GUIDELINES FOR
Mine Waste Dump and Stockpile Design

GUIDELINES FOR
Mine Waste Dump and Stockpile Design

EDITORS: MARK HAWLEY AND JOHN CUNNING

CSIRO
PUBLISHING

CRC Press
Taylor & Francis Group
Boca Raton London New York

CRC Press is an imprint of the
Taylor & Francis Group, an **informa** business

A BALKEMA BOOK

National Library of Australia Cataloguing-in-Publication entry

    Hawley, Mark, author.

    Guidelines for mine waste dump and stockpile design / Mark Hawley and John Cunning.

    9781486303502 (hardback)
    9781486303519 (ePDF)
    9781486303526 (epub)

    Includes bibliographical references and index.

    Waste products – Storage – Handbooks, manuals, etc.
    Mines and mineral resources – Waste disposal – Handbooks, manuals, etc.
    Mineral industries – By-products – Waste disposal – Handbooks, manuals, etc.
    Mineral industries – Waste disposal – Environmental aspects – Handbooks, manuals, etc.
    Australian.

    Cunning, John, author.

    622

Published exclusively in Australia and New Zealand by

CSIRO Publishing
Locked Bag 10
Clayton South VIC 3169
Australia

Telephone: +61 3 9545 8400
Email: publishing.sales@csiro.au
Website: www.publish.csiro.au

Published exclusively throughout the world (excluding Australia and New Zealand) by CRC Press/Balkema, with ISBN 978-1-138-19731-2

CRC Press/Balkema
P.O. Box 11320
2301 EH Leiden
The Netherlands
Tel: +31 71 524 3080
Website: www.crcpress.com

Front cover: Main waste dump at the Pierina gold mine, Huaraz, Peru. Photographed by R. Sharon. Courtesy Minera Barrick Misquichilca S.A.

Set in 10/12 Adobe Minion Pro and Optima
Edited by Joy Window (Living Language)
Cover design by James Kelly
Typeset by Thomson Digital
Index by Indexicana
Printed in China by 1010 Printing International Ltd

Original print edition:
The paper this book is printed on is in accordance with the rules of the Forest Stewardship Council®.
The FSC® promotes environmentally responsible, socially beneficial and economically viable management
of the world's forests.

Contents

Preface and acknowledgements

The Large Open Pit (LOP) project is an international research and technology transfer project focused on the stability of large slopes associated with open pit mines. It is an industry sponsored and funded project that was initiated in 2005 and managed by Dr John Read under the auspices of Australia's Commonwealth Scientific and Industrial Research Organisation (CSIRO). The project was renewed in 2014 under the leadership of Dr Marc Ruest and the University of Queensland. The sponsors have comprised a diverse group of multinational mining companies, joint venture partners and individual mines including Anglo American plc; AngloGold Ashanti Limited; Barrick Gold Corporation; BHP Chile; BHP Billiton Innovation Pty Limited; Corporación Nacional de Cobre Del Chile (Codelco); Compañía Minera Doña Inés de Collahuasi SCM; De Beers Group Services (Pty); Debswana Diamond Company; Newcrest Mining Limited; Newmont Australia Limited; Ok Tedi Mining Limited; Technological Resources Pty Ltd (RioTinto Group); Teck Resources Limited; Vale; and Xstrata Copper Queensland.

Among the initiatives mandated by the LOP sponsors is the development of a series of reference books on various aspects related to the design and stability of large slopes associated with open pit mines. The first of these books, *Guidelines for Open Pit Slope Design* (Read and Stacey 2009) covers the fundamentals of geotechnical investigation, analysis, design, and monitoring of open pit slopes. It represents the first comprehensive publication on this subject since the *Pit Slope Manual* was published by the Canadian Centre for Mining and Metallurgy (CANMET) in 1977. The second book, *Guidelines for Evaluating Groundwater in Pit Slope Stability* (Beale and Read 2013), covers the key influences of groundwater on the stability of open pit slopes and includes chapters on groundwater investigation, modelling of pore pressures and inflows, dewatering and depressurisation techniques and monitoring.

This book, *Guidelines for Mine Waste Dump and Stockpile Design*, is the third book in the series and focuses on the investigation, design, operation and monitoring of waste dumps, dragline spoils and stockpiles associated with large open pit mines. This book has been written by a consortium of geotechnical consultants and individuals including Piteau Associates Engineering Ltd, Golder Associates Ltd, Schlumberger Water Services, Sherwood Geotechnical and Research Services Inc., Dr. Oldrich Hungr and Dr Ward Wilson.

The next book in the series, *Guidelines for Open Pit Slope Design in Weak Rocks*, will focus on the unique aspects of open pit slope development in weak rocks and other materials whose characteristics and behaviour fall between rock and soil.

These reference texts are intended to help both geotechnical practitioners and non-specialists improve their understanding of the myriad of factors that can influence the stability of large, man-made slopes associated with open pit mines. They are intended to provide practical guidelines for both designers and operators. Another objective is to define the current state of practice and establish benchmarks or standards of practice that the mining industry can use to judge the suitability or acceptability of a given investigation, design or implementation approach.

While these works are considered to be comprehensive and current at the time of publication, it is recognised that there is ongoing research in a wide range of areas that relate both directly and indirectly to the subject matter. Existing technologies, techniques and methodologies continue to evolve, as do design objectives and tolerances, and practitioners are advised to consider these books as references and guidelines only; they are not a substitute for good engineering judgement and experience, and are not intended to be 'cookbooks'. Analytical and empirical techniques and methodologies that differ from those included in these references may also be valid and reasonable.

The editors acknowledge with thanks the dedication and hard work of all of the contributors to this book, especially Dr John Read, for his vision and encouragement, and the LOP sponsors for their support and patience. The editors are also very grateful to the following people for their important contributions:

Geoff Beale, Schlumberger Water Services, Shrewsbury, UK

Mike Bellito, Golder Associates Inc. Denver, Colorado, USA

Mike Bratty, Golder Associates Ltd, Vancouver, BC, Canada

Paolo Chiaramello, Golder Associates Ltd, Vancouver, BC, Canada

Leonardo Dorador, Golder Associates Ltd, Vancouver, BC, Canada

Jeremy Dowling, Schlumberger Water Services, Denver, Colorado, USA

Michael Etezad, Golder Associates Ltd, Mississauga, ON, Canada

Claire Fossey, Golder Associates Ltd, Vancouver, BC, Canada

Jason Garwood, Teck Coal Ltd Fording River Operations, Elkford, BC, Canada

Carlos Granifo Garrido, Schlumberger Water Services, Santiago, Chile

Brian Griffin, Golder Associates Inc., Houston, Texas, USA

Andy Haynes, Golder Associates Ltd, Vancouver, BC, Canada

James Hogarth, Piteau Associates Engineering Ltd, Vancouver, BC, Canada

David Holmes, Schlumberger Water Services, Shrewsbury, UK

Erin Holdsworth, Golder Associates Ltd, Vancouver, BC, Canada

Oldrich Hungr, University of British Columbia, Vancouver, BC, Canada

Fernando Junqueira, Golder Associates Ltd, Mississauga, ON, Canada

Simon Lee, Piteau Associates Engineering Ltd, Vancouver, BC, Canada

Kim McCarter, University of Utah, Salt Lake City, Utah, USA

Rowan McKittrick, Schlumberger Water Services, Shrewsbury, UK

Manuel Monroy, Golder Associates Ltd, Vancouver, BC, Canada

Humberto Puebla, Golder Associates Ltd, Vancouver, BC, Canada

John Rupp, Schlumberger Water Services, Reno, Nevada, USA

Eduardo Salfate, Golder Associates SA, Santiago, Chile

John Simmons, Sherwood Geotechnical and Research Services, Brisbane, Qld, Australia

Julia Steele, Golder Associates Ltd, Vancouver, BC, Canada

Björn Weeks, Golder Associates Ltd, Vancouver, BC, Canada

David Williams, University of Queensland, Brisbane, Qld, Australia

Ward Wilson, University of Alberta, Edmonton, AB, Canada

Robert Yarkosky, Golder Associates Inc., St Louis, Missouri, USA

Luca Zorzi, Golder Associates Ltd, Vancouver, BC, Canada

The book has been edited by Mark Hawley (Piteau Associates Engineering Ltd) and John Cunning (Golder Associates Ltd), with the assistance of an editorial subcommittee comprising Geoff Beale (Schlumberger Water Services), James Hogarth (Piteau Associates Engineering Ltd), Andy Haynes (Golder Associates Ltd), Peter Stacey (Stacey Mining Geotechnical Ltd), Stuart Anderson (Teck Resources Ltd) and Claire Fossey (Golder Associates Ltd). The assistance and encouragement of Briana Melideo, Lauren Webb and Tracey Millen from CSIRO Publishing are also gratefully acknowledged.

Mark Hawley and John Cunning
May, 2016

1

INTRODUCTION

Mark Hawley and John Cunning

1.1 General

In terms of both volume and mass, waste dumps associated with large open pit mines are arguably the largest man-made structures on Earth. Their footprints typically exceed the aerial extent and their heights often rival the depths of the open pits from which the material used to construct them is derived.

Figure 1.1 is a view of the East Dump at the Antamina Mine in Peru. This dump contains ~1 billion tonnes of material, covers an area of 240 ha, and has an overall height of more than 500 m. Figure 1.2 is a view of the waste dumps at Rio Tinto Kennecott's Bingham Canyon Mine in Utah, USA. This mine has a long development history spanning more than 100 years. The original dumps were constructed using rail haulage and tips, with subsequent expansions using truck haulage. Figure 1.3 is a view of a waste dump at a coal mine in the Elk Valley

region of British Columbia, Canada. A cumulative volume of waste rock of over 8.5 billion tonnes with overall dump heights of up to 400 m have been deposited in the Elk Valley area coal mines over ~45 years.

While most waste dumps worldwide have performed very well, there are many cases where they have been subject to large-scale instabilities with significant adverse consequences. Figure 1.4 illustrates one such failure that occurred in 1987 at the Quintette Coal Mine in British Columbia, Canada. This failure involved more than 5.6 million m^3 of material, and the runout distance exceeded 2 km (for additional details on this failure see BCMEM record #60 in Appendix 1).

Despite these metrics, the amount of effort expended on the investigation, design, implementation and monitoring of these massive structures is often small in comparison to the programs for their source open pits. Likewise, our understanding of their behaviour and our

Figure 1.1: East Dump at the Antamina Mine, Peru, ca. 2010. Source: M Hawley. Published with the permission of Compañia Minera Antamina S.A.

Figure 1.2: View of the Bingham Canyon Mine and associated waste dumps, ca. 2010. Source: Rio Tinto Kennecott Copper

Figure 1.3: View of waste dumps at a mine in the Elk Valley region of British Columbia. Note backfill waste dumps (active) in centre and reclaimed (inactive) waste dumps on right. Source: J Cunning

ability to model and reliably predict their stability is not as advanced as for open pit slopes and other large earth structures, such as tailings impoundments and water retention dams, and their design remains largely empirical.

1.2   Historical context

Some of the earliest work on developing a formal understanding of the mechanics of mine waste dumps was conducted in the early 1970s in response to the failure

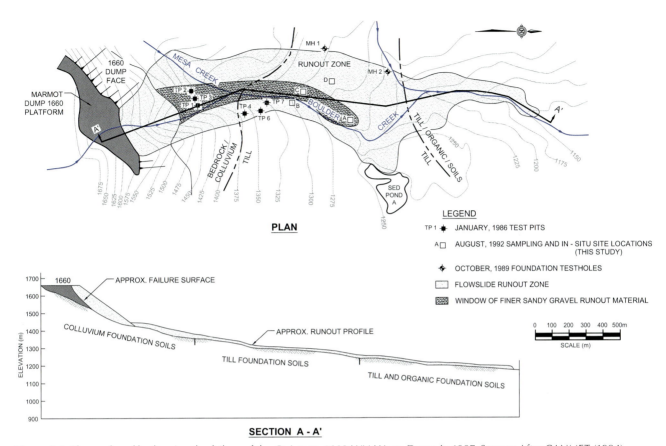

Figure 1.4: Plan and profile showing the failure of the Quintette 1660 WN Waste Dump in 1987. Source: After CANMET (1994). © Her Majesty the Queen of Canada, as represented by the Minister of Natural Resources, 2015

of a coal mine waste tip in Wales in 1966 (Fig. 1.5). Runout from this failure inundated a primary school and residential section in the town of Aberfan, killing 116 children and 28 adults. The failure was attributed to a build-up of pore pressure in the waste material due to heavy rains and natural springs in the foundation which triggered a liquefaction-type failure.

In 1975, the US Mining Enforcement and Safety Administration (MESA 1975) (predecessor to the current US Mine Safety and Health Administration) published a design manual for coal refuse disposal facilities. This manual was intended to provide guidelines and standards for open strip coal mine waste dumps being developed predominantly in the eastern United States (Virginia and Kentucky), and the design methodologies were based largely on classical soil mechanics approaches. The MESA manual was followed in 1977 by the *Pit Slope Manual*, published by the Canadian Centre for Mining and Metallurgy (CANMET 1977), which incorporated a chapter on waste embankments that included both tailings dams and waste rock dumps. In 1982, the US Bureau of Mines (USBM) published a comprehensive reference on the *Development of Systematic Waste Disposal Plans for Metal and Nonmetal Mines* (USBM 1982), and in 1985 the

Society for Mining and Metallurgy (SME) sponsored what appears to be the first focused workshop on the *Design of Non-impounding Waste Dumps* (SME 1985). In 1989, the US Department of the Interior's Office of Surface Mining published a new manual for the design and closure of spoils from surface coal mines (OSM 1989), replacing the earlier MESA manual. In 1991, as a follow-up to legislative changes flowing from the Aberfan disaster, the government of the United Kingdom published a *Handbook on the Design of Tips and Related Structures* (Geoffrey Walton Practice and Great Britain, Department of the Environment 1991).

In 1990, in response to a series of large waste dump failures at metallurgical coal mines in the Canadian Rocky Mountains, a committee composed of local mining companies, the Canadian Centre for Mineral and Energy Technology (CANMET) and the British Columbia Ministries of Environment and Energy, Mines and Resources (the British Columbia Mine Waste Rock Pile Research Committee [BCMWRPRC]) was formed to foster research on mine waste dumps. The outcome of this research included a series of interim guidelines and focused research reports, which are summarised in Table 1.1.

Figure 1.5: Aberfan coal tip failure, 21 October 1966. Source: M Jones and I McLean (n.d.) *The Aberfan Disaster.* <http://www.nuffield. ox.ac.uk/politics/aberfan/home2.htm>

While this work was based primarily on experience with large waste rock dumps associated with the surface metallurgical coal mines located in mountainous terrain in British Columbia and Alberta, Canada, the *Interim Guidelines* (reports #1 and #2 in Table 1.1) were generalised to include similar structures at other types of open pit mines. Following release of the *Interim Guidelines*, the BCMWRPRC sponsored a series of workshops throughout British Columbia that were intended to introduce the concepts and proposed classification and design methodologies and to solicit feedback from industry. After an introductory period, it was intended to update the *Interim Guidelines* and publish final versions. Unfortunately, the BCMWRPRC was unable to secure funding for this phase of the program, and the *Interim Guidelines* were never finalised. Nevertheless, they continue to be used as a practical reference by many practitioners and some regulators.

Since the mid-1990s there have been many individual contributions to the literature on waste rock dumps,

including papers describing advances in site investigation and materials testing, new analysis techniques and computer software codes, and case studies. In the Slope Stability 2000 conference sponsored by the SME in Denver, USA (Hustrulid *et al.* 2000), one session was dedicated to waste rock dumps, and in 2008 the Australian Centre for Geomechanics (ACG 2008) sponsored the First International Seminar on the Management of Rock Dumps, Stockpiles and Heap Leach Pads in Perth, Australia. In 2009 a second edition of the 1975 *Engineering and Design Manual – Coal Refuse Disposal Facilities* (MSHA 2009) was published. Several dedicated workshops, online courses and sessions associated with various conferences and symposia have also been held over the last several years.

Another good source for papers on mine waste dumps is the proceedings of the Tailings and Mine Waste Conference, which has been held annually in Fort Collins or Vail, Colorado, USA, Vancouver BC, Canada or Banff Alberta, Canada between 1994 and 2004, and between 2007 and 2016.

Table 1.1: Summary of BCMWRPRC and CANMET waste dump interim guidelines and related research reports

Report #	Report title	Date issued	Prepared by	Reference
1	*Investigation and Design Manual, Interim Guidelines*	May 1991	Piteau Associates Engineering Ltd	BCMWRPRC (1991a)
2	*Operating and Monitoring Manual, Interim Guidelines*	May 1991	Klohn Leonoff Ltd	BCMWRPRC (1991b)
3	*Review and Evaluation of Failures, Interim Report*	March 1992	Scott Broughton	BCMWRPRC (1992a)
4	*Runout Characteristics of Debris from Dump Failures in Mountainous Terrain Stage 1 Data Collection Volume I Text and Tables Volume II Drawings and Photographs*	March 1992	Golder Associates Ltd	BCMWRPRC (1992b)
5	*Methods of Monitoring Waste Dumps Located in Mountainous Terrain*	March 1992	HBT AGRA Limited	BCMWRPRC (1992c)
6	*Instability Mechanisms Initiating Flow Failures in Mountainous Mine Waste Dumps Phase I*	November 1992	University of Alberta	CANMET (1992)
7	*Liquefaction Flowslides in Western Canadian Coal Mine Waste Dumps Phase II Case Histories*	1994	CANMET and University of Alberta 1994	CANMET (1994)
8	*Consequence Assessment for Mine Waste Dump Failures, Interim Report*	December 1994	Golder Associates Ltd	BCMWRPRC (1994)
9	*Runout Characteristics of Debris from Dump Failures in Mountainous Terrain Stage 2 Analysis, Modelling and Prediction*	February 1995	Golder Associates Ltd and O. Hungr Geotechnical Research Ltd	BCMWRPRC (1995)
10	*Liquefaction Flowslides in Western Canadian Coal Mine Waste Dumps Summary Report Phase III Volumes I and II*	1995	University of Alberta	CANMET (1995)
11	*Rock Drain Research Program Final Report*	March 1997	Piteau Engineering Ltd	BCMWRPRC (1997)

Many of the above documents are available online at: http://www2.gov.bc.ca/gov/content/industry/mineral-exploration-mining/permitting/geotechnical-information

1.3 The Large Open Pit Project

The Large Open Pit (LOP) Project is an international research and technology transfer project focused on the stability of large open pit mines. It is an industry sponsored and funded program that was initiated in 2005. The LOP Project was initially managed by Dr John Read under the auspices of Australia's Commonwealth Scientific and Industrial Research Organisation. The project was renewed in 2014 under the leadership of Dr Marc Ruest and the University of Queensland. The sponsors comprise a diverse group of multinational mining companies, joint venture partners and individual mines.

One of the initiatives mandated by the LOP sponsors was the development of a reference book on the investigation, design, operation and monitoring of waste rock dumps, dragline spoils and stockpiles associated with large open pit mines. These *Guidelines for Waste Dump and Stockpile Design* are the result of this initiative and are intended to consolidate the historically important contributions detailed above, as well as the experience gained since the publication of the BCMWRPRC *Interim Guidelines*, into a single, concise, practical reference book. This work is not intended to be an exhaustive or detailed treatise of all the underlying science and engineering associated with these structures. Nor is it presented as an all-encompassing document that should be used as a framework for development of legislation or regulations. Its focus is the current state of practice, and its primary purpose is to provide insight and guidance to practitioners involved in the investigation, design, operation, monitoring and closure of waste dumps, dragline spoils and stockpiles and, most specifically, for those individuals that are responsible at the mine site level for ensuring the stability and performance of these structures.

1.4 Waste rock dump surveys and databases

1.4.1 1991 British Columbia waste dump survey

As part of the Canadian research effort in the 1990s, a multifaceted survey of waste dumps at all active mines in British Columbia was undertaken by Piteau Associates Engineering Ltd. Information was solicited on a wide range of factors, including waste material characteristics, dump configuration, foundation conditions, development and operation, and monitoring and stability history. Data were obtained on a total of 83 dumps from 24 mining operations, including eight coal mines, 15 metal mines and one asbestos mine. A complete listing of the survey results is included in an appendix to the 1991 *Investigations and Design Manual – Interim Guidelines* (BCMWRPRC 1991a).

1.4.2 Database of mine waste dump failures

Starting in the early 1990s, two separate databases were established for the study of mine waste dump failures in British Columbia, Canada. Golder Associates Ltd compiled a database containing records from 48 waste dump failures that occurred between 1968 and 1993 and prepared a series of reports for CANMET, which included a compilation of available data from each event and the nature of the runout of debris from each failure (reports #4 and #9 in Table 1.1). Scott Broughton later compiled a database containing records from of 44 waste dump failures that occurred in British Columbia between 1979 and 1991 and prepared a report for the BCMWRPRC (report #3 in Table 1.1). Appendix 1 includes key details from these two databases of waste dump failures.

1.4.3 British Columbia Ministry of Energy, Mines and Natural Gas database of waste dump incidents

The British Columbia Ministry of Energy, Mines and Natural Gas has continued to record and compile a database of details on reported waste dump incidents in British Columbia. This database, which is presented in Appendix 1, includes two parts. The first part comprises 73 incidents that were documented between 1968 and 1991, many of which are included in the failures that made up the Golder Associates Ltd database. The second part included 122 incidents that were reported between 1992 and 2005. It is important to note, however, that not all of these incidents resulted in failure or large-scale instability, unconstrained runout or adverse environmental impacts.

1.4.4 2013 Large Open Pit waste dump, dragline spoil and stockpile survey

A key component of the current work was a worldwide survey of existing and planned waste dumps, dragline spoils and stockpiles. The purpose of this survey was to develop a comprehensive database on the geometry, geotechnical and hydrogeological characteristics, design, operation and performance of these structures that could be used as a resource for this publication and for future research. The survey was developed using an online tool, and the scope and format were designed to limit the time and effort required to respond. Participation was solicited through the LOP sponsors, their consultants, and other interested parties. The survey was launched in April 2013 and closed in September 2013, and a total of 69 validated responses were received.

A compendium and analysis of survey responses is included in Appendix 2. At the request of the majority of the respondents, the identities of the mines and individual waste dumps have been excluded from the data summarised in Appendix 2.

1.5 Terminology

The terminology used to describe mine waste dumps, dragline spoils and stockpiles varies considerably from mine to mine and by jurisdiction. In North America, at conventional truck and shovel operations, these structures are most commonly referred to as waste rock dumps or waste rock storage facilities to differentiate them from tailings deposits. In Latin America, the generic 'botadero', which translates literally as 'refuse dump or landfill', is used. In coal mines in the central and eastern United States, the Canadian Prairies and Australia, they are traditionally referred to as 'spoils'. In the United Kingdom, waste rock deposits associated with both open pit and underground coal mines are historically referred to as 'tips'. The term 'tip' has also been used in many jurisdictions to describe waste deposits associated with hard rock underground mines. A distinction is also often made between waste rock dumps, mineral stockpiles, and overburden or topsoil stockpiles.

With the exception of mineral stockpiles, the common denominator for all of these structures is that they are composed of earth materials (rock and soil) that have been removed (mined or stripped) to expose ore. These materials are placed in heaps, piles or fills in areas peripheral to the ore deposit where they will not unduly restrict exploitation of the deposit. In an effort at a generic description that avoided the use of the terms 'waste' and 'dumps', both of which were felt to have negative connotations, the 1991 *Interim Guidelines* referred to these structures as 'mined rock and overburden piles'. In this current publication, all facilities intended for long-term containment of materials from stripping operations (including soils), run-of-mine and crushed waste rock, and residual materials from leaching operations (commonly referred to as 'ripios' in Latin America) are generically referred to as 'waste dumps' or simply as 'dumps'. The term 'stockpile' is also used and is intended to include all temporary storage facilities for natural and processed earth materials, such as run-of-mine and crushed ore, low or marginal grade ore, and overburden and topsoil materials stockpiled for later use in reclamation activities. The term 'spoil' is also used to describe dumps composed of waste materials generated by large-scale, relatively shallow stripping operations that primarily use draglines or bucket-wheels (Chapter 13). The terms 'embankment' and 'facility' are also generically used to describe the overall dump or stockpile structure. The design and operation of leach pads, dump leaching facilities and tailings or coal fines refuse are specifically excluded.

1.6 Waste dump and stockpile types

As shown in Fig. 1.6, most waste dumps and stockpiles may be characterised based on their intended purpose (i.e. permanent containment or temporary storage) and the materials used to construct them (e.g. rockfills, earthfills, mixed fills). Rockfills include mined rock and natural talus that is composed of coherent, angular rock particles with few fines. Earthfills include most overburden materials, residual soils, weak saprolitic materials and very weak rocks that disaggregate when excavated. Herein, the use of the term 'overburden' is limited to surficial soils (including topsoil) and other soil-like like materials, rather than the more general definition that is sometime applied to all materials (soil and rock) that overlie a mineral deposit or coal seam. Mixed fills are composed of a mixture of rockfill and earthfill materials.

Figures 1.7, 1.8 and 1.9 illustrate examples of different waste dump, dragline spoil and stockpile types.

Waste dumps and stockpiles may also be characterised on the basis of their overall configuration and topographic constraints as proposed by Wahler (1979) and illustrated in Fig. 1.10. As the name implies, 'valley fills' partially or completely fill a valley. The surface of the fill is typically graded to prevent impoundment of water at the head of the valley, or surface water is diverted around the fill in channels or under the fill via a flow-through rock drain. Valley fills that completely fill a valley are also known as 'head-of-hollow' fills and are common in the coal fields of the south-eastern United States.

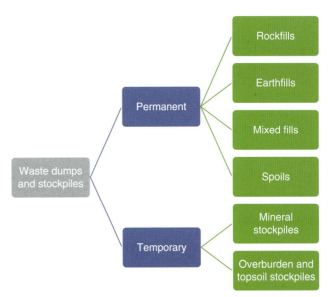

Figure 1.6: Basic waste dump and stockpile classification

Cross-valley fills are a variation of valley fills in which the structure spans the valley but does not fill it. Cross-valley fills may also be constructed to create causeways for haulroads, light vehicle access roadways, or conveyor or railway embankments. Cross-valley fills typically require construction of an engineered culvert or drainage structure, or a flow-through rock drain, to prevent impounding of water upstream of the fill. Sidehill fills are constructed on sloping terrain and typically do not block any major drainages. Slopes are usually inclined in the

Figure 1.7: Mixed fill waste dump; outer competent waste rock shells are designed to contain overburden in the core of the dump; East Waste Dump, Lagunas Norte Mine, Peru. Source: M Hawley. Published with the permission of Compañia Minera Barrick Misquichilca S.A.

Figure 1.8: Typical dragline spoil, Hunter Valley, Australia. Source: J Simmons

Figure 1.9: Low-grade ore stockpiles in Antamina Valley, Antamina Mine, Peru. Source: M Hawley. Published with the permission of Compañia Minera Antamina S.A.

same direction as the topography, and the toe of the fill is usually founded on flatter terrain in the valley bottom, or is buttressed against the lower slope on the other side of the valley. Ridge crest fills are a variation of the sidehill fill in which the fill spans a ridge crest and slopes are

established on both sides of the ridge. Heaped fills are founded on relatively flat or gently inclined terrain with fill slopes on all sides. Heaped fills are usually constructed from the bottom up in lifts. Figures 1.11 to 1.14 illustrate several of the basic waste dump and stockpile types.

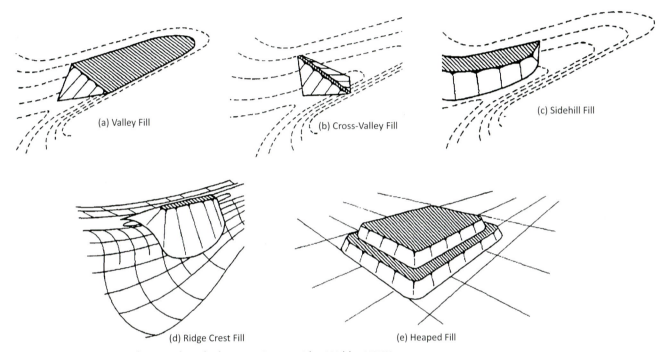

Figure 1.10: Basic waste dump and stockpile types. Source: After Wahler (1979)

Figure 1.11: Valley fill, Pierina Mine, Peru. Source: M Hawley. Published with the permission of Compañia Minera Barrick Misquichilca S.A.

Figure 1.12: Sidehill fills in the Tucush Valley, Antamina Mine, Peru. Source: M Hawley. Published with the permission of Compañia Minera Antamina S.A.

Figure 1.13: Ridge crest fill, East Waste Dump, Lagunas Norte Mine, Peru. Source: M Hawley. Published with the permission of Compañia Minera Barrick Misquichilca S.A.

Figure 1.14: Heaped fill at the Cerro Colorado Mine, Chile. Source: M Hawley. Published with the permission of BHP Billiton, Chile

2

BASIC DESIGN CONSIDERATIONS

Mark Hawley

2.1　General

Rational design of waste dumps and stockpiles requires consideration of a wide range of interrelated factors that may change throughout the life cycle of the mine. Site selection criteria must be defined and used to identify and rank alternatives at an early stage. Conceptual designs that respect the physical constraints imposed by the site, and also meet economic, geotechnical, social and environmental objectives, must be developed. Pre-feasibility-level studies then need to be undertaken to establish environmental baselines, investigate and characterise alternative sites, characterise fill materials, validate design concepts and narrow the site selection process. More detailed feasibility-level investigations and detailed design studies then follow to address any data gaps, refine designs, and develop detailed implementation plans, operational guidelines and controls and preliminary closure plans. During the operational phase, adjustments to the design will be necessary to accommodate inevitable changes to the mine plan and adapt to actual behaviour. Planning for closure must be a part of the design process from the beginning, and towards the end of mine life it may supersede other design objectives. Figure 2.1 illustrates the design process and key inputs.

The design process for these structures must also fit within the overall project development stage and objectives as summarised in Table 2.1.

This chapter provides an overview of the design process and main inputs. More detailed descriptions of each of the key components, including various methodologies and analytical techniques that can be applied to help develop rational and defendable designs, are provided in subsequent chapters.

2.2　Site selection factors

The first step in the design process is to select a site for the facility. This is a critical step that needs to be done methodically and carefully as it can have a major impact on successive steps. A poor site selection process can result in substantial delays in the design and permitting process and require costly and time-consuming iterations. Selecting the optimal site requires consideration and balancing of a wide range of often competing objectives. To minimise potential conflicts and delays, the perspectives of all key stakeholders must be considered at an early stage in the process.

For illustrative purposes, Fig. 2.1a indicates the output of the initial site selection process as an input at the start of the conceptual design phase. While this linear sequence would be an ideal application of the design process, in reality all of the information required as input to the site identification and selection process is typically not available at the outset. As a practical matter, much of the necessary data are collected at stages throughout the conceptual, pre-feasibility and feasibility phases; hence, site selection inevitably becomes an iterative process that must be refined and updated as the overall investigation and design study progresses.

As illustrated in Fig. 2.2, the key site selection factors can be divided into seven categories:

- regulatory and social
- mining
- terrain and geology
- environmental
- geotechnical
- fill material quality
- closure.

(a)

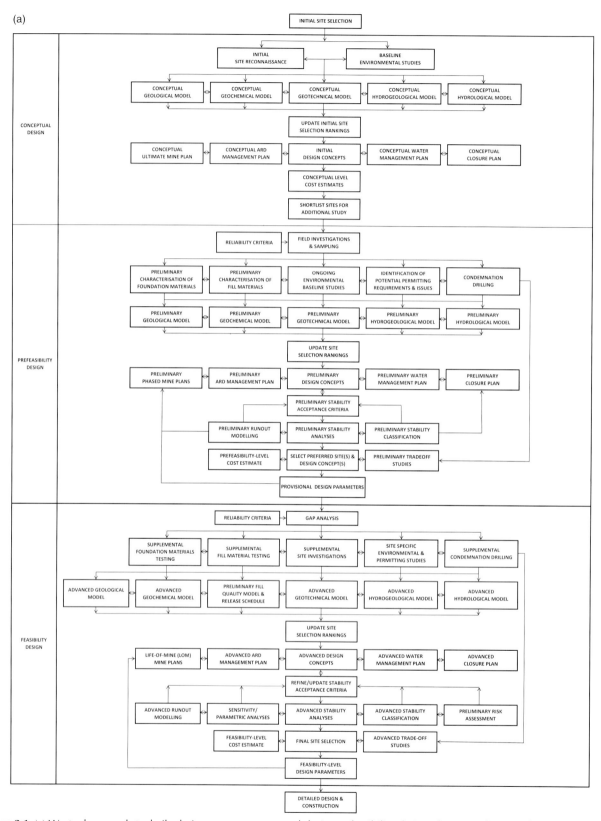

Figure 2.1: (a) Waste dump and stockpile design process: conceptual design to feasibility design, (b) Waste dump and stockpile design process: detailed design and construction to closure

(b)

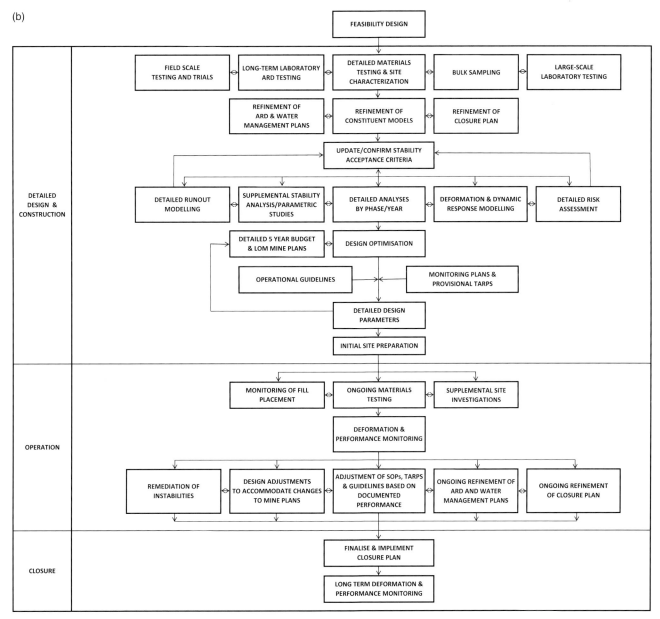

Figure 2.1: (Continued)

Each of these categories is described in more detail in the following sections. Some selection factors are relevant to more than one of these categories, so there is some redundancy in the following sections and in the suggested initial site screening tool described in Section 2.3.

2.2.1 Regulatory and social factors

Regulatory and social licensing needs must be met with assurance that the required approvals and permits can be obtained. Regulatory and social factors that should be considered in the site selection process include:

- permitting requirements
- regulatory standards

- land claims
- inhabitant relocation
- land and water use
- visual quality
- archaeology
- artisanal mining.

Most jurisdictions regulate the development of waste dumps, at least to some degree. Some have very well-defined and reasonably predictable permitting processes that set out the government's expectations with respect to the engagement of stakeholders, facility design, stability, environmental impact identification and mitigation, and closure and reclamation. Other jurisdictions are less

Table 2.1: Project development stages and objectives

Project stage	Objectives
Target identification and order of magnitude studies	Confirm that the identified mineralisation is sufficient in consistency and grade to warrant additional exploration and initiate project development plan. Provide an initial, high-level, order of magnitude financial analysis.
Conceptual design (Scoping)	Establish if project is worthy of further considerations and investigations based on regional and available data. Look for fatal flaws. The major drivers of the project should be defined, if possible.
Pre-feasibility	This level will test major operating options to establish a robust case for the project and develop a robust business case to support proceeding to feasibility. Pre-feasibility does not look to optimise the project, capital or operating plan. Initiate geotechnical site investigations. Assess options (if applicable) and select preferred option. The major drivers of the project need to be identified.
Feasibility	Confirm technical and financial feasibility including determination of the best operating configuration, which may need to be refined to optimise capital and operating costs compared to revenues. This stage will refine data and assumptions to enable the corporation to book reserves.
Detailed design and construction	Prepare detailed design and construction plans. Construct the project on time and within budget, and transfer skills to operations staff from construction staff. Pre-stripping would fall into this level. Develop the necessary action plans by discipline or function in enough detail to assure a smooth implementation from commencement of construction until early operations.
Operations	Execute mine production plans to the most efficient level possible, including health, safety, environment and loss prevention, while maximising return on investment.
Closure	Focus on the physical and chemical stability of materials and effluent products of the mining activity to ensure long-term care of human health and the environment. Develop specific objectives and completion criteria to reach a sustainable post-mining land use.

Source: Modified after Read and Stacey (2009)

transparent and less predictable, and some may not have any regulatory standards or a formal permitting process. Regardless of the sophistication of the jurisdiction, mine proponents are advised to research their specific regulatory environment, adopt a proactive approach to engage regulators and other interested parties and stakeholders, and set mutually agreeable targets and limitations very early in the project exploration and development process. Understanding the expectations and process at the outset is key to the success of the project.

As a minimum, most jurisdictions require the proponent to prepare and submit an overall project environmental impact assessment (EIA) that includes a description of plans for the disposition of waste rock and overburden materials. Such documents typically include baseline environmental data, assessments of potential

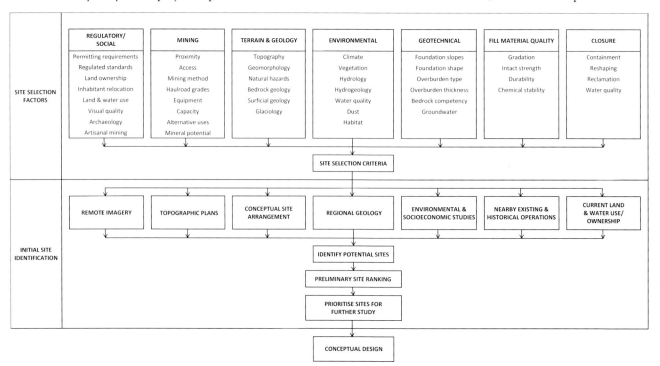

Figure 2.2: Initial site selection

environmental impacts, proposed measures to mitigate impacts, and environmental monitoring plans. Also typically included in EIAs are conceptual plans for reclamation, closure and long-term monitoring of waste dumps. Environmental compliance standards may be proposed as part of the EIA or may be negotiated during the EIA review process. While not cast in stone, once approved, EIA commitments may be difficult, time consuming and costly to modify or revisit. Consequently, it is important to establish reliable environmental baselines, and carefully evaluate alternative sites and design concepts, before submitting the EIA.

In recent years, the concept of 'social licensing' has become a key factor in the successful development of new or expanded mining ventures. In most jurisdictions, EIAs or similar documents must also include consideration of, and public consultation related to, social impacts, such as the need to relocate inhabitants, protect archaeological and culturally important sites, and respect traditional lifestyles. Traditional land and water uses and impacts to scenic vistas and touristic values also need to be considered. Legal entitlements to exploit the mineral resources may not be sufficient to guarantee a successful project. Unresolved indigenous land claims, 'squatter' rights, and artisanal and illegal mining operations may limit options and escalate costs, and their impacts need to be considered in the site selection process. Disputes between local, regional and federal governments, often over the distribution of revenues from taxation and royalties, can also impact site selection, particularly when sites span multiple jurisdictions. One issue that has become a lightning rod in some recent resource development disputes is the lack of benefits (perceived or otherwise) that flow back to local communities. In such circumstances, the mine proponent may be obliged to take on the role of benefactor in supporting and delivering services to local communities to obtain the support needed at the local level. Many of the recent successes in developing new mining ventures have established partnerships with local community and indigenous groups at an early stage to ensure that a portion of the jobs and other economic benefits generated by the project benefit these groups and promote sustainability of the local economy.

While many of these regulatory and social issues are overall project-level concerns, they may impact the siting and design of the waste dumps and stockpiles, and their impacts therefore need to be considered in the site selection process.

2.2.2 Mining factors

Mining-specific factors that should be considered in the site selection process include:

- proximity
- access
- mining method
- haulroad grades
- equipment options
- capacity
- alternative uses
- mineral potential.

The proximity of a site to the source waste or stockpile material may be a key factor in determining the site's economic viability. Haulage distances and cycle times in conventional truck and shovel operations make up a major component of the overall mining costs, and minimising these is often a mine planning priority. Sites that are close to the open pit are obviously more attractive than remote sites, all other things being equal. Phased mine plans often favour the development of smaller satellite facilities that are close to active phases, over a single large facility.

Ease of access is also an important consideration. Close-by sites that are difficult to access because of topography or other factors may be less attractive than more distant sites that are easy to access. Access to some prospective sites may only be feasible using tunnels, large fill causeways, or bridges, but the cost of these structures may be offset by proximity or other factors.

The location and configuration of the waste dump or stockpile may also dictate the mining method and the type of equipment used for construction. For locations that are far from the source material, in-pit or pit-rim crushing and conveying and spreading may be more economical than conventional truck haulage. The design of dragline spoils can be highly constrained by the mine plan, particularly with the configuration of ramps and equipment surcharge loading; consequently, these factors need to be considered early in the initial spoil planning process.

Haulroad grades may also be a factor. Allowable grades for most large haul trucks are typically limited to 8–10%, although flatter grades are generally preferred. As a general rule of thumb, for the same operating cost a loaded haul truck can travel about twice as far on a horizontal grade as it can uphill on a 10% grade. In addition, switchbacks tend to slow traffic and increase cycle times, so routings that minimise the number of switchbacks are generally preferred.

Truck size and haulroad width requirements may also need to be considered. Wide haulroads to accommodate very large trucks may be difficult or very costly to construct in steep terrain. If tunnelling is an alternative option to a long surface haul, there may also be a cost advantage to using smaller trucks or crushing and conveying.

The capacity of a given site, and its potential for future expansion, may also be important selection criteria. If capacity is limited and the required volume cannot be contained, it may be more attractive to develop an alternative, larger site that may require longer hauls or higher overall development costs. Few projects overestimate the ultimate requirements for management of waste rock and overburden materials, so options for expansion need to be factored into the site selection process.

Many mine sites are constrained by land ownership and topography, and there may be competition within the project for alternative land uses. A site that may seem ideal for construction of a waste dump or stockpile may be more valuable as a tailings storage facility, plant site or water reservoir. Haulroads may also have to detour around critical facilities, adding to both capital and operating costs. Site selection should not be conducted in isolation, but in concert with overall project planning.

The potential for economic mineralisation underlying a site that is being considered as a permanent waste dump also needs to be considered. Construction of a waste dump over a satellite deposit may limit future mining options. Condemnation drilling of prospective sites is often considered a low priority and may be deferred until later in the project development cycle. However, late identification of potentially economic mineralisation can cause significant delays and require fundamental changes to the EIA or mine plans. Preliminary geological investigations and condemnation drilling should therefore be conducted early in the project development cycle.

2.2.3 Terrain and geology factors

The terrain and geology category includes factors related to the overall geography and engineering geology of the site, including:

- topography
- geomorphology
- natural hazards
- bedrock geology
- surficial geology
- glaciology.

The topography and shape of the site can be favourable, but they may also be adverse and limit development options. Waste dumps and stockpiles situated on gentle terrain are typically easier to develop and more stable than those founded on steep slopes. Waste dumps and stockpiles that are confined by topography also tend to be more stable compared to those developed on open terrain.

Geomorphology is the study of landforms and the processes that shape them. A basic understanding of the geomorphology of a particular site can provide insight into potential development issues that may be of relevance in site selection. Adverse geomorphological features may include evidence of active or historical mass wasting

(landslides), avalanches or other natural hazards. Sinkholes may signify the presence of karst. Cirques, moraines and kame terraces are indicators of past glacial activity and may be associated with weak glacial lacustrine or glacial outwash deposits. V-shaped gullies and other erosional features may indicate deep weathering and weak residual soils. Steep, stable slopes and frequent bedrock exposures could indicate competent, resistive bedrock and favourable foundations.

A basic understanding of the bedrock geology of the site, including the lithology (rock type), alteration, weathering, stratigraphy (spatial relationship of the different geological units) and regional structure, may also be relevant to site selection. Sites underlain by competent, fresh bedrock with favourable stratigraphy and structure are preferable to those underlain by weak or altered/ weathered rocks, adverse stratigraphy, or structure that could present foundation stability issues.

A basic understanding of the surficial geology of the site is also important. The distribution and characteristics of both natural and anthropogenic (human-made) deposits may affect the suitability of a given site or how the site is developed. For example, a site underlain by a competent, dense basal moraine or thin, granular alluvium or colluvial deposits overlying bedrock would be preferred to a site with extensive and/or thick lacustrine or organic deposits or old tailings.

Glaciology is the study of glaciers and their associated processes. The presence of glaciers, icefields or periglacial (permafrost) terrain may restrict or preclude the use of a given site. In recent years, some jurisdictions have established restrictive laws to protect such areas from development.

2.2.4 Environmental factors

The key environmental factors that are important to consider in the site selection process include:

- climate
- vegetation
- hydrology
- hydrogeology
- water quality
- dust
- habitat.

While climate factors are often a common denominator when comparing alternative sites for a given project, there are cases where variations in temperature, precipitation, evapotranspiration and prevailing winds are materially different and relevant to site selection. Orographic effects associated with even small elevation changes can make substantial differences in cumulative precipitation, temperature and evaporation, and topographic divides and other geographic features can create unique microclimates.

A basic understanding of the distribution of biogeoclimatic zones (zones with similar climate, flora and fauna) within the project area may be useful in comparing climatic conditions of prospective sites. In general, drier sites are preferable to wetter sites, and warmer, more exposed sites with sparse vegetation and greater evaporation potential are preferable to cooler or protected sites and those that support dense vegetation.

The hydrology of a site will determine the need for diversions, flow-through rock drains and other surface water control measures. Sites with small catchments that contain only ephemeral (seasonal) streams and do not require construction of diversions are preferred over those with large catchments that support perennial streams or rivers that require complex diversions, or which contain lakes, ponds or wetlands. Major stream diversions or filling of lakes, ponds or wetlands should be avoided where possible as these activities can complicate EIA approvals and tend to be controversial. There are ample recent examples where projects have been stalled or failed to gain EIA approval or public acceptance due to proposals to divert major streams or infill lakes.

A basic understanding of the hydrogeology of the project area and prospective sites is also important. Sites located in groundwater discharge areas may simplify environmental containment and management, but may be problematic in terms of stability. Sites situated in groundwater recharge areas may be more stable, but it may be more difficult to manage or mitigate impacts to groundwater quality. Sites underlain by karst represent a special case that can require extensive investigation and mitigative measures to manage groundwater impacts.

Potential impacts to water quality, and the ability to contain and treat surface runoff and groundwater where necessary, are key considerations in site selection. Most modern waste dumps and stockpile designs incorporate downstream sedimentation ponds to manage suspended sediment loads. Some incorporate wetlands or other passive systems to reduce nitrates from blasting residuals and to treat low levels of heavy metals. Where acid rock drainage (ARD) is an issue, collection systems that separate unaffected flows from those needing treatment may be required and may include seepage recovery wells and sumps. Sites that offer easy containment of surface runoff and groundwater flows and opportunities for nearby gravity-fed treatment facilities (both sediment ponds and treatment plants) are better than those that are difficult to contain or where contaminated water must be pumped to remote treatment facilities. Where complex treatment plants are needed, it may be better to contain all potentially reactive materials in a single facility, even if this increases overall transportation costs.

Dust management can be an issue at some sites that are in close proximity to settlements or sensitive habitats. In rare cases, concerns over fugitive dust emissions accelerating the melting of adjacent protected glaciers and icefields must also be addressed. An understanding of prevailing wind directions and speeds, and how local topography may alter these, is necessary to assess the fate and impact of fugitive dust and gauge the need for extraordinary dust suppression measures.

Potential impacts to fish and wildlife habitat, including migration routes, can be very contentious issues, especially where endangered species are present. Sites that do not significantly impact fish and wildlife, or where the impacts can be easily managed, are clearly better than sites that present major impacts that are difficult or impossible to mitigate. Sites where development would disrupt important migration routes may be difficult or impossible to permit or to gain the acceptance of indigenous peoples or special interest groups. Similarly, sites containing unique or endangered plant species may require special management plans that may include transplantation or other offsets.

2.2.5 Geotechnical factors

Geotechnical factors include those aspects of the site that may impact the competency and stability of the waste dump or stockpile foundation:

- foundation slope
- foundation shape
- overburden type
- overburden thickness
- bedrock competency
- groundwater conditions.

The slope of the foundation has a direct influence on the global stability and general performance of the waste dump or stockpile. Embankments constructed on steep foundations are much more likely to be unstable or perform poorly compared to those constructed on flat foundations.

The shape of the foundation may also have an impact on the stability and general performance of the waste dump or stockpile. Facilities that are confined by natural topography tend to perform better that those that are not confined.

'Overburden' is a generic term that is used throughout this book to refer to any soil, fill or unconsolidated material that underlies a waste dump or stockpile, including residual soils derived from *in situ* weathering of bedrock. Overburden materials can have a substantial impact on the stability of a waste dump or stockpile. Overburden types may range from very weak soils, such as highly organic topsoils or very soft or sensitive clays, to moderately competent, dense, granular alluvium or colluvium, to very competent, dense glacial tills. Some soil types, such as those which may be susceptible to liquefaction, or frozen soils that are subject to cyclical thaw or creep, may be very problematic, and sites where these conditions exist need to flagged early in the selection process.

Overburden thickness can also play a role in stability. Where a weak overburden deposit forms a very thin veneer over more competent material, it may be displaced or scoured during fill placement and have negligible impact on stability. Where a waste dump or stockpile is constructed from coarse, angular rockfill, these materials may also penetrate the thin, weak layer and be keyed into the more competent underlying material. Thick deposits of otherwise competent soils may contain weak lenses or layers that could adversely affect stability.

The competency and structure of the bedrock underlying the foundation can also influence, and in extreme cases control, stability, particularly in the case of a very high fill founded on a weak bedrock foundation. Bedrock types may range from very weak, highly clay-altered or weathered, sheared or highly fractured rocks; to moderately competent, moderately fractured, slightly weathered or altered rocks; to very competent, very hard, unweathered, unaltered, massive bedrock. The potential for the occurrence of adversely oriented structural discontinuities (e.g. faults, contacts, bedding, joints) in the bedrock foundation should also be considered during site selection and investigation.

High groundwater levels and high pore pressures in the foundation can also adversely affect stability. As indicated above, waste dumps or stockpiles founded in groundwater discharge areas with upward gradients, such as valley bottoms or topographic lows, are more likely to experience elevated pore pressures and saturated conditions in the foundation than those founded in recharge areas with downward gradients, such as well-drained slopes or topographic highs.

2.2.6   Fill material quality factors

In this book, fill materials include waste rock, waste overburden, ore and topsoil stockpile materials, ripios (leached ore) and any other materials that will be placed in a waste dump or stockpile. Fill material quality factors include those aspects that may directly impact the internal stability of the waste dump or stockpile. In many cases, the quality of the fill materials may be a common denominator for all prospective sites and therefore may not be relevant to site selection. However, where the quality of the fill varies widely and where the fill materials can be practicably segregated, it could be highly relevant, as some sites may be better suited for the containment of poor quality or chemically reactive materials than others.

The key material quality attributes that should be considered in the site selection process include:

- gradation
- intact strength
- durability
- chemical stability.

The gradation or particle size distribution of the fill material is a key factor in determining the frictional component of its shear strength. Friction typically accounts for the majority of the shear strength of most waste rock, overburden and stockpile materials. In general terms, well-graded materials with a high percentage of coarse, angular particles and a low component of fines tend to have higher shear strength than poorly graded, fine-grained materials. The mineral composition of the fine-grained component may also be relevant. Fines composed of plastic clay minerals would be characterised by relatively low shear strength as compared to those composed of rock flour or non-plastic fines. Saturated fine-grained fill materials may also be susceptible to generation of excess pore pressures during loading and undrained failure. Fill materials composed of uniformly graded materials with a low clay content and rounded particles may also be susceptible to liquefaction and should be flagged for special consideration during the site selection process.

The shear strength of fill materials is also influenced by the intact strength and durability of the individual particles, particularly in high fills where inter-particle stresses may exceed the intact strength of the material and result in crushing. Material types may range from very weak, highly degradable materials derived from poorly indurated or weakly cemented rocks with very low intact strength; through materials derived from moderately strong rocks with moderate durability and limited susceptibility to freeze–thaw degradation; to very competent, highly durable materials derived from very strong, highly indurated rocks.

The intact strength of the fill materials may also degrade over time as a result of physical weathering or chemical processes, such as oxidation of sulphide minerals or dissolution of sulphates. In addition to the potential for generation of ARD, precipitates that may form as a result of these chemical processes may also impact the hydraulic conductivity of the fill. Materials with poor chemical stability should also be flagged for special consideration during the site selection process.

2.2.7   Closure factors

As indicated above, a closure mindset should be adopted from the outset of the design process, including during initial site selection. To be able to compare alternative sites from a closure perspective, it is first necessary to set out the expected closure requirements. The key site attributes from a closure perspective include:

- containment
- reshaping
- reclamation
- water quality management.

In this context, containment refers to the ability of the site to contain debris from future potential instabilities and runout. Even if the potential for large-scale instability following closure is very low or negligible, a site with good containment is preferable to an open site.

Reshaping refers to the effort required to modify the geometry of the waste dump to promote long-term stability, including erosion resistance. This may depend to a large degree on the construction method and sequencing; however, dumps constructed on confined, steep terrain from the top down are typically more difficult and costly to reshape than those constructed from the bottom up on gentle, open terrain.

Reclamation refers to surface treatments to limit the infiltration of water and oxygen (as an ARD management strategy), to promote revegetation and to control erosion. Reclamation challenges also depend largely on the construction method and sequencing, with dumps constructed from the bottom up on gentle, open terrain generally being easier to reclaim than those constructed from the top down on steep, confined terrain.

Water quality management includes requirements for collection and treatment of surface runoff and groundwater. While long-term sediment control may be appropriately managed by reclamation and associated erosion control strategies, long-term management of water chemistry may be more challenging, and some sites may be more amenable to collection and treatment than others.

2.3   Initial site identification

Having set out the various criteria that should be considered in the site selection process, the next step is to identify potential sites. This is ideally conducted early in the project based on a desk review of available remote imagery (air photos, satellite images), topographic plans, regional geology plans and reports, regional environmental and socioeconomic studies, and other available reports and supporting data. A basic understanding of the proposed project arrangement and the current land and water use and ownership is also needed. The experience of nearby existing and historical operations should also be considered. Communications with other groups within the project development team, such as those responsible for mine planning, sociology and environmental studies, are also advisable to ensure that all of the available information is considered.

2.3.1   Preliminary ranking of potential sites

Where multiple potential sites are identified, a method for comparison and ranking is needed. Several approaches can be used for ranking sites by their attributes, including simple checklists and qualitative comparisons, semiquantitative approaches that rely on subjective weightings and cumulative numerical ratings, and more detailed, quantitative approaches such as comparative impact analysis, cost–benefit analysis and trade-off studies. The approach taken will depend on the evaluation stage (conceptual, pre-feasibility, feasibility or detailed design) and on the information available.

For an initial comparison and ranking of multiple sites, a simple spreadsheet-based screening and ranking tool, such as illustrated in Fig. 2.3, may prove useful. This worksheet includes a suggested subjective rating scheme and a numerical weighting scale. Together, these can be used to derive a relative 'Site Selection Index', which can be used to compare alternative sites. The rating scheme and weighting scale can be customised to suit a particular site or emphasise a particular factor or group of factors.

The worksheet in Fig. 2.3 is intended to provide the user with a checklist of the main factors that should be considered during the initial site selection process and should only be used as an initial screening and ranking tool. It is not intended to replace more comprehensive classification and rating systems such as described in Chapter 3, or structured risk assessment studies such as described in Chapter 10.

In Canada, the federal government has published *Guidelines for the Assessment of Alternatives for Mine Waste Disposal* (Environment Canada 2011). It is a regulatory requirement of the Canadian environmental assessment process to provide an alternatives assessment where the placement of mine waste into a natural water body that is frequented by fish is contemplated. The guidelines outline a methodology that follows a 'multiple accounts analysis' process, and is useful for preparing and presenting a transparent method for site selection.

2.4   Conceptual design

Having identified potential sites, the next step in the process is to develop conceptual designs for each site. At a minimum, conceptual designs need to satisfy capacity requirements, either individually or in aggregate.

Access routes, site preparation requirements, equipment options and construction alternatives should also be considered at this stage. If not done during the initial site selection process, a preliminary site reconnaissance should be conducted to confirm the key assumptions used to develop the initial site ranking. Baseline environmental studies may also be initiated during the conceptual design stage.

PRELIMINARY SITE SELECTION WORKSHEET[1]

SITE DESIGNATION:

FACTORS	DESCRIPTION	RATING[2]	WEIGHTING[3]	WEIGHTED RATING[4]
REGULATORY/SOCIAL				
Permitting/EIA Requirements				
Regulated Standards				
Land Ownership				
Inhabitant Relocation				
Land and Water Use				
Visual Quality				
Archaeology				
Artisanal Mining				
MINING				
Proximity				
Access				
Mining Method				
Haulroad Grades				
Equipment				
Capacity				
Alternative Uses				
Mineral Potential				
TERRAIN & GEOLOGY				
Topography				
Geomorphology				
Natural Hazards				
Bedrock Geology				
Surficial Geology				
Glaciology				
ENVIRONMENTAL				
Climate				
Vegetation				
Hydrology				
Hydrogeology				
Water Quality				
Dust				
Habitat				
GEOTECHNICAL				
Foundation slopes				
Foundation Shape				
Overburden Type				
Overburden Thickness				
Bedrock Competency				
Groundwater				
FILL MATERIAL QUALITY				
Gradation				
Intact Strength				
Durability				
Chemical Stability				
CLOSURE				
Containment				
Reshaping				
Reclamation				
Water Quality				
SITE SELECTION INDEX[5]				

Notes:
1. This worksheet is intended to provide a checklist of the main factors that should be considered during the initial site selection process and should only be used as an initial screening and ranking tool. Complete a separate worksheet for each prospective site.
2. Suggested subjective rating scheme: 0 = Very Unfavourable; 1 = Unfavourable; 2 = Neutral; 3 = Favourable; 4 = Very Favourable.
3. Suggested weighting system: 1 = low importance; 2 = moderately important; 3 = very important.
4. Multiply the Rating by the Weight to obtain the Weighted Rating for each relevant factor.
5. The Site Selection Index is the sum of the Weighted Ratings for all relevant factors.

Figure 2.3: Preliminary site selection worksheet

Based on the results of the initial site selection desk study, an initial site reconnaissance and preliminary environmental baseline work may be undertaken, and conceptual-level geological, geochemical, geotechnical, hydrogeological and hydrological models may be developed. These constituent models would be used to update the initial site rankings and, in conjunction with the conceptual ultimate mine plan, to develop initial geometric design concepts and conceptual ARD management (if required), water management (surface and groundwater flows and water quality) and closure plans for each prospective site.

The main objective of the conceptual design stage is to develop a shortlist of viable sites and design concepts for more detailed investigations. Conceptual level estimates of development and operating costs may also be required at this stage, and may help to refine initial site rankings and focus ongoing studies on the most prospective and economic sites. Developing an early understanding of the geological, physical and climatic settings of the site, and of the geochemistry of the fill materials and their potential for the development of ARD, is of critical importance at this stage as subsequent stages in the design process build upon these conceptual models and design concepts.

2.5   Pre-feasibility design

Comprehensive field investigations and laboratory testing programs are usually initiated at the pre-feasibility design stage. The amount of effort that must be invested in these programs, and subsequent model development and analytical studies, will depend on the required reliability. In terms of capital and operating cost estimates, pre-feasibility studies have traditionally targeted an accuracy of ± 25–35%, although this can vary widely depending on the policies of the mine proponent. Designers should have a full understanding of the expectations of the mine proponent at all stages of the project so that they can set realistic budgets and schedules. Additional guidance regarding the suggested level of effort and target levels of data confidence by project stage is provided in Section 2.10.

Preliminary site investigations, such as surface mapping, test pitting, trenching and drilling, are undertaken to characterise the foundations and obtain samples for laboratory testing. A preliminary laboratory characterisation of the physical and chemical properties of the fill materials that will compose the waste dumps and stockpiles is also recommended at this stage. Environmental monitoring and testing should also be expanded to develop a reliable historical baseline. Information derived from these studies should be used to validate and enhance the conceptual constituent models. Site-specific permitting requirements and potential

social licensing issues should be clearly identified, and appropriate action plans developed to address them. Condemnation drilling is also typically undertaken at this stage in combination with ongoing ore reserve definition studies. If applicable, site selection rankings should be updated.

Conceptual designs and the ARD management, water management and closure plans should be refined based on the updated models and input from the mine planners. In addition to the ultimate project configuration, preliminary designs should consider the evolution of the mine plan and basic waste dump and stockpile development sequencing. Preliminary stability acceptance criteria should be established in consultation with the mine proponent, and with consideration of the consequence of instability and confidence in the input parameters and other factors as discussed in Section 8.2. Stability analyses should be conducted based on the updated models to validate the design concepts, identify design limitations and sensitivities and initiate the design optimisation process. Empirical runout assessments may also be needed at this stage to confirm the viability of a given site. At this point, there should also be enough information to prepare preliminary stability classifications and rankings (see Chapter 3) to facilitate detailed, objective comparison of alternative sites and development concepts.

Next, pre-feasibility capital and operating cost estimates should be developed and, together with the results of condemnation drilling and stability assessments, used to compare alternatives and select the preferred site(s) for consideration during the feasibility study. Provisional design parameters should then be developed for input into the pre-feasibility mine plan. If these parameters are not compatible with the mine plan, an iterative cycle may be required whereby the preliminary design concept, and potentially the ARD management, water management and closure plans, are modified and re-analysed. This cycle would be repeated until a compliant design is achieved.

2.6   Feasibility design

The feasibility stage should commence with a detailed gap analysis to identify deficiencies in the supporting data. Based on the outcome of this analysis, supplemental field and laboratory investigations, site- and value-specific environmental and permitting studies, materials testing and characterisation, and supplemental condemnation drilling may be required. As with pre-feasibility, the required accuracy of feasibility-level cost estimates will need to be considered during the gap analysis and may influence the scope of supplementary investigations and analyses. Traditionally, feasibility studies have targeted a capital/operating cost accuracy of ± 15–20%, again depending on the policies of the mine proponent.

Following completion of these supplemental studies, the various constituent models should be updated and refined, and site selection rankings updated as appropriate. Development of a preliminary fill material quality model and release schedule may be also be needed to support key design assumptions that may require placement of specific materials in designated areas within the waste dump. Refinement of design concepts and ARD management, water management and closure plans would flow from the updated models and would, in turn, lead to advanced stability analyses, runout modelling and parametric/sensitivity studies. Review and refinement of the stability acceptance criteria may also be appropriate. At this stage, designs should consider the waste dump development sequencing as contemplated in proposed life-of-mine mine plans. Preliminary stability classifications and rankings should be updated and preliminary risk assessments undertaken. The results of these analyses should be considered in conjunction with updated (feasibility-level) capital and operating cost estimates to confirm feasibility, support final site selection and develop feasibility-level design parameters. As with pre-feasibility design studies, if these parameters are not compatible with the mine plan, one or more iterative cycles may be required.

2.7 Detailed design and construction

Detailed design may require supplemental site investigations, bulk sampling and materials testing to refine the various constituent models, refine and update ARD management, water management and closure plans, and support detailed analyses. Adjustment of the stability acceptability criteria may also be warranted at this stage if there are material improvements in the confidence of the input parameters and changes in the potential consequences of instability. Detailed stability analyses may include numerical runout modelling, supplemental parametric/sensitivity analyses, detailed stability analyses by year or phase, deformation and dynamic response modelling and quantitative risk assessment. The results of these analyses would be used, with input from the mine planners, to optimise the design. Analyses should be focused on detailed validation of short- and medium-term (up to 5 years) detailed mine plans, as well as key long-range waste dump and stockpile configurations.

During this stage, preliminary operational guidelines, monitoring procedures and response plans (Chapters 11 and 12) would be developed as part of the detailed design. The detailed design parameters would be input into the mine plan, which would be adjusted as required to optimise the design.

This stage would also include site preparation activities, such as stripping and contouring of the foundation, construction of diversions and underdrainage systems (if required), installation of foundation instrumentation and development of access routes. Foundation preparation activities may afford opportunities to verify assumed foundation conditions and to identify the need for modifications to the design to accommodate changed conditions.

2.8 Operation

The design process would continue during the operating phase with ongoing monitoring of foundation preparation and material placement to confirm compliance with the design criteria, supplemental site investigations, field trials and material testing to verify design assumptions, and deformation and performance monitoring. Design adjustments will very likely be required throughout the mine's life to accommodate changes to the mine plan. Remedial designs may be needed to manage instabilities and performance issues, and operating and monitoring guidelines may need adjustment based on documented performance. Periodic updating of the ARD management, water management and closure plans will also be required.

Major waste dump and stockpile expansions and development of new sites that are required to accommodate expansions of the mine that were not contemplated during the initial mine development should be subjected to a site selection and design process similar to that described above.

2.9 Closure

Following completion of mining, or as portions of the waste dump(s) are completed, the closure plan for the facility must be finalised and implemented. Long-term deformation and performance monitoring requirements will also need to be established.

2.10 Study requirements

Table 2.2, which is patterned after Tables 1.2 and 8.1 in Read and Stacey (2009), is a summary of the suggested levels of effort needed to meet expected design confidence targets, and target levels of data confidence, for each project stage.

Table 2.2: Level of effort and confidence by project stage

Component	Project stage					
	Conceptual design (scoping)	Pre-feasibility design	Feasibility design	Detailed design and construction	Operations	Closure
Capital and operating cost estimate accuracy	± 35–50%	± 25–35%	± 15–20%	± 10–15%	± 10–15%	± 10–15%
Site investigations	Compilation and review of regional topographic and geological maps and available remote imaging. May include an initial site reconnaissance and initiation of baseline environmental studies.	Site reconnaissance. Bedrock geology, surficial geology, terrain, geohazard and hydrogeological mapping. Initial test pits/trenches with soil sampling and laboratory testing. Condemnation drilling.	Supplemental mapping, test pits and trenches. Drilling to characterise and sample deep soil deposits and bedrock. Detailed flora, fauna, and archaeological surveys.	Mapping and surveying of prepared foundation. Supplemental sampling of foundation soils. As-built documentation of constructed diversions and drains. Site-specific environmental studies.	Confirm characteristics of fill materials. Ongoing topographic surveys to verify compliance with design. Operational monitoring. Supplemental site investigations for optimisation and expansion. Investigation of instabilities.	Final as-built topographic surveys. Supplemental site investigations as necessary to support detailed closure design.
Characterisation of foundation materials	Identification of major soil types and aerial distribution. Initial estimation of properties based on basic field descriptions, literature values and experience.	Definition of soil units and basic stratigraphy. Laboratory index testing to establish range of properties.	Supplemental index testing to expand database and confirm soil units and properties. Shear strength testing. *In situ* hydraulic conductivity testing in boreholes. Surface percolation/infiltration testing.	Ongoing index and shear strength testing of samples obtained during foundation preparation activities to validate soil types and design properties and improve database reliability.	Laboratory testing of soil samples from expansion areas.	Supplemental sampling and testing as required to support closure.
Geotechnical characterisation of fill materials	Initial estimation of fill properties (density, shear strength, hydraulic conductivity) based on geological descriptions and reconnaissance-level geological mapping and rock/overburden characterisation, literature values and experience.	Preliminary evaluation of strength, durability and gradation of fill materials based on testing of core, geomechanical core logging, and field exposures. Preliminary index testing of overburden materials. Preliminary characterisation of fill according to quality.	Gradation and shear strength testing of small scale samples. Estimate gradation using empirical techniques. Durability testing. Preliminary fill quality model and release schedule.	Gradation and shear strength testing of large-scale, bulk samples. Supplemental durability testing. Refine fill quality model and release schedule.	Ongoing testing of bulk samples to verify gradation and shear strength assumptions. Field verification and as-built documentation of gradation, quality and distribution and design compliance.	Detailed assessment of long-term durability. Verify design compliance and variances with respect to spatial distribution according to fill quality. Laboratory testing of overburden materials for use in closure and reclamation.
Geochemical characterisation of fill materials	Initial geochemical characterisation of waste rock and stockpile types based on geological descriptions and available information from early exploration work. Review regional and district historical geochemistry issues.	Preliminary geochemical classification of fill material types as potentially acid generating, neutral or acid consuming. Define and initiate laboratory testing program; acid–base accounting (ABA) testing of all materials and static testing of potentially acid generating material types.	Static and kinetic testing on samples of all potential acid generating materials.	Ongoing static and kinetic testing; may include site trials.	Ongoing static and kinetic testing; may include site trials.	Supplemental sampling and field and laboratory testing to assess long-term *in situ* geochemical stability of fill materials.

(Continued)

Table 2.2: (Continued)

Component	Project stage					
	Conceptual design (scoping)	Pre-feasibility design	Feasibility design	Detailed design and construction	Operations	Closure
Geological model	Compilation and review of regional literature. Review of advanced exploration mapping and core logging data and preliminary geological model for ore body. Develop conceptual representative geological sections through prospective sites.	Initial 3D geological model of ore body and host rocks supported by outcrop mapping and exploration drilling. Preliminary engineering/surficial/ bedrock geology plans and sections of prospective sites.	Refined 3D geological model of ore body and host rocks (including lithology, alteration, mineralisation and major structure). Detailed engineering/surficial/ bedrock geology plans and sections of selected sites.	Continued refinement of 3D geological model of ore body and host rocks. Verification and refinement of engineering/surficial/ bedrock geology plans and sections of selected sites.	Ongoing validation and reconciliation of 3D geological model of ore body and host rocks based on mapping of benches, infill drilling and refinement of geological interpretations. Ongoing validation and refinement of engineering/surficial/ bedrock geology plans and sections.	Updating of engineering/ surficial/bedrock geology plans and sections based on supplemental investigations.
Geotechnical model	Conceptual 2D (sectional) geotechnical model based on topography, regional bedrock and surficial geology, and results of initial site reconnaissance. Identify potential instability mechanisms.	Preliminary geotechnical model with defined soil and bedrock units and stratigraphy. Properties based on preliminary index testing. Preliminary 2D limit equilibrium (LE) stability analyses of proposed ultimate configuration using bulk fill properties and including estimated phreatic surface based on preliminary hydrogeological modelling.	Advanced geotechnical model with detailed definition and characterisation of soil and bedrock units, stratigraphy and structure/fabric. Fill properties based on laboratory testing and zoned according to quality, as appropriate. Pore pressures incorporated based on advanced hydrogeological modelling. Advanced 2D and/or 3D LE analysis, and possibly numerical modelling/ deformation analysis, of ultimate and critical interim configurations.	Updated/refined geotechnical model based on as-built foundation geometry and conditions, updated shear strength parameters, advanced hydrogeological modelling, and refined fill quality model and release schedule. Supplemental stability analyses as required to validate final design.	Ongoing refinement of the geotechnical model based on updated material properties, as-built configurations, and updated hydrogeological modelling. Supplemental stability analysis (and back analysis) as required to support design modifications, expansions, and design of remedial stabilisation measures.	Updated geotechnical model based on final as-built configuration, documented distribution of fill materials and updated hydrogeological modelling. Supplemental stability analyses as required to support final closure design.
Geochemical model	Initial qualitative assessment of ARD potential. Develop conceptual geochemical model.	Evaluate results of initial laboratory testing and refine geochemical model.	Initial calibration of geochemical model based on preliminary results of longer term (kinetic) laboratory testing.	Ongoing refinement and calibration of geochemical model based on results of longer term (kinetic) laboratory testing and field trials.	Ongoing refinement and calibration of geochemical model based on results of longer term (kinetic) laboratory testing and field trials.	Update and calibrate geochemical model based on the results of laboratory testing and field trials.

Table 2.2: (Continued)

			Project stage			
Component	Conceptual design (scoping)	Pre-feasibility design	Feasibility design	Detailed design and construction	Operations	Closure
Hydrogeological model	Conceptual hydrogeological model based on regional climate/hydrological data, regional bedrock and surficial geology, topography and initial site reconnaissance.	Preliminary numerical groundwater model that incorporates preliminary bedrock and overburden units defined in the preliminary geotechnical model. Steady-state prediction of expected phreatic surface for ultimate dump configuration.	Advanced hydrogeological model that is aligned with the geotechnical model and incorporates detailed soil and bedrock units, stratigraphy, and major regional structures. Steady-state, pre-mining calibration based on available piezometric data, hydrogeological mapping and *in situ* testing. Transient modelling to predict foundation and fill pore pressures in ultimate and critical interim configurations.	Updated and refined hydrogeological model calibrated based on documented piezometric monitoring. Updated predictions of the distribution of transient pore pressures in the foundation and fill.	Ongoing refinement of the hydrogeological model based on piezometric monitoring data and surface flow monitoring. Updated predictions of transient pore pressures in the foundation and fill to support supplemental stability analyses, as required.	Updated hydrogeological model to incorporate closure and reclamation measures (revegetation, infiltration controls) and provide updated pore pressure predictions for post-closure stability analyses.
Hydrological model	Compilation of regional rainfall, temperature, prevailing wind, and streamflow data. Identification of watershed boundaries.	Preliminary hydrological model based on regional data with consideration of orographic effects. Preliminary infiltration and runoff estimates and water balance.	Advanced hydrological model based on regional climate data calibrated to limited site-specific climate and streamflow monitoring data.	Updated/refined hydrological model based on regional and site-specific climate and streamflow monitoring.	Ongoing calibration and refinement of hydrological model based on longer term site monitoring.	Updated hydrological model calibrated to long-term monitoring data to support development of detailed surface water management plans for closure.
Monitoring systems and Trigger Action Response Plans (TARPs)	Commission site climate station.	Shallow standpipe piezometers, stream flow gauging.	Establish long-term environmental monitoring and compliance sites and frequency.	Installation of foundation instrumentation. Establish detailed monitoring plan and TARPs.	Installation of piezometers in the fill to validate/calibrate hydrogeological model. Implement deformation and piezometric monitoring systems and adjust/modify systems and TARPs based on performance.	Develop and implement long-term (closure) groundwater, deformation and settlement monitoring plans and periodic visual inspection protocols.
Overall target levels of confidence	± 50%	50–65%	60–75%	70–85%	80–90%	> 90%

3

WASTE DUMP AND STOCKPILE STABILITY RATING AND HAZARD CLASSIFICATION SYSTEM

Mark Hawley

3.1 Introduction

The 1991 *Investigation and Design Manual – Interim Guidelines* (BCMWRPRC 1991a) introduced the first comprehensive stability and hazard classification system for waste rock dumps (Hawley 2000). Prior classification systems (e.g. Whaler 1979; USBM 1982; MESA 1975; OSM 1989) subdivided waste dumps into a few typical or characteristic types based on the overall foundation and dump configuration, and differentiated failure mechanisms and stability controls, but did not attempt to assign stability or hazard ratings or otherwise develop quantitative indices. The 1991 dump stability rating (DSR) system, which is reproduced herein as Table 3.1, established a numerical index based on consideration of 11 factors. Point ratings were assigned to each factor, weighted according to the authors' perception of their overall importance, and the sum of the individual point ratings was defined as the dump stability rating, or DSR. Higher DSR values inferred lower relative stability and vice versa. The range of possible DSR values (maximum possible value of 1800) was subdivided into four dump stability classes (DSCs), and each DSC was assigned a relative 'failure' or instability hazard descriptor (Table 3.2). For example, dumps with DSR ratings of less than 200 were assigned to DSC I, which was characterised as having 'negligible' instability hazard, and dumps with DSR values of more than 1200 were assigned to DSC IV and were characterised as having a 'high' instability hazard.

It is important to note that the DSR system was never intended as a stand-alone risk assessment tool. It did not include any explicit reference to exposure or the consequence of failure, which is a necessary component of any risk assessment whether it be qualitative or quantitative. Furthermore, the DSR system was introduced with the expectation that it would be subject to trials and

'calibration'. A follow-up program to engage industry practitioners to test and validate the system was envisaged, but unfortunately funding for this and several other related waste dump research initiatives was never secured, and the work was not advanced. Notwithstanding this, the DSR system has been adopted informally by some practitioners, and reporting of DSR values is required by at least one provincial mines regulator in Canada.

One of the key attributes of the DSR system was the recognition that many factors can impact the stability of a waste dump, and it served as a checklist for things to consider when evaluating stability issues and design options for such structures. The system also provided a qualitative way to assess how varying key design parameters or implementing different mitigative measures could affect stability and to compare the experience and behaviour of different waste dumps at a given site and from one site to another. Another application suggested in the *Investigation and Design Manual – Interim Guidelines* was to link the DSR to the level of effort required to investigate, design and construct waste dumps (Table 3.2). Waste dumps with higher DSR values, or that fall into a higher DSC, logically ought to require more investigative and design effort, and more care and monitoring during construction and operations, than dumps with lower DSR values or that fall into a lower DSC.

As part of the current effort to develop an updated and modernised reference text and guidelines for the investigation, design, construction, monitoring and closure of mine waste dumps, dragline spoils and stockpiles, it was decided to revisit and further develop the concept of a stability/hazard classification system or index based on the 1991 DSR system. The rest of this chapter lays out the basis, structure and attributes of the proposed new waste dump and stockpile stability rating and hazard classification (WSRHC) system.

Table 3.1: 1991 dump stability rating (DSR)

Key factors affecting stability		Range of conditions or description	Point rating
Dump height	Low	< 50 m	0
	Moderate	50–100 m	50
	High	100–200 m	100
	Very High	> 200 m	200
Dump volume	Small	< 1 × 10^6 BCM (bank cubic metres)	0
	Medium	< 1 × 10^6 –5 × 10^7 BCM	50
	Large	> 5 × 10^7 BCM	100
Dump slope	Flat	< 26°	0
	Moderate	26–35°	50
	Steep	> 35°	100
Foundation slope	Flat	< 10°	0
	Moderate	10–25°	50
	Steep	25–32°	100
	Extreme	> 32°	200
Degree of confinement	Confined	Concave in plan or section; valley or cross-valley fill, toe buttressed against opposite valley wall; incised gullies that can be used to limit foundation slope during development	0
	Moderately Confined	Natural benches or terraces on slope; even slopes, limited natural topographic diversity; heaped, sidehill or broad valley or cross-valley fills	50
	Unconfined	Convex slope in plan or section; sidehill or ridge crest fill with no toe confinement; no gullies or benches to assist development	100
Foundation type	Competent	Foundation materials as strong or stronger than dump materials; not subject to adverse pore pressure; no adverse geologic structure	0
	Intermediate	Intermediate between competent and weak; soils gain strength with consolidation; adverse pore pressures dissipate if loading rate controlled	100
	Weak	Limited bearing capacity, soft soils; subject to adverse pore pressure generation upon loading; adverse groundwater conditions, springs or seeps; strength sensitive to shear strain, potentially liquefiable	200
Dump material quality	High	Strong, durable; less than ~10% fines	0
	Moderate	Moderately strong, variable durability; 10–20% fines	100
	Poor	Predominantly weak rocks of low durability; greater than ~25% fines, overburden	200
Method of construction	Favourable	Thin lifts (< 25 m thick), wide platforms; dumping along contours; ascending construction; wrap-arounds or terraces	0
	Mixed	Moderately thick lifts (25–50 m); mixed construction methods	100
	Unfavourable	Thick lifts (> 50 m), narrow platforms (sliver fills); dumping down the fall line of the slope; descending construction	200
Piezometric and climatic conditions	Favourable	Low piezometric pressures, no seepage in foundation; development of phreatic surface within dump unlikely; limited precipitation; minimal infiltration into dump; no snow or ice layers in dump or foundation	0
	Intermediate	Moderate piezometric pressures, some seeps in foundation; limited development of phreatic surface within dump possible; moderate precipitation; high infiltration into dump; discontinuous snow or ice lenses in dump or foundation	100
	Unfavourable	High piezometric pressures, springs in foundation; high precipitation; significant potential for development of phreatic surface or perched water tables in dump; continuous layers of snow or ice in dump or foundation	200
Dumping rate	Slow	< 25 BCM/m of crest/d; crest advancement rate < 0.1 m/d	0
	Moderate	25–200 BCM/m of crest/d; crest advancement rate 0.1–1.0 m/d	100
	High	> 200 BCM/m of crest/d; crest advancement > 1.0 m/d	200
Seismicity	Low	Seismic Risk Zones 0 and 1[1]	0
	Moderate	Seismic Risk Zones 2 and 3[1]	50
	High	Seismic Risk Zones 4 or higher[1]	100

Source: After BCMWRPRC (1991a)
Notes: 1. Seismic Risk Zones based on Canadian and USA National Building Codes ca. 1990.

Table 3.2: Dump stability classes and recommended level of effort

Dump stability rating (DSR)	Dump stability class (DSC)	Failure hazard	Recommended level of effort for investigation, design and construction
< 300	I	Negligible	Basic site reconnaissance, baseline documentation; minimal laboratory testing; routine check of stability, possibly using charts; minimal restrictions on construction; visual monitoring only
300–600	II	Low	Thorough site investigation; test pits, sampling may be required; limited laboratory index testing; stability may or may not influence design; basic stability analysis required; limited restrictions on construction; routine visual and instrument monitoring
600–1200	III	Moderate	Detailed, phased site investigation; test pits required, drilling or other subsurface investigations may be required; undisturbed sampling may be required; detailed laboratory testing, including index properties, shear strength and durability testing likely required; stability influences and may control design; detailed stability analysis, possibly including parametric studies, required; Stage II detailed design report may be required for approval/permitting; moderate restrictions on construction (e.g. limiting loading rate, lift thickness, material quality); detailed instrument monitoring required to confirm design, document behaviour and establish loading limits
> 1200	IV	High	Detailed, phased site investigation; test pits, and possibly trenches, required; drilling and possible other subsurface investigations probably required; undisturbed sampling probably required; detailed laboratory testing, including index properties, shear strength and durability testing probably required; stability considerations paramount; detailed stability analysis, probably including parametric studies and full evaluation of alternatives probably required; Stage II detailed design report probably required for approval/permitting; severe restrictions on construction (e.g. limiting loading rate, lift thickness, material quality); detailed instrument monitoring required to confirm design, document behaviour and establish loading limits

Source: After BCMWRPRC (1991a)

3.2 Waste dump and stockpile stability rating and hazard classification system

The structure of the new WSRHC system is illustrated in Fig. 3.1. The system incorporates many of the concepts that were used in the 1991 DSR system; however, the structure is somewhat different. The new system requires evaluation of 22 key factors or attributes that are thought to affect stability (11 more than in the 1991 DSR system). These factors have been organised into seven groups. Numerical ratings are assigned to each factor, and the sum of these ratings defines the waste dump and stockpile stability rating (WSR). The maximum possible WSR is 100, with a higher rating indicating a more stable configuration. In the 1991 DSR system, the maximum possible rating was 1800, and a higher rating indicated a less stable configuration, so it was really an 'instability' rating. Similar to the 1991 DSR system, WSR values have been subdivided into several waste dump and stockpile hazard classes (WHCs). Waste dumps or stockpiles with a very high WSR rating (more than 80) are assigned to WHC I and are characterised as presenting a relatively very low potential for instability (i.e. a very low instability hazard). Conversely, waste dump or stockpiles with a very low WSR rating (less than 20) are assigned to WHC V and are characterised as presenting a relatively very high potential for instability (i.e. a very high instability hazard).

Intermediate classes (WHC II, III and IV) represent waste dumps and stockpile with an intermediate potential for instability/intermediate instability hazard. The architecture of the new system is similar in concept to the Rock Mass Rating (RMR) system (Bieniawski 1976) and the Geological Strength Index (GSI) classification (Hoek *et al.* 2002), both of which have gained wide acceptance and are well understood by most geotechnical practitioners in the mining industry.

The WSRHC system attempts to strike a balance between complexity and utility. While the number of factors that must be evaluated may at first seem daunting, many are based on objective parameters that are self-explanatory and should be relatively easy to obtain from plans or records, and detailed benchmark descriptions are provided to help the user select values for the more subjective factors. To simplify application, the system has been coded into an Excel workbook with each group of factors comprising a separate worksheet.[1] Key parameter ratings may be input directly into the individual worksheets, and the component indices and final stability rating value are calculated automatically.

The numerical values assigned to each factor and group have been weighted to reflect their relative importance. It is anticipated that, as experience using the system is gained, adjustments to these weightings, to the individual

---

[1] Copies of the Excel workbook may be obtained free of charge by contacting the author via email at info@piteau.com.

parameter values and ranges, and perhaps even to the factor hierarchy and content, may be required to refine and improve the system. In this regard, the author encourages direct feedback from practitioners regarding their experience using the system and suggestions for improvement. One of the goals of this work is to develop a comprehensive worldwide database of waste dump and stockpile designs and performance, and it is hoped that the WSRHC system will provide a consistent vehicle for gathering and cataloguing this data.

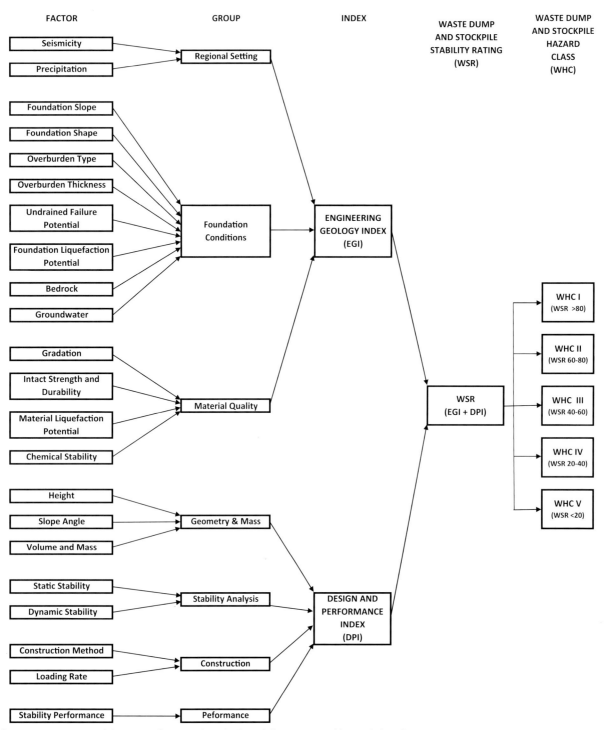

Figure 3.1: Structure of the waste dump and stockpile stability rating and hazard classification (WSRHC) system

The WSRHC system was initially intended to be applied only to conventional waste dumps and stockpiles. However, early trials indicate that it may also be useful in helping to characterise the stability and behaviour of dragline spoils. Future versions of the WSRHC system may include a specific dragline spoil adaptation.

The following sections provide detailed descriptions of the various factors that compose the WSRHC system, and guidance in evaluating individual factor ratings. They are organised into the seven groups as illustrated in Fig. 3.1. Tables summarising the factor descriptions and numerical ratings for each group, and corresponding to individual worksheets in the companion Excel workbook, are shown in Tables 3.3 to 3.9.

In many cases, more than one criterion or parameter is shown for a given factor. Where it is possible to evaluate more than one of these criteria or parameters, or the user cannot decide between values or ranges, an average rating value should be chosen.

Many of the factors that combine to define the WSR are subject to variability throughout the development cycle of a waste dump or stockpile, and the designer needs to be aware of this variability. As a general rule, individual factor rating values that are representative of the least favourable conditions throughout the life cycle of the structure should be chosen. Alternatively, a range of rating values can be calculated based on the range of input values for different factors to help understand this variability. Another approach would be to prepare stability ratings and hazard classifications for each major development phase. This information may help to evaluate how the stability of the waste dump or stockpile might vary throughout its development cycle, and to objectively identify critical phases.

3.2.1 Regional setting

The Regional Setting category (Table 3.3) includes factors that are related to the geographic location and climate of the site. The key factors in this category are seismicity and precipitation. Other regional/climate factors, such as temperature, humidity and wind speed/direction, may also be important in the design of waste dumps and stockpiles (e.g. to optimise dust suppression and for snow and avalanche management), but either do not impact stability (e.g. wind) or are considered indirectly in other groups. For example, hot and humid environments may accelerate degradation of dump or stockpile materials and impact the Chemical Stability and Intact Strength and Durability factors in the Material Quality group, and prolonged freezing could impact the Permafrost factor in both the Foundation Conditions and Material Quality groups.

Collectively, Regional Setting factors are weighted to account for a maximum of 10 points, or 10% of the maximum possible WSR (see Section 3.2.8).

Table 3.3: Regional setting factors and ratings

Factors[1]	Ratings				
Seismicity	Very High	High	Moderate	Low	Very low
Expected peak ground acceleration (g): based on 1:475 year return period event/ 10% probability of exceedance in 50 years	> 0.4	0.2–0.4	0.1–0.2	0.05–0.1	< 0.05
Expected peak ground acceleration (g): based on maximum credible earthquake (MCE)	> 0.6	0.4–0.6	0.2–0.4	0.1–0.2	< 0.1
Rating	0	0.5	1	1.5	2
Precipitation	Very High	High	Moderate	Low	Very low
Average annual precipitation: rainfall (mm)	> 2000	1000–2000	350–1000	100–350	< 100
Average annual precipitation: snowfall (cm)	> 200	100–200	35–100	10–35	< 10
Total annual precipitation: equivalent rainfall (mm)	> 2000	1000–2000	350–1000	100–350	< 100
Rating[2]	0	2	4	6	8
Regional Setting rating[3]				Maximum possible rating:	10

Notes:
1. Select a rating for each factor. Where more than one criteria is shown for a given factor, or you cannot decide between two ratings, select an average or intermediate rating.
2. For sites that experience intense seasonal rainfall or rapid runoff events, decrease the rating value by 1 point.
3. The sum of the ratings for the individual factors is the Regional Setting rating.

3.2.1.1 Seismicity

It is generally accepted within the geotechnical community that the design of certain types of earthfill and rockfill embankments, such as water retention dams and tailings dams, must take into account the potential impact of earthquakes, and this principal is enshrined in most mining codes worldwide. There is less agreement among practitioners as to the impact of earthquakes on waste dumps and stockpiles, and to author's knowledge there have been no documented case studies that convincingly link a large-scale instability of a waste dump or stockpile to a specific seismic event. Nevertheless, regulatory agencies in some jurisdictions, particularly those in high seismic risk zones (e.g. Chile and Peru), require that seismicity be explicitly considered in the design of all mine structures, including waste dumps and stockpiles. In view of this requirement, and to promote universal acceptance, it was felt that seismicity needed to be included as one of the key stability factors in the WSR. Further discussion regarding the selection of seismic design criteria and analysis of the impact of seismic events on waste dumps and stockpiles is given in Chapter 8.

The Seismicity rating is evaluated according to the expected ground acceleration at the site. Two alternative or complementary criteria are provided based on commonly used seismic design criteria: the expected peak ground acceleration based on the 1:475 year return period seismic event (also equivalent to the expected ground acceleration with a 10% probability of exceedance in 50 years), and the peak ground acceleration due to the maximum credible earthquake. Values for these parameters for a given site should be found in seismic risk studies that are routinely carried out to support the civil design components at the mine site, or as input for the design of tailings dams. Some jurisdictions and government agencies also provide maps and/or interactive websites that can be used to estimate these values at a given site (e.g. GSHAP 1999; NRC 2013; USGS 2015).

The Seismicity factor has a maximum rating of 2 points, or 2% of the maximum possible WSR. This relatively low individual weighting reflects the limited impact that seismicity is thought to have on waste dump and stockpile stability. It should be noted, however, that potential earthquake impacts are also indirectly included in the assessment of liquefaction potential under the Foundation Conditions and Material Quality groups, and directly in the assessment of the overall dynamic stability factor under the Stability Analysis group. When all of these factors are considered collectively, and depending on site-specific circumstances, high seismicity could have a significantly negative impact on the WSR.

3.2.1.2 Precipitation

Waste dumps and stockpiles constructed in wet environments have a higher incidence of failure and tend not to perform as well as those constructed in arid environments. This tendency is demonstrated in the results of the 1991 waste dump survey where surveyed waste dumps that reported poor or very poor performance (16 of 84) were subjected to annual precipitation rates in excess of 1000 mm, whereas the surveyed dumps that reported good or very good performance (62 of 84) were subjected to annual precipitation rates that were typically less than 1000 mm. The 2012 waste dump survey reported only one dump out of 49 validated results with poor performance, but that embankment was also subjected to an annual precipitation of more than 1000 mm.

The Precipitation factor is rated based on equivalent annual precipitation, including snowfall. This information is typically summarised in environmental impact assessments or general project descriptions, and most sites maintain climate stations that record precipitation data.

It is recognised that total annual precipitation alone is not sufficient to describe the potential impact of precipitation on waste dump and stockpile stability. Sites that are subject to strong seasonal variations (wet season/monsoons) with periods of intense rainfall, or rain-on-snow events during freshet periods that result in rapid snowmelt and high instantaneous runoff, are likely more susceptible to precipitation-related instability that those that receive an equal annual level of precipitation distributed evenly throughout the year. For sites subject to high rainfall intensity or runoff events, the Precipitation rating should be reduced by 1 point. For example, if the site is subject to a moderate level of annual precipitation of between 350 and 1000 mm (equivalent to a Precipitation rating of 4), but is also subject to rapid snowmelt and high runoff events during the freshet, the Precipitation rating should be reduced to 3.

The Precipitation factor has a maximum rating of 8 points, or 8% of the maximum possible WSR. This relatively high individual factor weighting reflects the significant potential impact that precipitation is expected to have on waste dump and stockpile stability. In addition, high precipitation levels tend to result in higher natural groundwater levels that could indirectly and adversely affect both the Groundwater and Liquefaction Potential factors in the Foundation Conditions group, and the Material Liquefaction Potential in the Material Quality group. This collective impact of precipitation on the WSR can be substantial and highlights the importance of this factor.

3.2.2 Foundation conditions

The Foundation Conditions group (Table 3.4) includes factors that are related to the physical attributes of the foundation or footprint of the waste dump or stockpile. The key factors in this group that are thought to potentially affect stability and are applicable to most waste dumps and stockpiles are the topography (Foundation Slope and Foundation Shape), the type and thickness of the foundation soils (Overburden Type), the competency and structure of the underlying bedrock, and groundwater conditions. Two other factors that could potentially have substantial negative impacts on stability but are applicable only to a relatively small subset of waste dumps and stockpiles are the Undrained Failure Potential factor and the Foundation Liquefaction Potential factor.

Collectively, Foundation Conditions factors are weighted to account for a maximum of 20 points, or 20% of the maximum possible WSR. Note that the Undrained Failure Potential and Foundation Liquefaction Potential ratings are negatively weighted, and hence only affect the WSR in unique cases where they may be applicable.

Table 3.4: Foundation conditions factors and ratings

Factor[1]	Ratings				
Foundation Slope	Very Steep	Steep	Moderate	Gentle	Flat; benched bedrock slope; pit backfills
Average overall foundation slope angle (°)	> 32	25–32	15–25	5–15	< 5
Rating	0	1	2.5	4	5
Foundation Shape[2]					
Section Shape	Convex on Steep to Very Steep slopes	Convex on Moderate slopes; Concave or Planar on Steep to Very Steep slopes	Convex on Gentle slopes; Planar or Concave on Moderate slopes	Planar or Concave on Gentle slopes	Planar or Concave on Flat or very irregular slopes
Plan Shape	Slopes with a pronounced convex plan shape ('nose')	Large radius convex slopes	Planar slopes with no lateral confinement	Concave slopes and wide valleys that provide limited natural confinement	Narrow valleys or gullies that provide substantial natural confinement
Rating	0	0.5	1	1.5	2
Overburden Type	Type I	Type II	Type III	Type IV	Type V
Description	Highly organic soils; very soft to soft silts and clays; very sensitive clays; other very weak soils	Soft to firm, weak or sensitive fine grained soils (e.g. lacustrine deposits, silts and clays, fine-grained residual soils)	Alluvial deposits; loose to moderately dense sands and gravels; mixed-grained colluvial deposits; sandy residual soils; stiff fine-grained soils	Highly weathered but coherent bedrock; competent talus deposits; moderately dense to dense, coarse-grained soils; moderately dense, mixed-grained moraine (glacial till)	Very dense, mixed-grained moraine (glacial till) and other very competent or hard soils; perpetually frozen soils with negligible potential for creep due to embankment loading; competent bedrock
Rating	0	1	2	3	4
Overburden Thickness (m)	> 5	3–5	1–3	0.3–1	< 0.3
Rating	0	0.5	1	1.5	2
Undrained Failure Potential[3]	Very High[5]	High[6]	Moderate	Low	Negligible
Description	Saturated, normally or underconsolidated, compressible Type I or II soils that have very low hydraulic conductivity and behave as low-strength S_u (frictionless) materials; very high potential for generation of excess pore pressures when loaded rapidly	Saturated, normally or slightly overconsolidated, compressible Type I or II soils that have low hydraulic conductivity and behave as Mohr-Coulomb (c-ϕ) materials; high potential for generation of excess pore pressures when loaded rapidly	Saturated, normally consolidated, mixed or fine-grained Type III or IV soils with low to moderate hydraulic conductivity and moderate potential for generation of excess pore pressures when loaded rapidly; unknown potential for undrained failure	Unsaturated, normally or overconsolidated, mixed or fine-grained Type III or IV soils with moderate hydraulic conductivity and low potential for generation of excess pore pressures when loaded rapidly; low potential for undrained failure but cannot be fully discounted	Heavily overconsolidated Type III, IV or V mixed-grained soils or competent bedrock or granular Type III or IV soils with high hydraulic conductivity, high strength and negligible potential for generation of excess pore pressures when loaded rapidly
Rating	–20	–10	–5	–2.5	0

(Continued)

Table 3.4: (Continued)

Factor[1]	Ratings				
Foundation Liquefaction Potential[4]	Very High[5]	High[6]	Moderate	Low	Negligible
Description	Very uniform; very loose; minimal plastic fines; open, clast supported structure; high void ratio; rounded clasts; saturated	Extra-sensitive clays and extremely weak soils	Moderate (or unknown) liquefaction potential	Low liquefaction potential but cannot be fully discounted	Well graded; dense; high content of plastic fines; matrix supported structure; low void ratio; angular clasts; dry
Rating	−20	−10	−5	−2.5	0
Bedrock	Type A	Type B	Type C	Type D	Type E
Competency[7]	Very weak and/or highly clay altered, sheared or highly fractured rocks; laminated, highly carbonaceous coal measures; phyllite; flysch; GSI/RMR < 20; Q < 1	Fine-grained sedimentary rocks; moderately weathered or altered rocks; moderately to intensely fractured; GSI/RMR 20–40; Q 1–4	Moderately competent; moderately fractured; slightly weathered/ altered; GSI/RMR 40–60; Q 4–10	Competent, hard, unweathered/ unaltered, blocky; GSI/ RMR 60–80; Q 10–40	Very competent, unweathered, unaltered, hard, massive; GSI/RMR > 80; Q > 40
Structure	Adversely oriented faults or shear zones; potential for structurally controlled foundation failure	Adversely oriented, continuous joints; potential for foundation failure on well-developed fabric anisotropy	Limited (or unknown) potential for foundation failure on major structure or moderately developed fabric anisotropy	Negligible potential for foundation failure on major structure or poorly developed fabric anisotropy	No adverse structure or fabric
Rating	0	1	2	3	4
Groundwater	High		Moderate		Low
Expected Groundwater Conditions	Groundwater table at surface; active discharge or seepage; strong upward gradients; potential for generation of high pore pressures in foundation due to embankment loading		Groundwater table > 5 m below ground surface; limited potential for development of adverse pore pressures in foundation due to embankment loading		Groundwater table at great depth; negligible potential for adverse pore pressures in foundation
Rating	0		1.5		3
Foundation Conditions rating[8]				Maximum possible rating:	20

Notes:

1. Select a rating for each factor. Where more than one criterion is shown for a given factor, or you cannot decide between two ratings, select an average or intermediate rating.

2. Choose the shape that best describes the geometry of the foundation in plan and section. Convex foundations steepen towards the toe in section and lack lateral confinement in plan; concave foundations flatten towards toe in section and provide lateral confinement in plan.

3. Evaluation of the potential for undrained foundation failure may require detailed *in situ* investigations and specialised laboratory testing. If unknown or unsure, use a default value of −5 and consult a geotechnical specialist.

4. Evaluation of the liquefaction potential of foundation soils may require detailed *in situ* investigations. If unknown or unsure, use a default value of −5 and consult a geotechnical specialist.

5. If the Undrained Failure Potential or Foundation Liquefaction Potential is judged to be Very High, the waste dump or stockpile should be classified as WHC V (Very High Hazard), regardless of the WSR.

6. If the Undrained Failure Potential or Foundation Liquefaction Potential is judged to be High, the waste dump or stockpile should be classified as WHC IV (High Hazard), unless the WSR is ≤ 20, in which case it should be classified as WHC V (Very High Hazard).

7. See Hoek *et al.* (2013) for a description of Geological Strength Index (GSI), Bieniawski (1976) for a description of RMR, and Barton *et al.* (1974) for a description of Q.

8. The sum of the ratings for the individual factors is the Foundation Conditions rating.

3.2.2.1 Foundation slope

The slope of the foundation of a waste dump or stockpile has a direct influence on the global stability and general performance of the structure. Waste dumps and stockpiles constructed on steep foundations are much more likely to be unstable or to perform poorly from a deformation perspective than those constructed on flat foundations. This tendency is demonstrated in the results of the 1991 waste dump survey, where four of 14 (29%) surveyed dumps that were constructed on steep slopes (i.e. foundation slopes greater than 25°) were characterised as having poor or very poor overall performance, whereas only 12 of the 70 (17%) remaining dumps in the survey that were constructed on slopes less than or equal to 25° reported poor or very poor performance.

The Foundation Slope factor is characterised based on the average overall foundation slope angle (Fig. 3.2). It is important to note that the average overall foundation slope angle for the ultimate dump may be flatter or steeper than average overall foundation slope angles during interim dump phases. While the Foundation Shape factor (see below) attempts to capture some of this variability, if the Foundation Slope factor varies widely by phase, it should be selected based on the least favourable overall slope angle throughout the development cycle of the waste dump or stockpile, not just on the ultimate configuration, unless the classification is being done for closure purposes only. Average overall foundation slope angles can be measured directly from dump sequence plans and representative sections.

3.2.2.2 Foundation shape

The shape of the foundation of a waste dump or stockpile may also have an impact on the stability and general performance of the structure. Waste dumps and stockpiles that are confined by natural topography tend to perform better that those that are not confined. Two criteria have been defined to help characterise the degree of confinement afforded by the shape of the foundation: the nature of the vertical profile of the foundation as it appears in cross-section (Section Shape) and the plan shape as represented by the topographic contours (Plan Shape).

The Section Shape rating is determined by comparison of the vertical profile of the foundation to three idealised shapes: convex, planar, and concave, and with reference to the overall steepness of the foundation slope. For slopes that have a similar overall foundation slope, concave shapes in which the foundation slope progressively flattens from crest to toe generally result in better long-term stability than those with convex shapes (foundation slope progressively steepens from crest to toe).

Planar foundation slopes would logically fall between the concave and convex cases; however, if the foundation slope is gentle, there may be little difference between the performance of waste dumps or stockpiles constructed on concave and planar foundations. In addition, foundations that have irregular sectional profiles with rolls or lateral ridges that disrupt potential dump-foundation shear interfaces tend to perform better than uniform foundations.

The Plan Shape parameter is characterised based on the shape of the foundation in plan or map view and the shape of the elevation contours. Waste dumps and stockpiles that are constructed in narrow valleys or gullies that provide natural lateral confinement tend to perform better than those founded on uniform slopes or slopes with convex plan shapes or 'noses' with little or no lateral confinement. Users should select a Foundation Shape

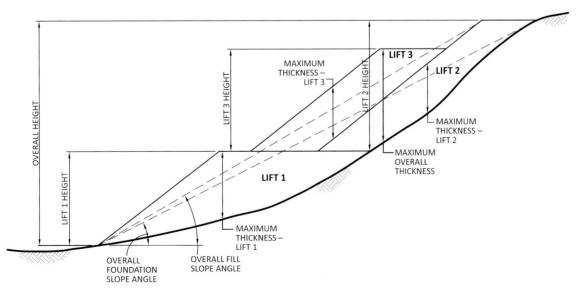

Figure 3.2: Geometry parameters

rating between 0 and 2 based on the descriptions of the Section Shape and Plan Shape parameters given above and summarised in Table 3.4.

3.2.2.3 Overburden type

In the context of the WSRHC system, the term 'overburden' refers to any soil or fill material that may underlie the waste dump or stockpile, including residual soils derived from *in situ* weathering of bedrock. Except as noted below, overburden does not include moderately weathered or fresh bedrock. For transitional materials, such as saprolite or 'sap-rock', the material should be characterised as overburden or bedrock depending on its expected behaviour.

Overburden materials can have a substantial impact on the stability of a waste dump or stockpile, and in some cases may control design. For the purposes of the WSR, five general overburden types have been defined:

Type I: Highly organic and very weak soils, such as amorphous peat and very soft and/or very sensitive clays with very little shear strength or bearing capacity (e.g. undrained shear strength typically less than 10 kPa).

Type II: Soft to firm, weak, fine-grained soils, such as lacustrine and glaciolacustrine deposits, alluvial silts and clays, and fine-grained residual soils (e.g. weak, clay-rich saprolites). These materials would be expected to have undrained shear strengths less than 25 kPa and/or low effective frictional strength (i.e. $\phi' < 15°$). Fibrous peat deposits and some 'topsoil' materials that exhibit some frictional shear strength might also be included in this group, depending on their composition and expected behaviour.

Type III: Most medium to coarse-grained alluvial deposits, such as loose to moderately dense sands and gravels, mixed-grained colluvial deposits, sandy residual soils (e.g. silty-sandy saprolites), and stiff fine-grained soils. These soils would be expected to behave as predominantly frictional materials with $\phi' > 25°$, or cohesive materials with undrained shear strengths of more than 50 kPa.

Type IV: Highly weathered but coherent bedrock or 'sap-rock'; competent angular talus deposits; moderately dense to dense, coarse-grained soils; moderately dense, mixed-grained moraine (i.e. glacial till); and very stiff, consolidated fine-grained soils. These soils would be expected to behave as predominantly frictional materials with $\phi' > 30°$, or as cohesive materials with undrained shear strengths of more than 200 kPa.

Type V: Dense, mixed-grained moraine (glacial till) and other very competent or hard soils, or perpetually frozen soils with negligible potential for creep due to

waste dump or stockpile loading. These soils would be expected to behave as predominantly frictional materials with $\phi' > 35°$, or as cohesive materials with undrained shear strengths of more than 500 kPa. As a default, Type V also includes cases where the dump or stockpile would be founded directly on competent bedrock. Overburden that is perpetually frozen may also be characterised as Type V; however, this should be supported by detailed *in situ* investigations and laboratory testing that conclusively demonstrates that the material will remain frozen and will not be subject to creep or stress-induced thawing under the expected loads, or subject to thawing as a result of changes to the groundwater flow regime.

Users should select an Overburden Type rating between 0 and 4 based on the descriptions given above and summarised in Table 3.4. If the Overburden Type varies throughout the footprint of the structure, an intermediate rating value should be chosen, weighted based on the expected distribution of the different overburden types. If the designer specifies that a weak surficial overburden layer is to be removed before waste dump or stockpile construction, then the Overburden Type rating should be based on the remaining overburden that underlies this layer.

3.2.2.4 Overburden thickness

The thickness of the overburden can also play a role in stability. Where a weak overburden deposit forms a very thin veneer over more competent material, it may be displaced during placement of the dump or stockpile material, or the dump or stockpile material may penetrate the thin, weak layer and be keyed into the more competent underlying material. It may also be feasible to remove weak layers if they are not too thick. On the other hand, even thick deposits of otherwise competent soils may contain weak lenses or layers that could adversely affect stability, and the thicker the deposit, the greater the likelihood that a weak layer may be present. As summarised in Table 3.4, the Overburden Thickness rating is directly linked to the average overburden thickness and is intended to recognise that the potential for instability increases incrementally with increasing overburden thickness, regardless of the Overburden Type. If the designer specifies that a weak overburden layer is to be removed before construction, then the Overburden Thickness rating should be based on the thickness of the remaining overburden.

3.2.2.5 Undrained failure potential

Undrained foundation failure can occur when weak foundation soils are loaded too quickly. Rapid loading of saturated (or near saturated) fine-grained soils with low hydraulic conductivity can increase pore pressures within the soil faster than they can be dissipated by normal consolidation processes. These types of failures can occur quickly once the undrained strength of the soils is exceeded, and displacements can be very large. A recent and well-publicised example of an undrained foundation failure is the 2014 failure of the Mount Polley tailings dam in British Columbia (Morgenstern *et al.* 2015).

Overburden types that are most susceptible to undrained failure include saturated, normally or slightly overconsolidated clays and silts. Mixed-grained and residual soils and saprolites whose shear strength properties are dominated by plastic clays with low undrained strength may also be susceptible and, under the right loading and groundwater conditions, even moderately strong overburden types that behave as Mohr-Coulomb materials and whose shear strength behaviour is composed of both cohesive and frictional components may be susceptible.

As indicated above, undrained failure is typically associated with a combination of factors (e.g. weak, fine-grained soils, adverse groundwater conditions, rapid loading), all of which are represented elsewhere within the WSRHC system. However, given its potentially severe impact on dump and stockpile design and construction, it was decided to incorporate a specific factor to emphasise the hazard associated with undrained failure. The default rating value for a negligible undrained failure potential was set at zero. At the other extreme, where the potential for undrained failure is very high, a rating value of –20 was assigned to ensure that such conditions are appropriately flagged. Users should select an Undrained Failure Potential rating based on the descriptions given in Table 3.4. To encourage practitioners to explicitly address the issue of undrained failure, if the potential for undrained failure is not known or has not been evaluated, a rating of –5 is assigned. Users that are not experienced geotechnical practitioners should consult a geotechnical specialist before assigning an Undrained Failure Potential rating.

3.2.2.6 Foundation liquefaction potential

In rare cases, it is possible that overburden deposits within the footprint of a waste dump or stockpile could be susceptible to liquefaction. Liquefaction may occur if the pore pressure in the material approaches or exceeds the overburden or confining stress. High pore pressures may be induced by cyclical shaking, such as may occur during an earthquake. Seismic shaking can also result in collapse of a loose, dispersed soil structure, which in turn can result in a sudden increase in pore pressure. Earthquake-induced liquefaction is also commonly referred to as dynamic liquefaction. Liquefaction may also be induced by high pore pressures due to artesian flow conditions, excess pore pressures induced by construction activities, or sudden loss of confinement (static liquefaction). Liquefied

material behaves more like a fluid than a soil and is subject to rapid loss of strength and large displacements when unconfined.

Overburden materials that are most susceptible to liquefaction have several common attributes. They are typically composed of very uniformly graded materials with rounded particles. They are also usually very loose, have a high void ratio, are saturated, and have a low clay content. Deposits of uniformly graded, loose, saturated silt and fine sand (e.g. some types of tailings deposits) are often highly susceptible to liquefaction, but coarser deposits of sand or gravel, and even cobble-sized material, may also be susceptible under certain conditions. Conversely, materials that are dense, well graded, unsaturated and contain appreciable amounts of clay and angular particles have a low susceptibility to liquefaction.

Because most overburden deposits upon which waste dumps and stockpiles are founded are not susceptible to liquefaction, the default rating value for a negligible liquefaction potential in the WSRHC system was set at zero. At the other extreme, where the liquefaction potential is very high, a rating value of –20 was assigned to ensure that such conditions are appropriately flagged. Users should select a Foundation Liquefaction Potential rating based on the material descriptions given in Table 3.4. To encourage practitioners to explicitly address the issue of liquefaction, if the liquefaction potential is not known or has not been evaluated, a rating of –5 is assigned. Users who are not experienced geotechnical practitioners should consult a geotechnical specialist before assigning a Foundation Liquefaction Potential rating.

3.2.2.7 Bedrock

The competency of the bedrock underlying the foundation can also influence or control stability, particularly in the case of very high waste dumps founded on weak bedrock foundations. The potential for the occurrence of adversely oriented structural discontinuities (e.g. faults, contacts, bedding, joints) in the bedrock foundation also needs to be considered.

Five bedrock foundation types have been defined and are characterised on the basis of general competency or strength and structural conditions as follows:

Type A: Very weak and/or highly clay altered, sheared or highly fractured rocks; laminated, weakly cemented sedimentary rocks (e.g. mudstones, claystones, siltstones, bentonitic or glauconitic sandstones, highly carbonaceous coal measures); low grade, clay- or mica-rich, poorly indurated, fissile metamorphic rocks (e.g. phyllites, mica schists, shales). Very poor-quality rock masses with a GSI (Hoek *et al.* 2013) of less than 20, an RMR (Bieniawski 1976) of less than 20, or a tunnelling quality index (Q) (Barton *et al.* 1974) of less than 1 would also be included in this type.

Bedrock foundations that contain discrete, persistent, adversely oriented faults or bedding/stratification joints or shears, or very persistent (slope-scale) joints, would also be classified as Type A. Adverse orientations would include structures that are parallel to or are undercut by the slope, or which in combination could form unstable planar, wedge or complex failure modes that could destabilise the foundation under expected dump or stockpile loading.

Type B: Fine-grained sedimentary rocks that are more competent than those that classify as Type A, and moderately weathered or altered and/or moderately to intensely fractured, poor-quality rock masses with GSI/RMR values in the range 20 to 40 or Q values in the range 1 to 4.

Bedrock foundations that do not contain discrete, persistent, adversely oriented structures as described above for Type A foundations, but are characterised by well-developed, adversely oriented structural fabric (e.g. discontinuous bedding or foliation joints, cross-joints, foliation joints, veins) with the potential for development of anisotropy-controlled foundation failure would be classified as Type B. Type B would also include bedrock foundations that are poorly understood, have not been investigated, or are situated in geological formations that could be subject to structurally controlled instability or low overall competency. This investigation-level default is intended to encourage designers to undertake at least a minimum level of *in situ* investigation, including condemnation drilling and reconnaissance mapping, so that the condition of the underlying bedrock can be rationally assessed.

Type C: Most moderately competent, moderately fractured and/or slightly weathered/altered rocks, and fair quality rock masses with GSI/RMR values in the range 40 to 60 or Q values in the range 4 to 10. The stratigraphy and structure of the foundation is reasonably well understood based on the results of condemnation drilling and reconnaissance mapping, and the potential for bedrock foundation instability is considered to be limited (but not negligible). Type C would also include cases where foundation investigation has been limited, but the underlying geological formations are expected to be reasonably competent, and the regional structure is expected to be favourable.

Type D: Competent, hard, unweathered/unaltered, blocky, good quality bedrock with GSI/RMR values in the range 60 to 80 or Q values in the range 10 to 40, and with negligible potential for foundation failure on major structure(s) or fabric anisotropy. As a further condition, Type D foundations must be supported by a detailed *in situ* investigation.

Type E: Very competent, very hard, unweathered, unaltered, massive, very good quality bedrock with GSI/RMR values higher than 80 or Q values higher than 40, and no adversely oriented discrete structures or fabric. As with Type D, Type E foundations must be supported by detailed *in situ* investigations.

Users should select a Bedrock rating between 0 and 4 based on the descriptions given above and summarised in Table 3.4.

3.2.2.8 Groundwater

Groundwater levels, hydraulic gradients, flow directions and the hydraulic conductivities of the materials that compose both the foundation and the waste dump or stockpile influence the pore pressures in the foundation. High groundwater levels and pore pressures in the foundation can adversely affect the static stability of the structure and can also create conditions (e.g. saturation) that can increase the potential for liquefaction failure. Waste dumps and stockpiles founded in groundwater discharge areas with upward gradients, such as valley bottoms or topographic lows, are more likely to experience elevated pore pressures and saturated conditions than those founded in recharge areas with downward gradients, such as well-drained slopes or topographic highs. Construction of waste dumps and stockpiles can also change the natural groundwater levels and flow regime by changing catchment areas or infiltration rates; hence, an evaluation of the groundwater conditions in the foundation needs to consider both the natural conditions and impact of dump or stockpile construction.

The potential impact of groundwater on the stability of the waste dump or stockpile is rated as either high, moderate or low by comparison of the known or expected conditions to the benchmark descriptions as detailed in following and summarised in Table 3.4:

■ **High:** A high groundwater condition would be characterised by the following criteria:
  → The natural groundwater table is at or near surface, or is expected to rise to or above the original ground surface during or following construction of the waste dump or stockpile.
  → The base or toe of the waste dump or stockpile is located in a discharge or seepage area, or encroaches on a seasonal or perennial wetland, lake or stream.
  → There is a strong upwards gradient in the base or toe of the waste dump or stockpile.
  → There is a potential for generation of high pore pressures in the foundation due to waste dump or stockpile loading.

■ **Moderate:** A moderate groundwater condition would be characterised by the following criteria:
  → The maximum seasonal elevation of the groundwater table is more than 5 m below the natural ground surface and is not expected to rise appreciably as a result of waste dump or stockpile construction.
  → Groundwater flow is either parallel to the slope or there is a downward gradient.
  → There is limited potential for development of adverse pore pressures in the foundation due to waste dump or stockpile loading.

■ **Low:** A low groundwater condition would be characterised by the following criteria:
  → The groundwater table is at great depth, below the base of the deepest potential critical failure surface.
  → There is a strong downward gradient.
  → There is negligible potential for adverse pore pressures in the foundation.

Users should select a Groundwater rating between 0 and 3 based on the descriptions given above and summarised in Table 3.4. An intermediate rating value between high (0) and moderate (1.5) should be selected if there is no evidence of discharge, seepage or standing water, but the depth to the groundwater table and potential for development of adverse pore pressures are unknown.

3.2.3 Material quality

The Material Quality group (Table 3.5) includes factors that are related to the physical attributes of the materials used to construct the waste dump or stockpile and collectively determine the shear strength, deformational behaviour and hydrological characteristics of the structure. The key factors in this category are the gradation (particle size distribution), intact strength and durability, and chemical stability. A fourth factor that could potentially have a substantial negative impact on stability, but is applicable only to relatively small subset of waste dumps and stockpiles, is the liquefaction potential of the material.

Collectively, Material Quality factors are weighted to account for a maximum of 20 points, or 20% of the maximum possible WSR. Note that the Material Liquefaction Potential ratings are negatively weighted and only affect the WSR in unique cases where this factor may be applicable.

Additional details on material properties and testing are given in Chapter 5.

Table 3.5: Material quality factors and ratings

Factors[1]	Ratings				
Gradation	Very Fine Grained		Mixed Grained		Very Coarse Grained
% Fines (passing # 200 Sieve; < 0.075 mm)	> 50%	25–50%	10–25%	5–10%	< 5%
% Greater than 75 mm	< 10%	10–25%	25–50%	50–75%	> 75%
Plasticity[2]	Highly plastic fines; LL > 50; PI > 20	Moderately plastic fines; LL 35–50; PI 10–20	Low plasticity fines; LL < 35; PI < 10	N/A	N/A
Rating	0	2	3.5	5	7
Intact Strength and Durability	Type 1	Type 2	Type 3	Type 4	Type 5
Intact Strength	Extremely weak to very weak rocks, R0-1 (UCS < 5 Mpa); Type A bedrock; Types I and II overburden	Weak rocks, R2 (UCS 5–25 Mpa); Type B bedrock; Types III, IV and most Type V overburden	Medium strong rocks, R3 (UCS 25–50 Mpa); most Type C bedrock; coarse grained alluvium and talus derived from hard rocks	Strong to very strong rocks, R4 (UCS 50–100 Mpa); most Type D bedrock	Very strong to extremely strong rocks R5–6 (UCS >100 Mpa); most Type E bedrock
Durability[3]	Material subject to breakdown during placement, time dependent degradation due to slaking or freeze–thaw, or crushing under anticipated static loading		Material subject to limited breakdown during placement, time dependent degradation due to slaking or freeze–thaw, or crushing under anticipated static loading		Material not subject to breakdown during placement, time dependent degradation due to slaking or freeze–thaw, or crushing under anticipated static loading
Permafrost[4]		Perpetually frozen Type 1 materials	Perpetually frozen Type 2 materials	Perpetually frozen Type 3 materials	Perpetually frozen Types 4 and 5 materials
Rating	0	2	4	6	8

(Continued)

Table 3.5: (Continued)

Factors[1]	Ratings				
Material Liquefaction Potential[5]	Very High[6]	High[7]	Moderate or Unknown	Low	Negligible
Description	Very uniformly graded; very loose; minimal plastic fines; open, clast supported structure; high void ratio; rounded clasts; saturated		Moderate (or unknown) liquefaction potential	Low liquefaction potential but cannot be fully discounted	Well graded; dense; high content of plastic fines; matrix supported structure; low void ratio; angular clasts; dry
Rating	−20	−10	−5	−2.5	0
Chemical Stability	Highly Reactive		Moderately Reactive		Neutral
ARD Potential	High potential for chemical breakdown/ oxidation/generation of ARD		Moderate (or unknown) potential for chemical breakdown/oxidation/ generation of ARD		Very chemically stable material with negligible content of reactive minerals
Impact of Precipitates	High potential for precipitates to fill voids, decrease hydraulic conductivity, and increase pore pressures over time		Limited (or unknown) potential for precipitates to fill voids, decrease hydraulic conductivity, and increase pore pressures over time		Negligible precipitates, or precipitates result in cementation and increase shear strength over time without adversely affecting pore pressures
Rating	−5	−2.5	0	2.5	5
Material Quality rating[8]				Maximum possible rating:	20

Notes:
1. Select a rating for each factor. Where more than one criteria is shown for a given factor, or you cannot decide between two ratings, select an average or intermediate rating.
2. LL = Liquid Limit; PI = Plasticity Index. These criteria apply only to cases where the fines content is > 10%.
3. Under expected ambient climatic conditions.
4. If material is perpetually frozen, select the next highest category to account for expected increased durability and strength.
5. Evaluation of liquefaction potential of waste dump and stockpile materials may require detailed investigations. If unknown, use a default value of −5 and consult a geotechnical specialist.
6. If the Material Liquefaction Potential is judged to be Very High, the waste dump or stockpile should be classified as WHC V (Very High Hazard), regardless of the WSR.
7. If the Material Liquefaction Potential is judged to be High, the waste dump or stockpile should be classified as WHC IV (High Hazard), unless the WSR is ≤ 20, in which case it should be classified as WHC V (Very High Hazard).
8. The sum of the ratings for the individual factors is the Material Quality rating.

3.2.3.1 Gradation

The gradation or particle size distribution of the waste dump or stockpile material is a key factor in determining the frictional component of its shear strength, which typically accounts for the majority of the shear strength of blasted waste rock and ore. In general terms, well-graded materials with a high percentage of coarse, angular particles and a low component of fines tend to have higher shear strength than poorly graded, fine-grained materials. Gradation is characterised using two indices: the proportion (% by weight) of the material with particle sizes of less than 0.075 mm (also known as the fines content), and the proportion with grain sizes larger than 75 mm (i.e. cobbles and boulders). As summarised in Table 3.5, very fine-grained materials would have more than 50% fines and less than 10% cobbles and boulders, whereas very coarse-grained materials would have less than 5% fines and more than 75% cobbles and boulders.

Another factor that may have a significant influence on the shear strength of fine-grained materials is the mineral composition of the fines. Are the fines composed of plastic clay minerals, which would be characterised by relatively low shear strength, or are they composed of rock flour or non-plastic fines, which tend to have higher shear strength? Two simple laboratory index tests are commonly used to help characterise the expected behaviour of fine-grained soils: the liquid limit (LL) test and the plastic limit (PL) test (Terzaghi and Peck 1967). The difference between the LL and PL of the material is defined as the plasticity index (PI). The LL, PL and PI (collectively known as the Atterberg limits) have been empirically correlated with the shear strength of fine-grained soils (Department of the Navy 1971). Fine-grained materials with a high LL (higher than 50) and high PI (higher than 20) typically have a high plastic clay content and lower strength. Fine-grained materials with a low LL (lower than 35) and a low PI (lower than 10) are typically composed of low plastic clays and non-plastic silts and tend to have a higher shear strength.

Most soils laboratories are capable of conducting Atterberg limits and standard particle gradation (up to cobble-sized particles) testing, although few may be equipped to test fully representative samples given that particle sizes can vary from fines to boulders larger than 1 m in diameter. Also, before commencement of mining, it may not be possible to obtain representative bulk samples. If reliable and representative laboratory index testing data are not available, visual estimates of gradation, together with detailed descriptions of the materials, including lithology and alteration, prepared by an experienced geologist or geotechnical specialist, should be used. Estimates of particle size distributions of waste rock and ore can also be based on semi-empirical fragmentation studies. Such studies are often conducted during feasibility work to help define blasthole spacing, burden and explosive loading factors, and for the design of crushers and mills. Photogrammetry techniques can also be used to estimate gradation of blasted waste and ore.

Users should select a Gradation rating between 0 and 7 based on an evaluation of the above factors. Note that if the fines content is less than 10%, the coarse fraction is assumed to control the behaviour of the material.

3.2.3.2   Intact strength and durability

The shear strength of the waste dump or stockpile material is also influenced by the intact strength and durability of the individual particles, particularly in high dumps where inter-particle stresses may exceed the intact strength of the material and result in crushing. Materials with poor mechanical durability could break down during placement or over time due to slaking or freeze–thaw processes. All of these processes tend to result in the generation of additional fines and change the gradation of the material.

To help characterise waste dump and stockpile materials on the basis of intact strength and durability, five material types have been defined as follows and are summarised in Table 3.5:

Type 1: Extremely weak to very weak, highly degradable materials. Poorly indurated or weakly cemented rocks with very low intact strength (unconfined compressive strength [UCS] less than or equal to 5 MPa; Field Hardness less than or equal to R1 [ISRM 1981a]) that break down easily when end-dumped on repose angle slopes, or under bulldozer or haul truck traffic, or that are susceptible to crushing under anticipated static loading, and rocks that contain expansive clay minerals and are highly susceptible to slaking or freeze–thaw degradation. Weak, fine-grained, plastic and/or highly organic soils. Includes materials derived from Type A bedrock (see Section 3.2.2.7) and Type I and II overburden (see Section 3.2.2.3).

Type 2: Weak, degradable materials. Rocks with low intact strength (UCS 5 to 25 MPa; Field Hardness R1 to R2) and/or low slake durability, and moderate susceptibility to freeze–thaw degradation, including Type B bedrock (see Section 3.2.2.7). Includes most mixed-grained soils and medium-grained alluvium (Type III, IV and V overburden [see Section 3.2.2.3]), except very coarse-grained alluvium and talus.

Type 3: Medium strength materials with moderate durability. Moderately strong rocks (UCS 25 to 50 MPa; Field Hardness R3), with moderate slake durability, limited susceptibility to freeze–thaw degradation, and limited potential for crushing under anticipated static loading. Includes most Type C bedrock (see Section 3.2.2.7) and very coarse-grained alluvium derived from durable rocks.

Type 4: Strong (UCS 50 to 100 MPa; Field Hardness R4), very durable rocks with low susceptibility to freeze–thaw degradation, including most Type D bedrock (see Section 3.2.2.7) and coarse, durable, angular talus composed of strong rock blocks.

Type 5: Very strong to extremely strong (UCS higher than 100 MPa; Field Hardness greater than or equal to R5), extremely durable rocks, and talus with similar strength and durability. These materials would not be susceptible to freeze–thaw degradation, mechanical breakdown during placement, or crushing under the anticipated static loading. Includes most Type E bedrock (see Section 3.2.2.7).

In rare cases where the embankment materials are subject to perpetual freezing, their effective intact strength and durability may be greater than comparable unfrozen materials. In these cases it may be appropriate to increase the strength and durability rating by one category to account for this improvement. Because Type 5 materials would be expected to be free draining, perpetual freezing is not expected to have any beneficial impact on their strength and durability. As with permafrost in overburden materials, increasing the strength and durability rating based on perpetual freezing conditions should only be considered if it is supported by detailed *in situ* investigations and laboratory testing that conclusively demonstrates that the material will remain frozen and will not be subject to creep or stress-induced thawing under the expected loads, or subject to thawing as a result of changes to the groundwater flow regime within the waste dump or stockpile.

3.2.3.3 Material liquefaction potential

Similar to overburden foundations, in rare cases it is conceivable that certain types of waste dump or stockpile materials could be susceptible to liquefaction. As noted in Section 3.2.2.6, materials that are most susceptible to liquefaction are typically composed of very loose, saturated, uniformly graded materials with a low clay content and rounded particles. Because most materials are not susceptible to liquefaction, the default rating value for a negligible liquefaction potential was set at zero. At the other extreme, where the liquefaction potential is very high, a rating value of –20 was assigned to ensure that such conditions are appropriately flagged. Users should select a Material Liquefaction Potential rating based on the material descriptions given in Table 3.5. To encourage practitioners to explicitly address the issue of liquefaction potential, if the liquefaction potential is not known or has not been evaluated, a rating of –5 is assigned. Users that are not experienced geotechnical practitioners should consult with a geotechnical specialist before assigning a Material Liquefaction Potential rating.

3.2.3.4 Chemical stability

The intact strength of the waste dump or stockpile materials may also degrade over time as a result of chemical processes, such as oxidation of sulphide minerals or dissolution of sulphates. Precipitates that may form as a result of these chemical processes may also impact the hydraulic conductivity of the structure. The Chemical Stability factor recognises the potential impact these chemical processes may have on stability. As summarised in Table 3.5, waste dump and stockpile materials are classified as Highly Reactive, Moderately Reactive (or unknown) or Neutral based on their potential susceptibility to chemical degradation and generation of precipitates that could reduce the bulk hydraulic conductivity of the material and increase saturation and pore pressures. Highly Reactive materials are assigned a rating value of –5 in recognition of the significant negative impact on stability that such processes may have over the long term, and to flag these materials for more detailed evaluation. Moderately Reactive materials, and materials that have not been thoroughly characterised according to chemical stability, are assigned a rating value of 0. Materials that have been definitively characterised as having negligible potential for long-term chemical breakdown are assigned a rating value of 5. As with liquefaction potential, this negatively weighted rating scheme is intended to encourage practitioners to carefully characterise the materials using appropriate field and laboratory tests designed to assess long-term chemical stability.

It is important to note that the WSRHC system is not intended to address environmental impacts, such as changes to surface water or groundwater chemistry that may occur as a result of leaching of heavy metals or blasting residues. A detailed discussion of such potential environmental impacts is beyond the scope of this guideline. It is, nevertheless, recognised that mitigation of environmental impacts may ultimately control design. An overview of potential environmental issues associated with waste dumps and stockpiles, current and emerging mitigation techniques, and references to the large body of work that has been conducted on this subject are included in Chapters 14, 15 and 16.

3.2.4 Geometry and mass

The Geometry and Mass group (Table 3.6) includes factors that are related to the size and shape of the waste dump or stockpile. Larger, higher waste dumps and stockpiles, and those with steeper overall slopes, tend to be more susceptible to instability, and such instabilities tend to be larger and have greater runout and more adverse impacts than smaller, lower and flatter structures. Of the 16 dumps in the 1991 survey that reported poor or very poor performance, seven (44%) had overall heights of more than 250 m, nine (56%) had overall slopes of more than 35°, and five (31%) exceeded a total mass of 2×10^8 t. The criteria chosen to characterise Geometry and Mass include various height and slope angle parameters and the bulk volume or mass of the waste dump or stockpile. Collectively, Geometry and Mass factors are weighted to account for a maximum of 10 points, or 10% of the maximum possible WSR.

3.2.4.1 Height

Three parameters (see Fig. 3.2) were chosen to characterise the height of the waste dump or stockpile: the overall (toe to crest) height, the maximum fill thickness and the maximum individual lift height. Where possible, the Height rating should be selected based on a balance of these three parameters.

3.2.4.2 Slope

The overall fill slope angle is the angle measured below horizontal of a straight line that connects the toe and the crest of the embankment (see Fig. 3.2). To avoid unduly penalising very low embankments constructed using repose angle lifts, where the overall slope height is less than 50 m, a minimum default rating of 2 should be assigned for the overall slope angle. Where the overall slope height is between 50 and 100 m, a minimum default rating of 1 should be assigned.

Table 3.6: Geometry and Mass factors and ratings

Factors[1]	Ratings				
Height	Very High	High	Moderate	Low	Very Low
Overall Height (m)	> 500	250–500	100–250	50–100	< 50
Maximum Vertical Thickness (m)	> 500	250–500	100–250	50–100	< 50
Maximum Individual Lift Height (m)	> 200	100–200	50–100	25–50	< 25
Rating	0	1	2	3	4
Slope	Very Steep	Steep	Moderate	Flat	Very Flat
Overall Fill Slope Angle (°)[2]	> 35	30–35	25–30	15–25	< 15
Rating	0	1	2	3	4
Volume and Mass	Very Large	Large	Medium	Small	Very Small
Volume (m³)	$> 1 \times 10^9$	$1 \times 10^8 – 1 \times 10^9$	$1 \times 10^7 – 1 \times 10^8$	$1 \times 10^6 – 1 \times 10^7$	$< 1 \times 10^6$
Mass (t)	$> 2 \times 10^9$	$2 \times 10^8 – 2 \times 10^9$	$2 \times 10^7 – 2 \times 10^8$	$2 \times 10^6 – 2 \times 10^7$	$< 2 \times 10^6$
Rating	0	0.5	1	1.5	2
Geometry and Mass rating[3]				Maximum possible rating:	10

Notes:
1. Select a rating for each factor. Where more than one criterion is shown for a given factor, or you cannot decide between two ratings, select an average or intermediate rating.
2. Overall fill slope angle is measured from toe to crest. Where the overall slope height is less than 50 m, use a minimum default rating of 2. Where the overall slope height is between 50 and 100 m, use a minimum default rating value of 1.
3. The sum of the ratings for the individual factors is the Geometry and Mass rating.

3.2.4.3 Volume and mass

The Volume and Mass factor should be estimated on the basis of the loose, or bulk, volume of the waste dump or stockpile. If bank volume (i.e. the volume of the material measured *in situ*, before blasting and excavation) is used, a bulking factor should be applied. Bulking factor is the ratio of the loose volume of blasted material to its *in situ* volume before blasting and excavation. Typical bulking factors for rock excavated using mass blasting techniques such as are commonly applied in open pit mines range between ~1.25 and 1.5. The ranges for the Volume and Mass factor shown in Table 3.6 were calculated assuming a loose wet density of the material of 2 t/m³, or an *in situ* wet density of 2.7 t/m³, and a bulking factor of 1.35. If the actual wet densities or expected bulking factors differ materially from these assumed benchmark values, then the Volume and Mass factor ranges should be adjusted accordingly, or the Volume and Mass rating should be based on the mass only, as mass is usually the most easily obtainable parameter.

3.2.5 Stability analysis

The Stability Analysis group (Table 3.7) is intended to directly capture and contrast difference in stability based on the results of objective stability analysis and the acceptability criteria upon which the design is based. As discussed in Chapter 8, a wide variety of analytical techniques may be used to analyse the stability of a waste dump or stockpile. These various techniques may be grouped into two basic categories: deterministic and probabilistic. Deterministic techniques are designed to calculate a specific index value that represents stability, such as factor of safety (FoS) or strength reduction factor (SRF). The FoS approach is most commonly associated with limit equilibrium (LE) analysis techniques familiar to most practitioners. The SRF approach is more commonly associated with numerical modelling techniques, such as finite element, distinct element and related techniques, which are becoming more popular.

While Table 3.7 infers an equivalency between FoS and SRF, in the author's experience this is not always the case. The FoS and SRF depend on many factors, including the specific analysis technique used, the complexity of the model, the critical failure path and, in the case of certain numerical models, the *in situ* stress conditions and the stress path. A detailed assessment of these differences is beyond the scope of these guidelines. As discussed in Chapter 8, FoS has been defined in this guideline as the primary deterministic stability index for the purposes of establishing acceptance criteria, and the stability index values shown in Table 3.7 are based on FoS. In applying Table 3.7, if results of stability analysis are presented in terms of SRF, the user must decide whether SRF is reasonably equivalent to FoS based on the specifics of the analysis, and is advised to consult with a geotechnical specialist to make this determination.

The results of some types of numerical analysis techniques can also be characterised on the basis of convergence. Numerical models may be characterised as 'stable' if they converge after a series of iterative cycles, or as 'unstable' if they do not converge. Convergence may be determined based on cumulative deformation, residual velocity of deformation, or other criteria specific to the numerical technique and experience of the modeller.

Probabilistic techniques calculate a probability of failure (PoF), which represents the likelihood that a failure of a given magnitude could occur based on the variability of the input parameters. The PoF is typically expressed as a percentage and is often favoured over deterministic indices for economic and risk-based design methodologies. Analyses of PoF can be framed using both LE and numerical modelling techniques.

The Stability Analysis factors have been grouped into two categories: Overall Static Stability, to capture the expected stability behaviour of the structure under conditions of normal static (gravity) loading, and Overall Dynamic Stability, to capture the expected stability behaviour under exceptional earthquake loading conditions. As discussed in Chapter 8, the typical acceptability criteria for these two cases are different. For both of these categories, rating ranges are expressed in terms of deterministic (FoS and SRF) values. In keeping with the acceptability criteria described in Chapter 8, probabilistic (PoF) criteria are only provided for static analysis cases.

The Stability Analysis rating accounts for a maximum of 10 points, or 10% of the maximum possible WSR.

3.2.5.1 Overall static stability

The Overall Static Stability rating should be determined according to Table 3.7 based on the results of either 2D or 3D LE or numerical modelling using a conventional analysis technique. This factor is heavily weighted (70% of the Overall Stability group) in comparison to a 30% weighting for the Overall Dynamic Stability factor to reflect the limited impact that seismicity is thought to have on waste dump and stockpile stability (see discussion in Section 3.2.1.1.). In the case where the stability has been evaluated using a numerical model and the results are only expressed in terms of convergence, a default value of 3.5 should be assigned for convergent behaviour, and a value of 0 should be assigned for non-convergent behaviour. To encourage designers to conduct stability analyses to support the design process, a low default value of 2 (out of a maximum of 7) is assigned if no stability analysis results are available.

Table 3.7: Stability analysis factors and ratings

Factors[1]	Ratings				
Static Stability[2]					
Factor of Safety (FoS) or Strength Reduction Factor (SRF)[3]	< 1.1	1.1–1.2	1.2–1.3	1.3–1.5	> 1.5
Probability of Failure (PoF)	> 20%	10–20%	5–10%	1–5%	< 1%
Other criteria[4,5]	Non-convergent Numerical Model	No Supporting Stability Analysis	Convergent numerical model		
Rating	0	2	3.5	5	7
Dynamic Stability[2,6]					
Factor of Safety (FoS) or Strength Reduction Factor (SRF)[2]	< 1.0	1.0–1.05	1.05–1.10	1.10–1.15	> 1.15
Other criteria[7,8]	Non-convergent numerical model	No supporting stability analysis	Convergent numerical model		
Rating	0	1	1.5	2	3
Stability Analysis rating[9]				Maximum possible rating:	10

Notes:
1. Select a rating for each factor. Where more than one criterion is shown for a given factor, or you cannot decide between two ratings, select an average or intermediate rating.
2. It is assumed that there is at least a moderate level of confidence in the input parameters and that the analysis results are credible and reasonable.
3. Stability index values shown are based on FoS. If the stability analysis results are presented in terms of SRF, the user must decide whether SRF is equivalent to FoS based on the specifics of the analysis and is advised to consult with a geotechnical specialist to make this determination.
4. If numerical analyses indicates convergent behaviour, a minimum default rating of 3.5 should be assigned. If numerical analyses indicated non-convergent behaviour, a default rating of zero (0) should be used.
5. If no supporting stability analyses are available, a default value of 2 should be assigned.
6. If pseudostatic analyses are used to evaluate dynamic stability, an appropriate reduction in the design peak horizontal acceleration should be applied.
7. If numerical analyses indicate convergent behaviour, a minimum default rating of 1.5 should be assigned. If numerical analyses indicate non-convergent behaviour, a default rating of zero (0) should be used.
8. If no supporting stability analyses are available, a default value of 1 should be assigned.
9. The sum of the ratings for the individual factors is the Stability Analysis rating.

The rating values given in Table 3.7 assume that there is a least a moderate level of confidence in the input parameters and that the analysis results are credible and reliable. If the confidence in the input parameters is low, or the results are judged to be unreliable or lacking in credibility, a lower rating value should be assigned. To qualify as having a moderate level of confidence, the results should be reviewed and accepted as reasonable and reliable by a geotechnical specialist. Guidance with respect to assigning a confidence level is provided in Section 8.2.

3.2.5.2 Overall dynamic stability

The Overall Dynamic Stability rating should also be determined according to Table 3.7 based on the results of either an LE or numerical modelling technique. Where pseudostatic analyses are used to evaluate dynamic stability, an appropriate reduction in the design peak horizontal acceleration, such as that suggested by Hynes-Griffin and Franklin (1984), should be applied to the peak horizontal acceleration.

In the case where the dynamic stability has been evaluated using a numerical model and the results are only expressed in terms of convergence, a default value of 1.5 should be assigned for convergent behaviour and a value of 0 should be assigned for non-convergent behaviour. To encourage designers to conduct stability analyses to support the design process, a low default value of 1 (out of a maximum of 3) is assigned if no stability analysis results are available.

As with Overall Static Stability, the rating values for Overall Dynamic Stability given in Table 3.7 assume that there is at least a moderate level of confidence in the input parameters and that the analysis results are credible and reliable. If the confidence in the input parameters is lower, or the results are judged to be unreliable or lacking in credibility, a lower rating value should be assigned. To qualify as having a moderate level of confidence, the results should be reviewed and accepted as reasonable and reliable by a geotechnical specialist. Guidance with respect to assigning a confidence level is provided in Section 8.2.

3.2.6 Construction

The Construction rating (Table 3.8) includes two roughly equally weighted factors, Construction Method and Loading Rate, and accounts for a maximum of 15 points, or 15% of the maximum possible WSR.

The construction sequence and rate of development can have a very significant impact on waste dump and stockpile stability and performance. All other factors being equal, waste dumps and stockpiles that are constructed slowly and from the bottom up in thin lifts are much more stable and perform much better than those that are constructed

rapidly using single, high lifts. In this context, performance would include both instability and settlement.

3.2.6.1 Construction method

The Construction Method factor is intended to capture the key differences in the sequencing and development of a waste dump or stockpile that can affect overall stability and performance. Five benchmark methods with associated descriptions have been defined as follows and are summarised in Table 3.8. Figure 3.3 provides idealised schematic illustrations of the five benchmark methods.

Method I: Descending construction sequence with single or multiple (wrap-around) very high (higher than 200 m) lifts constructed on very steep terrain (foundation slopes more than 32°). This construction method is generally accepted as the least favourable from a geotechnical stability perspective, but often represents the most economical approach due mostly to shorter haul distances.

Method II: Descending construction sequences with single or multiple (wrap-around) lifts constructed on steep terrain (foundation slopes 25–32°). Lift heights limited to less than 200 m. The flatter terrain and lower lift heights as compared to Method I should result in improved stability. The lower lift heights may increase incremental haul distances and costs in comparison to Method I.

Method III: Combination of descending and ascending construction sequence (also referred to as a hybrid construction method) with initial lifts developed on moderate terrain (local foundation slopes less than 25°) followed by wrap-around lifts founded on the lower lifts and steeper terrain (foundation slopes 25–32°). Lift heights limited to less than 200 m. This approach represents a compromise between more stable bottom-up construction and less expensive top-down construction.

Method IV: Ascending or bottom-up construction sequence with multiple lifts constructed on moderate terrain (foundation slopes 15–25°). Lift heights limited to less than 100 m. As indicated above, bottom-up construction is typically more stable than top-down, but can be more expensive if material must be hauled downhill from the pit crest. Where the topography beyond the pit crest is flat, there may be little cost differential between bottom-up and top-down.

Method V: Ascending (Method Va) or descending (Method Vb) construction sequence with multiple lifts constructed on gently sloping or flat terrain (foundation slopes less than 15°). Lift heights limited to less than 50 m. The gently sloping to flat foundation and low lift heights make this method the most stable.

Table 3.8: Construction factors and ratings

Factors[1]	Ratings				
Construction method[2]	Method I	Method II	Method III	Method IV	Method V
Description[3]	Descending sequence with single or multiple (wrap-around), very high lifts constructed on very steep terrain; lift heights > 200 m; overall foundation slopes > 32°	Descending sequence with single or multiple (wrap-around) lifts constructed on steep terrain; lift heights < 200 m; overall foundation slopes 25–32°	Hybrid (combination of descending and ascending) sequence designed to avoid founding lift toes on steep or very steep terrain; lift heights < 200 m; overall foundation slopes > 25°	Ascending sequence with multiple lifts constructed on moderate terrain; lift heights < 100 m; overall foundation slopes 15–25°	Descending or ascending sequence on gentle or flat terrain; lift heights < 50 m; overall foundation slopes < 15°
Rating	0	2	4	6	8
Loading Rate	Very High	High	Moderate	Low	Very Low
Volumetric Loading Rate[4] (m^3/d/m)	> 500	150–500	50–150	15–50	< 15
Mass Loading Rate[5] (t/d/m)	> 250	75–250	25–75	7.5–25	< 7.5
Crest Advancement Rate[6] (m^2/d)	> 500	150–500	50–150	15–50	< 15
Rating	0	2	3.5	5	7
Construction Rating[7]				Maximum possible rating:	15

Notes:
1. Select a rating for each factor. Where more than one criterion is shown for a given factor, or you cannot decide between two ratings, select an average or intermediate rating.
2. Select the method that best describes the development sequence. Where the construction sequence includes attributes of more than one method, choose an intermediate rating.
3. Descending sequence refers to embankments that are constructed from the top down using wrap-around lifts; ascending sequence refers to embankments that are constructed in lifts from the bottom up.
4. Volumetric Loading Rate = average daily loose volume (m^3/d) ÷ average active crest length (m). Includes bulking factor.
5. Mass Loading Rate = average daily mass of fill consigned to the waste dump or stockpile (t/d) ÷ average active crest length (m). Assumes a nominal bulk density of 2.0 t/m^3; adjust as necessary.
6. Crest Advancement Rate = average daily rate of crest advancement (m/d) × average lift height (m).
7. The sum of the ratings for the individual factors is the Construction rating.

Designers should compare the proposed construction method for their waste dumps and stockpiles with these benchmark descriptions and choose a rating value based on the description that fits the best. Where waste dump or stockpile development plans contain attributes of more than one of the benchmark methods, an intermediate rating should be selected.

3.2.6.2 Loading rate

The Loading Rate factor is intended to capture the impact on stability due to the rate of construction. Slow construction rates allow more time for both the embankment and foundation materials to consolidate and gain strength, and for construction-induced pore pressures to dissipate. Three alternative, and related, parameters have been defined to help characterise the rate of construction: Volumetric Loading Rate, Mass Loading Rate and Crest Advancement Rate. The Volumetric Loading Rate is defined as the average amount of loose

material (measured in cubic metres) placed on the waste dump or stockpile in a 24 h period, divided by the average active crest length (measured in metres). The Mass Loading Rate is defined as the average mass placed on the waste dump or stockpile in a 24 h period (measured in tonnes), divided by the average active crest length (measured in metres). The Crest Advancement Rate is defined as the average rate of active crest advancement in a 24 h period (measured in metres per day), multiplied by the average lift height (also measured in metres).

As for the Volume and Mass factor under the Geometry and Mass group (see Section 3.2.4.3), these three parameters are related by bulking factor and density. The ranges for the Volumetric Loading Rate and Crest Advancement Rate factors indicated in Table 3.8 were calculated based on the Mass Loading Rate assuming a loose wet density of 2.0 t/m^3, or an *in situ* (before blasting and excavation) wet density of 2.7 t/m^3, and a bulking factor of 1.35. If the actual wet densities or expected

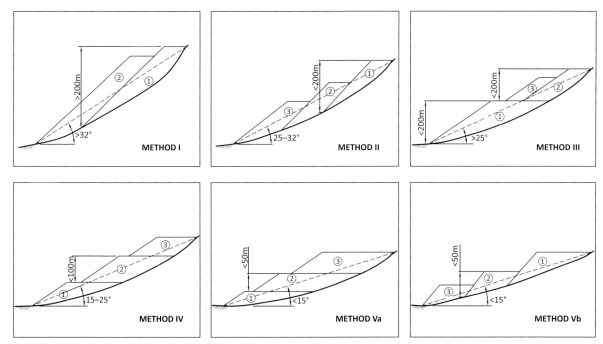

Figure 3.3: Construction methods

bulking factors differ materially from these assumed benchmark values, then the Volume Loading Rates and Crest Advancement Rate factor ranges should be adjusted accordingly, or the Loading Rate rating should be based on the Mass Loading Rate factor only, as the Mass Loading Rate is typically the most easily obtainable parameter.

3.2.7 Performance

The Performance rating is intended to capture the actual, documented stability performance of existing waste dumps and stockpiles. Five stability performance categories have been defined as described in the following list and are summarised in Table 3.9.

Very Poor: unstable. Waste dumps and stockpiles that have been subjected to large-scale (> 1 × 10⁶ t) instability, or single or multiple instabilities that have had a major impact on operations or have required substantial long-term closures or major remedial work.

Poor: metastable to unstable. Waste dumps and stockpiles that have been subjected to frequent short-term closures or frequent sliver, crest and/or local foundation instabilities, but no large-scale (> 1 × 10⁶ t) instabilities. Moderate impact on operations.

Fair: metastable to stable. Waste dumps and stockpiles that have been subjected to occasional closures due to deformation/settlement, or occasional small (< 1 × 10⁵ t) sliver failures. Impacts on operations have been limited. This category also includes waste dumps and stockpiles with no performance history.

Good: stable. Waste dumps and stockpiles that have been subject to minor deformation/settlement, rare closures, or rare small failures that have had negligible impact on operations.

Very Good: very stable. Waste dumps and stockpiles that have experienced negligible deformation/settlement, no closures, and no instabilities that have impacted operations.

Performance rating values range from –15 to +15. The negative rating for very poor and poor performance is intended to flag these structures for further evaluation.

3.2.8 Waste dump and stockpile stability rating

The aggregate of the rating values for each group of factors described above is defined as the waste dump and stockpile stability rating (WSR). Table 3.10 provides a convenient format for summarising and aggregating the ratings for the different indices, and is automatically populated in the Stability Rating worksheet in the companion Excel workbook.

In addition to calculating the overall WSR, the worksheet calculates two component indices: the Engineering Geology Index (EGI) and the Design and Performance Index (DPI). The EGI is equal to the sum of the Regional Setting, Foundation Conditions and Material Quality indices, and represents factors that are related to the site location and geological conditions, and which the designer has limited ability to influence. The DPI is equal

Table 3.9: Performance ratings

Factors[1]	Ratings				
Stability Performance	Very Poor[2]	Poor[3]	Fair	Good	Very Good
Description	Unstable; substantial long-term closures or major remedial work required; large-scale (> 1 × 10⁶ t) instability; major impact on operations	Metastable to unstable; frequent short-term closures; frequent sliver/crest failures and/or local foundation failures, but no large-scale (> 1 × 10⁶ t) instabilities; moderate impact on operations	Metastable to stable; occasional closures due to deformation/settlement; occasional small sliver failures (< 1 × 10⁵ t); limited impact on operations; waste dumps and stockpiles with no performance history	Stable; minor deformation and/or settlement; rare closures; rare small failures; negligible impact on operations	Very stable; negligible deformation/settlement; no closures; no failures; no impact on operations
Rating	−15	−7.5	0	7.5	15
Performance Rating[4]				Maximum possible rating:	15

Notes:
1. Select a rating for each factor. Where more than one criterion is shown for a given factor, or you cannot decide between two ratings, select an average or intermediate rating.
2. If the Stability Performance is judged to be Very Poor, the waste dump or stockpile should be classified as WHC V (Very High Hazard), regardless of the calculated WSR.
3. If the Stability Performance is judged to be Poor, the waste dump or stockpile should be classified as WHC IV (High Hazard), unless the WSR is ± 20, in which case it should be classified as WHC V (Very High Hazard).
4. The sum of the ratings for the individual factors is the Performance Rating.

to the sum of the Geometry and Mass, Stability Analysis, Construction and Performance indices, and represents factors over which the designer has some control. The maximum rating values for these two indices is balanced so that both represent 50% of the WSR.

3.2.9 Waste dump and stockpile hazard class

For descriptive purposes and to simplify comparison of different possible alternative configurations or design approaches for a given waste dump or stockpile, and in recognition of the somewhat subjective nature of the rating scheme, the possible range of WSR values has been subdivided in to five categories or waste dump and stockpile hazard classes (WHCs) as shown in Table 3.11.

A qualitative instability hazard description is associated with each class to help convey their relative potential for instability. These descriptions may be useful in qualitative and comparative risk assessments such as those discussed in Chapter 10.

Note that if either the Foundation Liquefaction Potential (Section 3.2.2.6) or the Material Liquefaction Potential (Section 3.2.3.3) is judged to be very high, or the Performance (Section 3.2.7) is judged to be very poor, the

Table 3.10: Waste dump and stockpile stability rating summary

Engineering Geology Index (EGI)	Regional Setting	Description	Very Adverse	Adverse	Neutral	Favourable	Very Favourable
		Rating	0	2.5	5	7.5	10
	Foundation Conditions	Description	Very Poor	Poor	Fair	Good	Very Good
		Rating	≤ 0	5	10	15	20
	Material Quality	Description	Very Poor	Poor	Fair	Good	Very Good
		Rating	≤ 0	5	10	15	20
Design and Performance Index (DPI)	Geometry and Mass	Description	Very Large	Large	Medium	Small	Very Small
		Rating	0	2.5	5	7.5	10
	Construction	Description	Very Unfavourable	Unfavourable	Intermediate	Favourable	Very Favourable
		Rating	0	5	7.5	10	15
	Performance	Description	Very Poor	Poor	Fair	Good	Very Good
		Rating	≤ 0	5	7.5	10	15
	Engineering Geology Index (EGI)			Design and Performance Index (DPI)			WSR

Table 3.11: Summary of waste dump and stockpile stability ratings, hazard classes and relative instability hazard

WSR	WHC	Instability Hazard
80–100	I	Very Low Hazard
60–80	II	Low Hazard
40–60	III	Moderate Hazard
20–40	IV	High Hazard
< 20	V	Very High Hazard

waste dump or stockpile should be classified as WHC V (Very High Hazard), regardless of the WSR. Likewise, if either the Foundation Liquefaction Potential or the Material Liquefaction Potential are judged to be high, or the Performance is judged to be poor, the waste dump or stockpile should be classified as WHC IV (High Hazard) unless the WSR is less than 20, in which case it should be classified as WHC V.

It may also be instructive to plot WSR results on the chart in Fig. 3.4. This chart illustrates the relative

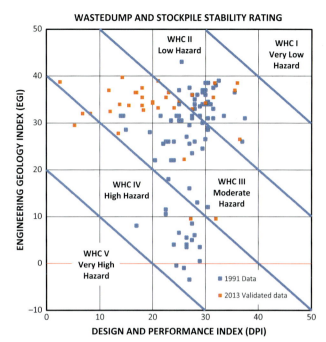

Figure 3.5: Waste dump and stockpile stability and hazard chart illustrating the results of the 1991 and 2013 surveys

Figure 3.4: Waste dump and stockpile stability rating and hazard class chart

weighting of the EGI and DPI indices and facilitates comparison of different waste dumps and stockpiles and possible alternative configurations or development phases for a given waste dump or stockpile. Figure 3.5 shows the results of the 1991 and 2013 surveys plotted on a waste dump and stockpile stability rating and hazard class chart.

Similar to the 1991 DSC, the WSRHC system can be used as a guide to the level of effort required to investigate, design and construct waste dump. Waste dumps and stockpiles with lower stability ratings, or that fall into higher hazard classes, logically ought to require more investigative and design effort, and more care and monitoring during construction and operations, than waste dumps and stockpiles with higher stability ratings, or that fall into lower hazard classes. Table 3.12 provides suggestion regarding the appropriate level of effort for the site investigation and characterisation, analysis and design, and construction and operation stages in the life cycle of a waste dump or stockpile based on WSR and WHC.

Table 3.12: Suggested level of effort based on waste dump and stockpile stability rating/hazard class (WSR/WHC)

Stability class		Level of effort		
Waste dump and stockpile hazard class (WHC)	Instability hazard	Investigation and characterisation	Analysis and design	Construction and operation
I	Very Low Hazard	Basic desktop studies to establish initial stability rating and hazard classification; basic site reconnaissance to confirm key assumptions from desktop studies and plan field investigations; limited mapping and test pitting to establish/verify subsurface conditions; material parameters based on literature/experience and validated with limited field and laboratory index testing; initiate limited baseline environmental monitoring; condemnation drilling	Simplified stability analyses to verify that stability does not influence design and potential impacts are minor; design by geotechnical specialist with peer review	Minimal site preparation; minimal restriction on construction; periodic visual monitoring; periodic inspection by geotechnical specialist
II	Low Hazard	Desktop studies to establish initial stability rating and hazard classification; site reconnaissance to confirm key assumptions from desktop studies and plan supplementary field investigations; mapping and test pitting as required to verify subsurface conditions; material parameters based on literature/experience and validated with field and laboratory index testing; initiate environmental baseline monitoring; condemnation drilling	Stability analyses to verify that stability has limited impact on design; design by experienced geotechnical specialist with peer review	Limited site preparation, may include minor diversions; limited construction constraints; standard instrument and visual monitoring with basic trigger action response plan (TARP); periodic inspection by experienced geotechnical specialist
III	Moderate Hazard	Comprehensive desktop studies to establish initial stability rating and hazard classification; detailed site reconnaissance to confirm assumptions from desktop studies; detailed mapping and subsurface investigations likely including test pitting/trenching and limited drilling and sampling; in situ instrumentation and testing and laboratory testing to verify foundation and fill material properties; initiate comprehensive baseline environmental monitoring; condemnation drilling	Comprehensive stability analyses, including consideration of runout potential; qualitative risk assessment; design moderately constrained by stability and potential impacts; design optimisation and impact mitigation studies; design conducted by experienced geotechnical specialist with peer review	Moderate site preparation, may include diversions and underdrainage; limited foundation instrumentation to verify performance; runout/rollout mitigation measures, if required; moderately constrained construction sequence; control of fill quality and placement as necessary; loading/advance rate restrictions; standard instrumentation and visual monitoring with well-defined TARPs; periodic (minimum annual) inspections by experienced geotechnical specialist
IV	High Hazard	Detailed desktop studies to establish initial stability rating and hazard classification; comprehensive site reconnaissance to confirm assumptions from desktop studies; detailed, phased mapping and subsurface investigations likely including test pitting/trenching, geophysics, specialised drilling and sampling; in situ instrumentation and testing and laboratory index and shear strength testing to establish foundation and fill material properties to a high degree of confidence; initiate comprehensive baseline environmental monitoring; condemnation drilling	Phased design study with detailed stability analyses of interim and final stages, including runout assessments; parametric studies; design constrained by stability and potential impacts; semi-quantitative risk assessment; optimisation, trade-off and mitigation studies; design by experienced geotechnical specialist with peer review; third party specialist review at critical stages in design	Moderate to extensive site preparation, may including underdrainage and diversions; foundation and fill instrumentation; runout/rollout mitigation measures; moderately constrained construction sequence with control of fill quality and placement; moderate to severe loading/advance rate restrictions; detailed instrument and visual monitoring with redundancy; well-defined/site-specific TARPs; frequent inspections and review by experienced geotechnical specialist; annual or more frequent review by third party specialist

(Continued)

Table 3.12: (Continued)

Stability class		Level of effort		
Waste dump and stockpile hazard class (WHC)	Instability hazard	Investigation and characterisation	Analysis and design	Construction and operation
V	Very High Hazard	Detailed desktop studies to establish initial stability rating and hazard classification; comprehensive site reconnaissance to confirm assumptions from desktop studies; detailed, phased mapping and subsurface investigations likely including extensive test pitting/trenching, geophysics, specialised drilling and sampling; comprehensive *in situ* instrumentation and testing and sophisticated and comprehensive laboratory index and shear strength testing to establish foundation and fill material properties to a high degree of confidence; extensive baseline environmental monitoring program; condemnation drilling	Phased design study with comprehensive and sophisticated stability analyses of interim and final stages; detailed runout assessments; parametric studies; design heavily constrained by stability and potential impacts; quantitative risk assessment; detailed optimisation, trade-off and mitigative studies; design conducted by experienced geotechnical specialist with peer review; review by expert review board at critical stages in design	Extensive site preparation, possibly including critical underdrainage and diversions; foundation and fill instrumentation; runout/rollout mitigation measures; highly constrained construction sequence with careful control of fill quality and placement; severe loading/advance rate restrictions; extensive (multi-system) instrument and visual monitoring with redundancy; well-defined/site-specific TARPs; frequent inspections and review by experienced geotechnical specialist; annual or more frequent review by expert review board

4

SITE CHARACTERISATION

Michael Etezad, John Cunning, James Hogarth and Geoff Beale

4.1 Introduction

This chapter describes the typical site characterisation and field investigation studies that are carried out to support the planning and design of mine waste dumps or major stockpile facilities. Site characterisation may include physiography, geomorphology, climate, surface water, groundwater, geochemistry and geotechnical conditions, all of which may be important during the site selection and facility design process.

The level of effort required to investigate and characterise a site is a function of the stage of the project, the potential for waste dump instability and the level of risk. Site characterisation studies should be advanced as the site selection and design process evolves, as illustrated in the flow chart in Fig. 2.1 (Chapter 2). Table 2.2 (Chapter 2) provides guidance on the level of effort required based on the stage of the project. In most cases, the more advanced the project and the greater the risk, the greater the required level of effort to characterise the site. Table 3.12 (Chapter 3) provides detailed guidance on the scope of site investigation and characterisation studies based on the potential for instability as represented by the waste dump and stockpile stability rating and hazard class. An initial classification of the site and proposed facility using the waste dump and stockpile stability rating and hazard classification system detailed in Chapter 3 can be helpful when planning the field investigations and site characterisation studies. The field investigations are usually scheduled to follow a phased approach, first obtaining preliminary data to support site selection and screening, followed by more advanced investigations to support detailed design.

The field investigations and site characterisation studies should cover the full footprint area plus any adjacent areas that could impact the site or be directly or indirectly impacted by the facility (e.g. downslope areas within possible runout zones and downgradient surface waters).

4.1.1 Conceptual studies

A typical scope would begin with a review of existing information to develop an initial understanding of the site setting, including the physiography, geomorphology, climate, surface water, groundwater, geochemical and geotechnical conditions. A project where a new (greenfield) mine site is being developed would typically require more effort than an existing site, and may require the preparation of extensive baseline studies to support the environmental assessment and permitting processes. For an existing waste dump expansion or brownfield site, there may be substantial historical site data, so a review of the existing records followed by a gap analysis should be carried out to define the scope of supplemental investigations and additional data needed to support site selection and advanced design stages.

At the conceptual design stage, the scope would include a desktop study to develop an initial understanding of the site setting. This may include aerial photo interpretation, review of available regional climate data, historical data review and topographic analysis. An initial site reconnaissance would usually be carried out to help validate the preliminary conceptual models. The initial site reconnaissance may include outcrop, terrain and stream mapping. For a greenfield site, a reconnaissance and review by an experienced practitioner may be needed to satisfy financial regulatory requirements if the mine waste management components will form a significant portion of the overall project costs. For a brownfield site, the actual performance data for existing waste dumps or stockpiles founded on similar terrain can be invaluable in supporting the characterisation of an expanded facility.

The physiography of the project area needs to be assessed in detail as part of the initial site characterisation studies. The physiography has a direct impact on the area available for waste dumps and stockpiles and also affects development costs. In mountainous terrain, sites could be limited by steep slopes and topographical constraints. Geological factors that could constrain development also need to be identified at an early stage. These may include weak bedrock, weak soil deposits or adverse geological structure. Potential natural hazards may include avalanches and landslides, permafrost and glaciers (which may be present in extreme northern or southern latitudes and/or at high altitudes) and seismic risk.

Geochemical characterisation and the need to manage potential acid rock drainage can have a significant impact on the design, operation and closure of waste dumps and major stockpiles, and an initial understanding needs to be developed early in the design process. Detailed procedures for geochemical characterisation of waste dumps and stockpiles are beyond the scope of these guidelines; however, an overview of the management of acid rock drainage is provided in Chapter 14. For a detailed discussion and guidance on the geochemical characterisation and management of mine waste materials, the reader is referred to the International Network for Acid Prevention's *Global Acid Rock Drainage (GARD) Guide* (INAP 2009).

4.1.2   Planning of field investigations

Following the completion of the conceptual studies, the field investigations and site characterisation would be progressively advanced to support the pre-feasibility, feasibility and detailed operation and closure design stages of the project. Field investigations are a key component of the site characterisation and design process and will normally be carried out at each stage of the project. The fieldwork may typically include geophysical surveying, test trenching, test pitting, drilling, *in situ* testing, installation of instrumentation, sampling and laboratory testing. At each stage of project development, it is necessary to identify data gaps, which can be filled during the subsequent stages of the field investigation. Where specific issues are identified, more advanced or specialised types of investigation and testing may be required to confirm that the characterisation studies are valid for specific areas being considered for waste dump and stockpile development, and to obtain site-specific information on soil and bedrock conditions, topography, surface water and groundwater.

The field investigations should be comprehensive enough to provide sufficient data to support the level of design of the waste dump or stockpile for each project stage, and should also be designed to capture all the site conditions that could impact stability (both positively and negatively). Inadequate site investigation may potentially result in delays in the design process, an overly conservative design or a high-risk design.

4.2   Site characterisation methods

The waste dump and stockpile design process includes development of the following models:

- geological
- hydrological (climate and surface water)
- hydrogeological (groundwater)
- geochemical
- geotechnical (foundation soil and bedrock).

As discussed in Chapter 1, the British Columbia Mine Waste Rock Pile Research Committee (BCMWRPRC) published a series of reports, including the 1991 *Investigation and Design Manual – Interim Guidelines* (BCMWRPRC 1991a), covering the investigation and design of mine waste dumps. The investigation and characterisation techniques associated with each of the key study areas presented in the *Investigation and Design Manual – Interim Guidelines* have been updated for this book to reflect the current state of practice. Table 4.1 is an update of Table 3.1 from the *Investigation and Design Manual – Interim Guidelines* and includes important characteristics, site selection and design implications, available sources of information and typical field investigation methods by design stage.

Characterisation of the surface water and groundwater setting is an essential component of any study for a mine waste dump and stockpile facility. Areas with high precipitation and shallow groundwater can be challenging in terms of water management. Catchment areas and the locations of streams, lakes and wetlands need to be considered. The depth of the water table and distribution of groundwater recharge and discharge zones may impact stability and water quality and need to be evaluated. Methods for characterising surface water and groundwater are discussed in Chapter 6.

4.3   Study areas

4.3.1   Physiography and geomorphology

Physiography refers to the physical features of the site surface. Undesirable physiographical conditions, such as steep slopes, lack of required storage capacity, low-lying areas subject to flooding and unfavourable drainage areas, may all affect the suitability of the available sites, which in turn influence the distance and elevation difference between the facility and the pit and the required haulage distance.

Table 4.1: Site characterisation and investigation methods by study area

Study areas	Important characteristics	Site selection and design implications	Available information sources	Investigation methods	
				Conceptual to detailed design	Detailed design to closure
Physiography and geomorphology	Site geology, location, topography (size, shape) Natural hazards (e.g. landslides, flood, debris flow, avalanche) Continuity and geometrical relationship between geomorphic units Landforms (plateaus, terraces, gullies) Seismicity (faults, ground motions) Glaciology; permafrost	Overall site suitability Topographic or geomorphic constraints Potential for stabilisation Dump type and construction methods Haul distance and grades	Geological maps Topographic maps Satellite imaging and air photos Seismic zoning maps Permafrost maps Glacier maps	Regional geology review Satellite image and aerial photo interpretations, terrain analysis LiDAR survey and interpretation Ground reconnaissance, terrain mapping	Ground surveys Site-specific seismic hazard study Fault identification
Climate and surface water (hydrology)	Climate information (precipitation in terms of rainfall and snowfall, evaporation rates, air temperature, relative humidity, prevailing wind speed and direction, solar radiation) Surface water runoff and infiltration characteristics Locations of perennial and ephemeral streams, lakes and other surface water features Size of contributing catchment basins Surface water quantity and quality	Direct precipitation Diversions, underdrainage and rock drains Snow accumulation Flooding potential Freeze–thaw degradation Potential for impact on downstream waters	Topographic maps Satellite imaging and aerial photos Published reports Government or regional climate station and flow gauging records Water licence records Terrestrial ecosystem mapping	Satellite imaging and aerial photo interpretation, terrain analysis LiDAR survey and interpretation Ground reconnaissance, stream mapping Stream flow measurements Installation of flow monitoring stations Baseline surface water studies	Detailed investigation programs, as described in Chapter 6
Groundwater (hydrogeology)	Upstream groundwater conditions Potential groundwater underflow Location of existing wells, springs, seeps Piezometric pressures Downgradient groundwater flow system Existing groundwater users Groundwater quantity and quality	Potential impact on downstream groundwater Impact of groundwater on foundation conditions for stability and runout Diversions, underdrainage, and rock drains Potential seepage	Topographic maps Satellite imaging and aerial photos Geological maps, published reports Exploration drilling records Water-level measurements Well logs Water licence records	Hydrogeological reconnaissance and mapping Geophysical surveys Inflows to trenches, test pits Piezometers In situ permeability testing Baseline groundwater studies	Detailed investigation programs, as described in Chapter 6

(Continued)

Table 4.1: (Continued)

Study areas	Important characteristics	Site selection and design implications	Available information sources	Investigation methods	
				Conceptual to detailed design	Detailed design to closure
Geotechnical: foundation soils	Soil types, distribution, stratigraphy Depth to bedrock *In situ* soil characteristics	Foundation preparation Characteristics of foundation soil materials for stability and runout Characteristics of soil waste/stripping materials	Surficial geology maps Soils maps Published reports Terrain, soils and ecosystem mapping Exploration drilling records	Satellite imaging and aerial photo interpretations, terrain analysis Ground reconnaissance, soils mapping Geophysical surveys Trenches, test pits, grab sampling, soft soil probing Boreholes with *in situ* testing: cone penetration testing (CPT), standard penetration testing (SPT), field vane testing (FVT), large penetration testing (LPT) and soil sampling	Specialised geophysical surveys *In situ* testing (e.g. field vane, shear wave velocity) Undisturbed soil sampling
Geotechnical: foundation bedrock	Geological structure *In situ* rock characteristics (lithology, competency, durability) Potential mineral resources (condemnation)	Characteristics of foundation bedrock material for stability and runout Impact on potential mineral resources Characteristics of waste rock materials	Regional geology maps and studies Published reports Exploration and condemnation drilling records	Regional geology review LiDAR and satellite imagining and aerial photo interpretations Ground reconnaissance, outcrop mapping Borehole and core logging, sampling and testing	Targeted boreholes and core logging, sampling and testing

Source: Updated from BCMWRPRC (1991a)

Geomorphology is the process through which the site surface and ground were formed. Understanding the geomorphology of the site is also an essential part of the design process. Lacustrine sediment or colluvial deposits could be sources of instability. Geomorphological hazards may include landslides and debris flows.

Information on the physiography and geomorphology of the site can be obtained from remote sensing and image analysis including light detection and ranging (LiDAR) surveys, site satellite imaging, aerial photographs, terrain analyses and geology reviews. These sources are useful for developing an initial understanding of the physical site and for planning of subsequent ground reconnaissance and field investigation programs.

4.3.1.1 *LiDAR*

LiDAR surveys measure ranges (variable distances) by illuminating a target with a laser and analysing the reflected light. The use of airborne laser scanning (ALS)-LiDAR-derived high resolution digital elevation models (HRDEM) for geological and geomorphological mapping, fluvial geomorphology and slope analysis (geomorphic analysis, landslide detection and characterisation, discontinuities and fracture analysis, and definition of deforming volumes) has gained in popularity due to the increased availability of the computing power required to manipulate the large amount of data included in a LiDAR survey. Figure 4.1 shows some example data from a LiDAR survey.

LiDAR can provide ~1 m accuracy in ground resolution and has the ability to penetrate vegetation canopies. A LiDAR-derived digital elevation model is a powerful tool

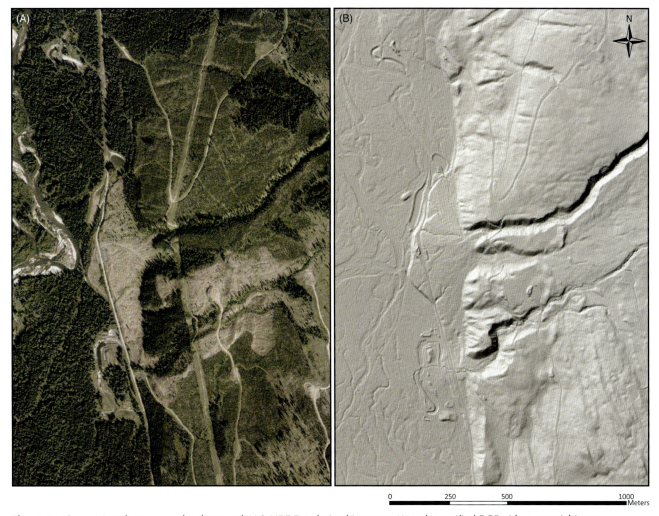

Figure 4.1: Comparison between orthophoto and ALS-HRDEM-derived imagery: (A) orthorectified RGB airborne aerial image, resolution: 1 m; (B) ALS-HRDEM-derived shaded relief image, ground resolution: 1 m. ALS-HRDEM data processing allows the extraction of bare ground elevation data, allowing a more detailed remote geomorphological and geological characterisation of an area.

that can be used to identify geomorphological and geomorphic features including the following:

- river terraces and river channels with dimensions as small as 1 m
- surface exposure of faults
- landslide related structures and geometries
- glacial and periglacial features (blockfields, rock glaciers, polygonal grounds, solifluction, etc.)
- areas affected by active erosion
- geohazards such as avalanche or rockfall hazard areas.

LiDAR datasets can be collected regularly during project development and used throughout the design life of the project, from initial planning to closure. Figure 4.2 shows shaded relief and 1 m topography contours from a LiDAR survey.

4.3.1.2   Satellite imaging and air photos

Satellite images are useful for defining geological features such as lithological contacts, faults and folds, the type and distribution of soils, drainage systems, and vegetation and land use. Air (sometimes aerial) photos can also assist with development of a preliminary understanding of geology and site features that are difficult to recognise from the ground or at small scale. Landforms and lineaments identified using satellite imaging techniques may indicate the presence of faults, deep soils or low strength materials. Topographic features such as ravines, sinkholes, stream channels and swamps, or drainage patterns can be mapped. Glacial landforms such as eskers, drumlins and kames may also be recognisable. The method is reliable, relatively inexpensive and easy to use, but weather conditions can affect the image quality. The Canada Centre for Mapping and Earth Observation, NASA and Google Earth are some sources of satellite images.

Aerial photos are sometimes used to supplement satellite images to provide larger scale and higher resolution images. The US Army Corps of Engineers technical manual on geology (USACE 2013) is an excellent source of soil and rock type interpretations from aerial

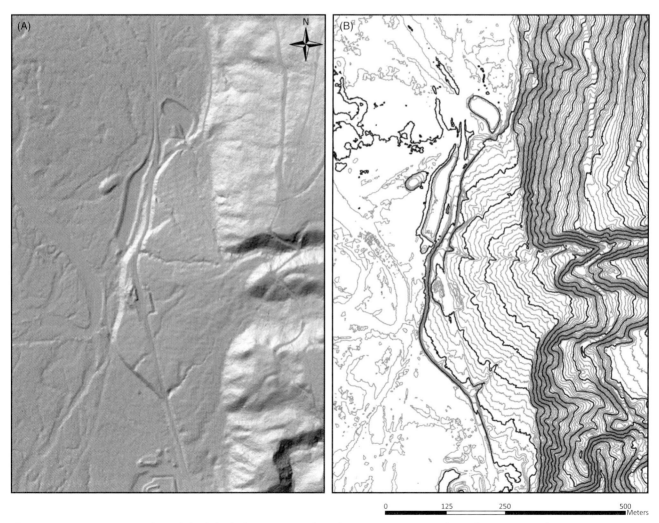

Figure 4.2: (A) ALS-HRDEM-derived shaded relief image, ground resolution 1 m; (B) LiDAR-derived 1 m topography contour map

photos. Features that can be seen in aerial photos, such as polygonal ground (distinct and often symmetrical geometric shapes formed by ground material in periglacial regions), thermokarst (very irregular surfaces of marshy hollows and small hummocks formed as ice-rich permafrost thaws) and pingos (dome-shaped mounds consisting of a layer of soil over a large core of ice), may indicate the presence of permafrost. Dark patches on aerial photos may be an indication of organic soils, a change in soil types or low-lying areas with poor drainage. The arrangement of vegetation boundaries may be related to change in geological conditions.

4.3.1.3 Photogrammetry

Photogrammetry is a mapping technique that uses stereographic photo images to generate a 3D image that can be viewed and used to develop orthographic images, digital terrain models or topographic plans. There are two types of photogrammetry: aerial and terrestrial. In aerial photogrammetry, stereo photos are taken from a plane or drone, and sometimes from a satellite. Aerial photogrammetry is most suitable for mapping large areas. In terrestrial photogrammetry, stereo photos are taken with cameras located on the ground. Terrestrial photogrammetry is the preferred technique for mapping small-scale features. Both photogrammetry techniques allow mapping of objects that are inaccessible on foot.

4.3.1.4 Topographic mapping

An understanding of the site topography is a fundamental requirement for site selection and design of mine waste dumps and stockpiles. Obtaining good topographic maps should be one of the first steps in every study. Ground relief, terrain types and landforms, drainage conditions and patterns can all be identified using topographic maps. Topographic maps may also provide insights into the underlying bedrock geology. For example, gently rounded hills may indicate weathered rock, and enclosed depressions may indicate sinkholes and the presence of karstic limestone. The topography of the site may also influence how the dump or stockpile is constructed. The topography of the site will also determine the available capacity.

At the outset of the project, large-scale topographic maps, often available from government, may be sufficient, but as the project advances, more detailed mapping is required. More detailed topographic maps may be obtained using aerial photogrammetric or LiDAR techniques as described above, or ground-based topographic surveys conducted using total station and/or global positioning system survey techniques. The use of drones to obtain high quality, georeferenced LiDAR or stereo photographic data that can be processed into very accurate topographic maps has recently become very popular. Drone surveys can be conducted very rapidly and are very economical in comparison with other mapping techniques.

4.3.1.5 Terrain classification and landforms

'Terrain' is a general term related to the geography of the area and refers to its physical features. A landform is part of terrain consisting of a specific geomorphic feature on the Earth's surface. Landform elements include hills, mountains, plateaus, canyons, peninsulas and bays. Terrain can be categorised based on slope steepness as summarised in Table 4.2.

The terrain categories in Table 4.2 are for unidirectional surfaces with a smooth longitudinal profile that is straight, concave or convex. In many cases, terrains consist of more complex landforms consisting of a multidirectional, non-planar surface as described below (Howes and Kenk 1997):

- **Undulating topography** – non-linear rises and hollows with slopes generally less than 15° (26%)
- **Rolling topography** – elongate rises and hollows with slopes generally less than 15° (26%)
- **Hummocks** – non-linear rises and hollows with many slopes steeper than 15° (26%)
- **Ridges** – elongate rises with many slopes steeper than 15° (26%)
- **Depressions** – hollows separated from an adjacent gentler surface by a marked break of slope.

Plains and plateaus may also be classified according to their origin and the materials that underline them (e.g. coastal, alluvial, glacial, lacustrine, sand, loess, volcanic and karst).

4.3.1.6 Glaciation

Glacial activity has resulted in glacial deposits that cover large areas of the Earth, mainly in the northern hemisphere and to a lesser extent in the southern hemisphere. There have been at least five major ice ages. The most recent one began about 2 million years ago during the Quaternary epoch. The last glacial retreat began with the start of Holocene epoch ~11 700 years ago

Table 4.2: Terrain category by slope

Terrain category	Slope	
	%	°
Plain (plateau)	< 5	< 3
Gentle	6–26	4–15
Moderate	27–49	16–26
Moderately steep	50–70	27–35
Steep	> 70	> 35

and is still ongoing. Between the 16th and 19th centuries, there was a cooling of the world's average temperature, resulting in some glaciers advancing. This period is known as the Little Ice Age.

During the Pleistocene epoch, much of North America, Northern Europe and Asia were covered by glaciers. The extent of the glacier ice during this period, however, fluctuated with changes in temperature. In North America, the Laurentide Ice Sheet was centred on Hudson Bay, covering most of the continent from the Rocky Mountains to north-central Canada. In Canada, as much as 97% of the land was covered in ice. The retreat of glaciers reduced their size from one-third of the Earth's surface to the current amount of ~10%.

Currently, the largest glaciers in the world are in Antarctica. Most of the Earth's current glacial ice resides within the Arctic and Antarctic regions, but significant parts of Alaska, Canada, Iceland, South America, Greenland and Russia are still covered with glaciers. Glaciers can also be found in many mountainous regions of the world. Significant retreat of glaciers has been observed since the 20th century. As an example, the glaciers on the eastern slope of the Rockies are 25–75% smaller than they were in 1850 (CCME 2003). Today, extensive continental ice sheets are only found in the Antarctic and Greenland. Alpine glaciers, which are much smaller than ice sheets, typically only exist at high elevations in areas with substantial precipitation. Glacial retreat is expected to accelerate with ongoing human effects on the environment that result in an increase of the Earth's temperature.

Glaciers

A glacier is composed of a body of dense ice that flows slowly downslope under its own weight. It exists in extremely cold regions when sufficient snow precipitation is present and forms by accumulation and compaction of snow over long periods. Glaciers can be continental or alpine. Continental glaciers, or ice sheets, cover large areas and form in high latitudes. Alpine glaciers form in mountains and can be subdivided to valley glaciers and piedmont glaciers. A valley glacier forms in an area eroded by stream action. When several valley glaciers merge at the foot of a mountain, a piedmont glacier forms.

Glacial action produces a range of soil deposits. Kehew (2006) presents a good summary of the types and characteristics of glacial deposits. Till is the direct deposit of glacier ice. Basal tills are typically dense and overconsolidated and have low permeability. Ablation tills tend to have less fines, are less dense and have higher permeability. Clay-rich glacial till may be fissured. Fissuring is more common in basal tills, but is also seen in ablation tills. The existence of fissures should be carefully investigated during the field program as fissures weaken

the till, resulting in lower strength and higher permeability of the soil. Glaciolacustrine sediments are normally dark coloured, are sometimes varved and generally have low bearing strength. Glaciolacustrine sediments may be sensitive (subject to rapid loss of strength when sheared) and susceptible to undrained shear failure. Drumlins are composed primarily of glacial till. Eskers and kames mainly consist of meltwater-transported sand and gravel, and are often prized as a source of granular borrow material.

Periglacial geomorphology

Periglacial is defined as the area around glaciers. Periglacial geomorphology is now more generally defined as the study of near-glacier locations or conditions including permafrost ground and seasonally frozen ground (active layer), and landforms and periglacial processes that are a result of freeze–thaw cycles. Periglacial environments include cold climate regions located in mid- to high-latitude or high-altitude climates that experience freeze–thaw cycles and in which permafrost ground exists.

Rock glaciers

Rock glaciers consist of a mass of debris composed of rock, soil and ice that moves slowly downslope through deformation of internal ice. They generally originate in mountainous areas with high topographic relief. Rock glaciers are normally either talus derived (formed by the periglacial processes involving freeze–thaw cycles) or glacier derived (formed in glacial environments with a high debris supply). They can also have a landslide origin. Rock glaciers are typically tongue-shaped and can be up to a several metres to tens of metres in thickness. Active rock glaciers typically flow downslope from a few centimetres to metres per year.

4.3.1.7 Permafrost

Permafrost forms when the ground remains at 0°C for at least two consecutive years. It exists in cold regions, when the precipitation is not high enough to allow the formation of glaciers. Air temperature is the main factor for the existence of permafrost, but other parameters such as vegetation, snow cover and terrain have influencing roles. The thickness of permafrost could be from less than 1 m to more than 1500 m. Currently, ~20% of the Earth's land surface is covered by permafrost. Permafrost covers ~50% of Canada and the former Soviet Union and ~85% of Alaska (Kehew 2006). The extent of permafrost has decreased over the last century. As a result of the climate change, the active layer of permafrost is expected to deepen over broad areas. Retreat of permafrost areas and thawing of frozen ground may result in less stable foundations and potential slope stability concerns.

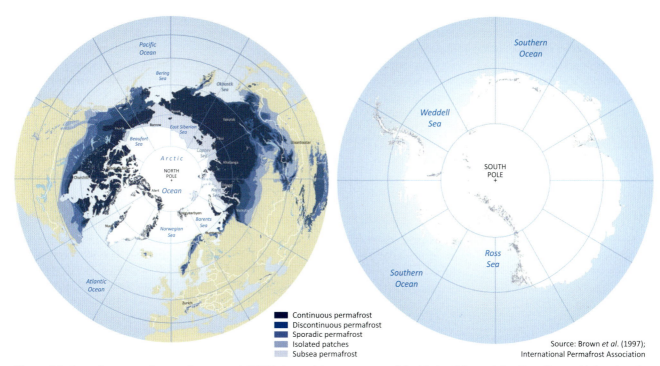

Figure 4.3: Permafrost map. Source: Brown *et al.* (1997). Image/photo courtesy of the National Snow & Ice Data Center, University of Colorado, Boulder, USA

Consequently, the long-term effect of climate change on the stability of the waste dump should be considered.

Permafrost areas are either continuous or discontinuous. Frozen and unfrozen surrounding ground is a sign of discontinuous permafrost. Permafrost is considered continuous when it covers ~90–100% of the landscape. Typically, permafrost extends to great depths at higher latitudes or elevations and gradually thins towards areas with higher ground temperature. A mean annual ground surface temperature of –5°C is generally the boundary between continuous and discontinuous permafrost. Permafrost can be found in Antarctica, the Antarctic Islands and high alpine or mountainous regions in the southern hemisphere. Figure 4.3 shows the extent of permafrost in the northern and southern hemispheres. The southern hemisphere contains less permafrost than the northern hemisphere, mainly due to less land area and more ocean area. The National Snow & Ice Data Center (http://www.nsidc.org) is a good source for maps and data on permafrost.

Above the permafrost layer is the active layer, which freezes and thaws seasonally. Localised unfrozen pockets called taliks may exist on top, underneath or within masses of permafrost. Ice-wedge polygons and pingos are landforms associated with permafrost.

Ground ice
Depending on the climate, soil condition and snow cover, ground ice of up to several metres can be found in frozen ground. The thickness of the active layer is controlled by the thermal and geotechnical characteristics of the near-surface soil, geothermal thermal gradient and atmospheric temperature. Layers of vegetation, snow cover and peat act as insulation and reduce the amplitude of seasonal temperature variation in the underlying soil. Disturbance of the ground due to removal of the vegetation cover, excavation into frozen ground, construction of dumps and stockpiles, and acid generation processes associated with some types of waste rock may lead to warming and increase the depth of the active layer.

Ground freezing can result in heaving of the surface. Silts are generally the most susceptible type of soils to frost heave. Thawing of the frozen ground may result in settlement and decrease in the strength of soils and can result in ground instability, landslides and changes in drainage.

The annual thaw depth can be estimated by measuring the ground temperature profile over the year or by test pits or holes conducted in the late summer or early autumn when a maximum thaw can be observed. Reduction of permafrost due to climate change should be considered, especially for the closure phase of projects.

4.3.2 Geology
Geology is the science that deals with the structure and origin of the Earth and the formation and alteration of the rocks of which it is composed. The geological makeup of a

site is a result of the complex interaction of many geological, environmental and tectonic processes over millions, and often billions, of years. An understanding of the site geology is important for every project, and through all project stages. Some mining activities may be located in geologically undesirable areas where topographic or hydrological constraints and distance from the pit may require construction of waste dumps and stockpiles in challenging settings.

The geology of the site influences foundation conditions, including the thickness and strength of the foundation materials, and their drainage characteristics. An understanding of surficial and bedrock geology is necessary to properly evaluate the foundation conditions, design constraints and potential hazards associated with the project. Features such as faults, bedding planes, joints, sinkholes and fissures may influence the drainage pattern and stability of the facility. The characteristics of dump and stockpile materials and the influence of local geology on construction methods should also be assessed. Lithology, structure, alteration, weathering and rock fabric can influence the durability, stability and permeability of the materials.

Sources of information on geology may include geological maps, reports published by governments, regional and local geological studies, academic theses and journal articles. These types of information are commonly available through government offices, universities, public libraries, journals and a wide range of online sources. A considerable amount of useful information can be obtained from geological maps. Features such as the stratigraphy, geological age and lithological character of the rock can be identified. Caution should be exercised when using this information, however, as many of the data are often interpreted with limited ground verification.

The key components of a geological desktop study can be summarised as follows:

- review of published and unpublished geological information (both regional and site specific), including results of previous studies; this may also include reviewing information obtained using various remote sensing techniques
- assessment of the available geological data and preparation of a gap analysis, including an evaluation of the potential impact of geology on the project
- assessment of potential failure mechanisms that could occur during and following construction of the waste dump or stockpile, as they pertain to site geology.

The results of the desktop study should be used to develop the scope for subsequent site investigations to validate the data and interpretations and establish a comprehensive understanding of site conditions.

Geological data from the site mineral resource model should also be reviewed and used as input.

4.3.3 Natural hazards

Natural hazards could result in catastrophic slope failure and environmental consequence. Therefore, their existence should be identified and potential mitigation measures assessed as part of the design. The assessment of the hazards should not be limited to the area of the facility but should extend to the area upstream or upgradient of the site, where existing conditions could eventually affect the stability or operation of the structure. Hazards may be related to site topography, geology and climate conditions. Geological maps, terrain maps, satellite imaging, HRDEM and aerial photos are useful sources for locating natural hazards. Waste dump or stockpile construction in areas subject to potential natural hazards should be avoided. If construction in these areas is unavoidable, mitigative measurements and protection works should be considered.

4.3.3.1 Landslides

Landslides involve mass movement of soil or rock in sloping terrains. They sometimes occur in conjunction with other hazards such as earthquakes and floods. Landslides can also be caused by excessive precipitation, weathering or progressive weakening of the slope over time, and human activities such as loading of natural slopes with mine waste or stockpile materials, ponding of water behind the crest of the slope, removal of vegetation, and excavation at the toe of the slope. Existing landslides can often be identified by interpretation of aerial photos and terrestrial photogrammetry, HRDEM interpretation, LiDAR-derived data or visual investigation. The Resources Inventory Committee (RIC 1996) described various mapping methods for landslide hazards. The presence of a landslide can be identified by the following:

- cracks at or near the top of the slope
- areas of anomalous vegetation (e.g. unvegetated areas of soil or rock, trees that lean in a different directions or have bent lower tree trunks, areas where trees are younger or display a difference in the density of the vegetated area)
- hummocky ground, steep ground, curved scarps or internal drainage areas.

It is generally not desirable to construct waste dumps or stockpiles in areas that are prone to landslides. If designing close to a landslide is unavoidable, the design should assume the residual shear strength for the landslide soil and include a comprehensive instrumentation program consisting of surveying, piezometers, settlement gauges and inclinometers to investigate any excessive ground movements. Blake *et al.* (2002) described a procedure to evaluate the shear strength of soil in landslide areas. Variations in the groundwater regime and the effect of freezing and thawing of the ground surface should also be studied. Additional control measurements such as buttressing and monitoring of surface drainage may also be needed.

4.3.3.2 Deep-seated gravitational slope deformation

Deep-seated gravitational slope deformations (DSGSDs) are a type of mass movement whose depth is a function of the slope size, and whose displacement is small in comparison with the depth of the displaced mass (Agliardi *et al.* 2012; Zorzi *et al.* 2014). These types of mass wasting phenomena are commonly defined by low to extremely low deformation rates, induced by a gravity-driven creep deformation mechanism. Typical landforms associated with potentially ongoing DSGSD were described by Agliardi *et al.* (2001) and Wolter *et al.* (2013) and include the following:

- double ridges
- scarps
- counterscarps
- trenches (tension cracks).

High-relief slopes, mainly, but not exclusively, in de-glaciated terrain, especially in areas characterised by a complex structural setting (i.e. faulting, active tectonism, highly deformed metamorphic rocks), are prone to be affected by ongoing deep-seated deformation. Despite the overall low rate of deformation, DSGSD may evolve into catastrophic collapse, such as rock slides (e.g. the Vajont rock slide in Italy (Wolter *et al.* 2013 and 2016)) and/or rock avalanches (e.g. the Flims rock avalanche in Switzerland (Pollet and Schneider 2004)). In addition, small-scale slope failures affecting mainly the toe of the deforming slope are common.

If the waste dump or stockpile is to be constructed in mountainous areas, especially in high mountainous areas, detailed desktop topographic analysis using an HRDEM is desirable to help define the potential existence of this type of slope instability. If detected, investigations should be conducted to assess the potential impacts that the waste dump or stockpile might have on the deforming system.

4.3.3.3 Snow avalanches

Avalanches develop on slopes that are shallow enough to allow snow to accumulate but steep enough for the snow to accelerate once the snow load exceeds its strength. Events such as rainfall, earthquakes, rockfalls and human activities could trigger an avalanche. Avalanches involve rapid sliding or flowing of a large amount of snow and ice, as well as incorporated rock, soil, debris and vegetation. Avalanches normally occur in convex-shaped slopes with little roughness when snow temperature is ~0°C at the snow–soil interface and on slopes at angles in the range of 15–55°. Slopes greater than 55° usually do not allow the accumulation of enough snow to trigger an avalanche. McClung and Schaerer (2006) presented a classification system for avalanches according to their size and destructive potential. Statistical and dynamic models are available to estimate the avalanche runout. Lied and Bakkehöi (1980) and McClung *et al.* (1989) developed statistical methods based on the data of 192 Norwegian avalanche paths and paths located within the areas of the Rocky and Purcell mountains, respectively. McClung (2001) provided a detailed comparison of available empirical models. Artificial barriers such as snow nets, snow fences, deflection dykes or other landscaped barriers can be effectively used to mitigate the effect of avalanches.

4.3.3.4 Debris torrents and debris flow

A debris torrent is defined as 'a mass movement that involves water-charged, predominantly coarse-grained inorganic and organic material flowing rapidly down a steep, confined, pre-existing channel' (VanDine 1985). A debris flow generally describes a similar mass movement on a planar, unconfined slope. A debris torrent can occur in creeks located in relatively high relief terrain, and where there is high precipitation and abundant loose sediment, soil or weathered rock. Jakob and Hungr (2005) provided a discussion of various methods available to predict the runout distance and extent of debris torrents, which are summarised in Table 4.3.

One approach to mitigation of debris torrents and debris flows is to control the drainage, thus decreasing the energy of the flow mass or debris. It is also possible to construct structures to divert or absorb the impact of debris. For example, Geobrugg AG of Switzerland markets a flexible debris flow barrier system (see http://www.Geobrugg.com). Further discussion on debris flows and countermeasures is available from Takahashi (2007).

The review of potential for debris flows or debris torrents is required as part of site characterisation as a possible impact on the development of a mine waste dump or major stockpile facility. The runout analysis as it applies to the failure of waste dumps is described in detail in Chapter 9.

4.3.3.5 Seismic hazards

Seismic hazards for mine sites and mine waste dump and stockpile facilities usually include ground motions (shaking effects) and permanent deformation of the ground due to surface fault rupture. Seismic hazard studies generally include determining the levels of ground

Table 4.3: Overview of runout prediction methods for debris torrents

General approach	Keywords to characterise method	Main reference
Total travel distance (entire path length)		
Travel distance and event	Travel angle	Corominas (1996)
Magnitude	Volume and descent height	Rickenmann (1999)
Volume balance	Without entrainment With entrainment	Cannon (1993) Fannin and Wise (2001)
Mass point models	Iverson approach	Lancaster et al. (2003)
Limiting criteria	Critical slope and junction angle	Benda and Cundy (1990)
Runout length (depositional part of flow)		
Critical slope and deposition on fan	Empirical methods	VanDine (1996)
Volume balance	Deposition area and flow cross-section	Crosta et al. (2003)
Analytical approaches	Mass point models Constant discharge model	Körner (1980) Perla et al. (1980) Hungr et al. (1984) Takahashi (1991)
Runout distance (entire path or depositional part only)		
Continuum based simulation	Various constitutive equations	Iverson (1997) McDougall and Hungr (2003)

Source: After Jakob and Hungr (2005)

acceleration for a range of structural responses, understanding the potential for surface fault rupture and range of potential deformations, and understanding of the impact of earthquake-induced landslides and resulting surface settlements.

This section presents an overview of the methods and approaches commonly used to define seismic hazards for mine sites with waste dump and stockpile facilities. Detailed coverage of the theory and procedures for quantifying seismic hazards is beyond the scope of this guideline. The information provided is based on current developments for characterising seismic hazards and on recent project experience in seismically active and stable continental regions. It is strongly recommended that the definition, characterisation and quantification of seismic hazards be performed by experienced specialists with a background in seismology, geology, soil mechanics and structural dynamics.

Seismic ground motions

Seismic ground motions for the design of waste dumps and stockpiles are generally described in terms of parameters such as peak ground acceleration and spectral acceleration. These ground motion values are necessary to estimate the potential seismic response of the facility and its foundation, including the potential for liquefaction, slope displacements and instability. Empirical ground motion prediction equations (GMPEs) are used for determining seismic ground motions. They describe the rate of dissipation of seismic energy with distance from the source of the earthquake. Ground motion prediction equations utilise data recorded from strong motion accelerographs and are commonly updated and improved after the occurrence of large earthquakes.

The acceleration time history data used to developed GMPEs are separated into groups that share similar attributes, such as local site conditions, type of faulting (e.g. strike-slip, normal-slip and reverse-slip) and tectonic regime (subduction or crustal). Soil conditions are usually classified in terms of shear wave velocity values.

Ground motion prediction equations require the magnitude or size of the earthquake event, a measure of distance from the earthquake source, type of faulting and soil conditions. Recent GMPEs use the moment magnitude as a metric for the size of the earthquake. However, for stable continental regions, it is still common to find GMPEs based on surface wave magnitude, body wave magnitude or local magnitude. The distance from the earthquake source is usually expressed as the epicentral distance (the distance from the point on the Earth's surface immediately above the focal point of the seismic event), hypocentral distance (the distance from the focal point of the seismic event) or shortest distance to the fault rupture plane.

Recorded ground motion data show large levels of scattering. This is mainly caused by directivity or path effects from the earthquake source to the site as well as the interaction of seismic waves as they propagate to the site from multiple locations along the fault rupture plane. The scattering of the data is reflected as uncertainty in the values given by the GMPEs and quantified in the GMPEs by values of normalised standard deviations from the median. The uncertainty term is typically a function of earthquake magnitude.

The selection of GMPEs for estimating design ground motions should be based on criteria that include origin, quality and extent of the available seismic data. Because many mining projects and related facilities are located in regions with scarce or non-existent seismic records, it is common practice to select available regional GMPEs developed for tectonic conditions similar to those at the project site. Where possible, it is recommended that design ground motions be estimated as the geometric mean of at least three representative GMPEs.

Once the GMPEs are identified and selected, design-level ground motions can be assessed by using well established deterministic and probabilistic methods. A brief description of each of these methods follows.

A deterministic approach, or a deterministic seismic hazard analysis (DSHA), uses available historical seismic and geological data to generate discrete, single-value earthquake events to model earthquake ground motions at the site. A DSHA requires identification and characterisation of the major seismic sources likely to contribute to the earthquake shaking at the site. Earthquakes generated at these sources are evaluated to identify the largest earthquakes that will produce the strongest earthquake ground motion at the site. These maximum earthquakes are specified by magnitude and source-to-site distance. Usually, the earthquakes are assumed to occur on the portion of the seismic source closest to the site.

A key feature of DSHA is that it does not explicitly consider the recurrence interval of the maximum earthquakes that are used to develop design-level ground motions. The DSHA method generally results in a conservative assessment of maximum ground motions at a site because near-maximum parameter values are used and return periods are not considered. A DSHA is commonly used for estimating design ground motion values for the assessment of slope stability and other hazards related to liquefaction for tailings storage facilities located in seismically active regions.

Probabilistic seismic hazard analysis (PSHA) uses the seismic source identification and characterisation elements of the DSHA and adds an assessment of the likelihood that ground motions of specified amplitude will occur at the site. The probability or frequency of occurrence of different magnitude earthquakes on each significant seismic source and inherent uncertainties are directly accounted for in the PSHA. The possible occurrence of each magnitude earthquake at any part of a source, including the closest location to the site, is also incorporated in the PSHA.

The results of the PSHA are used to select the design earthquake ground motion parameters based on the probability of exceeding a given parameter level during the service or design life of the structure or for a given return period. Through a technique known as 'hazard disaggregation', the results of the PSHA can also be used to identify which combinations of magnitudes and distances and/or specific seismic sources are the largest contributors to the hazard, as well as the number of standard deviations a level of ground motion is away from its median value.

A comprehensive PSHA incorporates not only the inherent randomness, or variability, of earthquake occurrence and seismic wave propagation, but also the uncertainty associated with the choice of particular models and model parameters for characterising seismic sources and estimating ground motions (epistemic uncertainty). Accordingly, no additional conservatism is normally added to the seismic hazard values from a comprehensive site-specific PSHA. The degree of conservatism is explicitly defined by the choice of annual exceedance probability. For example, some tailings storage facilities located in stable continental regions are designed for seismic ground motions with an annual exceedance probability of 1 in 2500 years or 1 in 10 000 years to take into account the consequence of failure of the facility and the acceptable risk. For mine waste dump and stockpile facilities, which many practitioners consider to be much less susceptible to failure during a seismic event than tailings facilities, the use of an annual exceedance probability of 1 in 475 years is common.

Probabilistic seismic hazard analysis typically forms the basis for the development of building code seismic hazard maps. Therefore, initial values for the seismic analyses of waste dumps and stockpiles can be obtained from such maps, when available, up to the return periods required. If not available, a site-specific study should be undertaken. Additional discussion on the selection of seismic design criteria for mine waste dumps and stockpiles is included in Chapter 8.

Surface fault rupture

Surface fault rupture can be an important consideration for waste dump and stockpile facilities located in seismically active regions because of the instability and failure it can impose on the facility. Identification of active faults is usually based on information from satellite imaging, aerial photos and fault trenching.

A fault is characterised by its location, orientation, width of fault rupture zone, the amount and direction of potential fault displacement and level of activity. Generally, a fault is considered active if it can be demonstrated to have displaced the ground surface during the Holocene epoch (i.e. within the past 11 700 years). Due to the large uncertainties in fault characterisation, the evaluation of fault activity inevitably implies judgement and subjectivity on the part of the geologist.

If a characteristic earthquake and associated recurrence interval can be established through field studies and historical seismicity, the expected fault displacement may be estimated as the cumulative slip between the design earthquakes (i.e. the slip rate per year times the recurrence interval). Alternatively, fault displacement may be established from relationships published in the available literature.

4.3.4 Climate

An understanding of climatic data is a fundamental part of the site characterisation process. The level of detail required will depend on the site setting and topography, the planned size of the facility and the prevailing climate.

4.3.4.1 Types of data

The following types of climate data are usually important:

- Mean monthly precipitation, by type (i.e. rain and snow) and evaporation:
  - → These data are used for early-stage planning, screening level assessments and alternatives analysis.
  - → For early-stage planning, before establishment of an on-site weather station, the data will usually need to be obtained from regional stations.
  - → In some dry climates, monthly precipitation and evaporation data may be all that are required.

- Daily (and sometimes hourly) precipitation:
  - → Daily precipitation data are required to support the peak rainfall and runoff analysis necessary for the planning and design of upgradient surface water diversions and to support the surface water management plan for the facility.
  - → Daily precipitation data are also required as input to infiltration and recharge modelling for the dump or stockpile surface.
  - → For early-stage planning, unless the facility is part of an established site, it is usually necessary to use published data as input to the studies.

- Daily evaporation:
  - → Evaporation is spatially less variable than precipitation, and it is normally possible to provide reasonable estimates using regional data, provided that the effects of topographic elevation are understood.
  - → Topographic aspect (or direction of dump or stockpile face with respect to the sun) may also be important for establishing local-scale variability.
  - → The use of solar radiation, wind speed and direction and other micro-climate data obtained from an on-site weather station are important for calculating evaporation to support detailed infiltration and recharge studies for the dump or stockpile surface, should these be required.

- Ambient air temperature:
  - → Air temperature data are important in colder climates for determining how much of the precipitation may occur as snow, and over what period of time the snow may accumulate.
  - → Air temperature data are important for calculating evaporation and estimating the near-surface water balance of the facility.
  - → In colder climates, a knowledge of air temperature may be used to assess the potential for seasonally frozen ground in the upgradient catchment area, or freezing on the outer slopes of the dump or stockpile (seasonal or permafrost).
  - → Daily, monthly and annual records may be available from regional stations before the establishment of an on-site weather station.

- Wind speed and direction:
  - → An understanding of the wind is important for assessing variations in precipitation and evaporation over the surface area of the facility.
  - → Wind data are important for estimating areas of potential snow accumulation and/or variations in sublimation over the surface of the facility, both of which may greatly influence the spatial distribution of recharge to the facility.
  - → Wind data may also be important for estimating the potential for dust erosion either from operating surfaces or from reclamation covers.
  - → Records may be available from regional stations before the establishment of an on-site weather station.

- Snowfall and snowpack:
  - → For cold climates, analysis of snowfall and snowpack data may be important for assessing the potential amount of accumulating snow on the facility or in the upgradient catchment area.

Depending on the proximity of a project site to populated areas, the number and type of existing weather stations will vary. The available data from existing stations may range from daily precipitation or daily temperature only, to a comprehensive suite of measurements that may include continuous precipitation and temperature, barometric pressure, wind speed and direction, humidity and solar radiation. Access to available data varies around the world; however, data from government sources, meteorological organisations and airports can typically be accessed via the internet. It is usually beneficial to establish a site-specific weather station as early as possible in the project development process, particularly if the closest climate stations are located a significant distance away or at different elevations. By comparing site-specific climate data with data from nearby weather stations, it may also be possible to establish empirical relationships between the site data and regional stations where the period of record is longer. This can be important at sites located at different elevations from the nearest weather stations and/or where orographic effects may result in different weather conditions as elevation changes.

4.3.4.2 Requirement for a daily record of climatic variables

In some cases, usually for advanced planning and design, it may be necessary to derive a daily record of climatic variables to allow the following:

- a reliable assessment of peak rainfall and surface water runoff from the upgradient catchment area and/or from the surface of the facility
- more reliable estimates of surface infiltration and recharge to the facility to support modelling and pore pressure studies or, more commonly, for design of reclamation covers for concurrent or final closure.

Modelling of the flux of water into the exposed surfaces usually requires a daily climatic dataset that is used as a direct input to the upper layer of numerical soil flux or unsaturated flow model (Chapter 6), which simulates the daily (and sometimes hourly) water balance for the near-surface layer based on precipitation, temperature and evaporation (which may need to be derived from temperature, solar radiation, relative humidity and other parameters).

For the majority of sites (and particularly for early-stage studies), there are typically insufficient site-specific data to allow the development of a long-term dataset of daily climatic variables, so it is necessary to develop a synthetic record of precipitation, temperature and evaporation, usually over a 100-year period. When developing this record, it is important to ensure that the dataset includes peak rainfall events, wet cycles, dry cycles and other extreme conditions or trends.

4.3.4.3 Influence of freezing conditions

In many regions, some or all of the precipitation may occur as snow. Most climate model inputs assume that precipitation at mean daily temperatures above freezing occurs as rain while precipitation at mean daily temperatures below freezing is stored on the surface as snow. Snowmelt can be computed using standard procedures, for example the Snow Accumulation and Ablation Model, SNOW-17 (NWSRFS 2016). Using this procedure, the melt process is divided into precipitation that occurs during non-rain periods and precipitation that occurs as rainfall. Rain-on-snow melt is computed using an energy balance approach. Non-rain melt is computed using air temperature as an index to energy exchange across the snow–air interface.

For any analysis, it is important to realise the uncertainty that may result from the following:

- the redistribution of snow by the wind, which may have a major impact on the soil moisture balance of both the natural upgradient catchment area and the surface of the facility; the effects of wind can greatly influence the actual flux of water that penetrates below the near-surface layer and moves deeper (as recharge) into the natural ground or the dump/stockpile materials
- the influence of frozen surfaces on the infiltration flux, considering that the insulating effect of snow may either prevent the onset of freezing of the surface or may cause the materials beneath snow banks to remain frozen when the surrounding surfaces have thawed.

As with any modelling, the overriding assumptions that are used to develop the input parameters are usually more important than the selection of model code, procedures or other modelling details.

4.4 Field investigations for geotechnical conditions

4.4.1 Planning of geotechnical field investigations

The key objectives of a geotechnical field investigation are to develop the material properties and lithology of the subsurface conditions, including groundwater conditions, to support analyses and design. Preliminary site characterisation should be carried out in advance of field investigations and can be used to assist in the planning of the field investigations. Typically, excluding initial reconnaissance, the first field investigation would take place to support the pre-feasibility study, and at this stage the conceptual models developed for the site geological, hydrological, hydrogeological, geochemical and geotechnical models should be reviewed. Planning would include development of the field investigation methods specific to the locations of testing, sampling and reconnaissance. Overall, the objectives for data collection should all be established before the commencement of the site work.

Information from the field investigation should be sent to the design engineer as soon as it becomes available at the site. This allows the investigation program to be modified as new subsurface information is obtained. If significant deposits of problematic soils or features such as faults and sinkholes are found, additional subsurface investigations, including geophysics, test pits and/or boreholes, may be required. Following initial investigations, more targeted investigations would be carried out to collect advanced data to support more detailed design.

Sources of data for both site characterisation and for planning of field investigations are presented in the following sections.

4.4.2 Foundation investigations

Waste rock and stockpile materials commonly exhibit high frictional strength (Chapter 5), and in many cases the stability of the dump or stockpile is controlled by the strength and pore pressure conditions in the foundation. Foundation characteristics, including strength and hydraulic conductivity, will have an important effect on the design of the facility.

The scope of a foundation investigation will vary depending on whether the key concern is soil or the bedrock, and may include geophysics, test pits, trenches and/or boreholes and require *in situ* testing and collection of samples for laboratory classification and testing.

A poorly designed or inadequate subsurface investigation that does not yield the data required to fill knowledge gaps may require the use of more conservative design criteria. Where the foundation geology of a waste dump or stockpile cannot be adequately investigated, it may not be feasible to ascertain or assess all potential modes of instability.

4.4.2.1 Overburden soil

Characterisation of the foundation soil should include the origin, nature and distribution of soil units within the site. The characteristics and extent of soft or problematic soils should be identified. The depth to bedrock or competent foundation soil should be identified. It is important that boulders are not confused with bedrock, so drilling normally extends to at least 3 m into the competent rock to make sure that the bedrock is correctly detected.

There are many soil classification systems referenced by various standards associations and used by practitioners around the world. These systems can be broadly divided into two types: those which classify based on particle size and plasticity (e.g. the Unified Soil Classification System, ASTM D2487 [ASTM 2011a]; AS1726 [Standards Australia 1993]) and those which classify based on bulk behaviour (i.e. cohesive or non-cohesive and particle size distribution, e.g. British Standard BS5930 [British Standards Institution 2015]; Dumbleton 1981).

Soil classification carried out in the field is typically confirmed with the subsequent laboratory testing. Records of field observations and conditions during site investigations should be recorded in sufficient detail and documented to support design stages and any additional investigation stages.

Material characterisation and laboratory testing are described in detail in Chapter 5.

4.4.2.2 Soil types and stratigraphy

The type of site investigation will be influenced by the soils which are expected to be encountered. To select the most suitable methods of investigation, a general understanding of the soil type and thicknesses is required.

Soils are generally categorised as gravel, sand, silt or clay depending on the particle size and the plasticity. Sands and gravels are categorised as coarse-grained soils and have a grain size larger than 0.075 mm. Silts and clay are grouped as fine-grained soils, having a grain size of less than 0.075 mm. According to the Unified Soil Classification System (ASTM 2011a), soils can be grouped into gravel (76.2 to 4.75 mm), sand (4.75 to 0.075 mm) and silt and clay (less than 0.075 mm). 'Clay' is sometimes used to mean soils that have particle sizes of less than 0.002 mm. However, clay classification should be based on plasticity limits and not according to the grain size. The engineering behaviour of coarse-grained soils is mainly influenced by grain size and density, whereas that of fine-grained soils is mostly influenced by mineralogy of the soil.

Soil deposits form under a large range of natural conditions. Some of the most common soil types and their modes of formation are listed below:

- **Organic** soils develop by accumulation of organic materials from growth and decay of plants. Peat is organic soil derived from the decomposition of organic materials normally found in lowland areas with high water table. Peat is usually encountered in coastal areas and glaciated regions and has dark colour and odour.
- **Residual soils** are developed by *in situ* weathering of parent rock and cover the bedrock surface from which they are produced.
- **Colluvial deposits** include talus, hill-wash and landslide deposits, which are transported by gravity.
- **Aeolian deposits** include loess and dune sands, which are deposited by wind action. These deposits are typically sand and silt with uniform grading.
- **Alluvial or fluvial deposits** are soils transported by rivers and streams.
- **Glacial soils** are material deposited by glacier action. They generally consist of a mixture of different soils, including cobbles or boulders, gravel, sand, silt and clay. Some of the main glacial deposits are as follows:
  - → **Glacial till** is a mixture of different soil particles and mainly consists of coarse material.
  - → **Glaciolacustrine deposits** are fine-grained soils deposited from glacial meltwater into ancient lakes. Seasonal winter freezing of the lake during deposition can contribute to stratified or varved silt and clay deposits.

→ **Glaciofluvial deposits** are transported and deposited by streams of meltwater from glaciers. Kames and eskers may be present.

▦ **Volcanic materials** such as volcanic ash are deposited by volcanic action.

▦ **Marine soils** are transported and deposited in a marine environment.

Many soils exhibit special characteristics that can prove problematic in geotechnical engineering. Recognition of these soils is a key to a successful design. Some of the problematic soils are described below:

▦ **Peat, organic silt and clay** are weak soils consisting of organic materials that can undergo large settlement as a result of loading. If these soils contain an appreciable amount of fibrous peat, testing may indicate relatively high shear strength. However, the shear strength of these types of soils will decrease over time as the fibrous parts decompose, so it is usually advisable to assume a lower strength for design. Peat should be probed to determine its depth and extent under the foundation. ASTM D4544 (ASTM 2012a) provides a guideline for estimating the thickness of surficial peat deposits using the probing technique.

▦ **Expansive soils** are soils (mostly clays) that undergo a large volume increase due to wetting. Certain types of clays, including montmorillonite and smectite, are especially susceptible to expansion/swelling, and any spoils containing appreciable amounts of these clays should be flagged for special consideration due to the potential for strength reduction upon swelling.

▦ **Collapsible soils** are soils that undergo a large decrease in volume upon saturation. They are mostly found in arid or semi-arid environments. Certain types of aeolian and residual soils exhibit collapsible behaviour.

▦ **Sensitive clays** are soils that experience a large strength loss when disturbed. The remoulded strength could be 25% or less of the undisturbed strength. Embankments on sensitive clays can fail with only small displacements. A typical indication of the possible presence of these soils is a water content that is equal to or greater than the liquid limit. A sensitivity of more than 16 (i.e. a residual shear strength less than 1/16 of the undrained shear strength) suggests the presence of 'quick' clays. Quick clays are a type of clay that collapses completely when remoulded (disturbed), resulting in a reduction of the shear strength to almost zero. Quick clays are typically found in the previously glaciated areas of North America and Scandinavia, but they have also been found in Japan and Alaska.

▦ **Liquefiable soils** are soils that experience significant strength loss when subjected to changes in internal stress conditions. Liquefaction can be triggered by static loading (e.g. excessive rate of loading) or by dynamic loading (e.g. earthquake shaking). Liquefied soils have extremely low shear strength and can behave more like a fluid than a soil.

4.4.2.3   Investigation requirements and techniques

The scope of an initial investigation will be determined during the desktop study stage of the project together with visual inspection/reconnaissance of the site. Attention should be given to any social, geological, environmental and hydrogeological features that could be unfavourable to the construction of the facility. The scope of the investigation would include initial estimates of the number and location of boreholes, test pits, field tests and instrumentation. As part of the planning for investigations, in addition to determining the scope, the required field equipment, manpower, schedule, cost and access to the site should be defined. An appropriate quality control program should be in place for each phase of the investigation. The field investigation should be designed to provide the following information:

▦ the suitability of the site for waste dump or stockpile construction, and whether the risks associated with the site are at an acceptable level

▦ the potential impact of the facility on the environment and the surroundings

▦ an understanding of the subsurface soil and stratigraphy

▦ design parameters for the foundation materials

▦ the elevation of the groundwater surface, including aquifer condition and seasonal variation of the ground-water surface

▦ any critical feature in the foundation (fissures, tension cracks, weak zones, remoulded zones)

▦ requirement of any ground improvement or additional structures (e.g. grouting, stripping, diversion dams).

Geophysical investigations

Geophysical methods are capable of exploring large areas rapidly and economically. They can be used to identify zones where further investigation is needed and, hence, to optimise the investigation program. Many of the geophysical testing techniques are non-intrusive. Geophysical surveys can provide useful information regarding subsurface conditions along the surveyed alignment, but need to be verified by drilling boreholes and excavating test pits. Information that can be obtained from geophysical surveys includes general soil stratigraphy and depths, bedrock depth, water table depth and the location of foundation features such as cavities, fault zones and buried river channels.

A wide range of geophysical methods are available. Some can be conducted at ground surface, and others require excavation of drill holes. Each method is applicable to a particular physical subsurface property and has limitations in terms of equipment, site and subsurface condition and signal noise. Naval Facilities Engineering Command's *Soil Mechanics Design Manual DM 7.01* (NAVFAC 1986) describes the applicability and limitations of different geophysical methods. ASTM D6429 (ASTM 2011b) describes procedure for selecting surface geophysical methods. Seismic and electric resistivity methods are routinely used in geotechnical investigation. Seismic methods are used extensively to estimate the profile of shear wave velocity required for seismic site studies and liquefaction assessment. ASTM D5753 (ASTM 2010a) provides guidelines for planning and conducting geophysical borehole logging. ASTM D5777 (ASTM 2011c) provides information on the seismic refraction method for ground investigation, and ASTM D4428 (ASTM 2014a) discusses the cross-hole seismic testing method. More information about various available geophysical testing methods can be obtained from Sabatini *et al.* (2002).

Test pits and trenches

Test pits and trenches are relatively inexpensive means for visual inspection of shallow subsurface conditions. Soil features such as slickensides and fissures can be identified, and both disturbed and undisturbed samples can be collected. The location of the water table should always be reported, and the location of all completed test pits and trenches should be surveyed. Logs and photographs of the walls of the test pit or trench provide information on the soil stratigraphy (layering) and competency. Estimates of relative density of soils can be made based on the level of effort required to dig the test pit or trench.

The depth of test pits and trenches is limited by the maximum reach of the excavation equipment, which generally ranges from ~3–7 m. In soft subsurface soils and when a high water table is present, the excavated walls may become unstable, which in turn limits the excavation depth. Even in competent soils, sudden collapse of test pit or trench walls can occur. Hence, extreme caution and suitable safe work procedures prepared by experienced personnel must be considered and used by all personnel involved in excavating, mapping, sampling and photographing around test pits and trenches. Some jurisdictions prohibit entry of personnel into open test pits or trenches that exceed a certain depth (typically greater than ~1 m) unless they are appropriately sloped or supported. In deep or unstable test pits or trenches, it may be necessary to sample the materials remotely using the excavator. It is also advisable (and required by many jurisdictions) to fill in test pits as soon as possible following completion.

Boreholes

Boreholes are the only method that can provide soil samples and subsurface information both near the surface and at depth. *In situ* testing and collection of samples for laboratory tests to determine the engineering properties of the subsurface soil such as permeability, consolidation and shear strength are generally carried out in concert with drilling.

Drilling fluid (water or drilling mud) may be used when the borehole walls are prone to sloughing or heave at the bottom of the hole, or to cool down the drilling equipment. Positive head provided by the drilling fluid also reduces the tendency for instability. Any significant loss of drilling fluid during drilling should be recorded, as this may be an indication of defects such as faults or cavities in the borehole.

The position of the water table should be reported during drilling and upon completion of the borehole. Granular soils tend to have high permeability, and the groundwater level in these soils will typically reach equilibrium soon after the drilling is completed. However, in finer grained soils, it may take a long time for the groundwater level to stabilise in the borehole, and water level measurements obtained just after completion of drilling in these spoils could be misleading. The best way to reliably establish long-term static water levels is to install piezometers in the completed borehole. Piezometers also allow long-term monitoring of water levels and assessment of seasonal changes in the water table. A more detailed discussion of groundwater monitoring is provided in Chapter 6.

To protect aquifers and limit the potential environmental impact of boreholes, many jurisdictions require that boreholes be fully grouted and sealed after completion and installation of any instruments. The collar locations of all completed boreholes should be surveyed.

The choice of boring method depends on cost, equipment availability, foundation condition, required field testing and sampling objectives. Table 4.4 shows common boring methods and their applicability. Most of the table is adapted from Naval Facilities Engineering Command DM 7.01 (NAVFAC 1986). ASTM D1452 (ASTM 2009a) provides guidelines for soil investigation and sampling using auger boring techniques.

Selection of test pit and borehole locations

The number and depth of test pits and boreholes depends on the stage of the project (i.e. pre-feasibility, feasibility or detailed design), soil type, subsurface soil variability, groundwater condition, height of the waste dump or stockpile and the amount of existing information. In general, the number of boreholes or test pits required increases with the complexity of the project, with the instability hazard and with the level of risk. For the preliminary stage of design, wide borehole or test pit spacing may be acceptable, but for detailed design, closer spacing may be necessary. Less developed sites may need more boreholes and test pits because there may be little available information.

As a general rule of thumb, the required depth of the exploration can be estimated as the depth where the increase in vertical effective stress in the ground due to the applied waste dump weight is ~10% of the applied load, or less than 5% of the current soil effective stress, unless competent bedrock is encountered. Penetration into sound

Table 4.4: Common boring methods

Boring method	Procedure	Applicability
Auger boring	Hand- or power-operated augering with periodic removal of material In some cases a continuous auger may be used, requiring only one withdrawal Changes indicated by examination of material removed Casing generally not used	Ordinarily used for shallow explorations above water table in partly saturated sands and silts and soft to stiff cohesive soils Can clean out hole between drive samples Fast when power-driven Large-diameter bucket auger permits hole examination Not suitable for soft and sandy soils below water table
Hollow-stem flight auger	Power operated Hollow stem serves as a casing	Access for sampling (disturbed or undisturbed) or coring through hollow stem Should not be used with plug in granular soil Not suitable for undisturbed sampling in sand and silt below water table
Wash-type boring	Chopping, twisting, and jetting action of a light bit as circulating drilling fluid removes cuttings Changes indicated by rate of progress, action of rods, and examination of cuttings in drill fluid Casing may be needed to prevent caving	Used in sands, sand and gravel without boulders, and soft to hard cohesive soils Usually can be adapted for inaccessible locations, such as on water, in swamps, on slopes, or within buildings Difficult to obtain undisturbed samples
Rotary drilling	Power rotation of drilling bit as circulating fluid removes cuttings from hole Changes indicated by rate of progress, action of drilling tools and examination of cuttings in drilling fluid Casing usually not required except near surface	Applicable to all soils except those containing large gravel, cobbles, and boulders (where it may be combined with coring) Difficult to determine changes accurately in some soils Not practical in inaccessible locations for heavy truck-mounted equipment (track-mounted equipment is available) Soil and rock samples usually limited to 150 mm diameter
Sonic drilling	Dual-cased single tube core barrel system that uses high frequency mechanical vibration to produce cutting action at the bit face and obtains continuous core samples of up to 200 mm diameter	Applicable to almost any formation Limited penetration into very competent bedrock formations Provides continuous core samples with good recovery rate and minimal contamination Drilling method typically results in disturbed soil core Allows easy installation of piezometers Method slows with increasing depth
Percussion drilling (churn drilling)	Power chopping with limited amount of water at bottom of hole Water becomes a slurry that is periodically removed with bailer or sand pump Changes indicated by rate of progress, action of drilling tools, and composition of slurry removed Casing required except in stable rock	Not preferred for ordinary exploration or where undisturbed samples are required because of difficulty in determining strata changes, disturbance caused below chopping bit, difficulty of access, and usually higher cost Sometimes used in combination with auger or wash borings for penetration of coarse gravel, boulders, and rock formations Could be useful to probe cavities and weakness in rock by changes in drill rate
Rock core drilling	Power rotation of a core barrel as circulating water or mud removes ground-up material from hole Water/mud also acts as coolant for core barrel and bit Generally hole is cased to rock	Used alone and in combination with other types of boring to drill weathered rocks, bedrock, and boulder formations
Wire-line drilling	Rotary type drilling method, where coring device is an integral part of the drill rod string, which also serves as a casing Core samples obtained by removing inner barrel assembly from the core barrel portion of the drill rod Inner barrel is released by a retriever lowered by a wire-line through drilling rod	Efficient for deep core holes

Source: After NAVFAC (1986)

bedrock should be at least 3 m. Where the presence of large boulders is geologically likely, the depth of the penetrating into rock should be increased. For embankments, CEN (European Committee for Standardization) (2007) recommends a minimum investigation depth of 0.8 to 1.2 times the embankment height or 6 m, whichever is more.

In situ *testing*

When representative, undisturbed samples can be obtained and transported to the laboratory without disturbance, laboratory testing can often be the most reliable and efficient method for obtaining the key physical and mechanical properties of the soil. However, for many soils it can be difficult to obtain and transport undisturbed samples for laboratory testing. In these cases, *in situ* testing may provide a convenient alternative. *In situ* testing may also be performed in conjunction with the laboratory testing to obtain geotechnical parameters for soils. Properties including shear strength, bearing capacity and overall behaviour of a soil can be determined using *in situ* testing techniques.

The suitability of *in situ* testing depends on ground conditions, surficial geology, soil stratigraphy and soil type. Several *in situ* testing techniques have been developed. The *Canadian Foundation Engineering Manual* (Canadian Geotechnical Society 2006) provides a good summary of common tests and references. Table 4.5 summarises the common *in situ* tests and their applicability for different ground conditions.

The standard penetration test (SPT) is one of the most common methods for investigating soil conditions; however, several correction factors should be applied when interpreting the results. Kulhawy and Mayne (1990) and the *Canadian Foundation Engineering Manual* (Canadian Geotechnical Society 2006) discussed factors that could affect the accuracy of SPT results. When testing below the groundwater table, maintaining the hydrostatic pressure is required for soft or granular soils. Otherwise, soil at the bottom of the hole may become quick and transform into a loose state, resulting in lower SPT values. The typical sampling interval is 1.5 m or at the interface of zones of different materials. The smallest recommended SPT sampling interval is 0.75 m, and may be carried out at locations where close attention to soil properties and layers is required, such as near the ground surface.

The SPT is generally not suitable for carrying out investigations in gravelly soils as the larger particles in these soil types can result in penetration refusal or block the SPT sampler. Large penetration testing (LPT) is similar to SPT, but has a larger sampler size and is used for gravelly soils. The Becker penetration test (BPT) is another large-scale penetration test and is commonly used to investigate deposits of gravel to cobble size with a Becker hammer drill.

The field vane test (FVT) is the most common test used to estimate the undrained shear strength of cohesive soils. Cone penetration testing (CPT) is another testing method commonly used to characterise soft, cohesive soils, but is not used for dense soils due to the limitation of this

Table 4.5: Summary of *in situ* testing techniques

In situ testing method	Properties that can be determined	Applicability		References
		Best suited	Not applicable	
Standard penetration test	Compactness Comparison of subsoil stratification	Sand	Clay	ASTM D1586 (ASTM 2011d)
Cone penetration test	Continuous evaluation of density and strength of sands and undrained strength in clay	Sand, silt and clay	Gravel	ASTM D3441 (ASTM 2005); ASTM D5778 (ASTM 2012b)
Dynamic cone penetration test	Compactness Comparison of subsoil stratification	Sand	Clay	ISSMFE Technical Committee on Penetration Testing of Soils–TC 16 (1989)
Field vane test	Undrained shear strength	Clay	Gravel and sand	ASTM D2573 (ASTM 2015a)
Flat dilatometer test	Ko, overconsolidation Undrained shear strength	Sand and clay	Gravel	ASTM D6635 (ASTM 2015b)
Becker penetration test	Compactness Comparison of subsoil stratification	Sand and gravel	–	Harder and Seed (1986); Sy and Campanella (1994)
Large penetration test	Compactness Comparison of subsoil stratification	Gravel	–	Kaito et al. (1971); Crova et al. (1993); Daniel et al. (2003)
Permeability tests	Evaluation of coefficient of permeability	Sand and gravel	–	ASTM D5126 (ASTM 2010b); ASTM D4043 (ASTM 2010c); ASTM D4044 (ASTM 2015c); ASTM D4050 (ASTM 2014b)

Source: After Canadian Geotechnical Society (2006)

equipment. Correlations with FVT can be developed to help estimate the undrained strength of soil from CPT results for large projects. Mayne (2007) and Robertson and Cabal (2015) provided detailed information regarding CPT interpretation. The main drawback of the CPT method is that it does not allow sample collection and visual observation. Consequently, boreholes drilled adjacent to CPT sites or other sampling techniques should be used to confirm the interpretation of soil layers.

Dilatometer testing or pressuremeter testing can provide information on soil type and a range of parameters such as overconsolidation ratio, horizontal stress magnitudes, friction angle and undrained shear strength. These tests can be performed in most soil types except for very dense, cemented or gravelly soils. Dilatometer and pressuremeter testing require specialised equipment, but are relatively simple and quick to perform and are relatively inexpensive. However, unlike CPT, which can provide continuous downhole information, they provide results at discrete intervals and are more difficult to correlate. Similar to CPT, no soil samples are obtained using these techniques.

Sampling of soils for laboratory testing

Soil sampling is carried out for visual identification and to obtain specimens for laboratory testing. The sampling program should be comprehensive so that all information required for characterisation of the foundation soils is obtained. The type and number of samples should be based on the stage of the investigation, complexity of the structure and geology of the site. Samples are usually taken at any change of soil stratum or every 1.5 m. Samples should be representative and should contain only the soil type of the stratum that they are representing. The sample container should not deteriorate over time or react with

the sampled soil. The moisture content of the samples should be preserved. If salt is present in the soil, sampled soil properties may change if water is added or removed from the sample. Organic samples should not be dried at high temperatures (i.e. higher than 60°C) to avoid burning of organic matter. Sample size should be consistent with the requirements of the specific tests being carried out.

Laboratory testing complements the field testing program, and the selected samples should be scheduled for testing as soon as possible. ASTM D4700 (ASTM 2015d) provides guidelines for sampling in the vadose (unsaturated) zone and ASTM D4220 (ASTM 2014c) provides recommendations for preserving and transporting soil samples.

Samples are classified as either undisturbed or disturbed. Recompacted, remoulded and reconstituted samples are all forms of disturbed samples. CEN (European Committee for Standardization) (2007) provides a convenient method for classifying sample disturbance and specifies suitable laboratory testing techniques.

Disturbed samples

Disturbed samples are generally only suitable for soil description, soil classification and laboratory index testing. Index testing includes water content, Atterberg limits, particle size distribution and specific gravity.

Reconstituted samples can be used for strength and permeability testing of fill materials (Chapter 5).

Undisturbed samples

Undisturbed samples are required to estimate hydraulic conductivity, strength and deformation characterisation of the soil. Obtaining undisturbed samples of granular soil is difficult and expensive. Therefore, soil parameters of

Table 4.6: Common samplers for disturbed and undisturbed sampling

Sampler	Disturbed/ undisturbed	Appropriate soil types	Method of penetration
Split-barrel (split spoon)	Disturbed	Sands, silts, clays; gravels may invalidate the data	Hammer driven
Shelby tube	Undisturbed	Cohesive fine-grained or soft soils; gravelly soils may crimp the tube	Mechanically pushed
Piston	Undisturbed	Soft to medium clays and silts	Hydraulically pushed
Pitcher	Undisturbed	Stiff to hard clay, silt, sand with cementation	Rotation and hydraulic pressure
Denison	Undisturbed	Stiff to hard clay, silt, sand with cementation	Rotation and hydraulic pressure
Continuous auger	Disturbed	Most of the soils, except those that are too hard to dig	Rotation
Bulk	Disturbed	Gravels, sands, silts, clays	Hand tools, bucket augering
Block	Undisturbed	Cohesive soils, residual soils	Hand tools
Cold Regions Research and Engineering Laboratory barrel	Undisturbed	Coring in fine-grained frozen ground and permafrost soils	Rotation

granular soils are normally estimated using *in situ* testing techniques such as those described above. For cohesive soils, it may be possible to obtain high quality, undisturbed block samples from test pits. Relatively undisturbed samples can also be obtained from test pits and boreholes using a thin-walled tube sampler. ASTM D1587 (ASTM 2015e) is a commonly referenced standard of practice for thin-walled tube sampling.

Radiography (X-ray analysis) of thin-walled tubes before removing the samples (extruding) from the tube can be very useful to assess sample disturbance.

Undisturbed samples must be treated with care during sampling, handling, transportation, storage and extrusion. Sample disturbance can significantly alter soil parameters.

Table 4.7: Causes of soil disturbance

Before sampling	During sampling	After sampling
Base heave	Failure to recover	Chemical changes
Piping	Mixing or segregation	Migration of moisture
Caving	Remoulding	Changes of water
Swelling	Stress relief	content
Stress relief	Displacement	Stress relief
Displacement	Stone along cutting edge	Freezing
		Overheating
		Vibration
		During extrusion
		During transportation and handling
		Due to storage
		During sample preparation

Source: After USACE (2001a)

Table 4.8: Effect of sample disturbance on laboratory testing

Factor	Effect
Physical disturbance from sampling and transportation	Effects on shear strength: Reduces UU and UC strength Increases CU strength Little effect on DS Decrease cyclic shear resistance Effects on consolidation test results: Increases C_r Reduces C_c and C_α Reduces C_v in vicinity of σ'_p and at lower stresses
Change stress conditions from *in situ* to ground surface locations	Similar to physical disturbance but less severe
Contamination of sand from drilling mud	Greatly reduces permeability of undisturbed samples

Source: After USACE (2001a)
Note: DS: drained direct shear test; CU: consolidated undrained triaxial test; UU: unconsolidated undrained triaxial test; UC: unconfined compression test; σ'_p: preconsolidation pressure; C_r: recompression index; C_c: compression index; C_α: secondary compression index; C_v: coefficient of consolidation.

Terzaghi *et al.* (1996) and Lunne *et al.* (1997, 2006) developed methods based on a rating system to estimate the amount of the disturbance of the samples. Terzaghi *et al.* (1996) suggested that reliable estimation of engineering parameters can only be achieved if the sample quality is in the 'B' (good to fair) category or better. Low-quality, disturbed samples should not be used for testing other than index properties.

Sampling methods and causes of sample disturbance are summarised in Table 4.6 and Table 4.7, respectively. Table 4.8 summarises how sample disturbance can affect the results of laboratory testing.

4.4.2.4 Permafrost foundations

The characterisation of permafrost or frozen foundation conditions can be obtained using a combination of the following:

- evaluation of available maps showing permafrost zones to confirm if site located in a zone of continuous, discontinuous or sporadic permafrost
- recording ice content and conditions as encountered in the soil or bedrock based on samples obtained from drilling or test pits
- installation of downhole thermistor strings to collect ground temperature data at intervals below ground with time
- evaluation of regional- and site-scale data on ground temperatures, solar aspect and variations across the footprint area of the site
- installation of vibrating wire piezometers equipped with thermistors or dedicated, multilevel thermistor strings that can measure temperature within the footprint area and downgradient.

Many waste dump sites include a range of solar aspects and topography, so there is likely to be a variation in ground temperature profiles, the thickness of the active zone and the base of the permafrost over the planned dump footprint area. There must be an adequate number of instruments to assess variations in the ground temperature profile across the site, and the thermistor strings should normally be deep enough to determine temperature gradient in the foundation to make an estimate of the base of the permafrost. The characterisation must include an assessment of warming trends obtained from either regional or on-site thermistors, as this may be important for the assessment of long-term (post-closure) changes in foundation conditions and potential long-term changes in the hydrogeology and seepage condition in the dump.

4.4.2.5 Bedrock foundations

Rock (bedrock) is made of naturally occurring minerals that are either tightly compacted or held together by a cement-like mineral matrix and cannot be readily broken by hand. Almost all rocks contain features that influence the geotechnical behaviour of the rock. The engineering properties of rock are controlled by both discontinuities in the rock mass and the characteristics of the intact rock. The origin of weak zones could be related to tectonic plate movements, earthquakes, volcanic eruptions and excessive heat. Some soluble rocks, such as limestone, contain minerals that can be dissolved by groundwater. Channels, caverns and sinkholes may be present in these kinds of rocks.

Bedrock underlying a proposed waste dump or stockpile should have sufficient strength and competency to support the anticipated load without excessive deformation or foundation instability. Shear strength and bearing capacity of the bedrock may decrease over time if the bedrock is susceptible to weathering and degradation and should be taken into consideration. Appropriate site investigations should be conducted to characterise the rock types (lithology); depth, type and impact of weathering and alteration; stratigraphy, structural conditions and the strength and durability of the intact rock; and the strength and competency of the rock mass. Field investigations may include geological mapping of exposures on surface or in test pits, trenches or road cuts, and core drilling.

Methods for investigating, classifying, testing and characterising bedrock and rock masses are extensively covered in many well known references, including *Guidelines for Open Pit Slope Design* (Read and Stacey 2009), *Rock Slope Engineering* (Wyllie and Mah 2005), *Engineering Rock Mass Classifications* (Bieniawski 1989) and *Introduction to Rock Mechanics* (Goodman 1989), and the reader is referred to those works for additional detailed discussion on this subject.

4.4.3 Errors and deficiencies in geotechnical site investigations

Neglecting problems that can arise during the execution of geotechnical site investigations can lead to challenges in interpretation of the data collected and so to errors in the design. Good engineering practices that include careful planning and execution of the program should be followed. During planning for geotechnical site investigations, the team should identify ranges for possible ground conditions to be encountered and consider challenges to the data collection processes based on the experience of site investigation in similar materials.

During execution, the program should be flexible to adapt to the conditions being encountered and when significant changes to the plan are encountered, be prepared to carry out additional investigations. Unidentified foundation conditions can lead to deficiencies in the geotechnical models, which can affect the degree of reliability of the results of the design analyses. Fell *et al.* (2005) have identified a list of common problems during geotechnical site investigations and present this along with a summary of some key consequences and possible remedies.

The use of experienced geotechnical staff can be key to the success of a geotechnical investigation program. Testing equipment should be calibrated at regular frequencies and the testing should be carried out according to standard procedures. Detailed photographic records of the daily events and samples taken during a site investigation program can assist in documenting details. The collected samples should be tested as soon as possible following the investigation. Care should be taken to minimise sample disturbance (refer to Table 4.8 for common causes of the sample disturbance).

5

MATERIAL CHARACTERISATION

Leonardo Dorador, John Cunning, Fernando Junqueira and Mark Hawley

5.1 Introduction

This chapter describes material characterisation carried out to support the planning and design of mine waste dumps or major stockpile facilities. Characterisation of the materials upon which the dump or stockpile will be founded and the fill materials that will be used to construct the dump or stockpile is undertaken to assemble a knowledge base of the geotechnical properties needed to support analyses and the design process.

Site characterisation studies and field investigations, discussed in Chapter 4, are key components of the material characterisation process. Site characterisation establishes the overall setting and context and is used to assist in determining the material properties. Field investigations are undertaken to collect geotechnical data on the *in situ* conditions and to collect samples for laboratory testing. For brownfield sites, the performance of existing foundations and waste or stockpile fill materials can provide important geotechnical data to support design and analyses. For greenfield sites, material characterisation data may need to be collected to support baseline reports for permitting processes or for project decision studies in addition to supporting design and analyses.

The required scope of the material characterisation and laboratory testing programs is a function of the stage of the project, the potential for instability and the level of risk. Table 2.2 in Chapter 2 provides guidance on the level of effort and reliability required for material characterisation studies based on the stage of the project. The waste dump and stockpile stability rating and hazard classification system presented in Chapter 3 can be used to evaluate the level of hazard and provides guidance on the detail of investigations that may be required to adequately define material characteristics (see Table 3.12). For very low instability hazard class dumps and early project stages, properties can be inferred and validated with limited field investigations. For more advanced stages of design, and for moderate to very high instability hazard class dumps, more detailed and targeted field investigations and comprehensive laboratory testing programs would typically be carried out.

The results of site and material characterisation studies are used for, and evolve through, the site selection, design and analysis stages. An understanding of the material properties is needed to support the various physical models described in Chapter 2 and to define the required input parameters for subsequent analyses and design (presented in Chapters 8 and 9).

5.1.1 Definitions

Mine waste dumps and stockpiles can be founded on overburden soils which overlie bedrock, or directly on bedrock foundations. The materials that make up these structures may be composed of overburden soils removed during pre-stripping operations, blasted run-of-mine waste rock or rockfill materials from the mining process, or a mixture of soil and rockfill. Dumps can also be composed of residual materials from leaching operations (ripios). Stockpiles can be composed organic or mineral overburden soils and crushed or run-of-mine ore or low-grade rockfill materials.

In this chapter, material characterisation is described separately for soils and rock for both foundations and fill materials, and for mixed fills. The *Canadian Foundation Engineering Manual* (CGS 2006) provides the following definitions to distinguish soil from rock:

- Soil is 'that portion of the Earth's crust that is fragmentary, or such that some individual particles of a dried sample may be readily separated by agitation in

water; it includes boulders, cobbles, gravel, sand, silt, clay and organic matter'.
- Rock is 'a natural aggregate of minerals that cannot be readily broken by hand and that will not disintegrate on a first wetting and drying cycle'.

5.2 Foundation materials

The distribution, type, stratigraphy and groundwater conditions of the foundation materials may have a controlling impact on the stability and performance of a mine waste dump, dragline spoil or stockpile. Together with the slope and shape of the foundation, these make up the Foundation Conditions group of the waste dump and stockpile stability rating and hazard classification system as described in Section 3.2.2. Some of the key properties of the foundation that are needed to support the constituent models described in various chapters in this book are listed below:

- soil profile, depth to bedrock, bedrock stratigraphy, position of the water table
- *in situ* soil conditions (*in situ* density and moisture content)
- the presence and distribution of soft and dispersive soils, quick clays, permafrost, residual soils, peat and bog soils and soils susceptible to liquefaction
- soil classification (index properties, particle size distribution)
- rock classification (geology, rock mass classification)
- bedrock structures (joints, faults, contacts, dykes and sills, karst)
- presence of potential planes of weakness in the soil or bedrock (laminations, shear zones, faults)
- shear strength, hydraulic conductivity, compressibility and durability of the soils and bedrock.

5.3 Foundation soils

Foundation soils include the overburden materials left in place within the footprint of the dump or stockpile; these may include fills, soils and residual soils. The properties of the overburden soils within the footprint are important as they may control the overall stability of the dump and performance during operations and can influence closure considerations.

The constituents of foundation soil are the various particles that make up the soil mass. These include the following types of materials:

- organic matter (not mineral in origin)
- fine-grained particles (silts and clays)
- coarse-grained particles (sands and gravels)
- cobbles (75–300 mm size) and boulders (greater than 300 mm size).

Foundation soils can be subdivided by behaviour into cohesive soils (those that remain intact when rolled or that stick together when moist) and non-cohesive (cohesionless) soils that fall apart when not confined.

The following sections present methods for the characterisation of foundation soils.

5.3.1 Soil description versus classification

It is important to understand the distinction between soil classification and soil description as used in this chapter:

- Soil classification involves grouping soil into categories based on selected parameters. Classification can be achieved through visual or tactile methods from *in situ* data and/or laboratory testing.
- Regardless of the classification system employed, classification alone does not capture all of the important engineering information needed to support analysis and design.
- Soil classification is a component of a soil description.
- Soil description provides other important information, such as behaviour, density, consistency, field moisture condition, colour and structure of the soil.

5.3.2 Soil description

Description of the foundation soils is usually conducted in the field based on a visual examination and mapping of natural exposures, test pits, trenches and boreholes. Results of any geophysical surveys that may have been conducted and laboratory index testing undertaken are used to supplement and confirm field descriptions. This information allows the various soil units to be identified and characterised.

Visual descriptions of soil characteristics should be based on the applicable standards for the jurisdiction in which the project is located, some examples of which are listed below:

- *Geotechnical Site Investigations* – AS 1726:1993 (Standards Australia 1993)
- *Code of Practice for Site Investigations* – BS 5930:2015 (British Standards Institution 2015)
- *Geotechnical Investigation and Testing – Identification and Classification of Soil – Part 1* – BS EN ISO 14688–1 and Part 2 – BS EN ISO 14688–2 (British Standards Institution 2002, 2004)
- *Canadian Foundation Engineering Manual* (CGS 2006)
- *Eurocode 7: Geotechnical Design Part 1: General Rules* – EN 1997–1:2004 (CEN 2004) and *Part 2: Ground Investigation and Testing* EN 1997–2:2007 (CEN 2007)
- *Standard Practice for Classification of Soils for Engineering Purposes (Unified Soil Classification System)* – ASTM D2487–11 (ASTM 2011a)
- *Standard Practice for Description and Identification of Soils (Visual-Manual Procedure)* – ASTM D2488–09a (ASTM 2009b)

5.3.3 Soil index properties

For low instability hazard dumps or very early stages of the design, the index properties of a soil can be used to provide a preliminary estimate of the expected performance of the foundation soils in terms of strength, hydraulic conductivity and compressibility. As the instability hazard increases, and for later stages of project development, additional data are collect and sample testing completed to refine these initial estimates. A summary of typical index properties that should be evaluated for cohesionless and cohesive soil types is presented in Table 5.1.

5.3.3.1 Particle size distribution

A particle size distribution (PSD) (also referred to as a grain size distribution curve) presents the distribution of the dry mass of a soil based on particle size. The division between categories of particle sizes varies with the different classification systems. Typically, the fines content (or fines) is defined as the percentage passing the No. 200 sieve (0.0029-inch openings) or the 0.075 mm sieve size. Particle size distribution can be determined using ASTM D422–63(2007)e2 and C136/C136M (ASTM 2007a, 2014d).

In cohesionless soils, the PSD can provide information regarding the competency and shear strength of the soils. In cohesive soils, the PSD provides the proportions of the sample that are fine and coarse grained and can assist with understanding the soil behaviour.

The PSD may be characterised as well graded (comprising an evenly distributed range of particle sizes), or poorly graded (comprising a uneven distribution of particle sizes), uniformly graded with a limited range or single particle size, or gap graded (with gaps in the size distribution). The shape of the PSD curve can be further characterised and objectively quantified using the coefficient of uniformity (C_U) and coefficient of curvature (C_C) indices, which are defined as shown in Eqns 5.1 and 5.2:

$$C_U = \frac{D_{60}}{D_{10}} \quad C_C = \frac{D_{30}^2}{D_{10} \cdot D_{60}} \qquad \text{(Eqn 5.1, Eqn 5.2)}$$

where:
  D_{60} = particle diameter (in mm) corresponding to 60% of the sample passing by weight
  D_{30} = particle diameter (in mm) corresponding to 30% of the sample passing by weight
  D_{10} = particle diameter (in mm) corresponding to 10% of the sample passing by weight.

A soil with a low C_U, for example close to 1, would indicate a uniform gradation or a soil with only one grain size. Soils with C_U between 2 and 3 are poorly graded, and soils with high values of C_U (greater than 10) are composed of a wide range of particle size or are well graded. A soil with a C_C between 1 and 3 is considered to be well graded,

Table 5.1: Typical soil index properties

Cohesionless soils	Cohesive soils
Particle size distribution and shape	Particle size distribution
Water content	Water content
Specific gravity	Specific gravity
Void ratio or porosity	Void ratio or porosity
In situ density	Atterberg limits
Compactness	Clay content and mineralogy
	Sensitivity
	In situ density
	Consistency

provided the C_U is over 4 for gravels and over 6 for sands. Well-graded and coarse-grained soils tend to have higher shear strength than poorly graded and fine-grained soils.

Estimates of hydraulic conductivity can also be derived from the PSD.

5.3.3.2 Particle shape

The shape of the coarse-grained particles should be described in terms of both their sphericity and angularity, each of which can affect the shear strength behaviour of a soil (Mitchell and Soga 2005). Under low to medium confining pressures, angular particles are expected to result in greater interlocking among particles and, therefore, a stronger soil mass. Under high confining pressures, the angularity is less important due to particle breakage and rearrangement. Minimum and maximum densities of a soil also depend on the particle shape, especially in sands (Biarez and Hicher 1994).

Several authors have published visual charts that can be used to classify the shape of the particles (e.g. Powers 1953; Krumbein and Sloss 1963), and commercial software programs are available that use digital imaging to define PSD and also provide indications of particle shape (Maerz *et al.* 1996). Figure 5.1 presents typical soil particle shapes.

5.3.3.3 Water content

Water or moisture content in both soil and rockfill is defined as the mass of water divided by the mass of dry solids, and is usually expressed as a percentage. Moisture content can be determined using ASTM D4643–08 (ASTM 2008a) or ASTM D2216–10 (ASTM 2010d). In saturated cohesionless soils, the *in situ* moisture content, if accurately collected during sampling, can provide an indication of the void ratio and density of the soil, which along with the results of other laboratory tests can assist in the understanding of shear strength, compressibility and liquefaction potential. *In situ* water content in fine-grained soils can be compared to the Atterberg limits (Section 5.3.3.6) and provide an indication of likely soil behaviour, including shear strength, consolidation and deformation properties.

Rounded   Subrounded

Subangular   Angular

Figure 5.1: Example of particle shape descriptions

5.3.3.4   Specific gravity

Specific gravity (Gs) is the density of the solid soil particles. It is a dimensionless coefficient and is expressed as a ratio of the density of soil solids to the density of water. Different tests are carried out to measure specific gravity for fine and coarse particles. For fine-grained particles, specific gravity may be determined using ASTM D854–14 (ASTM 2014e). For coarse-grained particles, specific gravity may be determined using ASTM C127–15 (ASTM 2015f). Specific gravity is required in the calculations of results for many soil laboratory tests, including the determination of density from moisture content and degree of saturation. It can be estimated for initial studies, and should be measured for more advanced studies.

5.3.3.5   Void ratio and porosity

Void ratio (e) is the dimensionless ratio of the volume of voids in a soil mass to the volume of the solids. Porosity (n) is the ratio of the volume of voids to the total volume and is expressed as a percentage. Porosity (n) and void ratio (e) are related, and one can be calculated directly from the other using Eqns 5.3 and 5.4:

$$n = e \,/\, (1+e) \text{ or } e = n \,/\, (1-n) \qquad \text{(Eqn 5.3, Eqn 5.4)}$$

Void ratio (or porosity) can be used in conjunction with specific gravity (Gs) of the soil particles and water content to calculate the *in situ* wet and dry densities or unit weights of the soil, or if density or unit weight, water content and specific gravity are known, then void ratio and porosity can be calculated.

5.3.3.6   Atterberg limits

The plastic limit (PL), liquid limit (LL), shrinkage limit (SL) and plasticity index (PI) of a soil (collectively known as the Atterberg limits) provide an indication of expected behaviour of a fine-grained soil under different moisture contents. Water contents below the plastic limit indicate that the soil will behave as a solid. When the water content is between the plastic limit and the liquid limit, the material will display plastic behaviour and be mouldable. The plasticity index is defined as the difference between the liquid and plastic limits and represents the range of water contents over which a soil behaves plastically. Above the liquid limit, the material will behave as a liquid and flow or creep. The shrinkage limit is the water content below which further loss of water by evaporation does not result in a reduction in volume. The plasticity index and liquid limit are required to represent the state of the soil on a plasticity chart and for classification purposes (Fig. 5.2). Atterberg limits should be determined using the laboratory testing procedures defined in ASTM D4318–10e1 (ASTM 2010e).

5.3.3.7   Clay content and mineralogy

Clay content may have a strong influence on the behaviour of a fine-grained soil. As noted by Mitchell and Soga

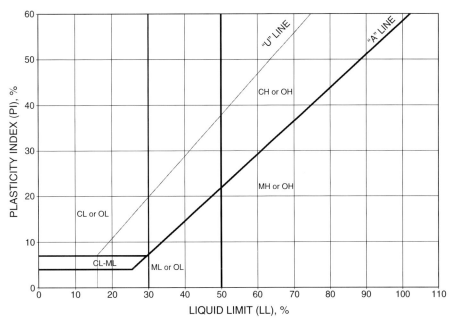

Figure 5.2: Plasticity chart for classification of fine-grained soils.
Source: After Casagrande (1948); Notes: CL = clay of low plasticity.
OL/OH: organic silt or organic clay. Low or high depends on the limits relative to the 'A' line (see ASTM D2487–11 [ASTM 2011a]).
CH: clay of high plasticity; CL-ML silty-clay; ML = silt of low plasticity; MH = silt of high plasticity.

(2005), the more clay in a soil and the higher its plasticity, the greater the potential for shrinkage and swelling, the lower the hydraulic conductivity, the higher the compressibility, the higher the cohesion, and the lower the internal friction angle. Hydrometer testing (ASTM D422–63(2007)e2 [ASTM 2007a]) should be carried out on the fine fraction of the soil (finer than 2 mm or passing the No. 10 sieve) to determine the percentage of silt and clay-sized particles present in the soil.

In addition to the percentage of clay-sized particles, the mineralogy of the clay particles is also important. Some types of clays (e.g. smectite clays such as montmorillonite) are highly plastic and have very low friction angles (in some cases less than 10°), whereas other clay minerals (e.g. kaolinite and illite) have much lower plasticity and exhibit friction angles in the 10–20° range. Clay-sized particles composed of finely ground hard minerals (i.e. 'rock flour') may have very little true cohesion and relatively high friction angles (greater than 30° in some cases).

The Atterberg limits test described in Section 5.3.3.6 can be used to provide some insight into the nature of the clay-sized fraction, but more sophisticated testing, such as X-ray diffraction or scanning electron microscope testing, may be needed to identify which clay minerals are present and their relative importance. If smectite or other highly plastic or weak clays are suspected, clay mineralogy testing should be conducted to help characterise the soil. In addition to a low friction angle, montmorillonite clay (also known as bentonite) has a high swelling capacity.

5.3.3.8 Sensitivity

In some cases, fine-grained soils can lose their strength and stiffness after being disturbed. The degree of potential strength loss can be measured using the sensitivity index, which is defined as the ratio between the undisturbed (peak) and disturbed (remoulded) strengths of the soil. Soils with low to medium plasticity and over-consolidated clays tend to be insensitive. Clays with a sensitivity index less than 2 are characterised as having low sensitivity. Soils composed of normally consolidated, medium plasticity clays typically have medium sensitivity. Clays with a sensitivity index in the range of 4 to 8 are characterised as moderately sensitive. Normally and under-consolidated, highly plastic clays with a sensitivity index greater than 16 are characterised as highly sensitive or quick clays. Marine clays typically have a high sensitivity.

5.3.3.9 In situ *density*

The *in situ* density of the foundation soil is another key parameter in determining the strength of the soil as well as its compressibility. Density can vary depending on the type of soil. For cohesionless soils, density will depend mainly on the gradation and shape of the particles and deposition history. For cohesive soils, water content is necessary to determine the bulk density of a soil. For cohesive soils, in addition to density, the pre-consolidation condition should be understood to be able to determine the available strength under a range of loading conditions.

Table 5.2: Typical values for soil density

Material	Density (kg/m³) Bulk density (assumed saturated)	Dry density
Sands and gravels: Very loose	1700–1800	1300–1400
Loose	1800–1900	1400–1500
Medium dense	1900–2100	1500–1800
Dense	2000–2200	1700–2000
Very dense	2200–2300	2000–2200
Poorly graded sands	1700–1900	1300–1500
Well-graded sands	1800–2300	1400–2200
Well-graded sands/gravel mixtures	1900–2300	1500–2200
Clays: Unconsolidated muds	1600–1700	900–1100
Soft, open-structured	1700–1900	1100–1400
Typical, normally consolidated	1800–2200	1300–1900
Boulder clays (over consolidated)	2000–2400	1700–2200
Red tropical soils	1700–2100	1300–1800

Source: After Carter and Bentley (1991)

Wet or bulk density accounts for the moisture content of the soil. Dry density is defined as the total dry weight of solids divided by the total volume. Dry density can be determined by dividing the wet or bulk density by 1 plus the water content in percent. When the soil is 100% saturated, the wet or bulk density is also known as the saturated density. Typical values of natural densities for a variety of soil types are given in Table 5.2.

5.3.3.10 Compactness

The compactness of the foundation soil can also influence shear strength and bearing capacity. Methods commonly used for evaluating the compactness of cohesionless soils include the *in situ* standard penetration test (SPT), cone penetration test (CPT), Becker penetration test and dynamic cone penetration test. Typical descriptions of soil compactness based on SPT N value data are summarised in Table 5.3.

When using SPT blow counts for engineering assessments, including liquefaction analyses, numerous

corrections are required as discussed in ASTM D6066 – 11 (ASTM 2011e).

5.3.3.11 Consistency

The term 'soil consistency' is used to describe the degree of adhesion between the soil particles in fine-grained soils and the resistance offered against forces that tend to deform or rupture the soil aggregate (Terzaghi *et al.* 1996). Table 5.4 relates qualitative descriptions of consistency with SPT N, unconfined compressive strength (q_u), undrained shear strength (c_u) and wet (bulk) density.

5.3.4 Soil classification

Soil classification is an important step in the geotechnical characterisation of foundation soils. There are many soil classification systems referenced by various standards associations and used by practitioners around the world. These systems can be broadly divided into two types: those which classify based on particle size and plasticity (e.g. the Unified Soil Classification System [USCS] [ASTM 2011a]), and those which classify based on bulk behaviour (i.e. cohesive or non-cohesive) and particle size distribution (e.g. British Standards Institution Standard BS5930:2015 [British Standards Institution 2015]; Dumbleton 1981).

The USCS provides a convenient framework for classifying soils based on visual and manual testing procedures and for assigning group names and group symbols to natural soils. The USCS is described in detail in the *Engineering Geology Field Manual* (US Department of Interior Bureau of Reclamation 1998), and standard procedures for classifying a soil according to the USCS method are described in ASTM D2487–11 (ASTM 2011a).

Table 5.3: Correlations of soil compactness with standard penetration test data

Compactness	SPT 'N' (blows per 0.3 m)
Very loose	0–4
Loose	4–10
Compact	10–30
Dense	30–50
Very dense	> 50

Note: SPT 'N' determined in accordance with ASTM D1586–11 (ASTM 2011d).

Table 5.4: Ranges of soil consistency

Consistency	Very soft	Soft	Medium	Stiff	Very stiff	Hard
SPT, N blows per 0.3 m	0–2	2–4	4–8	8–16	16–32	> 32
Unconfined compressive strength (q_u) (kPa)	0–25	25–50	50–100	100–200	200–400	> 400
Undrained shear strength (c_u) (kPa)	0–12.5	12.5–25	25–50	50–100	100–200	> 200
Wet density (kg/m³)	1602–1922	1762–2082			1922–2243	

Source: After Cheney and Chassie (2000)

Using these procedures, soils are classified into groups based on index properties, including grain size, and Atterberg limits. The plasticity chart presented in Fig. 5.2 is used in the USCS system to classify fine-grained soils in terms of their plasticity.

All records of test pits, test trenches and boreholes should include a reference to the soil description and classification systems used, and laboratory testing results should clearly reference the guiding system and standards.

5.3.5   Shear strength

An understanding of the shear strength of foundation soils is critical to the evaluation of the bearing capacity and slope stability for waste dumps and major stockpiles. The most common constitutive model for soils used to quantify shear strength is based on Mohr-Coulomb failure criteria. This model defines the shear strength of cohesionless soils based solely on the friction between the soil particles. For cohesive soils, Mohr-Coulomb defines the shear strength based on a combination of inter-particle bonding (cohesion) and friction between the soil particles.

For initial studies, estimates of shear strength can usually be obtained from existing site data or from the soil

index properties. For conceptual and pre-feasibility project stages, Table 5.5 provides guidance on the range of shear strength by material type. Additional guidance on estimating geotechnical material properties of soils based on index properties can be found in Koloski *et al.* (1989), Carter and Bentley (1991), Das (2002) and Hough (1957).

For advanced project stages, laboratory testing should be carried out to determine shear strength properties. Typical tests used to determine shear strength properties in the laboratory include the following:

- direct shear under consolidated and drained conditions (ASTM D3080/D3080M–11 [ASTM 2011f])
- triaxial testing, consolidated drained (ASTM D7181–11 [ASTM 2011g]), and consolidated undrained (ASTM D4767–11 [ASTM 2011h])
- cyclic triaxial testing (ASTM D5311/D5311M–13 [ASTM 2013a]) or cyclic simple shear for liquefaction and post-earthquake assessments (Matsuda *et al.* 2011, 2012).

Testing should be carried out at as close to the *in situ* conditions as can be simulated in the laboratory. This can be achieved by either collecting undisturbed samples (difficult in many soils) or by reconstituting samples to

Table 5.5: Typical shear strength parameters for soils

Material	Unified Soil Classification System (USCS) group symbol	Cohesion (kPa)	Effective friction angle ϕ' (°)
Gravels, gravel with sand, alluvial deposits (high energy), well-graded sand, angular grains	GW, GP, GM, SW	0	30–45
Outwash (glacial), volcanic soil (lahar)	GW, GP, GM, SW, SP, SM	0–50	25–40
Alluvial (low energy), uniform sand, round grains	SW, SM, SP, ML	0–25	15–30
Glaciolacustrine	SP, SM, ML	0–140	15–35
Lacustrine soil (inorganic)	SP, SM, ML	0–10	5–20
Silty sand	SM	0	30–34
Till, silty clays, sand-silt mix	SM, ML	0–200	34–45
Clayey sands, sand-clay mix, volcanic soil (tephra)	SC, SM, ML	0–50	20–35
Silt (non-plastic) clayey silts	ML	0–30	30–35
Sandy clay, silty clay, clays (low plasticity)	CL, CL-ML	0–20	18–34
Clays (high plasticity), clayey silts	CH, MH	0	19–28
Silt loam, clay loam, silty clay loam	ML, OL, CL, MH, OH, CH	0–20	18–32
Lacustrine soil (organic)	OL, PT	0–10	0–10

match field conditions. Measurements of *in situ* density and moisture are needed when reconstituting samples.

For some advance stages of investigation where it may not be possible or practical to remove weak fine-grained soils from the foundation before construction of the waste dump or stockpile, *in situ* testing to determine the available shear strength may be required. *In situ* testing methods include field vane shear testing (ASTM D2573/D2573M–15 [ASTM 2015a]) and cone penetration testing (ASTM D5778–12 [ASTM 2012b]).

5.3.5.1 Drained shear strength

The drained shear strength of a soil is the effective strength due to the inter-particle contact with negligible excess pore pressure and is typically expressed in terms of Mohr-Coulomb friction and cohesion. The effective shear strength parameters summarised in Table 5.5 are only applicable in the case of drained loading conditions in which any excess pore pressures induced by loading are quickly and fully dissipated by internal drainage.

5.3.5.2 Undrained shear strength

In a saturated soil mass, an undrained condition occurs when the rate of loading is greater than the rate at which the pore water is able to drain out of the soil. As a result, most of the load is taken by the pore water, resulting in an increase in pore water pressure and decrease in effective stress. If the pore pressure rises to the point at which the effective stress becomes zero (i.e. equivalent to confining stress), fully undrained conditions occur and the shear strength of the soil is governed by its undrained strength. The pore pressure parameter B is used in laboratory tests to confirm saturation and can also be used in stability analyses to investigate the effects of partially drained behaviour on the foundation soils. B is defined as follows:

$$B = \frac{\Delta u}{\Delta \sigma_c} \qquad \text{(Eqn 5.5)}$$

where Δu is the change in pore pressure and $\Delta \sigma_c$ is the change in confining stress.

The undrained shear strength of fine-grained soils is well correlated with the liquidity index (LI) (Vardanega and Haigh 2014), which relates the natural moisture content (w) to the plastic limit (PL) and liquid limit (LL) of the soil as follows:

$$LI = (w - PL)/(LL - PL) \qquad \text{(Eqn 5.6)}$$

Measurements of the undrained shear strength of a saturated, fine-grained soil can be obtained using conventional *in situ* testing techniques (e.g. vane shear, standard penetration or cone penetration testing) or in the laboratory using direct shear, consolidated undrained or unconsolidated undrained triaxial tests (ASTM D2850–15 [ASTM 2015g]).

In cases where liquefaction susceptible soils may be subjected to seismic loading or pore pressure changes that could induce liquefaction (see Chapter 8), cyclical triaxial testing and other more sophisticated laboratory testing may be needed. In such cases, standard penetration or cone penetration testing is also typically conducted to help evaluate the liquefaction potential. Guidelines for investigation and design of foundations subject to seismic shaking can be found in Kramer (1996).

5.3.6 Hydraulic conductivity

Simple groundwater flow equations (such as Darcy's law) can be used to calculate the flow of water in saturated soils. Darcy's law is based on a linear relationship between the rate of flow and driving forces. Hydraulic conductivity, commonly referred to as permeability, is an important parameter in waste dump and stockpile design as it controls the rate of drainage and pore pressure dissipation in both the waste material and the foundation soil and bedrock. The hydraulic conductivity of a soil is highly dependent on the gradation of the material and, in particular, on the proportion of clay and silt. Figure 5.3 summarises typical ranges of hydraulic conductivity for a variety of different soil types.

Characterisation of groundwater is discussed in Chapter 6. Field testing to obtain *in situ* permeability can

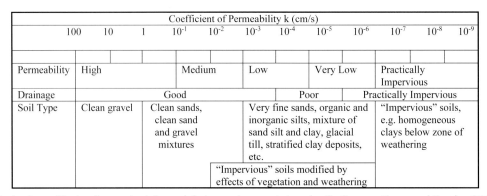

Figure 5.3: Typical values of hydraulic conductivity and drainage characteristics of soils. Source: Carter and Bentley (1991)

involve falling head or rising head tests in groundwater wells and packer testing in boreholes. Laboratory techniques for determining hydraulic conductivity in granular soils can be determined following ASTM D2434 (ASTM 2006b).

5.3.7 Consolidation and creep

Deformation will occur when load is applied to a soil material. Vertical deformation is also referred to as settlement. For soils, the total (ultimate) deformation will be the sum of the immediate (elastic) settlement, consolidation settlement (which occurs over time) and secondary settlement (which also occurs over time). Secondary settlement is also referred to as creep.

Consolidation of soil results from expulsion of water from pore spaces and causes an increase in the bulk soil density. Consolidation may also increase shear strength and reduce permeability. The rate of consolidation is controlled by the permeability of the soil. In fine-grained soils (silts and clays), deformation under loading will be slower due to their relatively lower permeability in comparison to coarse-grained soils. A detailed discussion of consolidation theory can be found in Lambe and Whitman (1969). The laboratory procedure to determine consolidation properties follows ASTM D2435/D2435M–11 (ASTM 2011i).

5.3.8 Permafrost and frozen ground

Thermal conditions, including expected ground temperatures over time, are important considerations in the design of waste dumps and stockpiles founded on permafrost or ground subject to seasonal freezing. If the site is located in an area that contains permafrost or ground that is subject to seasonal freezing, special characterisation of the soils subject to freezing may be necessary. Site characterisation (Chapter 4) should identify areas of continuous and discontinuous permafrost, and soil horizons that are subject to seasonal freezing. General procedures for description of frozen soils are based on visual examination and should follow ASTM D4083 – 89(2007) (ASTM 2007b), NRCC-7576 (Pihlainen and Johnston 1963) or similar protocols.

In frozen soils, total moisture content can be an indicator of whether the frozen soil is ice rich or ice poor. In addition to using a visual ice classification system, the total water content, including any excess water liberated from the sample as it thaws, would be included in the water content calculation.

From a geotechnical perspective, a frozen foundation soil at a low temperature will have high strength. However, if the frozen soil is ice rich and warms to near thawing conditions, creep associated with the ice content can reduce foundation strength and result in stability or deformation issues. Creep testing may be required for ice rich permafrost to determine strength and deformation

properties over a range of temperatures. ASTM D5520–11 (ASTM 2011j) can be used to characterise the creep properties of frozen soils.

5.4 Foundation bedrock

The character of the waste dump or stockpile foundation bedrock can also be an important factor in site selection, design, operation and closure. Even though a waste dump may be founded on soil, the competency and structure of the underlying bedrock may control stability. In some cases, weak soils may be stripped from beneath dump or stockpile footprints, directly exposing the underlying bedrock. In other cases, the dump or stockpile may be founded directly on bedrock (e.g. in-pit backfill dumps).

The geotechnical characterisation of bedrock foundations needs to consider the engineering geology of the rock mass beneath the foundation, including the orientation and characteristics of faults, bedding planes and joints, and the groundwater flow system and surface drainage patterns. Chapter 4 provides a summary of site characterisation studies that would include investigation and characterisation of the foundation bedrock based on ground reconnaissance, geological mapping, drilling, geophysical surveys and other techniques appropriate to the stage of the study, the potential for instability and level of risk.

Methods for classifying, testing and characterising bedrock and rock masses are extensively covered in many well-known references, including *Guidelines for Open Pit Slope Design* (Read and Stacey 2009), *Rock Slope Engineering* (Wyllie and Mah 2005), *Engineering Rock Mass Classifications* (Bieniawski 1989) and *Introduction to Rock Mechanics* (Goodman 1989), and the reader is referred to those works for additional detailed discussion on this subject.

The following sections provide a general overview of bedrock characterisation methods.

5.4.1 Rock characterisation standards and methods

Key ASTM standards for rock characterisation include the following:

- *Standard Test Method for Determining Rock Quality Designation (RQD) of Rock Core* – ASTM D6032–08 (ASTM 2008b)
- *Standard Practice for Rock Core Drilling and Sampling of Rock for Site Exploration* – ASTM D2113–14 (ASTM 2014f)
- *Standard Test Method for Determination of the Point Load Strength Index of Rocks, and Application to Rock Strength Classifications* – ASTM D5731–08 (ASTM 2008c).

Key International Society for Rock Mechanics (ISRM) methods for rock characterisation include the following:

- *Suggested Methods for the Quantitative Description of Discontinuities in Rock Masses* (ISRM 1978a)
- *Rock Characterisation Testing and Monitoring* (ISRM 1981a)
- *Basic Geotechnical Description of Rock Masses* (ISRM 1981b)
- *Suggested Methods for Determining Point Load Strength* (ISRM 1985).

5.4.2 Bedrock geology and rock types

Basic information on bedrock geology and rock types can usually be obtained from regional and site geological maps, sections, stratigraphic columns and geological reports. Core samples from boreholes can be used to characterise the geological conditions at depth. Basic geological descriptions of the rock types, alteration and surficial weathering can be used in the initial project stages to estimate rock properties including strength and durability.

5.4.3 Intact rock strength

A rock mass is composed of an assemblage of intact blocks separated by discontinuities such as fractures or joints, faults, bedding or foliation joints and veins. The strength of the intact rock blocks is dependent on a wide range of factors including mineral composition, grain size, particle bonding and cementation, alteration type and intensity, weathering, and origin and stress history of the rock. Intact rock strength is a key input parameter in most commonly used rock mass classification systems (Bieniawski 1989) and in most commonly used methods for the determination of rock mass strength (Read and Stacey 2009).

For conceptual and preliminary designs, initial estimates of intact rock strength obtained from the literature (e.g. Goodman 1989) may be sufficient. Initial estimates may also be obtained using simple field tests such as the ISRM field strength test (ISRM 1981a) or a similar procedure defined in the British Standard BS 5930:2015 (British Standards Institution 2015).

For more advanced studies, field and laboratory testing of intact rock samples is usually conducted to refine preliminary estimates of intact rock strength. Typical methods for obtaining intact rock strength from field and laboratory tests on core samples are described in the following standards:

- *The Complete ISRM Suggested Methods for Rock Characterisation, Testing and Monitoring: 1974–2006* (ISRM 2007)
- Rock characterisation testing and monitoring: *ISRM Suggested Methods* (ISRM 1981a)

- *Standard Test Method for Compressive Strength and Elastic Moduli of Intact Rock Core Specimens under Varying States of Stress and Temperatures* – ASTM D7012–14 (ASTM 2014g)
- Splitting tensile strength: ASTM D3967–08 (ASTM 2008d)
- Brazilian tensile strength: ISRM (1978b).

5.4.4 Alteration and weathering

The physical and chemical properties of rock can change as a result of a wide range of geological processes such as hydrothermal alteration, contact metamorphism and changes in stress that alter the mineralogy and fabric of the rock. Chemical processes such as oxidation and sulphidation can also alter the properties of the rock. In addition, near-surface rocks may be subject to physical weathering processes such as cyclical wetting–drying, freeze–thaw, heating–cooling, exposure to intense sunlight and high humidity. Exposure to meteoric waters can also result in chemical and physical changes to the rock. The development of near surface saprolitic and lateritic horizons in some rock types in tropical climates is a good example of how environmental factors can change the properties of rocks, including their intact strength, density, modulus and hydraulic conductivity. The degree of weathering and alteration of the foundation bedrock should be qualitatively classified using methods such as described in ISRM (1981a).

5.4.5 Discontinuities and fabric

The occurrence and nature of discontinuities such as faults or shears, joints or fractures and veins can have a significant impact on the strength of the rock mass. To characterise a discontinuity, its shear strength, orientation, position and length should be obtained from surface mapping and drill core. Where multiple discontinuities with a common orientation and character occur (i.e. discontinuity sets), the spacing between them also needs to be understood.

For smooth, planar discontinuities, it may be sufficient to characterise their shear strength using a simplified Mohr-Coulomb criteria that neglects cohesion and relates shearing resistance to normal stress as illustrated in Eqn 5.7.

$$\tau = \sigma_n \cdot \tan \phi_b \qquad \text{(Eqn 5.7)}$$

where:

τ = shearing resistance
σ_n = normal stress
ϕ_b = basic friction angle of a smooth diamond saw-cut surface.

Equation 5.7 uses the basic friction angle, which is different from the peak and residual friction angles. In the case of natural discontinuities which exhibit surface asperities, discontinuity strength can be quantified using several criteria, including Mohr-Coulomb with inclusion of cohesion, Patton's bilinear failure envelope (Patton 1966) and the approach proposed by Barton and Choubey (1977), which is summarised in Eqn 5.8:

$$\tau = \sigma_n \cdot \tan\left(\phi_b + JRC\,Log_{10}\frac{JCS}{\sigma_n}\right) \qquad \text{(Eqn 5.8)}$$

where:

    JRC = joint roughness coefficient
    JCS = joint wall compressive strength.

The parameters JRC and JCS can be obtained from tables and correlations in Barton and Choubey (1977).

Special consideration must be given to filled discontinuities as the available shear strength may be reduced when part or all of the surface is not in intimate contact but instead is covered by soft filling materials such as clay gouge. In addition, the presence of water pressure can change the effective normal stress (Hoek 2007) and reduce the available shearing resistance. Measurements of the shear strength of discontinuities can be obtained from direct shear testing using procedures such as those described in ASTM D5607–08 (ASTM 2008e) or ISRM's *Suggested Methods for Determining Shear Strength* (ISRM 1974).

'Fabric' refers to the special arrangement of the mineral aggregates and other geological features in a rock, such as bedding in sedimentary rocks or foliation or cleavage that forms in some types of metamorphic rocks. Fabric can impart anisotropy in which the compressive and shear strength of the intact rock and rock mass varies depending on the direction of the applied stress. In the case of highly anisotropic rocks, the bearing capacity and stability of the foundation can be adversely affected. Strength anisotropy due to fabric can be evaluated using many of the field and laboratory testing techniques described previously by varying the direction of the applied loads and stresses.

5.4.6 Rock mass classification

As discussed above, a rock mass is composed of intact rock blocks separated by discontinuities, and its behaviour under applied stress is controlled by the intact strength of the blocks and the frequency, orientation and characteristics of the discontinuities.

Rock mass classification provides a practical way of evaluating the overall quality and competency of a rock mass and comparing rock mass conditions. In addition, as discussed in Section 5.4.7, rock mass classification is an integral part in the evaluation of rock mass strength. Bieniawski (1989) provides a comprehensive discussion of a variety of rock mass classification systems and their application in engineering analysis and design. One of the most popular systems used for classification of rock masses for open pit slope and shallow bedrock foundation

design is the rock mass rating (RMR) system. This system was originally proposed in 1976 (Bieniawski 1976) and requires the selection of numerical values for five indices based on the following:

- unconfined compressive strength of the intact rock
- rock quality designation (Deere 1963)
- discontinuity spacing
- conditions of discontinuity surfaces
- groundwater conditions.

The RMR, which has a maximum value of 100, is the sum of these five indices. Once the RMR has been determined, adjustments can be applied to account for the orientation of discontinuities. Several modifications to the original RMR system have been proposed since it was originally introduced, but the underlying fundamentals have remained unchanged.

Another popular rock mass classification system is the geological strength index (GSI) originally proposed by Hoek *et al.* (1995). This system requires selection of two parameters: one related to the size of the blocks and one related to surface condition of the discontinuities. As for RMR, the maximum possible value of GSI is 100.

5.4.7 Rock mass strength

As discussed in Read and Stacey (2009), there are many different methods for assessing rock mass strength, and their usage depends on the nature of the rock mass and preference of the individual practitioner. The Mohr-Coulomb criterion (Eqn 5.9) that is commonly used for soils is also often used for weak rock masses such as saprolites and other highly weathered or altered rocks.

$$\tau = Si + \tan\phi \qquad \text{(Eqn 5.9)}$$

where:

    ϕ = angle of internal friction
    Si = shear strength intercept of the Mohr-Coulomb envelope.

For stronger rock masses where the shear strength is controlled by both the intact strength of the rock and the shear strength of the discontinuities, the non-linear Hoek-Brown criterion (Hoek and Brown 1997; Eqn 5.10) is often used.

$$\sigma_1' = \sigma_3' + \sigma_c\,(m_b\,\frac{\sigma_3'}{\sigma_c} + s)^a \qquad \text{(Eqn 5.10)}$$

where:

    σ_1' and σ_3' = the maximum and minimum effective principal stress at failure
    σ_c = uniaxial compressive stress of intact pieces of rock
    m_b, s and a = constants which depend on the composition, structure and surface conditions of the rock mass.

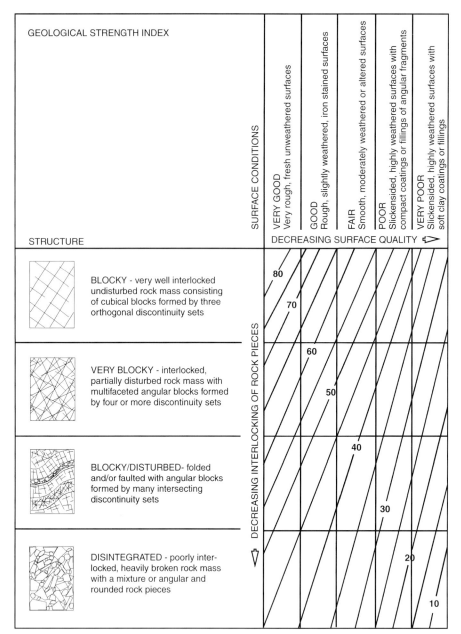

Figure 5.4: Estimating the Geological Strength Index from geological descriptions. Source: Hoek and Brown (1997)

In addition to shear strength, the deformation modulus of the rock mass (a required input parameter for most numerical stress and deformation analysis codes) can be estimated using Hoek-Brown rock mass strength parameters (Hoek and Diederichs 2005).

There have been many modifications to the Hoek-Brown criterion since it was first introduced in the 1980s (Hoek and Brown 1980), including formulations that include the Geological Strength Index (GSI) (Fig. 5.4) and consider disturbance due to blasting and excavation. The reader is referred to Hoek and Brown (1997), Hoek *et al.* (2002), Hoek (2007) and Read and Stacey (2009) for

additional discussion of the theoretical basis and application of the Hoek-Brown failure criterion.

5.4.8 Mineralogy and petrography

Rocks consist of an assemblage of minerals, so mineralogy (the study of the chemical and physical properties of minerals and their crystalline structure) and petrography (the study of the mineral composition and structure of rocks) are important components in understanding the mechanical and chemical properties of rocks. Minerals in rocks determine physical properties such as specific gravity, hardness, cleavage and fractures. Other simple tests such as reactivity to acid and colour provide more information about the types of minerals in rocks. An understanding of the mineral composition of a rock can also provide insight into potential problems that may emerge during development and operation of waste dumps and stockpiles. For example, Goodman (1989) identified the following problem minerals:

- **Soluble minerals** – calcite, dolomite, gypsum, anhydrite, salt (halite) and zeolite
- **Unstable minerals** – marcasite and pyrrhotite
- **Potentially unstable minerals** – nontronite (iron-rich montmorillonite), nepheline, leucite and micas rich in iron
- **Minerals whose weathering releases sulfuric acid** – pyrite, pyrrhotite and other sulphides (ore minerals)
- **Minerals with low friction coefficients** – clays (especially montmorillonite), talc, chlorite, serpentine, micas, graphite and molybdenite
- **Potentially swelling minerals** – montmorillonites, anhydrite and vermiculite.

5.4.9 Durability

Change in the rock properties over time can be produced by weathering processes such as exfoliation, hydration, decrepitation (slaking), solution, abrasion and other process (Goodman 1989). Degradation impacts bedrock strength and bearing capacity. Laboratory techniques that are commonly used to assess the potential for rock degradation include the slake durability test (ASTM D4644–08 and ASTM D5312/D5312M–12 [ASTM 2008f, 2013b]), the sulphate soundness test (ASTM C88–13 [ASTM 2013c]) and the Los Angeles abrasion test (ASTM C131/C131M–14 and ASTM C535–12 [ASTM 2006a, 2012c]).

5.4.10 Hydraulic conductivity

An understanding of the hydraulic conductivity of a bedrock foundation is needed to assess whether pore pressures will develop in the foundation during loading by the dump or stockpile, and how such pore pressures could affect stability. Hydraulic conductivity, groundwater flow, pore pressures and their impact on stability are discussed in Chapter 6.

5.5 Waste dump and stockpile fill materials

Most mine waste dumps and mineral stockpiles are composed predominantly of rockfill; however, many dumps also incorporate soils from stripping operations. In addition, many operations are required to stockpile overburden materials for use in reclamation. The physical properties of the various fill materials and their

distribution within the waste dump or stockpile will influence, and in some cases control, the overall stability of the structure. Careful characterisation of these materials is therefore required to support site selection, stability analysis, design and closure. For the purposes of the following discussion, fill materials have been subdivided into rockfill and overburden.

5.5.1 Rockfill

5.5.1.1 Material quality

An understanding of the quality and variability of the rockfill materials is important when evaluating their shear strength, deformational behaviour and hydrogeological properties. The geological setting and source of the waste rock may provide insight into the physical characteristics of the material. As discussed in Chapter 2 (Section 2.2.6), key attributes that can affect the quality of the fill materials include gradation, strength, durability and chemical stability. Many of these factors are interrelated and are influenced by the way the material is blasted, excavated, transported and placed in the dump or stockpile.

As suggested in the *Investigation and Design Manual – Interim Guidelines* (BCMWRPRC 1991a) and in Chapter 3, rockfill material quality can be qualitatively characterised based on strength, durability, and particle size distribution of the rockfill particles as follows:

- poor quality – weak rocks with low durability and a fines (% passing No. 200 sieve) content above 25%
- moderate quality – intermediate strength and durability rocks with 10–25% fines
- high quality – strong and durable rocks with less than 10% fines.

5.5.1.2 Geochemistry

The geochemistry of the rockfill and the need to manage potential acid rock drainage can have a significant impact on the design, operation and closure of waste dumps and major stockpiles, and an initial understanding needs to be developed early in the design process. Detailed procedures for geochemical characterisation of rockfill are beyond the scope of these guidelines; however, an overview of the management of acid rock drainage is provided in Chapter 14. For a detailed discussion and guidance on geochemical characterisation and management of mine waste materials, the reader is referred to the *Global Acid Rock Drainage (GARD) Guide* (INAP 2009).

Some mines have different geochemical classes of materials, and these need to be separated and stored in different areas of the waste dump, as discussed in Chapter 11.

5.5.1.3 Particle size distribution

The PSD of the rockfill can control the *in situ* density of the dumped or placed material and influence its hydraulic conductivity and shear strength. The bulk gradation of rockfill materials can be difficult to obtain given the maximum size of the particles that are often produced by the mining operations (often greater than 1 m). Representative gradations of the material as placed in the dump can also be difficult to obtain due to the large degree of particle segregation that occurs during development. Figure 5.5 illustrates some typical ranges of PSD for rockfill materials.

Gradation curves for rockfill materials should be developed following recognised standard procedures such as ASTM D5519–15 (ASTM 2015h) in which the upper limit of the particle size measured is limited by the available sieve size and ASTM D6913–04(2009)e1 (ASTM 2009c) for the particle sizes below 75 mm (3 inches). Gradation of rockfill material is typically characterised by the maximum particle size, the proportion with grain sizes greater than 75 mm (i.e. cobbles and boulders content) and the proportion (% by weight) of the material with particle sizes passing the 0.075 mm sieve size (fines). As discussed in Chapter 3, a fine-grained waste material would have more than 50% fines and a low proportion (typically less than 5–10%) of cobbles and boulders, whereas very coarse-grained materials would have very low proportion of fines (less than 5–10%) and more than 50% cobbles and boulders.

Mechanical sieve analyses can take significant quantities of rockfill material and require large areas of laboratory space. Digital image processing is a powerful, low cost tool that can be used to estimate particle size distribution of rockfill materials using high quality digital photographs. Several systems such as Fragscan (Schleifer and Tessier 1996), WipFrag (Maerz *et al.* 1996) and Split (Kemeny *et al.* 1993; La Rosa *et al.* 2001) are commercially available.

5.5.1.4 Block shape

The shape (angularity and sphericity) of the blocks that comprise the largest size particles of the rockfill is an intrinsic property of the material and should be considered in its geotechnical characterisation. Block shape will have an influence on the *in situ* density, void ratio and shear strength of the rockfill. The degree of angularity is sensitive to fragmentation resulting from the original blasting and excavation process, secondary degradation that occurs during initial placement of the material in the dump, and subsequent breakdown of particles due to loading and changing stress conditions within the dump. Weathering of rockfill, including geochemical degradation, can also change the block shape and, hence, the strength properties of the material.

5.5.1.5 Water content

The water content of the fill materials can be important when these materials contain a large amount of fines and

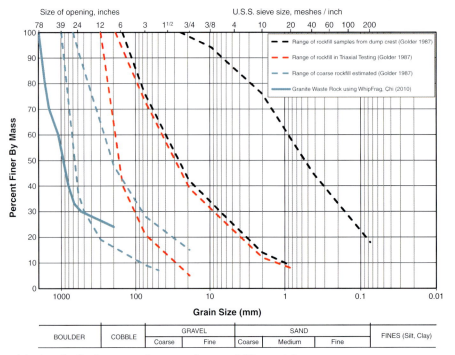

Figure 5.5: Typical particle size distribution ranges for waste dump rockfill materials

can influence strength, consolidation behaviour and hydraulic conductivity. For rockfill materials, the water content is typically estimated based on the minus 19 mm (0.75 inch) component, as the water content of material that is larger than 19 mm is generally negligible (Breitenbach 1993).

5.5.1.6 Void ratio, porosity and density

The void ratio of a rockfill directly affects its hydraulic conductivity and is needed for the design of rock drains (Chapter 7). The void ratio of a rockfill can be very difficult to determine due to the large size and range of particles and the significant level of particle segregation that can occur during dump or stockpile construction. As for soil, void ratio (e) and porosity (n) are related, and one can be calculated directly from the other (see Section 5.3.3.5). The void ratio or porosity can be directly calculated if the water content, specific gravity of the solids, degree of saturation and *in situ* bulk density are known. For some rockfills, it is possible to obtain a direct measurement of *in situ* density and to derive void ratio or porosity using the procedures described in ASTM D5030/ D5030M–13a (ASTM 2013d).

Published values of the void ratio of rockfills are limited, and are expected to range greatly within a rockfill dump as a result of segregation. Linero *et al.* (2006) reported a void ratio of 0.46 for waste rock material from the Andina Mine (Chile). Dawson *et al.* (1998) reported a range of field void ratios between 0.4 to 0.9 for heaped waste rock, and Golder Associates Ltd (1987) reported field measurements of void ratio varying from 0.25 to 0.3.

Rockfill PSD, particle shape and segregation in the waste dump or stockpile will affect the *in situ* density and void ratio. Rockfill materials will consolidate as the dump

Table 5.6: Typical ranges for dry density and porosity of intact rock by rock type

Rock type	Density (g/cm³)	Porosity, n (%)
Granite	2.5–2.8	0.5–1.5
Dolerite	3.0–3.1	0.1–0.5
Limestone	2.5–2.8	5–20
Dolomite	2.5–2.6	1–5
Quartzite	2.65	0.1–0.5
Sandstone	2.0–2.6	5–25
Shale	2.0–2.7	10–30
Coal – anthracite	1.3–1.6	–
Coal – bituminous	1.1–1.4	–
Sediments	1.7–2.3	–
Metamorphic rocks	2.6–3.0	–

Source: After Peng and Zhang (2007)
– = not provided

height increases, resulting in an increase in density and decrease in void ratio. Valenzuela *et al.* (2008) presented data based on triaxial and large oedometer testing (Fig. 5.6), showing the relationship between *in situ* void ratio and dump height. These authors indicated that for waste rock dumps higher than 100 m, particle crushing can be significant, resulting in finer particle size distribution and lower void ratio.

Table 5.6 provides a summary of typical ranges for intact rock dry density and porosity by rock type, which can be used along with PSD, block shape and void ratio to estimate rockfill density.

5.5.1.7 Intact strength

The intact strength of the individual rock blocks influences the shear strength of the rockfill. Rockfills composed predominantly of weak or soft rocks tend to have lower shear strength than those composed of strong, hard rocks. Methods for determining the intact rock strength of rockfill blocks are the same as those discussed in Section 5.4.3.

5.5.1.8 Alteration and weathering

Alteration and weathering of the rock blocks that compose the rockfill can result in a reduction in durability, shear strength and hydraulic conductivity. Methods for evaluating the impact of alteration and weathering on the properties of the rockfill are the same as those discussed for bedrock foundations in Section 5.4.4.

5.5.1.9 Shear strength

The shear strength of a rockfill is perhaps the most important physical property that can affect the stability of the waste dump or stockpile. It is also one of the most difficult parameters to measure due to the wide range in particle sizes and shapes, anisotropy that develops due to

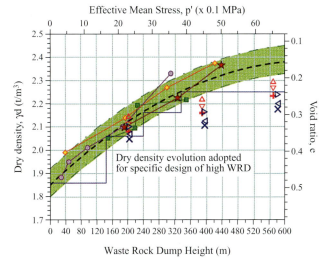

Figure 5.6: Effective mean stress versus void ratio for waste rock. Source: Valenzuela *et al.* (2008)

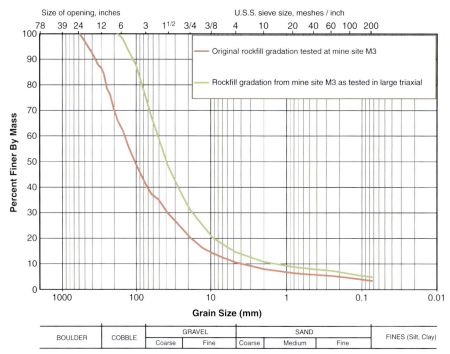

Figure 5.7: Example of the application of a parallel gradation method from a rockfill laboratory testing program

segregation during placement, and changes in shear strength that occur over time due to consolidation and degradation. Unlike some soil deposits in which shear strength can be measured *in situ* using special testing equipment (Section 5.3.5), there are no reliable methods for directly measuring the shear strength of rockfills *in situ*. It may be possible to empirically correlate rockfill shear strength with penetration rates for certain types of drill rigs (e.g. the Becker hammer); however, the authors are not aware of any published data from penetration rate studies done on waste dumps that support this approach.

The most reliable method of assessing bulk rockfill shear strength is through back-analysis of instabilities, although this approach is highly site and material specific and not available for greenfield sites or sites than have not experienced instabilities.

Small-scale laboratory testing

Due to the large maximum particle size of some waste dump materials (sometimes up to 2 m), in most cases it is not possible or practical to conduct laboratory testing on the original whole gradation of the material. In addition, large-scale tests, such as discussed below, can be very expensive and time consuming to complete. An alternative is to conduct laboratory testing on a finer gradation sample of the material.

Various methods of testing to obtain the shear strength using the fine fraction of waste dump and stockpile material have been used in the past, including the matrix method (Siddiqi 1984), scalping method (Al-Hussaini

1983), scalping–replacement method (Donaghe and Torrey 1979) and parallel gradation method (Lowe 1964). Verdugo and De la Hoz (2006) described a parallel gradation method for testing coarse granular materials which consists of scalping the oversized particles and preparing a new, finer gradation with a grain size distribution curve parallel to the original. The maximum allowable grain size

Table 5.7: Large triaxial devices

Laboratory	Specimen diameter (mm)	Country
University of Cataluña	300	Spain
Geodelft	400	Netherlands
University of Karlsruhe	800	Germany
University of California - Berkeley	915	USA
University of Nantes	1000	France
Missouri Institute of Science and Technology	420	USA
University of Nottingham	300	UK
University of Chile (Idiem Institute)	1000	Chile
Norwegian University of Science and Technology	500	Norway
National Laboratory of Civil Engineering	300	Portugal
Building and Housing Research Center, Tehran	300	Iran
Snowy Mountains Engineering Corporation, Cooma	570	Australia

Figure 5.8: Large triaxial equipment at: (a) Idiem Institute (Chile) and (b) National Laboratory of Civil Engineering (Portugal). Source: (a) Linero *et al.* (2006), (b) Araújo and Correia (2011)

of the finer gradation will be determined by the maximum diameter of the specimen that can be accommodated by the testing apparatus. Typically, a ratio of the specimen diameter to the maximum grain size of six is used. However, some key characteristics of original whole gradation sample must also be maintained in the finer gradation sample, such as mineralogy, hardness, shape and

roughness of the particles, and density (De la Hoz 2007). The finer gradation should not contain more than 10% fines due to the importance of the fines in the mechanical response of the material (Verdugo and De la Hoz 2006). Figure 5.7 presents an example of a rockfill material gradation as tested at the mine site and the gradation of the material as prepared for testing in a large triaxial cell.

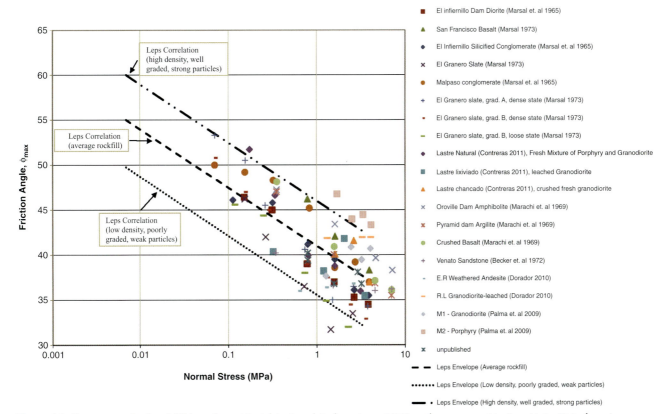

Figure 5.9: Shear strength of rockfill from large triaxial testing data from Leps (1970) and more recent testing. Note: Data from Leps (1970) with recent testing data added – see references in Table 5.8.

Large-scale laboratory testing

The most direct method to obtain the shear strength properties of rockfill materials is through large direct shear tests or large triaxial tests. Large direct shear apparatus that can test specimens of 300 mm in diameter are available commercially; however, large triaxial equipment is still relatively uncommon.

Table 5.7 is a summary of the main large triaxial devices that are reported as being used for research purposes around the world, and Fig. 5.8 shows two examples of large triaxial equipment.

Leps (1970) published the results of large-diameter triaxial testing on 15 different rockfill materials that represent a range of grain size distribution, particle strength and density. The range of confining stresses considered by Leps established a key understanding that the shear strength curve for rockfill is non-linear, particularly at low confining stresses. This non-linearity needs to be considered in selecting appropriate shear strength parameters for use in the stability analyses for waste dumps and stockpiles.

Leps identified a range of rockfill material strength envelopes depending on the density, gradation, shape and intact strength of the particles. High-density, well-graded materials with angular, high intact strength particles had higher shear strength envelopes than materials composed of lower density, uniform or gap-graded materials with rounded, lower intact strength particles. Figure 5.9 shows the results compiled by Leps along with additional test data published up to 2011. The source of data used to prepare Fig. 5.9 is summarised in Table 5.8.

Shear strength function

Based on the work by Leps (1970) and others (De Mello 1977; Charles and Watts 1980; Maksimović 1996; Douglas 2002; Linero *et al.* 2006; De la Hoz 2007), a non-linear shear strength function (Eqn 5.11) that depends on normal stress was developed for coarse granular soils (including waste rock and stockpile materials). Research on a range of rockfill materials shows that the constants *a* and *b* depend not only of the intrinsic parameters of the blocks (e.g. strength, angularity, gradation), but also on state parameters of the material such as density.

$$\tau = a\sigma^{-b} \qquad \text{(Eqn 5.11)}$$

where:

τ = shearing resistance

σ = normal stress, and

a and *b* are constants.

Figure 5.10 illustrates how this function can be used to develop non-linear shear strength – normal stress failure envelopes that are fitted to the same test results plotted in Fig. 5.9.

Table 5.9 presents a summary of suggested non-linear shear strength functions for high and moderate quality rockfill based on the triaxial tests reported by Leps (1970), and includes an upper bound function for the shear strength behaviour for high-density, well-graded, strong particles, and a lower bound function for low density, poorly graded, weak particles. The selection of which shear strength function to apply in a given case depends on the method of dump construction, the quality of the rockfill materials and the potential for instability.

For poor-quality fill and soil-like materials, shear strength parameters should be chosen on the basis of conventional laboratory direct shear and/or triaxial testing of the finer component of the fill.

5.5.1.10 Durability

As discussed in Section 5.4.9, changes in the properties of the individual rock blocks composing the rockfill can change over time as a result of weathering and alteration, and similar laboratory techniques as discussed in that section are also applicable in helping to assess the durability of rockfill.

Table 5.8: Summary of details for rockfill materials presented in Fig. 5.9

Rockfill name	Rock type	Reference
El Infiernillo	Diorite	Marsal *et al.* (1965)
San Francisco	Basalt	Marsal (1973)
El Infiernillo	Silicified conglomerate	
El Granero	Slate	
Malpaso	Conglomerate	
El Granero A (dense)	Slate	
El granero B (dense)	Slate	
El granero B (loose)	Slate	
Fresh waste rock	Mixture of porphyry and granodiorite	Contreras (2011)
Leached waste rock	Granodiorite with some degradation by leach	
Crushed waste rock	Fresh granodiorite	
Oroville dam	Oroville dam material – amphibolite	Marachi *et al.* (1969)
Pyramid dam	Pyramid dam material – argillite	
Crushed material	Basalt	
Venato	Sandstone	Becker *et al.* (1972)
E.R.	Weathered andesite	Dorador (2010)
R.L.	Granodiorite with some degradation by leaching	
M1	Granodiorite	Linero *et al.* (2009)
M2	Porphyry	

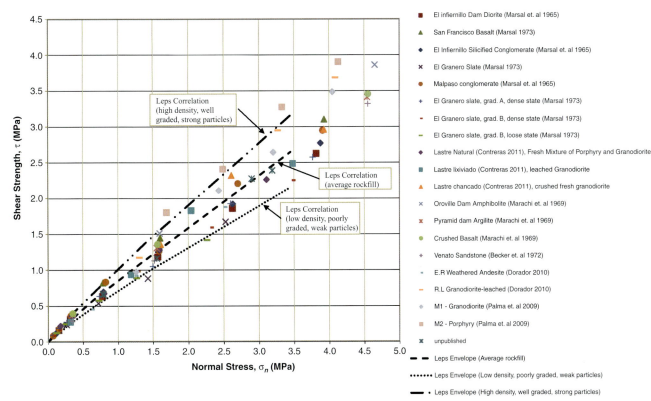

Figure 5.10: Shear strength of rockfill plotted using non-liner shear strength function with data from large triaxial tests. Data from Leps (1970) with recent testing data added – see references in Table 5.8

5.5.1.11 Hydrogeological properties

An understanding of the hydraulic conductivity of the rockfill materials is needed to assess the nature of groundwater flow through the waste dump or stockpile and the potential for development of a phreatic surface or perched water tables and the distribution of pore pressures within the structure. Methods for evaluating hydraulic conductivity and related hydrogeological parameters of importance in rockfills are discussed in Chapter 6.

5.5.2 Overburden and mixed fills

Methods for assessing the characteristics of overburden soils that may be incorporated in a waste dump, or that

may compose overburden stockpiles, are the similar to those presented in Section 5.3 for foundation soils.

As discussed in Chapter 1, mixed fills are composed of mixtures of rockfill and overburden. The properties of mixed fills will depend on the relative proportions of overburden and rockfill. Mixed fills composed predominantly of rockfill will behave more like rockfills, provided there is a high level of contact between individual rockfill blocks and that the overburden component is contained within voids in the rockfill. In this case, the compaction/settlement and density characteristics of the mixed material may be better than the original rockfill, although the bulk hydraulic conductivity may be lower.

Table 5.9: Suggested rockfill shear strength functions for high and moderate material quality rockfill

Waste dump Construction method	Shear strength function (τ and σ in kPa)	Leps (1970) envelope (incorporating recent data) (Figs. 5.9 and 5.10)	Waste dump instability hazard
Top down spoil construction	$\tau = 1.446\sigma^{0.898}$	Lower bound	Very high to high (WHC IV and V)*
Top down spoil construction	$\tau = 1.576\sigma^{0.899}$	Lower quartile, between low and average rockfill (typical to model top down construction greater than 75 m high)	Moderate to low (WHC I to III)*
Bottom up spoil construction	$\tau = 1.899\sigma^{0.898}$	Upper quartile, between average and high rockfill (typical to model bottom up construction)	Very high to high (WHC IV and V)*
Bottom up spoil construction	$\tau = 2.067\sigma^{0.899}$	Upper bound	Moderate to low (WHC I to III)*

* See Chapter 3 for a description of the waste dump and stockpile hazard class (WHC)

On the other side of the spectrum, mixed fills composed predominantly of overburden materials are likely to exhibit properties more representative of the overburden materials, especially if there is no contact between the rockfill blocks. When characterising the properties of mixed materials, the designer is cautioned to be conservative and consider the relative proportions. Wherever possible, property selection should be based on objective laboratory testing of a representative mix of the materials.

6

SURFACE WATER AND GROUNDWATER CHARACTERISATION

Geoff Beale

6.1 Introduction

The assessment of surface water and groundwater is an important component for the planning, design and operation of any mine waste dump or major stockpile facility. This chapter describes the characterisation work that may need to be carried out to evaluate the following:

- upstream hydrological conditions and the need for surface water diversions
- hydrogeological characteristics of the foundation materials and the groundwater underflow system
- hydrogeological conditions within the waste rock dump or stockpile itself, including:
  → zones of saturation and pore pressure
  → seepage that may occur from the base of the facility
  → seepage faces that may occur in the side slopes or downgradient toe area
- potential impacts downgradient of the facility.

Section 6.2 describes the general planning process for surface water and groundwater characterisation and is divided into studies carried out to determine upgradient conditions, conditions beneath the footprint area of the facility, the hydrology and hydrogeology of the facility itself, and downgradient conditions. Section 6.3 describes the main components of the conceptual hydrogeological model pertaining to the facility. Section 6.4 describes the surface water characterisation. Section 6.5 describes the near-surface water balance of the facility together with infiltration and recharge modelling. Section 6.6 describes the hydrogeological modelling for the overall facility and includes pore pressure modelling of the waste or stockpile materials, and Section 6.7 describes the modelling of the foundation and underflow system.

The collection of climate data required to support the hydrology studies is described in Chapter 4. Hydrogeological instrumentation and monitoring that may be needed to support operations and closure planning are discussed in Chapter 12. The hydrological input to the reclamation and closure studies is included in Chapter 16. Water chemistry is not discussed in detail in these guidelines, but a brief discussion of water chemistry in the context of acid rock drainage prediction and mitigation is provided in Chapter 14.

6.2 Investigation of surface water and groundwater

One of the goals of the Large Open Pit Project has been to provide guidance on the level of effort required at each stage of project development and implementation. This guidance may be equally applied to greenfield sites, brownfield sites or to the expansion of an existing operation. As discussed in Chapter 2 and summarised in Table 2.2, as the developing project progresses from the concept stage through to operations and eventually closure, there should be a corresponding increase in the confidence in the data and designs. This level of confidence should, in turn, be commensurate with the overall level of certainty required for the project as a whole. In addition, as discussed in Chapter 3 and summarised in Table 3.12, the level of effort required also depends on the relative instability hazard and level of risk, particularly with respect to the environmental protection of water.

6.2.1 Components of the investigation program

The key components of the investigation program are typically as follows:

- **Upgradient conditions**, which may include:
  - → surface water conditions upstream of the planned footprint area, the potential for surface water run-on to the facility and the requirement to divert surface waters; the requirement to construct and maintain large surface water diversions sometimes represents one of the major long-term risks for a waste dump
  - → groundwater conditions upgradient of the planned facility, and the potential for groundwater to flow into the footprint area and below the base of the facility (groundwater underflow).

- **Conditions beneath the footprint area of the facility**, which may need to involve:
  - → the hydrogeology of the foundation materials (which should incorporate the assessment of upgradient surface water and groundwater flow)
  - → the nature of any groundwater underflow, and how this may be disturbed by the construction of the facility
  - → the potential for the foundation materials to develop undrained pore pressures as a result of increasing total stress due to loading as the waste dump or stockpile is constructed.

- **The hydrology and hydrogeology of the facility itself**, including:
  - → rainfall and runoff from the surface of the waste dump or stockpile, and the extent to which this may cause:
    - · infiltration and recharge into the waste dump or stockpile materials
    - · erosion of the waste dump or stockpile materials
    - · potential downstream impacts to surface water and/or groundwater
  - → unsaturated and saturated groundwater movement within the facility, the build-up of pore pressure within saturated zones and the potential for internal erosion (piping) to occur behind the side slopes
  - → the potential to generate undrained pore pressures in low permeability layers within the waste dump or stockpile where the increase in total stress due to loading occurs at a greater rate than pore pressures can dissipate as the materials consolidate.

- **Downgradient conditions**, which may need to incorporate:
  - → seepage that could occur from the base of the facility and/or from the lower toe areas, and the potential for this to impact downstream surface water and/or groundwater
  - → runoff that may leave the facility footprint and may potentially impact the downgradient drainage area

  - → sediment that may become transported into the downgradient surface water system either from facility runoff or from diversions.

The surface water and groundwater studies are usually interlinked. For any given site, there may be a considerable amount of surface water–groundwater interaction, both in the surrounding hydrological system and within the waste dump or stockpile facility itself.

6.2.2 Planning considerations

For surface and groundwater studies, the planning process must pay careful attention to local regulatory requirements and compliance standards, including how these may evolve with time and potentially become more stringent during the planned life of the facility and following closure. The hydrology program must also consider the mining company's internal corporate standards and environmental governance. Depending on the overall layout of the mine site, it is usually necessary to integrate the specific waste rock hydrology program into the site-wide water management plan. Most waste dump and stockpile facilities need to meet downstream water quality and sediment loading criteria. One or more downgradient compliance points may need to be established for both surface water and groundwater.

As part of the early studies it is usually advisable to obtain a good surface water and groundwater baseline that is focused on providing an adequate characterisation of naturally elevated chemical parameters and any pre-existing disturbance to the natural system. A good baseline characterisation is important for helping to demonstrate to the public, regulators and other stakeholders that changes to the hydrological system may result from natural variance, or from other developments external to the mine site, rather than from the waste dumps or stockpiles themselves.

6.2.3 Investigation of conditions upgradient and beneath the footprint of the facility

The investigation of upgradient surface water and groundwater conditions, and the hydrogeology of the foundation materials, is usually carried out early in the planning process (well before construction) as part of the alternatives analysis and staged design process (see Chapter 2). The hydrology data collection process is often tied to the site characterisation program (see Chapter 4).

For each potential facility location, there may be a need to determine the following:

- the requirements for surface water diversions and whether they will create significant, long-term hazard/risk issues and maintenance requirements (extending into the post-closure period)

- the hydrogeology of the foundation materials to assess:
  - → the potential for build-up of pore pressure beneath the facility
  - → the potential for groundwater seeps or springs to enter the base of the facility, and how this water may contribute to instability and/or potential downgradient impacts
  - → the implications of any foundation excavations on the groundwater underflow system
  - → the potential for loading of the facility to change the hydraulic properties of the foundation materials which may, in turn, affect the groundwater underflow system.

The conceptual design stage usually includes a walk-over survey and desktop study to define alternative dump locations and would normally identify the items listed in Table 6.1.

Pre-feasibility work would be carried out for the preferred facility location and potentially one or two alternative sites. Typical hydrological and hydrogeological studies carried out as part of the pre-feasibility investigations are listed in Table 6.2. Important data gaps would be identified as part of the pre-feasibility studies.

The feasibility study would be carried out for the preferred facility location identified in the previous studies. A typical program for feasibility studies is summarised in Table 6.3. The development of a conceptual hydrogeological model should be

carried out as early in the program as possible and should evolve as the project progresses. By the time of the feasibility study, the conceptual hydrogeological model should be of sufficient detail to support an analytical or numerical analysis (as necessary) and a pre-construction water balance of the footprint area.

The feasibility study will also need to provide sufficient supporting data to allow planning and design of the following:

- any surface water diversions
- any upstream groundwater cut-off requirements
- any required dewatering of foundation materials
- environmental protection and mitigation requirements for surface water and groundwater (during operations and closure).

It is necessary to ensure there is a sufficient number of surface water monitoring sites and environmental monitoring wells for the establishment of a surface water and groundwater baseline. The environmental baseline usually requires a minimum of 12 months of monitoring before the start of facility construction, with a minimum of quarterly sampling for water quality and suspended sediment load (often tied to the seasonality of the hydrological system). The specific baseline for the waste dump/stockpile facility will need to be integrated into the site-wide environmental baseline program.

Characterisation and monitoring of the groundwater system may utilise the following:

Table 6.1: Surface water and groundwater reconnaissance studies for conceptual design

Upgradient and beneath the planned facility footprint	
Surface water	**Groundwater**
Obtain available local and regional climatological data (see Section 4.3.4).	Obtain any existing groundwater reports that may be available.
Determine major upgradient catchment areas, including the flow in any streams that may require diversion.	Determine any important groundwater units that may need to be protected.
Assess minor upgradient topographic catchments that may contribute overland (non-channelled) run-on to the facility.	Determine any existing (and potential future) water supply sources that may need to be protected.
Map the upgradient drainage divides, and ideally place the facilities as close to the drainage divides as possible in the initial design and site layout.	Carry out field reconnaissance of seeps, springs and other potential groundwater discharge features.
Determine the location of any lakes or wetland areas and how these may influence the facility footprint. There may be a need to maintain habitat and/or riparian flows and preserve surface water flows and quality.	Obtain information on groundwater levels that may be available from any existing wells in the area or from historical mineral exploration holes.
Within the planned waste dump and downgradient	
Assess the facility location relative to its surrounding topography and upgradient areas and whether there is potential for enhanced recharge into the waste dump/stockpile materials in the long term.	
Evaluate the physical nature of the waste dump/stockpile materials and whether there is potential for pore pressure to build up in the interim or final downgradient toes.	
Assess whether there is potential for the facility to produce seepage to the downgradient environment in the short or long term.	
Evaluate the likely water quality of any seepage and the implications of this for long-term planning and environmental risk.	

- condemnation drill holes and/or geotechnical drill holes and test pits:
  - → *in situ* permeability testing (falling and rising head tests), groundwater level monitoring, groundwater chemistry sampling, testing to characterise the permeability and porosity of identified hydrogeological units
  - → installation of piezometers and monitoring wells to determine the phreatic surface (water table), any perched groundwater zones, vertical hydraulic gradients and any confined groundwater units
  - → sampling and laboratory testing to characterise the water content of identified hydrogeological units above the phreatic surface

- dedicated hydrogeology drill holes: where layering is expected to influence the hydrogeology, there will probably be a need to install multilevel standpipe piezometers (for sampling) and/or multilevel vibrating wire piezometers
- mapping of rivers, lakes, springs and seeps: the mapping data will need to be integrated with the interpretation of the phreatic surface, including any perched groundwater zones.

Surface geophysical studies can be considered to assist with groundwater mapping where the underlying hydrogeology is expected to be of concern, and particularly where the depth to water is within 20–30 m of the ground surface.

Table 6.2: Pre-feasibility surface water and groundwater studies

Upgradient and beneath the planned facility footprint	
Surface water	Groundwater
Establish an on-site rainfall station, and potentially a full climatological station, depending on the quality of climatological data that are already available for the mining district (see Section 4.3.4).	Carry out hydrogeological logging, groundwater level monitoring, *in situ* permeability testing and possibly downhole geophysical surveys in condemnation drill holes, geotechnical holes and/or test pits. Initial permeability testing will usually involve rising or falling head tests carried out at several depths within the hole during drilling (Beale and Read 2013).
Establish flow measurement stations in any major or locally important streams.	Use the condemnation and geotechnical programs (drill holes and test pits) to establish piezometers for the hydrogeological program and monitoring wells for the environmental program (Beale and Read 2013).
Carry out synoptic flow surveys to assess the spatial distribution of stream flow, including any surface water losses and/or gains to groundwater.	Carry out *in situ* permeability testing in the completed piezometers.
Carry out preliminary surface water quality sampling and initiate a baseline study for selected facility location(s).	Use the drilling and test pit results to prepare an evaluation of the superficial and alluvial hydrogeology above the bedrock surface and, in particular, the influence of geological layering on the groundwater flow system.
Sample the sediment loading at the flow measurement stations, and potentially other locations, for a range of flow conditions.	Use the site investigation results (including mapping data) and any published data to assess the bedrock hydrogeology, incorporating the lithology, degree of weathering, primary geological structures and any voids, cavities, sink holes or other karst features.
Carry out initial planning for surface water management, possibly including diversions, rock drains, sediment control and water quality management.	Identify groundwater units that may affect the foundation stability and/or facility design.
Prepare an initial assessment of any potential impact to downgradient streams or other surface water features.	Identify any groundwater units that may require environmental protection.
Within the planned facility	
Analyse climatological data to evaluate potential infiltration and recharge into the surface of the facility.	
Assess whether zones of saturation may develop within the dump/stockpile and how these may interact with the underlying and downgradient groundwater system.	
If the potential for saturation of the dump/stockpile materials has been identified, determine what mitigation measures may be required for reducing pore pressures or for seepage control.	
Evaluate whether and how the proposed fill placement strategies may impact the hydrogeology of the facility in terms of the build-up of pore pressure and long-term environmental risk.	
Downgradient of the planned facility	
Assess how facility runoff and seepage will be incorporated into the overall site water balance for operational and closure planning.	
Evaluate detailed planning and topography data for the facility and how the design of the facility may affect the local-scale hydrogeology.	
Compare and rank alternate facility locations for short- and long-term risk in terms of pore pressure, environmental protection and cost of long-term mitigation.	

The groundwater interpretation of the foundation and underflow system will need to include the following:

- estimation of the groundwater flux moving within the foundation materials and the direction of flow (underflow)
- the interaction of the near-surface hydrogeology with any surface water flows
- the relationship between the near-surface hydrogeology and mapped seeps and springs
- the interaction of the near-surface groundwater system in the superficial deposits and alluvium with the underlying bedrock hydrogeology
- an assessment of the groundwater chemistry.

Care should be taken to distinguish any zones of perched groundwater from the main phreatic surface

Table 6.3: Feasibility surface water and groundwater studies

Upgradient and beneath the planned facility footprint		
Surface water	**Groundwater**	**Operational planning**
Establish additional surface water flow monitoring stations, as required. A minimum of one upstream and one downstream station is normally required for each facility for the baseline studies.	Install any dedicated hydrogeology holes and piezometers and/or environmental monitoring wells.	Determine the phreatic surface and pore water pressures within the underflow system as input to the geotechnical stability analysis.
Analyse the requirements for surface water diversions, including seasonal and peak flows and how surface water may influence closure planning.	Ensure there is a sufficient number of baseline monitoring wells.	Determine how any planned excavation may affect the hydrogeology and potentially disturb the baseline groundwater flow system.
Analyse the sediment loading of required surface diversions.	Carry out surface geophysical surveys, as necessary, to assist the groundwater investigation and, in particular, to identify voids, cavities or other features if karst conditions may be encountered.	Evaluate whether the foundation materials may become compacted and develop undrained pore pressures under loading, therefore reducing the effective stress in finer grained materials.
Obtain sufficient surface water information to allow design of any required diversions.	Install and test additional piezometers within geotechnical drill holes and/or test pits.	Determine the potential for underlying groundwater or upstream surface water to enter the waste dump/stockpile materials, either during operations or during closure.
Obtain sufficient surface water information to allow an assessment of potential downstream changes in the flow regime and/or the water chemistry.	Characterise the hydraulic properties of all important groundwater units in terms of their porosity, drainable porosity, specific retention, horizontal and vertical permeability (K_h:K_v), degree of hydraulic layering and/or vertical interconnection, surface water-groundwater losses and gains, and interaction with the wider scale groundwater flow system.	Assess the deeper hydrogeology and the requirement to drill any dedicated hydrogeological investigation holes, and the need to carry out any wider scale testing (such as prolonged airlift testing or pumping tests).
Within the planned facility		
Assess the types of materials to be placed in the facility and their likely spatial distribution, including an evaluation of how the materials may fragment or otherwise degrade during or following placement.		
Assess the initial water content of the placed materials.		
Develop predictions of how the materials may become segregated or layered during deposition, as discussed in Section 6.3.1.		
Estimate the saturated and unsaturated flow properties of the materials.		
Assess how water will enter the facility and at what rate. Sources of water that may enter the facility are discussed in Section 6.3.3.		
Evaluate the water balance of the facility, including redistribution of the water content and saturation with time, and whether (and how) water within the facility may discharge to the underlying or downgradient environment.		
Assess the potential for any saturated zones to develop, as discussed in Section 6.3.4.		
Carry out analytical or numerical modelling of pore pressures within the saturated zones if they are considered to be significant, with the results incorporated into the geotechnical analysis.		
Predict seepage flows (and variability with time) from any identified saturated zones and whether these flows could have the potential to create internal erosion (piping) or external erosion of the materials below the seepage locations.		
Downgradient of the planned facility		
Develop predictions of short- and long-term (operational and closure) seepage flows and chemistry and how the discharge may possibly affect downstream surface and groundwater flow rates, downstream water quality, downstream sediment loading, existing water supply sources, riparian wetland sites or other downgradient receptors.		
Develop mitigation plans for management of the water, as required, considering that the physical and chemical nature of the placed materials, and hence the water quality, may evolve with time.		

(water table), considering that zones of perched water may sometimes be seasonal and the main phreatic surface may show a large seasonal variation or may respond to occasional recharge events.

The type and amount of data required will depend on the planned size of the facility, the nature of the hydrogeology and, in particular, the depth of the water table. An assessment of the surface water and groundwater risks should be carried out as part of the feasibility study. This should include an evaluation of the potential for certain layers in the foundations to generate undrained pore pressure.

Except where there are specific surface water and/or groundwater concerns, the following general guidelines may apply:

- If there is groundwater within ~20–30 m of the ground surface within the footprint of the facility, there should be a minimum of 15 locations where permeability testing has been carried out and groundwater levels have been measured, and there should be a minimum of six piezometer installations (preferably multilevel installations), of which at least three should be standpipe completions to allow groundwater sampling.

If there is groundwater within ~30–50 m below the ground surface, there should be a minimum of 10 locations where permeability testing has been carried out and groundwater levels have been measured, and there should be a minimum of three piezometer installations, of which at least two should be standpipe completions to allow groundwater sampling (upgradient and downgradient). The amount of detail required for the detailed design and construction stage will depend on the nature of the site and the importance of surface water and groundwater. The following may need to be considered:

- *Surface water*:
  - → filling in any data gaps in the surface water study
  - → ongoing monitoring of surface water (established stations or synoptic surveys)
  - → construction and monitoring of any required surface water diversions for interim or final footprint layouts.

- *Groundwater*:
  - → ongoing monitoring of groundwater (geotechnical piezometers and environmental monitoring wells)
  - → further characterisation of any important groundwater units
  - → additional dedicated deeper hydrogeological test holes, as required
  - → implementation of any required groundwater cut-offs (for interim or final footprint layouts)
  - → installation of any required dewatering or pumpback wells in the toe area, either for

excavations related to construction or for facility operation.

The nature of any ongoing studies and monitoring during operations will depend on the importance of surface water and groundwater on the facility design and/or for environmental control. The work may include the following:

- inspection and maintenance of surface water diversions
- monitoring of pore pressures in foundation and/or toe area piezometers to ensure that design criteria are being met
- surface water quality monitoring if dedicated stations have been installed
- groundwater quality monitoring in environmental wells
- sediment monitoring.

Details of the ongoing monitoring program are included in Chapter 12.

6.2.4 Investigation of conditions within and downgradient of the facility

The hydrogeological characterisation of the waste dump and stockpile materials themselves is often carried out later in the overall program than the investigation of the upgradient conditions and facility foundations. Characterisation requires an understanding of the geology of the various material types that will be mined and placed in the waste dump/stockpile. Details are often not available in the early stages of the investigation, so much of this work is usually carried out during feasibility and detailed design studies, and concurrently with mine operations.

For each potential facility location there may be a need to determine the following:

- how long the facility water balance may take to reach a steady-state condition, and whether seepage rates may increase with time post-closure as the dump materials gradually reach field capacity (i.e. the maximum amount of water that a given material can retain in its pore spaces before downward gravity drainage occurs)
- the seepage water chemistry, and how this may change with time (in the operational and post-closure periods)
- how the dump or stockpile may impact the downstream surface water and groundwater environment.

In drier climates, the steady-state condition may take many hundreds of years to be achieved and, in some cases, the surface recharge may be so low that seepage may never occur (e.g. at many mine sites in northern Chile and other desert regions).

Conceptual level studies usually include a preliminary (broad-scale) estimation of the hydrology and water balance of the facility itself, often with particular emphasis on the suitability of the location for closure planning and potential post-closure risks. Early-stage investigations may include those items listed in Table 6.1.

Pre-feasibility level studies often include an interactive analysis by hydrogeologists, geotechnical engineers, mine planners and environmental staff regarding how the facility design may be modified to better suit the ambient hydrological conditions. This may include modifying the footprint to allow placement of materials closer to watershed divides or varying the design geometry to minimise the accumulation of wind-blown snow on leeward slopes.

The typical requirements of the pre-feasibility level studies are included in Table 6.2. Data gaps would be identified following the pre-feasibility study and addressed at the outset of the feasibility study. The required level of detail is less in situations where the waste dump/stockpile materials have been identified as being chemically inert, of high strength with good free-drainage potential, less prone to fragmentation and less prone to physical or chemical breakdown with time.

The preferred facility location(s) will usually have been identified before commencing feasibility studies, so the hydrogeology investigation would focus on the preferred site layout and could include those items listed in Table 6.3.

The detailed design and construction stage often includes the installation of pore pressure and/or soil moisture monitoring systems that may be within the foundations or initial lifts. The monitoring system would typically be designed during the feasibility studies, with the design modified as necessary during the detailed design and construction stage. The operational stage would include monitoring of one or more of the components listed in Chapter 12.

6.3 Conceptual hydrogeological model

This section describes the components of a conceptual hydrogeological model of a waste dump or stockpile (Fig. 6.1). The conceptual model may be described as a realistic and defensible, albeit simplified, representation

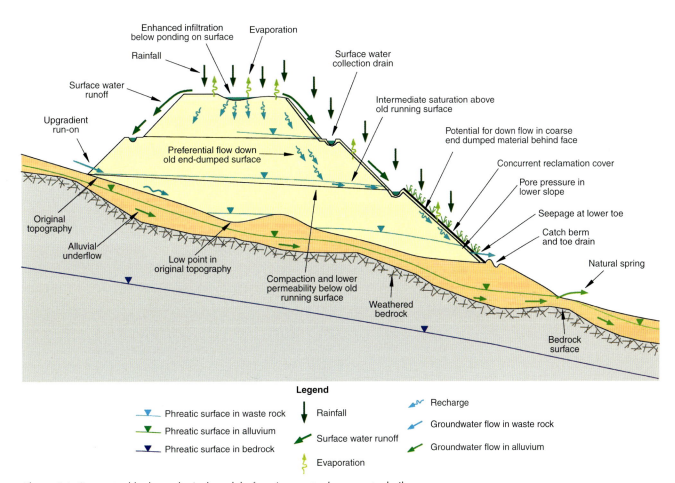

Figure 6.1: Conceptual hydrogeological model of a mine waste dump or stockpile

of the hydrogeological system. For a waste dump or stockpile, the model must consider the following:

- the geology and nature of the waste dump or stockpile materials through which the flow of water will take place (Section 6.3.1), including changes with time (Section 6.3.6)
- the initial water content of those materials (Section 6.3.2), considering that new material (and therefore pore water) is being continually added to any active waste dump or stockpile
- the water that may enter the facility from external sources (recharge; Section 6.3.3)
- the ongoing flow and redistribution of water within the facility (Section 6.3.4)
- the water that leaves the facility (discharge; Section 6.3.5).

The level of accuracy for predicting future hydrogeological conditions within the facility is entirely dependent on the conceptual understanding of the above processes. There is little value in preparing detailed predictions of future conditions if the recharge processes or the nature of the materials is poorly understood.

The conceptual model must assess how the hydrology of the facility interacts with the groundwater underflow system. In some cases, rather than downward discharge from the base of the facility, there may be the potential for upflowing groundwater to enter the base of the facility from the foundation.

In addition, the conceptual model needs to consider the changing dimensions of the facility and the changes to the hydraulic properties that occur with time as the facility is developed (Section 6.3.6).

For existing or active waste dumps and stockpiles, it is usually beneficial to evaluate the available water chemistry

Figure 6.2: Internal preferential flow pathways as a result of segregation during end dumping and subsequent internal erosion (piping). Source: G Beale

data (e.g. using samples from seepage faces or from downgradient groundwater) to help support the development of the conceptual model. Variations in the seepage water chemistry compared with geochemistry testing results (Section 6.3.8) can help determine the extent to which preferential flow pathways are developing within the dump materials.

6.3.1 Hydrogeological characterisation of waste dump and stockpile materials

The physical characterisation of waste dump and stockpile materials is discussed in detail in Chapter 5. Characteristics that may be important in terms of the conceptual hydrogeological model may include the following:

Figure 6.3: Segregation on (a) outer end dumped face and (b) external discharge (springs) due to preferential flow behind the face. Source: G Beale

- the geological, geotechnical and hydraulic properties of the materials that will be placed within the facility
- the likely spatial distribution of the material types (lateral and vertical)
- how the materials will fragment upon blasting, transport and placement, and how these processes may affect the particle size distribution, with an emphasis on understanding the content of the fine-grained component after the materials are placed
- how placement of the different material types will be sequenced over the life of the facility
- how the materials may become segregated or layered during placement; segregation may lead to:
  - → internal preferential flow pathways that may develop because of end dumping of individual lifts; end dumping may lead to alternating and intermixed coarse and fine layering, which may subsequently become enhanced because of internal erosion (Fig. 6.2)
  - → preferential flow pathways that may develop behind end dumped outer faces (Fig. 6.3)
  - → layering that develops between lifts caused by the active running surfaces (Fig. 6.4)
  - → a coarse layer at the base of individual lifts within the facility (depending on the height and amount of segregation that occurs on the end dumped faces).

The initial planning tools for the waste rock hydrogeology assessment are the geological block model for the mining area, the mine plan and the waste rock sequencing over the planned life of the facility. The domain size of the block model will need to be assessed. The block model is often more focused on ore assemblages rather than waste rock, so the evaluation may need to use drill logs and other primary data sources to characterise waste rock materials outside of the block model domain.

6.3.2 The initial water content of the placed materials

The initial water content of the placed materials can be an important factor for the hydrogeological model. In drier climates, the initial water content may be the dominant variable that controls when and whether the materials reach saturation. Where the initial water content is low, the materials may never reach field capacity, so there may be no potential for the development of pore pressure or for any discharge from the facility. The following factors may need to be evaluated:

- the water content of the materials before mining
- changes to the water content that occur during drilling, blasting and transport
- the drainable porosity and total porosity of the placed materials, including changes that may occur during placement
- the initial (unsaturated) water content of the materials immediately following placement
- the specific retention properties of the placed materials
- the unsaturated permeability of the placed materials.

These parameters are discussed further in Section 6.6.
An issue for waste dump and stockpile facilities located in seasonal or cold climates is the snow that may be incorporated within the waste dump or stockpile materials as the waste dump or stockpile is constructed. The presence of snow that becomes entrained as the materials are placed can sometimes have a large impact on the initial water content. For example, seasonal melting of snow layers within the Quintette dumps in British Columbia was thought to have been an important contributing factor to the large-scale instabilities that occurred during the mid-1980s, and was also considered to have contributed to the long runout distance.

Figure 6.4: Development of internal layering on old running surfaces. Source: G Beale

6.3.3 Recharge entering the waste dump

The main sources of water that may recharge the waste dump or stockpile facility are:

- infiltration of incident precipitation and runoff on the active or final waste dump/stockpile surfaces (which may include snow blown onto the facility from the surrounding area)
- surface water run-on from the surrounding (upgradient) slopes
- groundwater that may enter the base of the facility because of buried seeps or springs, reduced permeability in foundation materials as a result of compaction during loading, or other disturbance.

The factors that need to be considered for assessing infiltration and recharge are discussed further in Section 6.6. They may include atmospheric variables, the water balance of the near-surface zone and spatial variations across the facility footprint due to topography, solar aspect and prevailing wind direction.

Fragmentation and breakdown of materials at the dump surfaces is an important factor for determining the amount of recharge that may occur. Fragmentation can generate a large amount of fine-grained material, which can create a high water retention capacity in the near-surface layer and may therefore reduce the potential for recharge (Fig. 6.5). The hydraulic properties of waste rock are controlled primarily by the fraction that is less than 5 mm in diameter, provided this fraction comprises at least 30–35% of the material by volume (Yazdani *et al.* 2000). As for any natural system, recharge is often concentrated in discrete areas where water may accumulate rather than occurring globally across the footprint.

6.3.4 Flow pathways through the dump materials

The redistribution of water and changes to the saturation state may occur very slowly in waste dump/stockpile facilities located in arid and temperate regions, but may occur rapidly in high rainfall tropical settings where

Figure 6.5: Illustration of fragmentation on an end dumped surface. Source: G Beale

saturated flow pathways may develop within the first few years of placement.

Recharge from the surface of the waste dump/stockpile will move downward through the pile as unsaturated or semi-saturated flow. At any given depth, percolating water that enters the pore spaces from above will initially be retained within the pore spaces but will drain downward once the water content has increased sufficiently to overcome the suction forces. Waste rock dumps containing clastic materials may retain water within the pore spaces between the grains and within the grains themselves. This may be particularly evident where the source rock has significant primary porosity and where the fragmentation is poor. The water content will increase within the unsaturated materials but will not reach saturation because of the ongoing gravity drainage. Where active downward gravity flow is occurring but the placed materials remain unsaturated, slight negative (suction) pore pressures will develop, typically –10 to –35 kPa or slightly greater in materials with a high silt or clay content (Fig. 6.6).

The ability of the material to retain water in its pore spaces (field capacity) is strongly related to its particle size. Water is retained in sandy materials mostly by capillary binding, and the material will release much of the water under low suction forces. Water is retained in clay materials due to adhesive and osmotic binding and will only be released under higher suction. The field capacity will be higher if the material is highly fragmented and contains a large proportion of fines, such as commonly occurs within the near-surface zone.

If the unsaturated permeability is great enough to transmit all the recharge through the pile by gravity drainage, then no zones of saturation will develop. In this case, any discharge from the facility may potentially occur as a result of diffuse seepage from the base. However, unsaturated flow will usually occur more rapidly through spatially distinct pathways. It is a dynamic and variable process, and the flow pathways may change as the water content changes, which may itself be a function of loading and total stress. Consequently, the ability to predict variations in unsaturated flow through the facility is usually limited.

Saturation is reached when the water content becomes equal to the porosity (considering that there may still be entrained air in some of the pore spaces). Once saturation occurs, a phreatic surface will develop, and pore pressures will become positive within the saturated zone. Although most saturated flow tends to occur through the larger pore spaces between the grains of rock, the grains themselves may be porous and allow the passage or retention of water, particularly the larger grains. Zones of saturation tend to develop preferentially above barriers to vertical (downward) flow, which are usually fine-grained or compacted layers. The barriers may be:

- the underlying natural topography at the base of the facility, where compaction has further reduced the vertical permeability of the natural foundation materials
- old running surfaces or lifts internal to the facility, where the materials have become compacted by operating equipment (Fig. 6.4)
- layers of finer-grained material that have been placed within the facility.

Initial saturation may also develop preferentially along more permeable materials, and saturated flow pathways may sometimes occur when the surrounding materials remain unsaturated.

The most common zone of saturation is at the base of the facility, immediately above the natural topography

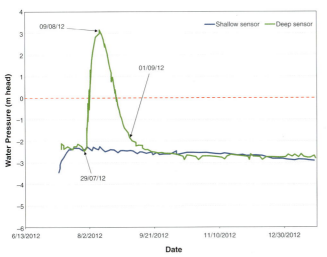
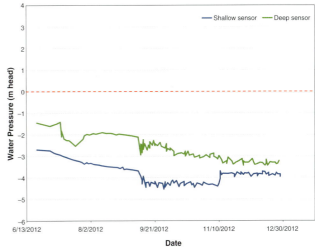

Figure 6.6: Illustration of slight negative pore pressures in a waste dump subject to ongoing recharge. The pore pressures are typically –10 to –35 kPa under active recharge. The left-hand graph shows temporary positive pore pressure after a recharge event

(particularly if a prepared surface is present at the base of the facility). Saturation may preferentially occur within hollows (depressions) in the original topography (Fig. 6.1). If there is a gradient in the original topography (as there is for most waste dumps), the lower sections of the waste dump above the downstream toe may be the first zones to reach saturation. If saturation develops above the downstream toe, then pore pressures in the toe area will be positive and external seepage may occur from the toe.

The time for the placed materials to reach saturation will vary widely depending on material type, spatial location within the facility and the rate at which new materials are placed on top of the existing waste dump or stockpile. Depending on the recharge, changes in water content usually occur more quickly in fine-grained soft rocks, and these usually have a higher water content. Fine-grained materials may also have a lower permeability and may be more compressible. The fine-grained fraction will determine the matric suction at which residual saturation is achieved.

The following factors are often of importance:

- how the dumping pattern may influence the development of preferential flow pathways (Figs. 6.2 and 6.3)
- whether internal erosion (piping) may occur and remove the fine-grained materials from the preferential flow pathways, further increasing the difference in hydraulic properties between the main preferential flow paths and the surrounding materials (Fig. 6.7)
- the role of layering within the facility that arises because of secondary dozing of materials or running surfaces between lifts; in wetter climates, it is common for saturation to develop above compacted low permeability zones between lifts
- the magnitude of unsaturated permeability $kx:ky:kz$ (and spatial variations)
- time-dependent changes in permeability due to loading and compaction from overlying lifts.

Preferential flow will often occur along the outer surfaces of end dumped faces because of the segregation of coarser materials. Short-circuiting of recharge behind outer faces is a frequently observed phenomenon (Fig. 6.1) and can lead to active seepage at the toe, even if the inner dump materials are unsaturated (Fig. 6.3b). This may be particularly evident on the leeward slopes in temperate or cold climates, where accumulating snow banks can produce strong localised recharge that may flow down the coarser outer materials immediately behind the dump face. End dumping may also produce a coarse rubble zone at the base of the rock pile.

6.3.5 Discharge of water from the facility

Discharge of water from the waste dump or stockpile may preferentially occur from saturated portions of the facility. It may be a result of the following:

- general downward seepage through the facility that leads to 'global' saturation at the base; in this case, discharge may occur from the entire footprint area and from unsaturated mateirals
- localised zones of saturation above low points in the original (underlying) topography, where discharge is limited to only those locations
- saturation and discharge from only the lower toes (either as downward seepage to the foundations or as external seepage to the lower faces above the toes)
- intermediate-level saturation above lower permeability layers that develop higher in the waste dump/stockpile materials, and discharge as external seepage at mid-points on the faces as a result of the saturation
- discharge at the toe as a result of preferential unsaturated flow in the coarser materials immediately behind the waste dump/stockpile faces that creates localised zones of saturation behind the outer toe areas.

The early-stage conceptual model must provide a sufficient level of detail to identify which of the above

Figure 6.7: Illustration of piping along internal preferential flow pathways. Source: G Beale

factors may be important for the particular setting and to broadly quantify the important processes to determine the implications for the facility design and the requirement for downstream water management controls.

6.3.6 Changes with time

Hydrogeological conditions within an operating waste dump or stockpile are always dynamic. The conceptual model will need to address changes that may occur as the dimensions of the facility change, and changes to the hydraulic properties with time as the facility is constructed. This may include the following:

- how the height of the facility may vary for each phase of operation
- how the footprint area of the facility may vary with time
- how the proportional area of the side slopes versus upper surfaces may vary with time
- how the angle of the side slopes may vary
- any variations (trends) in placed material types and hydraulic properties
- changes to placement techniques that may occur with time
- any physical (or chemical) changes that may be implemented for the upper layer to minimise long-term infiltration into the reclaimed facility.

The ongoing placement of materials (new lifts) alters both the recharge pattern and the dynamic flow pathways. If the new material is placed with a lower water content than the existing materials, it will adjust to the prevailing recharge conditions before transmitting any water down to the underlying materials. Therefore, the ongoing placement of new material may reduce the downward flow of water through the pile.

The waste dump/stockpile materials will also become compacted by the placement of new (overlying) materials. Compaction tends to reduce both the permeability and porosity of the placed materials, but tends to increase their water content (measured as a percentage). In some cases where the materials are compressible, compaction may cause unsaturated materials to achieve saturation and thus develop pore pressure.

The effect of temperature profiles within the facility may also need to be considered. A higher temperature may accelerate the breakdown of materials and the change in hydraulic properties. There is often complex and strong coupling between physical and chemical processes, and the generation of heat can result in thermally driven air circulation, which may have the potential to impact the unsaturated flow pathways. Temperature may also affect the solubility of the materials and hence the chemistry of the percolating water.

6.3.7 Approach for characterisation studies

The hydrogeological characterisation of the waste dump and stockpile materials is often carried out using a two-step process:

1. **Screening assessment** of the material types to be placed – this usually includes both the physical and chemical properties.
2. **Detailed assessment** of the materials – the requirement for a detailed assessment will depend on whether there is potential for pore pressures to develop and on the predicted flow rate and chemistry of any discharge from the facility (determined by the screening assessment).

The **screening assessment** will normally consist of the following:

- an evaluation of the **site setting** and the planned location of the waste dump relative to the surrounding surface water and groundwater hydrogeology
- an evaluation of the **material types** that will be placed on the dump surfaces and their basic geomechanical properties, including grain size distribution, uniaxial compressive strength, fragmentation studies and initial water content; this information will be used to provide a first-pass estimate of the hydraulic properties
- an initial estimate of the **infiltration rate** into the surface of the facility, the time required for the waste materials to reach saturation and the potential for (and timing and magnitude of) any zones of saturation, pore pressure development and/or and seepage discharge from the facility
- an assessment of the **geochemistry laboratory testing data** to determine the potential for any poor-quality runoff or seepage water.

For many projects, a screening hydrology assessment may be sufficient to support the decisions that need to be made throughout all stages of study, including detailed construction and operation of the waste dump. This may be the case at sites where:

- the mine is located in a dry climate, where the potential for saturation, the development of pore pressures and seepage from the facility is low
- the facility will contain mostly high strength materials, and the data indicate that the waste rock will be free draining with little potential for saturation or the development of pore pressure
- the waste rock facility will contain mostly chemically inert materials.

In cases where a **detailed assessment** is required, it may be necessary to carry out analytical or numerical modelling of the facility, as described in Sections 6.5 and 6.6. Literature values of hydraulic parameters can be obtained for given waste rock materials, or information can be gathered from one or more surrogate sites. However, where the waste dump already exists (or a surrogate site exists), a site-specific hydrogeological characterisation program may be carried out. This would typically include field testing and sampling of the materials using test pits and/or boreholes to derive both saturated and unsaturated flow properties. The following types of testing may be applicable:

- laboratory testing of field samples from test pits, which may include particle size distribution, the development of saturated and unsaturated permeability, and specific retention testing
- large-scale empirical column testing to evaluate saturated flow and development of pore pressure (the tests may be linked to geochemical characterisation program)
- *in situ* field tests using shallow permeameters, borehole permeameters or tension infiltrometers, with the field data used to calculate permeability values (Fig. 6.8)
- installation of monitoring boreholes and piezometers into any saturated zones to determine the saturated hydraulic properties.

Obtaining reliable and representative site-specific estimates of hydraulic parameters of dump materials can be difficult. A wide range of values may be typical and they may change with time due to fines migration, weathering and/or loading. However, a good interpretation of the fine-grained component of the waste rock materials is essential if any meaningful analysis or modelling is to be carried out. The proportion of fine-grained material will be the primary control of the water content versus tension relationships. The presence of non-plastic fines may be particularly important, since this component plays a key part in controlling the unsaturated flow properties, generally having the effect of enhancing water retention capacity.

6.3.8 Chemical characterisation

A similar two-step process can be adopted for the chemical characterisation. However, in many countries, a detailed geochemical characterisation is required for environmental permitting even in situations where the materials to be placed in the facility are chemically inert or there is little potential for mobilisation of chemical constituents because of low precipitation. Nonetheless, where the screening level assessment shows that the reactivity of the

Figure 6.8: Illustration of field testing for unsaturated flow properties. The main photograph shows a typical test array. The inserts show a single ring infiltrometer (right insert) and a double ring infiltrometer (left insert)

waste rocks is low, the number of required samples and the intensity of the testing program can usually be reduced.

The goal of the chemical characterisation program is to provide the basis for predicting the following:

- the chemistry of any saturated zones and variations with time
- the chemistry of any external seepage and variations with time
- runoff water chemistry, seasonal variations and longer term trends
- how the waters will react as they enter the downgradient environment and mix with other downgradient waters.

For the dump design, the characterisation program must also allow a determination of the following:

- whether any of the dumped materials may be susceptible to weathering or chemical breakdown, which could change the physical and/or hydraulic properties of the material
- whether secondary chemical precipitation may occur; this may also influence the physical and/or hydraulic properties.

The chemical characterisation of waste rock can be a complex process involving a considerable amount of testing and modelling, particularly for early-stage investigations related to permitting of the facility. In-depth geochemistry studies are beyond the scope of this book; therefore, only brief guidelines for the geochemistry characterisation are provided. The reader is referred to *The Global Acid Rock Drainage (GARD) Guide* (INAP 2009) for guidance on geochemical characterisation.

6.4 Surface water characterisation

6.4.1 Introduction

This section describes the characterisation of surface water and the planning and site water management that are typically required for waste dumps and stockpiles. Surface water controls may include the following:

- large-scale diversions of upgradient streams
- diversion of local upgradient runoff (overland flow or drainage channels)
- management of the water on the surface of facility.

The amount of data and engineering studies required will depend on the size of the upstream drainage area, the site setting and topography, the planned size of the facility and the ambient climatological conditions. Surface water may be less of a concern where the mean annual precipitation is less than ~250 mm per year. However, surface water management can still be a major issue at sites with very low mean annual precipitation but where the majority of rainfall occurs in a small number of discrete events (e.g. one event in five years).

Any design studies should carefully evaluate specific local runoff conditions, and also potential sediment loading and how this may affect the performance of diversion structures during peak flow events. The ability of the operation to implement a maintenance program for diversion structures is also an important decision factor in the design process. Surface water diversions that carry only occasional flows are notoriously prone to failure, so it is necessary to specify a regular maintenance and cleaning program immediately before, and regularly throughout, the period of the year where rainfall events may occur. The regular removal of sediment from ditches and control sumps needs to be factored into the maintenance plan, particularly in tropical areas where the materials into which the diversion channel is excavated may themselves be weathered and erodible.

6.4.2 Estimating the magnitude of runoff events from small upgradient catchments

Where there is sufficient knowledge of site conditions, conventional surface water modelling methods may be used to estimate peak runoff intensity. One approach is to use the United States Department of Agriculture Natural Resource Conservation Service, formerly the Soil Conservation Service (SCS), distribution curves to model rainfall runoff intensity and duration (USDA NRCS 2004). Different curve numbers are available for different climatic conditions. Using the runoff curve approach, modelling of design storms can be carried out using the US Army Corps of Engineers Hydrologic Engineering Center Hydrologic Modelling System (HEC-HMS; USACE 2001b). Where a watershed can be subdivided into subdrainages composed of similar runoff conditions, watershed response can be modelled for each subdrainage area using SCS curve numbers to estimate the quantity of direct runoff. Curve numbers account for differences in runoff resulting from infiltration into different soils and depression storage, for example, and are used to model runoff response. Using the curve numbers, runoff coefficients, representing the percentage of runoff that will report to the subdrainage/watershed outlet, can be calculated using the formula:

$$\text{Runoff (mm)} = (P - 0.2S)^2/(P + 0.8S) \qquad \text{(Eqn 6.1)}$$

where:

P = precipitation (mm)

S = potential maximum water retention (mm) = $254*((100/CN) - 1)$

CN = SCS curve number.

Using the initial SCS curve numbers chosen as described above, runoff coefficients for each of the subdrainages can be calculated. Direct runoff response can then be modelled using the SCS unit hydrograph. Model parameters can also be verified against recorded rainfall and flow records for the site, if available, to confirm whether an appropriate CN is being used. Curve number values can be modified to account for topography and size of the upgradient catchment area, the nature of the materials in the catchment area, soil groups, land use and antecedent moisture conditions. Some caution should be exercised when selecting the CN value as the technique was originally developed for agricultural lands. It is often best to apply a runoff coefficient based on empirical evidence or on experience at mine sites in similar settings. An experienced hydrologist should be engaged to help define appropriate CN values and to assist with developing the hydrological model.

Potential trends in climate may also influence predictions. Normal-year peak runoff events may occur due to seasonal temperature increases and melting of the snowpack, and have the potential for significant contact with near-surface materials. Extreme-year runoff events may occur as a result of rain on snow over frozen ground. The following factors should be considered when selecting the design return period:

- the amount of data and knowledge that exists on the ambient rainfall patterns in the area
- the nature of the contributing catchment area and its applicability for modelling
- the ability to clean and maintain the diversion, and the degree to which it is necessary to over-design the system to account for sediment accumulation

- the planned life of the facility and whether the design needs to be maintained for long-term closure
- applicable local regulations.

Alternatively, a risk-based approach may be used to size diversion channels, considering the following design parameters:

- The **potential magnitude of peak runoff flows**, and the extent to which these events are locally understood – in remote, drier climates, there are often limited data to assess the magnitude of peak rainfall events. Where the consequences of failure are high, it is often prudent to incorporate substantial contingency into the design.
- The **nature of the materials that will form the bed of the diversion structure**, and the need to line the structure – lining may minimise erosional scouring, thus maintaining the integrity of the structure.
- The **extent to which sediment and debris may be washed into the diversion** during peak events, consequently reducing its capacity – in some desert regions, there may be a large amount of debris in runoff flows during peak rainfall events, making the implementation of surface water diversions very difficult and therefore requiring a large design contingency. Sediment transport during extreme events may be so great in some situations that it severely affects the design or, in some cases, precludes the construction of a diversion altogether. Transport of large rocks and boulders may create damage to the channel, liner or structures placed within the diversion.
- Whether **chemical precipitates** have the potential to reduce the carrying capacity of the diversions themselves or associated infrastructure.

Figure 6.9: Example of an inert waste rock berm used for stormwater management in a low rainfall area. The berm is used to help protect the main waste dump from occasional high runoff events. It represents a short haul for waste material and can therefore be oversized

- The **extent to which snow may accumulate** and become compacted within the diversion structure – the accumulation of windblown snow may render the diversion channel ineffective to carry a large runoff event, particularly rain-on-snow events that may occur during the time of spring breakup (freshet).
- The **ability to provide ongoing access** to allow routine cleaning and maintenance – any diversion will require regular maintenance, such as clearing of sediment and debris to prevent blockage of the system. If the maintenance is not carried out on a routine basis, it may compromise the entire diversion structure. Systematic cleaning is often required immediately before, and regularly throughout, the wet season.
- The **risk and consequences of failure** or overtopping of the diversion channel.

The engineering aspects of surface water diversions are discussed in Chapter 7 of these guidelines. In some cases, it may be more appropriate to use runoff diversion berms in preference to (or in conjunction with) excavated channels (Fig. 6.9). Diversion berms may be less susceptible to capacity reduction or damage due to sediment and may be less prone to overtopping due to accumulating compacted snow than excavated channels. If berms are used, construction materials should be chemically inert.

6.4.3 Estimating the magnitude of runoff from the waste dump or stockpile

The prediction of runoff volumes from waste dumps and stockpiles depends on several key variables, including the following:

- the relative proportion of flat surfaces versus sloping faces
- the geology of the near-surface materials and the amount of fine-grained material on the surface of the facility
- the degree to which the dumped materials have been dozed to remove hummocky surfaces
- the layout and distribution of the haul ramps and the extent to which these may form drainage arteries
- the presence of any step-out berms and the extent to which these may promote the collection of runoff
- the percentage of surfaces/platforms that may have been concurrently reclaimed
- whether crest berms are present
- the antecedent rainfall conditions, including any ponded water already present on the surfaces/platforms
- the amount of evaporation that has occurred since the previous rainfall event
- the presence of any frozen surfaces
- the rainfall amount and duration of specific events.

Factors that tend to increase runoff are rapid drainage along haul ramps, working surfaces that are sloped towards the ramps and the installation of surface water drainage channels. Dump surfaces that include a high proportion of fine-grained materials with a low proportion of hummocky or blocky areas will also have an increased potential for runoff.

Although the CN approach can be used for assessing the magnitude of runoff from the waste dump/stockpile, the above factors must be carefully considered, and caution should be used when selecting the CN value because of the complex geometry and variable surfaces of most operating dumps. The dynamic nature of a working waste dump or stockpile means that, for any given rainfall intensity and duration, the proportion of the rainfall that contributes to runoff can vary with time as the facility evolves. Consequently, the development simple runoff coefficients (with an envelope for uncertainty) is often appropriate.

Surface water management for operating dumps is discussed in Chapter 11 of these guidelines. If predictions of peak flow rates are to be used for the design of operational facilities, an experienced hydrologist should be engaged to help define appropriate design flows.

6.5   Infiltration and recharge

This section describes the near-surface water balance of a waste dump or stockpile facility together with studies that may be required for infiltration and recharge modelling. It is normal practice to perform the recharge analysis before and independently of any numerical modelling studies for the facility has a whole (described in Section 6.6).

6.5.1   Near-surface water balance

Depending on the intensity of the precipitation event (P) and the antecedent water condition of the near-surface materials, a proportion of the precipitation will run off from the surface of the facility (Ro), a portion will evaporate from the surface (E_s) and a proportion will infiltrate to the near-surface layer.

The infiltrating water temporarily increases the water volume stored in the near-surface layer of the waste dump/stockpile material (ΔSW). The water held in storage in the near-surface layer is subsequently removed by evapotranspiration (ET) or by downward percolation that recharges the facility (R), until a new equilibrium is reached or another infiltration event occurs.

The simplistic near-surface water balance is shown in Fig. 6.10 and can be written as follows:

$$\Delta SW = P - Ro - E_s - ET - R \qquad \text{(Eqn 6.2)}$$

where:

ΔSW = change in water storage of the near-surface materials

P = precipitation

Ro = surface water runoff

E_s = evaporation from surface

ET = evaporation/evapotranspiration from the near-surface materials

R = downward percolation to form recharge to the facility.

Storage changes due infiltration in the near-surface zone (ΔSW) is an important consideration for the water balance. Water withdrawn back to the atmosphere by evaporatranspiration (ET) does not contribute to deeper recharge. The maximum depth from which evapotranspiration can remove subsurface pore water is called the extinction depth. Any water that percolates downward below the extinction depth recharges the facility (R).

The soil moisture deficit (SMD) is the amount of water that is needed at any given time to bring the water content of the near-surface zone back to field capacity. If the SMD has not been satisfied, unsaturated conditions continue to occur, and most of the water remains stored in the near-surface horizon, held in place by surface tension on the rock and soil particles. If surface infiltration continues to occur, the SMD of the near-surface zone reduces until the field capacity is reached (the storage capacity is used up), and any excess water entering the soil drains downward by gravity and may become recharge to the facility.

The surface of most waste dump or stockpile facilities has a significant proportion of fine-grained porous material. The processes that contribute to this condition include the following:

- fragmentation/degaradtion of the material during mining and transportation
- fragmentation/degradation during placement
- potential further disturbance by dozing
- additional accumulation of wind-blown material.

In cases where rainfall and evapotranspiration are evenly distributed throughout the year, the effective rainfall may be low, and the the SMD may remain high. In this case, most of the water that infiltrates into the dump surface is subsequently removed by evaporation and only a limited amount of water percolates downward to become recharge. Penman type estimates can provide realistic predictions of evaporation from the surface of a dump provided that a method of limiting actual evaporation is employed (Carey *et al.* 2005). The solar aspect of the facility may have a large influence on the near-surface water balance, and faces that are shaded from the sun and wind may experience less evaporation and may therefore be slower to develop an SMD. Wind is often an important factor in determining the evaporation potential from waste dump and stockpile surfaces. Wind blowing over the exposed dump surface can potentially create advective air movement within the facility.

6.5.2　Spatial variations

For the majority of waste dump and stockpile facilities, the spatial distribution of recharge varies across the footprint area as a result of the topography of the site, the surface topography of the facility itself, solar aspect, prevailing wind direction, variable near-surface materials and potentially variable vegetation. Several conditions are commonly encountered:

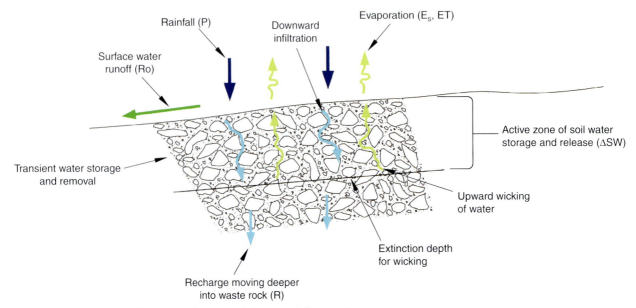

Figure 6.10: Illustration of infiltration, soil moisture balance and dump recharge

- For mines located in areas of high rainfall and/or low evaporation, there may be a considerable amount of areal or 'blanket' recharge across the entire surface area of the facility caused by the incident precipitation. Runoff can reduce the recharge rate on the side slopes compared with the upper surfaces. However, for end dump slopes, the presence of coarser materials may increase the rate of recharge on the side slopes.
- In temperate climates, recharge may be enhanced on windward slopes that receive a greater proportion of the rainfall.
- In colder climates, accumulation of snow (and therefore recharge) may be increased on the leeward slopes (particularly north-east facing leeward slopes in the northern hemisphere).
- There may be enhanced evaporation and less recharge on the south facing slopes (in the northern hemisphere), or more sublimation of snow on south and south-west facing slopes.
- Recharge can also vary with the surface topography of the facility, and there may be an increased potential for accumulation of water and recharge in any low points on the facility surface.
- For end dumped facilities, water may preferentially infiltrate and move downward by gravity within the segregated coarser materials behind the face of the facility, leading to seepage at the toe, even if most of the facility remains unsaturated.
- In colder regions, a portion of the waste rock surface may freeze during part or all of the year, and this may locally inhibit infiltration.

- Where there is significant snow accumulation, sublimation can significantly reduce the volume of water available for infiltration (particular at sites that are at a higher topographic elevation).

Thus, when carrying out water balance and recharge studies for waste dumps and stockpiles, the following factors may need to be considered:

- atmospheric variables and their variation across the facility footprint (prevailing wind direction, solar aspect (Fig. 6.11)
- exposure and orientation of the faces (windward or leeward slopes) and how this may affect wind exposure and evaporation from the surface of the waste rock, and also the accumulation of snow
- how the soil moisture balance may vary between the upper surfaces, side slopes and lower slopes/toe areas
- internal facility topography, including any low spots
- any surface water run-on from surrounding (upgradient) natural slopes.

Assessment of the local-scale variations and influence of precipitation, snow accumulation and evaporation is often subjective, and it is important to use site observations to indicate how and where snow accumulations occur (Fig. 6.12). Wind conditions and solar aspect will greatly affect snow depths across the surface of the dump. Aspect causes variations in the rates of both evaporation and sublimation.

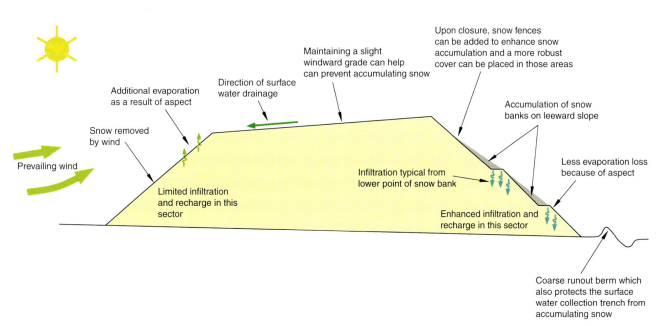

Figure 6.11: Importance of prevailing wind direction and solar aspect in waste dump planning and design. The wind direction can have a large influence on evaporation from the surface of the waste rock and the accumulation of snow

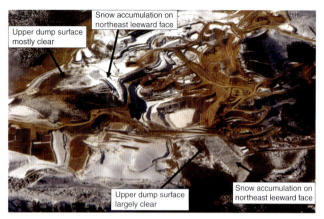

Figure 6.12: Illustration of snow accumulation on leeward faces. Source: G Beale

6.5.3 Effective rainfall

To gain a general understanding of the recharge potential to the facility, it is often useful to calculate the effective rainfall, which is defined as the difference between total rainfall and actual evapotranspiration.

The assessment of the mean effective rainfall on a monthly basis can provide a useful rule-of-thumb method for estimating the recharge rate to the facility. Some of the effective rainfall may be lost to runoff, so an estimate of runoff needs to be subtracted from the monthly effective rainfall amounts (Table 6.4). The analysis may be refined by including water storage in the near-surface layer above the extinction depth. For most climatic settings, effective

Table 6.4: Simple method of estimating recharge to a waste dump based on effective rainfall. The analysis may be refined by including water storage and extinction depth

	Mean precipitation (mm)	Mean potential evaporation (mm)	Estimated runoff (mm)	Recharge approximation (mm)
January	76	10	28	38
February	57	8	24	25
March	64	20	26	18
April	54	26	19	9
May	66	54	12	0
June	32	70	0	0
July	28	86	0	0
August	18	93	0	0
September	30	61	0	0
October	97	49	46	2
November	102	20	55	27
December	80	12	39	29
Total	704	509	249	148[1]

[1] 21% of mean annual precipitation

rainfall is only positive for certain months of the year, and potential recharge occurs only during those months. In the absence of rainfall, evaporation removes water stored in the near-surface layer, increasing the SMD.

6.5.4 Modelling of infiltration and recharge

There are several methods for estimating the water balance and infiltration potential of the near-surface zone. In each case, the goal is to estimate the amount of water that percolates downward below the extinction depth and therefore becomes recharge to the dump. Wherever possible, water balance estimates should be empirically derived from site data before numerical flow modelling is completed.

6.5.4.1 Analytical water balance

Adequate estimates of recharge can often be obtained with a simplistic daily water balance (Eqn 6.2) in a spreadsheet approach. A key assumption in the analysis is the runoff percentage from the surface of the facility during high rainfall events. It may be necessary to estimate this by inspecting the area and using judgement to make empirical estimates, or by using a simple runoff model, for example the TR-55 method (USDA NRCS 1986) assigning a suitable CN.

The daily water balance analysis assumes a volume of material above the extinction depth and applies the difference between the field capacity and interconnected porosity to determine an active storage volume. Once the storage volume is exceeded, the balance of the water becomes recharge.

6.5.4.2 One-dimensional numerical analysis

A 1D numerical simulation may be developed, again using a daily time step, with the addition of specific retention and unsaturated flow properties to better quantify the active storage and water retention properties of the near-surface materials.

Downward water flow though the unsaturated dump materials can be characterised as either matrix flow, where water moves under capillary action and is defined by the Richards equation, or by macropore or non-capillary flow, where the water is more rapid and channelised. The degree to which each mechanism is important depends on the properties of the waste rock (particularly the fine-grained fraction) and the timing and magnitude of recharge events. Software codes such as UNSAT2 (Davis and Neuman 1983), SVFlux (SoilVision 2015), VADOSE/W (Geo-Slope International 2015), HYDRUS (PC Progress 2011) or SOILCOVER (GeoAnalysis 2000) that solve the Richards equation for saturated/unsaturated flow systems typically include a function for specifying the flux in the near-surface zone. The total parameter set may include the following:

- saturated and unsaturated permeability
- total porosity, drainable porosity
- specific retention, field capacity
- van Genuchten parameters
- time-dependent decreases in permeability due to loading and compaction
- material variability
- anisotropy.

However, it is usually possible to perform a reasonably accurate water balance simulation using rainfall, runoff, evaporation, porosity and extinction depth.

The soil moisture retention curve can be approximated with established models such as Brooks Corey or van Genuchten, both of which have empirical constants. These parameters can be assigned in a variety of ways, including laboratory fit to field samples, using estimates based on soil properties (i.e. grain size distribution), or from literature based on soil classification. The water retention curve defines the relationship between the water that is retained in the material and water removal under increasing suction pressure (often termed matric potential in soil science). It is strongly dependent on the fine-grained fraction of the material. Water retention curves are non-linear and are relatively difficult to obtain accurately because samples must be disturbed during collection for laboratory testing. Laboratory testing for soil moisture retention also indicates saturated, drainable and non-drainable porosity.

Any numerical analysis must incorporate a valid daily climatic dataset, as described in Section 4.3.4, and the dataset must include extreme events, dry cycles and wet cycles. The analysis must consider that, in some climates, there may be no recharge to the facility during an average year, and all the recharge may be derived from a limited number of isolated, high intensity precipitation events that may have a return period of several years. The estimate of water lost to runoff during these events may be subjective.

It is difficult to capture the spatial variability in surface conditions across the facility within a single numerical simulation. Therefore, a common approach to infiltration modelling is to divide the surface of the facility into several distinct spatial areas and to set up a 1D model that is representative of each area. The output of each model can then be summed in proportion to its representative area to provide an estimate of the total infiltration and recharge. The facility may be spatially subdivided based on criteria such as the following:

- active or non-active deposition
- reclaimed or non-reclaimed surface
- near-surface material type
- flat surface (reduced surface water runoff) or side slope (greater surface water runoff)

- windward or leeward slopes
- north facing or south facing slopes.

In colder climates, where melting snow banks account for much of the infiltration and recharge, the facility surface may be further subdivided into four zones: negligible snow accumulation, low snow accumulation, normal snow accumulation and high snow accumulation. The fourth zone may include snow that is transferred by the wind from adjacent zones of lower snow accumulation (windward slopes) within the facility footprint and snow that is blown onto the facility from outside of the footprint area. The majority of near-surface moisture may be held as storage within the snowpack and may become lost to sublimation. In the northern hemisphere, the potential for sublimation is likely to be less on the north and north-east facing slopes (vice versa in the southern hemisphere) so, in these areas, the snowbanks may persist for longer and the eventual recharge may be greater. Another important consideration is the water content of the snow at the time of snow melt. Although there are several methods for estimating this, it is usually preferable to obtain field measurements of snow depth and water content, either from the facility itself or from similar settings.

6.5.4.3 Two-dimensional numerical analysis

If the input parameters and conceptual model are known with a reasonable degree of certainty, a 2D model can be used to simulate layering and lateral or downslope flow, if these processes are important to the analysis. If layering is included within the model, it may be necessary to input both unsaturated and saturated hydraulic values for each layer. A 2D model may also allow different snow accumulation zones to be incorporated within a single overall model domain, instead of using multiple models.

There are many variables controlling the soil moisture balance and recharge, and engineering judgement is usually required. The degree to which fragmentation and the development of fine-grained materials will occur in the near-surface material is often difficult to predict in advance of facility construction, making the modelling results for new facilities somewhat interpretive and subjective. As for most types of model, the overriding control on the model performance is the assumptions used to represent the system, rather than the choice of code or other modelling details, such as grid sizing or time steps selection.

The modelling results will be more realistic if they can be validated with seepage (outflow) measurements (i.e. measured toe seepage or lysimeter data). Opportunities for analogues using data that may exist from operational or reclaimed facilities that are located nearby, or at other locations with similar geological and climatic settings, should be pursued wherever possible. However, when comparing model results with seepage or outflow measurements from similar facilities, it is necessary to consider that many active waste dump and stockpile facilities (particularly those in drier climates) will not have reached field capacity and steady-state conditions.

6.6   Hydrogeological modelling of the waste dump/stockpile facility

This section describes the modelling for the overall waste dump or stockpile facility and includes pore pressure modelling of the materials.

6.6.1   Objectives

Flow modelling of a waste dump or stockpile facility may be carried out to achieve three possible objectives:

- **To predict zones of saturation that may develop at the base of the facility (and particularly the downgradient toe areas) or at higher levels within the facility caused by layering** – the goal is usually to predict how pore pressures may develop within the saturated zones and to assess the impact of pore pressure on stability. The output of the saturated flow model can be used as input to a geotechnical stability model. In climates with seasonally variable precipitation, it may be necessary to evaluate seasonal (transient) saturation of otherwise unsaturated materials or seasonal variations in the phreatic surface and pore pressure profile.
- **To predict the locations and flow rates of any seepage that may occur from the base of the facility to the underlying groundwater system** – saturated conditions may occur at the base of the facility over the entire footprint area, only at low points within the base of the facility, or only in downgradient parts of the footprint area. The goal is to predict the amount of water that may enter and therefore have an impact on the groundwater system beneath and downgradient of the facility.
- **To predict the locations and flow rates of any external seepage flows that may occur from the saturated zones that develop within the facility**, usually at the downgradient toes but also potentially at discrete intermediate layers within the facility. The goal is to predict the amount of water that may enter the downgradient surface water system. In addition, any external seepage flows may be a concern for internal erosion behind the face (piping).

Zones of saturation and pore pressures are more likely to develop in facilities where there are layers of weaker materials that have low initial permeability or may compact so that free drainage does not occur, and/or in facilities located in wetter climatic settings with a greater rate of infiltration and downward flux of water through the facility.

6.6.2   Consideration of transient conditions

Prior to any numerical simulation, it is essential to develop a conceptual model to gain an understanding of important factors such as the following:

- how quickly the placed materials may reach field capacity and whether zones of preferential saturation may initially develop
- how the ongoing placement of materials and the increasing material volume placed within the facility may influence the time for the lower levels to reach field capacity
- whether there may be any intermediate zones of saturation within the facility.

In concept, seepage will occur from the base of the facility when the lower levels reach field capacity. The ongoing placement of new material usually results in progressively expanding footprint areas and/or increasing thicknesses. The newly placed material will itself need to reach field capacity before any water percolates downward into the older materials. These transient conditions have a strong influence on the time taken for any zones of saturation to develop or for seepage to occur. For facilites located in dry regions, it may take many hundreds of years for the materials at the base of the dump to achieve field capacity, if they reach it at all.

6.6.3   The influence of loading on hydraulic properties

Ongoing loading of the facility may affect the hydraulic properties of the materials as follows:

- **Foundation materials** – compaction of the foundation usually leads to a reduction in porosity and permeability, particularly where soils or soft deformable rocks are present below the facility footprint. The consequences of this may be:
  - → an increase in the underlying piezometric surface, which may potentially reach the base of the facility and cause an upward flow of water into the facility
  - → the development of pore pressures in low permeability layers, which may also be the most susceptible to undrained failure.
- **Basal layer** – the original ground surface often forms a low permeability basal layer, particularly if an engineered surface has been prepared before construction of the facility. Compaction may further reduce the permeability of the basal layer and act to decouple any saturation in the facility from the underlying groundwater.
- **Waste dump/stockpile materials** – compaction of the waste dump/stockpile materials themselves may lead to

a reduction in porosity and permeability, particularly if unconsolidated or soft materials are present in the initial lifts. Again, undrained pore pressures may develop in any low permeability and deformable materials. Ongoing compaction may also cause the materials in the lower lifts to reach field capacity at a lower water content than when they were initially placed (and, if certain materials are compressible, some water maybe be 'squeezed out').

Each of the above factors will very likely be progressive as a result of the ongoing loading of the facility. In addition to the effects of increasing total stress, unfavourably oriented layers of low permeability materials may be affected by shear displacements, which may also cause an undrained response with a local increase in pore pressure. In some cases, the material may initially be unsaturated, but the pore pressure may respond to the applied stress. Such increases in pore pressure are rarely measured in the field, particularly when the movement is progressive. The surrounding more permeable materials may potentially show a different behaviour, and it is conceivable that piezometers located away from the immediate area of shear movement may not show any significant response to the locally increased pressures.

It may also be necessary to consider the possible change in hydraulic properties that may result from oxidation, weathering or chemical breakdown of the materials. Again, these would normally be expected to cause softening, leading to an increase in primary porosity, a reduction in permeability and often an increase in compressibility. Secondary chemical precipitation resulting from percolating water may also reduce the pore spaces between rock fragments and soil grains, and therefore decrease both porosity and permeability.

6.6.4 Modelling approach

Some form of pore pressure prediction is usually required as part of the stability analysis during the design and construction of the waste rock facility. The goal is to predict the development of saturated zones and pore pressures in the lower footprint area, and particularly in the area(s) of the lower toes. The analysis may also need to consider the potential for decoupled zones of saturation and pore pressure that may develop above low permeability layers at intermediate locations within the facility (sometimes at several intermediate locations).

The prediction of saturated conditions and pore pressures can use one of three approaches:

- rules of thumb
- an analytical approach
- a numerical analysis that may consider:
  - → only the saturated portion of the waste rock, or
  - → variable saturation that requires more complex modelling.

As described in Section 6.5, it is usually advisable to estimate recharge using a stand-alone analysis of the near-surface layer, rather than attempting to model recharge as part of an overall facility model. This same principle applies to groundwater models of natural sites, where the recharge rates are typically estimated using an independent analysis and then applied to the upper surface of the groundwater model. Of particular importance for waste rock models is a consideration of the spatial and temporal variability of the recharge over the footprint area, and how the variability of recharge may influence the development of saturated zones.

Because recharge varies seasonally and with changing climatic cycles, and also because of the changing footprint area and waste rock topography, it is usually necessary to use averaged or 'smeared' recharge values for waste rock studies to avoid the requirement to simulate short stress periods (time steps). If there are monitoring data to allow calibration of the model (site-specific or from surrogate sites), then it may be possible to adjust the recharge input during the process of model calibration.

It is also necessary to decide whether the model should include only the waste rock facility itself, or whether the foundations should be included within the model domain. Where continuous saturation is anticipated between the lower levels of the waste rock and the underlying groundwater system, then the foundations should be incorporated. A further complexity may be the vertical groundwater flow paths that could have been created in the foundations as a result of condemnation (sterilisation) or geotechnical drilling carried out in the footprint area before dump construction.

6.6.5 Rules of thumb

Rules of thumb can be applied as a screening tool to determine the potential for saturated zones (and therefore positive pore pressure) to develop. For example, the mean annual or mean monthly recharge rate (e.g. metres per day) can be estimated and then compared with the vertical permeability of the facility materials (using the same units) to determine whether the water will pass through the materials without causing saturation. If the permeability of the materials in the facility is higher than the infiltration rate, then zones of saturation may not develop. An example of this, carried out during a field inspection in Botswana, is described in Chapter 11.3.1.1. Discrete layers of lower permeability material can be incorporated into the analysis as appropriate.

For facilities with relatively low rainfall, or composed predominantly of strong and/or poorly fragmented waste rock, it may become clear that zones of saturation are unlikely to develop. However, it is also necessary to consider the natural (or prepared) ground surface at the base of the facility and whether saturation may develop above this layer.

During the early stages of project development, it is preferable to validate any predictions by benchmarking with other facilities in similar environments, particularly for the alternatives analysis. Where there is limited site-specific information for the materials, an acceptable approach for estimating the zones of saturation and phreatic surfaces may be to use actual piezometer data from facilities located in similar settings.

It is also necessary to consider the potential for surface water run-on or groundwater underflow entering the base of the facility, as these factors may create or contribute to zones of saturation. These factors may be significant downsides in the alternatives analysis and site selection.

6.6.6   Analytical approach

A simple analytical model can be developed during the early stages of a project when the reliability of the input parameters is low. The potential for saturation to develop in the facility can be estimated by this simple water balance equation:

$$(\text{Inflows}) - (\text{Outflows}) = (\text{Change in water content [storage]})$$

Inflows may occur from recharge to the upper surface of the facility, surface water run-on or groundwater underflow entering the base of the facility.

Outflows may occur as discrete seepage from the toe of the facility (or possibly at intermediate levels higher in the side slopes) or as diffuse downward seepage from the base of the facility.

If an analytical model is to be considered, it is usually best carried out using a 2D cross-section, as illustrated in Fig. 6.13. The model considers the zone of saturation at the base of the facility, as follows:

- The recharge reaching the saturated zone is estimated. Applied recharge can be constant or can be varied spatially along the cross-section (continually, or in bands, as shown in the figure). Assumptions will need to be made regarding how quickly the recharge will reach the saturated zone at the base of the facility.
- The permeability of the saturated materials can be estimated based on the known rock types and rock properties, or potentially by carrying out hydraulic tests in boreholes drilled into the saturated zone of a surrogate waste dump or stockpile.

Simple groundwater flow equations (such as Darcy's law) can be used to calculate the hydraulic gradient at various points along the cross-section (usually coincident with the recharge bands) and therefore the thickness of the saturated zone at the base of the facility. This, in turn, can be used to estimate the phreatic surface along the section. The recharge rate, facility profile and permeability of the materials can all be varied with time to provide a phreatic surface that changes with time.

Such analytical methods may be adequate for early-stage planning when the controlling processes are poorly understood and there are no field measurements or monitoring results to justify the use of a more complex approach.

6.6.7   Numerical analysis

A 2D numerical model of the saturated zone can be used to estimate the level of the phreatic surface at the base of the facility, and potentially at intermediate levels. With this approach, it is often expeditious to make the same assumptions regarding recharge and permeability as discussed in Section 6.6.6. The numerical simulation can be carried out rapidly and the sensitivity of varying the

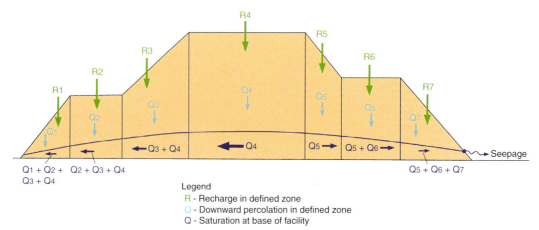

Figure 6.13: Illustration of simple 2D analytical model to estimate pore pressure development in the base of a waste dump. In this example, the dump has been divided into seven recharge zones. The flow at the base of the dump is calculated by summing the recharge. Using assumptions for the permeability of the materials, the height of the water table can be calculated using standard groundwater flow equations. The calculation assumes a steady-state condition, so only applies to dumps in wet climates and/or freely draining materials

input parameters can be easily investigated to produce a range of possible phreatic surfaces.

Although it is sometimes easier to discretise the model to complex geometries with finite element codes such as SEEP/W (Geo-Slope 2012) and FEFLOW (DHI 2016), in many applications an evenly spaced grid provides the most practical approach to avoid numerical instabilities and re-gridding of the model. The stress period (time step) should be based on the modelling objectives but, for most applications, the use of 3–5 year stress periods (time steps) is sufficient for any pore pressure or seepage analysis (depending on the life of the facility). Since the stability analysis and facility designs are often carried out using a 2D limit equilibrium approach, it is usually adequate to provide

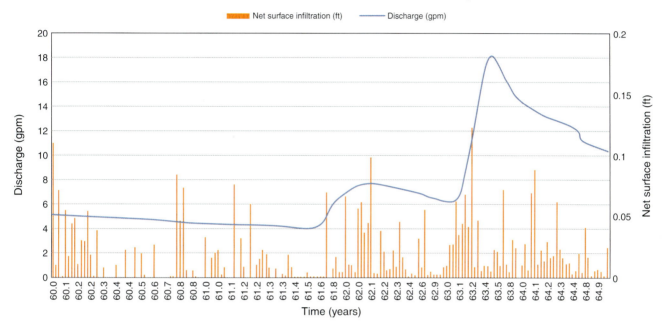

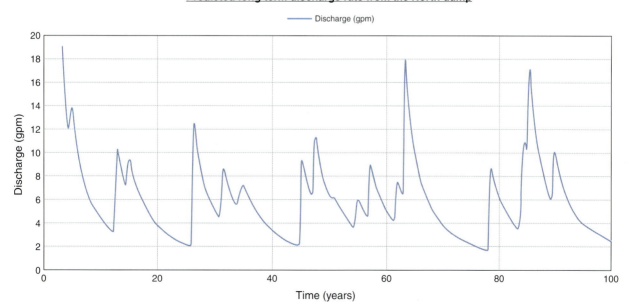

Figure 6.14: Predicted discharge from a 2D waste dump model using 100 year synthetic climate data as input. The upper graph shows the surface infiltration and seepage rate for the wettest 5 year period. The lower graph shows the predicted rate for the full 100 year period

the pore pressure input using a 2D analysis. The model can also be used to predict the variation in discharge rates from the facility with time, as illustrated in Fig. 6.14.

For most new or planned future facilities, it is acceptable to use phreatic surfaces as input to the stability analysis rather than pore pressure profiles that consider the development of vertical pore pressure gradients within each saturated zone. In most circumstances, large vertical pore pressure gradients do not develop within individual saturated zone(s) because of the relatively small flux of water passing through the facility in relation to its footprint area. Where vertical gradients do occur within the materials, they tend to be downward, in which case the use of a phreatic surface would be conservative (i.e. would tend to overestimate pore pressures). However, upward gradients can develop within the foundation materials, and these may require more detailed analysis (Section 6.7).

Two-dimensional models that include both saturated and unsaturated flow conditions are sometimes required if it is expected that significant pore pressures may develop, or where the presence of intermediate saturation may need to be simulated. However, attempting to compute the process of water movement through the unsaturated zone to the phreatic surface will add a significant amount of complexity to the model and is usually only appropriate only where the variables can be understood and where there is a reasonable basis for predicting the unsaturated flow properties of the materials.

If a more complex approach is to be considered, the purpose and objectives of modelling need to be clearly understood. FEFLOW (DHI 2016), VADOSE/W (Geo-Slope International 2015) and MODFLOW-SURFACT (HGL 2013) can simulate variable saturation, but model results can be misleading unless the limitations of the input parameters are clearly defined.

The modelling steps may be as follows:

1. prediction of recharge and its spatial variability; definition of recharge zones applied to the model (Section 6.5)
2. model conceptualisation, including:
   - → determination of the overall model domain: the upper boundaries of the model will usually be the side slopes and surface of the facility; the lower boundary may be the original topographic surface, or it may be the underlying groundwater system if the foundation materials are also included (Section 6.7)
   - → assessment of whether discrete hydrogeological zones within the model domain can be justified based on the material characterisation data, or whether all materials will be 'lumped' and hydraulic properties held constant over the entire domain

- → definition of anisotropy, which may be a k_h:k_v ratio to simulate layering or directional anisotropy to simulate end dumping
- → input of any preferential zones of high permeability, or layers of low permeability (such as running surfaces between lifts) into the model domain, as justified
- → input of unsaturated and saturated hydraulic properties, with variations for the different hydrogeological zones (as justified) and for preferential flow pathways or low permeability barriers
- → selection of starting water content for the initial conditions model
- → assessment of the need to vary the hydraulic parameters with time (between time steps) to simulate the effects of compaction
3. selection of an appropriate code to simulate either saturated only or both saturated and unsaturated conditions
4. determining whether the model will be run for a fixed period and facility geometry or whether the transient development of the facility will be included by adding to the model domain for each time step; if the model parameter distribution is to be dynamic, then material that is added to the model during each successive step may need to be included as a set of sequenced transient model runs
5. including the influence of upgradient surface water run-on, if this may be of concern.

With such a large number of contributing variables, the prediction and modelling of pore pressures will inherently rely on a series of key assumptions. It is often preferable to carry out several runs with varying input parameters, and to develop a bracketed range of model results, rather than attempting to predict absolute conditions. The bracketed results can be used to develop an instrumentation and monitoring plan, and potential mitigation measures, as required (see Chapter 12). Again, there will be more confidence in the model predictions if a comparison can be made to actual conditions within a similar existing facility. However, even when the facility is operational, it can be very difficult to understand the internal processes in detail, even if the facility is well instrumented.

The use of a range of pore pressure conditions as input to the stability analysis also allows the design engineer to evaluate the importance of water in the overall facility design. This allows the subsequent work programs to be focused. It may be beneficial to use the pore pressure model to investigate the possible use of mitigation measures, which may include the placement of alternative (more permeable) material types on the downstream faces or close to the downstream toe areas.

It is generally easier to predict hydraulic behaviour of waste dumps and stockpiles in wetter climates because the time to reach saturation and near steady-state conditions is usually quicker, so there is more information on seepage flows that can be used as the basis for calibrating the model. The inherent uncertainty of model predictions has been illustrated at a number waste dumps in wet climates where saturation and elevated pore pressure were predicted by modelling, but where the subsequent monitoring data did not show any evidence for elevated pore pressures.

Development of a 3D numerical model of a waste dump or stockpile can also be carried out and may be appropriate for the situations below:

- A relatively large part of the facility materials is saturated and it is necessary to design dewatering measures to lower the phreatic surface in the facility.
- The materials are of low permeability, thus creating saturated conditions, and the geometry of the facility is such that a 3D pore pressure grid is required to support the geotechnical analysis.
- The recharge and/or hydrogeological properties are spatially variable and are known in sufficient detail,

with adequate supporting and monitoring data to support a 3D model.
- The valley-fill geometry may be more suited to 3D analysis (Fig. 6.15).

Three-dimensional models can also be carried out using conventional saturated groundwater flow codes such as FEFLOW (DHI 2016) or MODFLOW-SURFACT (HGL 2013) (the example in Fig. 6.15 used FEFLOW). If it is desirable to include the entire facility in the model domain, then codes that can handle both unsaturated and saturated flow will probably be needed. Again, it is necessary to exercise caution because of the large number of assumptions required to underpin the model and the inherent uncertainty in the model input parameters. The use of a model to obtain a bracketed range of results is usually more pragmatic than trying to obtain absolute answers.

6.6.8 Prediction of seepage chemistry

The chemistry of the saturated pore water and potential seepage outflows can be predicted based on (1) the estimation of the seepage flux rates through the

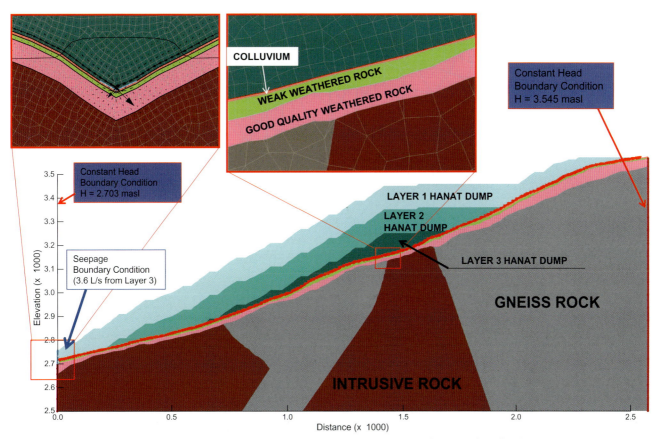

Figure 6.15: Example of 3D seepage model used to predict saturation (pore pressure) in the base of a valley fill dump and seepage outflow at the lower toe

facility; (2) the distribution of material types within the facility; and (3) the geochemical test results for the various material types.

It is beyond the scope of these guidelines to describe the geochemical analysis and prediction of seepage chemistry in detail. Readers are referred to the *The Global Acid Rock Drainage (GARD) Guide* (INAP 2009) for this topic. For a given seepage flow rate, the rate of chemical dissolution from each material type can be predicted from the kinetic test results. The contact water chemistry is then estimated using an equilibrium thermodynamic model (most often the PHREEQC code [Parkhurst and Appelo 1999]) to simulate chemical reactions between the mobile water and different material types. As for the physical modelling, the prediction of seepage water chemistry is highly dependent on several key assumptions:

- the spatial distribution of infiltration into the surface of the facility
- the wide range of preferential flow pathways that may occur, and therefore the different leaching rates of the placed materials
- variance in the oxidation state of the water within the facility and the rate at which chemical parameters may become dissolved or re-precipitated
- the occurrence of reactions that cause secondary chemical precipitation which may either (1) reduce the water-rock/soil contact and affect the rate of dissolution or (2) change the distribution of the preferential flow pathways of the percolating water.

Chemistry predictions will be more reliable if they are supported by sampling data obtained from large-scale field testing or actual seepage from waste dumps or stockpiles with similar materials. Once actual seepage occurs from the facility, it is usually preferable that water quality data from actual samples are used in preference to (or in conjunction with) the modelling data. However, the influence of preferential flow pathways and short circuiting along the outer (end dumped) faces must still be considered, together with longer term reactions and changes in seepage water chemistry with time.

6.7 Modelling of the foundation materials

This section describes the hydrogeological modelling of the foundation materials and considers the groundwater underflow system, the possible contribution of seepage from the overlying waste dump or stockpile and the potential for undrained pore pressures to develop within the foundations.

6.7.1 Characterisation

During the planning and design stages, the hydrogeological characterisation of the foundation materials can normally be integrated with the geotechnical studies. With careful planning, the required hydrogeology data can often be obtained solely from the geotechnical drill holes and test pits (Section 6.2.3). However, in cases where the underlying groundwater is identified to be of concern during the early phase studies, it may be necessary to install dedicated hydrogeological test holes for the feasibility study. Depending on the depth to bedrock, the hydrogeological study may need to include the superficial and/or alluvial materials and the underlying bedrock. It must consider the changes that may occur if near-surface materials are excavated as part of the foundation design. The following information is typically required to support the hydrogeological analysis.

- **Bedrock**:
  - → a bedrock surface elevation map (if the bedrock is within ~50 m of the ground surface within the proposed facility footprint)
  - → the thickness of the weathered bedrock zone
  - → a structural geology map of the bedrock
  - → definition of any continuous geological structures, cavities, solution channels or other karst features.

- **Superficial/alluvial materials**:
  - → detailed geology of the materials present above the bedrock surface, and particularly those within 30 m of the ground surface within the footprint area (focusing in particular on geological layering)
  - → any prominent depressions, sink holes or other unusual topographical features.

Characterisation of the groundwater system will typically use the approach outlined in Section 6.2. The information from the field programs must be sufficient to allow the following:

- a reliable estimation of the flux and direction of the groundwater underflow
- an assessment of the potential for build-up of pore pressure beneath the facility and, if required, modelling of foundation pore pressures
- evaluation of the potential for groundwater to enter the base of the facility
- an understanding of the implications of any foundation excavations on the groundwater underflow system
- changes to the permeability or pore pressures that may occur as the foundation materials compact under loading.

It may also be necessary to characterise the deeper zones of the bedrock as part of the pore pressure analysis for the foundations. As described in Section 6.2, this may require dedicated hydrogeology drill holes, which are installed to a greater depth than the geotechnical drill holes. Where karst conditions are expected within the bedrock, and the bedrock is within ~20–30 m of the ground surface, surface geophysical surveys may be advisable to determine the presence of cavities or other karst features.

6.7.2 Pore pressure modelling

Provided that the materials have been adequately characterised before facility construction, the confidence level for hydrogeological modelling of the foundations is usually higher than that for the modelling of the facility itself. The need for a numerical analysis for the foundations will depend on the depth to groundwater, the presence of perched zones and their relationship to the main phreatic surface, and the permeability of the underlying materials. A model will typically be required when:

- there is groundwater within 20–30 m of the original topographic surface
- saturated conditions are present along the predicted geotechnical failure surface below the facility, or
- the permeability of the foundation materials is below ~10^{-8} m/s and there is potential for undrained pore pressures to develop.

If numerical modelling is carried out with the objective of estimating pore pressures, the analysis is often carried out using a 2D approach, and the phreatic surfaces are input directly into a 2D geotechnical model along the same section line. The model may also be used to investigate the implications of changing hydraulic properties during loading and compaction. If there is a significant groundwater underflow rate and the formation permeability is reduced because of compaction, it is possible that the piezometric surface may rise into the base of the waste dump or stockpile. It may therefore be appropriate to change the hydraulic properties within the successive model time steps. The input of pore pressures to the geotechnical analyses is described in Chapter 8.

Pore pressure modelling is typically carried out using a groundwater flow code for saturated flow, most often SEEP/W for 2D analysis, or FEFLOW or MODFLOW-SURFACT for 2D or 3D analysis. The upgradient model boundary may be a fixed head or constant flux boundary if the underflow rate can be reliably estimated. The boundary needs to be far enough upgradient so that changes to the groundwater system as a result of facility operations do not extend to the boundary. The model domain usually needs to extend a greater distance downgradient than upgradient, so that downgradient boundary conditions do not influence the model results within or immediately downslope of the facility footprint. The upper surface of the model may be the

ground surface or, if continuous saturation is expected between the foundations and the overlying waste rock, the pore pressure model for the foundation materials may be integrated with any overlying waste rock model. The lower surface of the model will normally represent a low permeability layer, below which there is minimal potential for active groundwater flow to occur. The required number of model layers will depend on geological layering below the footprint area and on the amount and quality of monitoring information available to support the model. For most waste dumps and stockpiles, an annual stress period (time step) is adequate for modelling of the foundations, but shorter stress periods (time steps) may be required to account for seasonal changes or to allow simulation of discrete rainfall and recharge events that may affect the underflow system.

An undrained or coupled analysis may be required in situations where the permeability of either the facility and/or foundation materials is less than ~10^{-8} m/s, and where compressibility of the materials may create a reduction in porosity such that there a tendency for the water pressure to increase. Undrained pore pressures may develop solely as a result of loading and compaction caused by the progressive development of the facility, regardless of the location of the phreatic surface. Where the materials are located above the water table, knowledge of the initial water content is usually important for predicting the undrained response. It should be noted that undrained pore pressure may develop in discrete, low permeability layers while the surrounding materials have low pore pressure.

Hydrogeological modelling of facility foundations is most often carried out using a 2D approach. A 3D model of the groundwater system may be required in situations when:

- a 3D geotechnical analysis requires pore pressures as input
- excavations for the foundations of the facility will disturb the groundwater underflow system, and/or temporary or permanent dewatering of the foundations is required
- a model of the wider groundwater flow system is required to simulate the potential impact of seepage flows.

A description of the broader procedures for groundwater flow modelling is beyond the scope of this book. A detailed discussion of 3D model development can be found in Beale and Read (2013). Most commonly used groundwater flow codes can be used for modelling of the foundations and the wider groundwater setting.

Where the foundation materials occur within a permafrost zone, this zone may be temporarily disturbed during initial dumping, but freezeback into the base of the dump may occur after the placement of the initial lifts, depending on the temperature of the permafrost within the foundations, the ambient air temperature and the ability of the advective air to circulate through the placed materials. The presence of permafrost may prevent any thaw settlements in the foundations and provide an effective barrier for any seepage that passes to the base of the dump. If the thermal conditions are suitable, the seepage itself may freeze and build up a frozen base to the facility and add to the hydraulic containment system, again depending on air termperatures and the effect of the airflow through the waste rock. Groundwater flow codes can be used to help simulate freezeback in waste dumps and stockpiles underlain by permafrost, but monitoring of thermistors installed at the base of the dump and within the foundation materials is advisable at the earliest opportunity to verify the model assumptions.

7

DIVERSIONS AND ROCK DRAINS

James Hogarth, Andy Haynes and John Cunning

7.1   Introduction

Mine waste dumps and major stockpiles can cover large areas, and measures may be required to control surface runoff to prevent saturation of slopes, prevent development of a phreatic surface within the dump or stockpile, convey flows under dumps or stockpiles, manage site water quality and minimise surface erosion or development of localised failures on dump or stockpile surfaces. These measures may be required before dump construction, during operations of the dump and, in some cases, in closure after mining operations are completed. Separation of contact (mine-affected) and non-contact water is also required on some sites as part of the mine site's water quality management plan. The main water quality issues can be either physical (sediment loading) or chemical (leaching or exposure of water to oxidised materials).

To minimise contact water, surface water from catchment areas outside the dump or stockpile footprint should be collected and diverted around the dump and stockpile area using diversion channels. Runoff from direct precipitation onto the dump or stockpile surface is typically contact water and may require collection for treatment in a sediment pond system or in some cases for chemical treatment. Some waste dumps or stockpiles require existing streams or diverted water to be conveyed through or beneath the dump or stockpile in properly engineered and constructed rock drains or underdrainage systems. This chapter provides an overview of diversion channels, rock drains and other drainage systems for dumps and stockpiles.

7.2   Diversion channels

Diversion channels are often feasible for sidehill fill and heaped dumps, but are usually difficult to incorporate into valley or cross-valley fills unless topography and gradients are such that the majority of the stream flows can be intercepted upstream of the dump and channelised on the valley slope beside the dump.

It is necessary to understand the site hydrology and service life so that diversion channels can be designed for suitable flows. Local regulatory requirements and compliance standards need to be reviewed in determining design requirements. Drainage areas should be divided into subdrainages composed of similar foundation conditions with similar runoff characteristics.

Surface water characterisation is presented in Chapter 6, in Beale and Read (2013) and in Haan *et al.* (1994). Depending on the site geometry, rainfall runoff modelling could use either the rational method or the HEC-HMS system discussed in Section 6.4. Where possible, model results should be verified against recorded rainfall and flow records, verified results incorporated into the model, and revised runoff coefficients for the design storm determined. Using the calculated runoff coefficients and the appropriate distribution curve, peak flow values for each subdrainage and the entire watershed may be calculated.

Channel design should follow standard practices which are well documented (e.g. OSM 1982, Haan *et al.* 1994 and Hustrulid *et al.* 2000).

7.3   Rock drains

In open pit mining in mountainous terrain there is often a need, due to economic and space limitations, to place waste materials in valleys. These valleys invariably contain streams (perennial or ephemeral) that must be conveyed either around the facility in a diversion channel or along the bottom of the dump in a rock drain. A rock drain may be the preferred option where a diversion channel is difficult to construct and maintain or where it is not feasible to intercept sufficient flow for the diversion channel to be fully effective.

A rock drain can be defined as a zone of coarse, durable rockfill that is capable of transmitting flows with low impedance. A rock drain may be constructed either as a pre-placed structure from select rockfill or allowed to form naturally, through the process of segregation, during construction of the dump or stockpile.

End dumping is commonly practised at most large mines where haul trucks are used to transport waste materials and ore to dumps and stockpiles. As the truck backs up to the dump crest, the box is raised and material slides out and directly down the face of the dump. The end dumping method aids in the segregation of the material as the momentum developed by the material as it slides out of the truck box helps carry larger rocks to the bottom of the dump face. A rock drain created by the end dumping method is naturally graded with the largest rocks at the base of the drain and the smallest at the top (Fig. 7.1).

Push dumping, also referred to as 'dump-short-and-push', is used in some cases where, for safety or other reasons, trucks are kept away from the dump crest. The material is first dumped onto the platform and then pushed over the crest of the dump or stockpile by a bulldozer. In push dumping operations there is no vertical momentum gained by the bulldozer push. Consequently, the larger rocks tend to hang up within the finer material at the top of the dump, reducing the potential for good segregation to occur.

Adequate segregation will typically occur for repose dump heights of greater than ~20 m. Where the source rock material is generally coarse and competent, rock drain development via natural segregation is generally the preferred method of construction. This is because it is usually impractical or uneconomic to haul rockfill in the volume, quality and size required to pre-construct a rock drain that will be as effective as one developed by natural segregation. A rock drain developed through segregation of relatively well-graded waste rock is also naturally filtered through the segregation process.

It is important to note that while a 20 m dumping height is typically adequate to cause segregation, a greater height may be required to produce a rock drain zone of sufficient capacity to convey the required flow (Claridge *et al.* 1986). For example, if a 5 m thick rock drain is required, then a minimum dumping height of 25 m would be required. Furthermore, if the material is not composed of a significant portion of good quality rock, a greater height may be required. Where there is insufficient good quality material present within the material being dumped, pre-construction of rock drains may be necessary.

7.3.1 CANMET rock drain research program (1992–97)

In 1992, the Canadian Centre for Mineral and Energy Technology (CANMET) and Manalta Coal Ltd sponsored a research project to evaluate issues related to the long-term performance of rock drains in the mining industry and to establish guidelines for data acquisition and analysis to support the design and construction of rock

Figure 7.1: View of segregation in a nominally 50 m high waste rock dump, with coarser material at toe. Source: J. Hogarth

drains. One of the outcomes of this research was development of a guideline for the design and construction of rock drains (BCMWRPRC 1997). The study identified rock drains that had been designed to convey flows up to 38 m³/s. The primary focus of the program was a large rock drain in the Line Creek Valley near Sparwood, British Columbia, Canada. Construction of the Line Creek rock drain commenced in 1989 and the rock drain research program started in December 1992.

The experience gained as part of the rock drain research program indicates the following advantages of using rock drains:

- may result in lower overall surface area disturbance in relation to waste rock volumes
- more convenient and economical waste disposal locations
- avoidance of expensive stream diversions and erosion control measures
- minimal construction costs, other than requirement for selective disposal of waste rock
- attenuation of flooding effects downstream related to major rainfall events.

At the time of the rock drain research program, there were only a few instances of valley dumps crossing significant streams. Regulatory agencies and environmental groups were concerned with the following issues:

- the effect of different drain construction methods on rock gradation and durability
- ponding at the inlet of the rock drain and the potential for plugging of the inlet
- variations in the water level within the drain and effect on dump stability
- sediment deposition within the rock drain
- conveyance capacity
- diminishing capacity over time
- the effects on sediment regime and substrate at the outlet of the drain
- the effect on water quality
- biological impacts upstream and downstream of the drain.

The rock drain research program investigated these issues by means of field monitoring, data analysis and flow modelling. The results of the study were summarised in a series of annual reports and a final report issued in 1997 (BCMWRPRC 1997).

7.3.1.1 Summary of major findings

Rock drain construction and properties
At the Line Creek waste rock dump, heights of at least 20 m generally provided sufficient segregation to create a rock drain with a highly transmissive layer of coarse durable rock at its base. The end dumping method of rock drain construction was effectively used to create a drain of coarse, d_{50} (particle diameter corresponding to 50% of the sample passing by weight) greater than 0.5 m rock at the base, consisting of durable, hard sandstone with lesser amounts of hard siltstone. Poorer quality waste rock may not segregate as well and evaluation of drain rock parameters is advisable before construction. The main rock drain in the Line Creek Valley was constructed of durable, non-acid producing rock with generally not more than 10% of the rock being less than 30 cm in diameter.

Where waste rock is coarse and competent, a single end dumped lift will generally create a superior and more economic rock drain than pre-construction of a drain with selected material. The total dump height should include the depth required for segregation (minimum 20 m for good quality rock) plus the required depth of drain material.

Although the sandstone rocks at the inlet and outlet of the Line Creek rock drain were exposed to repeated wetting and drying, there was no evidence of deterioration of the rock strength during the study period. Minor loss of strength and fissuring were observed in some exposed siltstone rocks. Freeze–thaw effects are expected to be small or non-existent within the Line Creek rock drain, and degradation of siltstone within the rock drain is expected to be limited.

Flow-through characteristics
The construction of the Line Creek waste rock dumps and associated rock drains caused significant changes in the hydrology of Line Creek. A widely used flow-through velocity equation, the Wilkins equation (see Section 7.3.3), was found to provide an upper bound on measured flow velocities.

The quality of the rock in a drain is of great importance in determining its flow-through characteristics and structural integrity. End dumped drains tend to exhibit significant gravity-related sorting (segregation) of particles, with the larger, more durable rocks reaching the bottom of the drain while the smaller particles remain near the top. This natural sorting creates a high porosity zone at the base of the drain, provided sufficiently large material is present in the dump material. If excessively fine material is used for drain construction, then poor hydraulic characteristics will result. With respect to the Line Creek rock drain, the durability of material exposed at surface varied considerably.

Measured flow-through velocities were less than predicted by Wilkins's turbulent flow equation (Wilkins 1956), ranging from 43 to 67%. Increasing the dump height from 35 to 50 to 170 m did not result in improved flow-through properties. An interpretation of tracer test data and piezometer levels in the rock drain suggests that, above a minimum level, flow-through velocities are likely to be relatively constant. No measureable reduction in flow-through capacity due to plugging with sediment was evident during the study, even following a large flood event. However, small drops in flow velocity, possibly due to deposition of sediment in the drain, were measured. Sediment deposited in the inlet pond to the rock drain during extreme flow events was consistent with expected values.

Due to the lower than predicted velocities measured, a reduction of the Wilkins factor from 5.24 to 4.2 $m^{0.5}/s$ was required to fit the modelling results to observed data. This adjustment compensates for increased friction losses throughout the rock drain due to the larger effective size of the drain rock.

Environmental effects of rock drains

The following environmental observations were noted during the study:

- small magnitude of changes in water temperature
- no effect on mean water temperature range
- diurnal variability reduced, but water temperatures within pre-drain temperature range
- reductions in water temperatures that are unlikely to negatively impact salmonid species (e.g. trout).

In the years since the study was completed, it has been observed that water passing through the Line Creek waste dumps and reporting to rock drains in this geographic area can contain selenium and nitrates, and it is therefore recommended that designers engage qualified geochemists and water quality specialists to assist with assessments regarding environmental aspects of waste dumps with rock drains.

7.3.2 Alignments

Rock drains should be designed to capture and convey foundation seepage and meteoric water that infiltrates into the embankment as quickly and efficiently as possible. Alignments should take advantage of natural terrain features to help minimise the need for excavations. Where necessary to drain a relatively large area, a series of smaller secondary drains arranged in a dendritic or herringbone pattern draining into a larger central drain (Fig. 7.2) can help reduce excavation costs where rock drains are excavated into the foundation, while providing some redundancy in the system by reducing the amount of flow any one secondary drain would need to carry. Drains should be as steep as possible; the greater the gradient, the smaller the cross-sectional area required to transmit a given flow and the lower the likelihood that the drain will be affected by sedimentation.

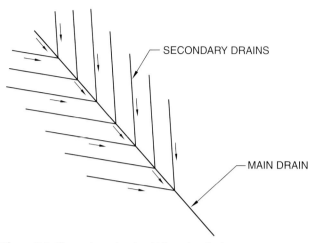

Figure 7.2: Illustration of a dendritic or herringbone type pattern

7.3.3 Design flows

Rock drains should be designed to convey at least the 1:200 year 24 h storm flow during operations, or the more conservative probable maximum flood, without consideration of the reduced infiltration rate provided by reclamation activities such as capping of the dump with a low permeability layer. This approach provides an additional factor of safety against long-term plugging of the drains due to sedimentation or precipitates over the long term following closure. As much as 60% of precipitation onto a waste dump/stockpile surface can infiltrate the dump and report to the rock drains in closure.

Where post-closure flows are expected to be significantly lower than during operations, consideration can be given to the use of lower expected post-closure flows for design of the rock drains. To accommodate higher peak flows that could occur during operations, it may be necessary to incorporate perforated high-density polyethylene (HDPE) pipes in the base of the rock drains to increase the conveyance capacity of the drains. However, the additional capacity afforded by perforated pipes should not be relied on over the long term following closure, as the pipes may eventually be crushed by the dump/stockpile. The use of perforated HDPE pipes to supplement rock drain flows may be applicable where flow-through rock drains are not required for the long term and only capture seepage flows in the long term.

It may also be necessary to consider additional flow volume due to groundwater discharge to the underdrainage system. This may require modelling of the hydrogeological conditions in the vicinity of the dump or stockpile.

The minimum cross-sectional area of each rock drain is derived from the flow velocity through the void spaces in the drain (i.e. void velocity). The void velocity depends primarily on the hydraulic gradient of the drain, which is simply the slope of the foundation, and the gradation of the drain rock. The void velocity can be calculated using the Wilkins equation for turbulent flow through a rock drain (BCMWRPRC 1997):

$$V_{void} = W(m^{0.5})(i^{0.54}) \qquad \text{(Eqn 7.1)}$$

where:

V_{void} is the average velocity of flow through the rock drain

W is an empirical constant (Wilkins constant) that accounts for the shape and roughness of the rock drain material

m is the hydraulic radius of the rock drain material

i is the average gradient along the rock drain.

For coarse angular rock, a Wilkins constant of 4.2 to 5.24 $m^{0.5}$/s is appropriate, while polished marbles have a Wilkins constant of 7.33 $m^{0.5}$/s. The hydraulic radius is a measure of the size of the interconnected pore spaces through which water flows based on monosized particles (Leps 1973).

For rock drain design, it is important to consider how the Wilkins equation is applied to well-graded rockfills. As was noted by Leps (1973), the determination of the hydraulic radius for clean, monosized drain rock is fairly reliable. However, for well-graded or non-homogeneous rockfill, it is necessary to consider the particle size that controls the flow, as flow through a porous media is generally governed by the smallest particles. The hydraulic conductivity of soil in the Darcy equation for flow through porous media can be approximated by using the Hazen equation as follows:

$$k = C(d_{10})^2 \qquad \text{(Eqn 7.2)}$$

where:

k = coefficient of permeability (cm/s)

C = Hazen's constant = 0.4 to 1.2, typically 1.0

d_{10} = grain size of 10% passing by weight (mm).

This equation empirically relates the d_{10} of a soil to its hydraulic conductivity. Even though the flow through the void space in a rock drain is turbulent, d_{10} is considered to best represent the flow in a rock drain. Based on the results of full-scale tests carried out for the CANMET rock drain research program, the use of the d_{10} was determined to be reasonable and was substantiated by dye tracer tests conducted for that study. Where well-graded drain rock is employed for rock drain construction, the use of the d_{50} in the Wilkins equation, as described by Leps (1973), increases the conveyance capacity in the rock drain many fold, but this was not observed in the field.

It is also important to note that much of the research into drain rock flow characteristics has utilised relatively uniform drain rock; therefore, the d_{50} would represent the approximate average particle size. However, waste rock and quarried rockfill, unless processed to provide a uniform particle size, are typically relatively well graded. For coarse, uniformly graded drain rock, the particle size associated with the d_{10} is therefore recommended for design, which is consistent with the findings of Abt *et al.* (1991).

The cross-sectional area required to convey the design flow is calculated using the following formula:

$$A = \frac{Q}{(V_{void})\,(n)} \qquad \text{(Eqn 7.3)}$$

where:

Q = the design flow rate (m³/s)

n = porosity.

An appropriate factor of safety, which can range between 2 and 10 depending on site conditions and proposed construction methods, should be applied to the calculated area.

7.3.4   Inlet capacity

For flow-through rock drains, the capacity of the drain inlet to transmit water is controlled by the dump slope, dump height, amount and nature of sediments accumulating upstream of the inlet and potential obstructions, such as may be formed by debris from landslides and organic materials. In some cases, an inlet pond may be required at the inlet of a flow-through drain, as illustrated in Fig. 7.3. The purpose of this pond is to attenuate storm flows, allowing suspended sediment to settle out. It also allows steam flows to transition to flow through the rock drain. The pond should have sufficient capacity to contain the design flood while the flow transitions from open channel flow to flow through the rock drain. This can be further assisted by a culvert installed in the inlet of the drain (Fig. 7.3). The installation of a culvert will increase the effective cross-sectional area of the inlet and allow more rapid infiltration. As illustrated in the figure, the culvert can also be perforated by cutting evenly spaced holes along its length to increase the discharge rate to the rock drain.

In the long term, there is a possibility that alluvium, which would normally be transported downstream during the freshet and storm events, will completely fill the inlet pond. This impact on long-term performance of the dump and rock drain should be evaluated as part of the design process. Where sedimentation of a flow-through rock drain is a significant concern, particularly during operation, it may be necessary to construct a separate sedimentation pond immediately upstream of the inlet of the drain to help protect against sedimentation, as illustrated in Fig. 7.4.

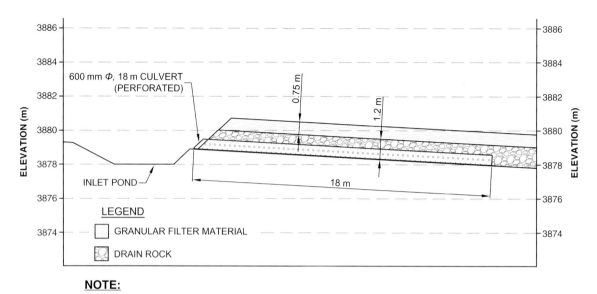

Figure 7.3: Schematic illustration of an inlet pond for a flow-through rock drain

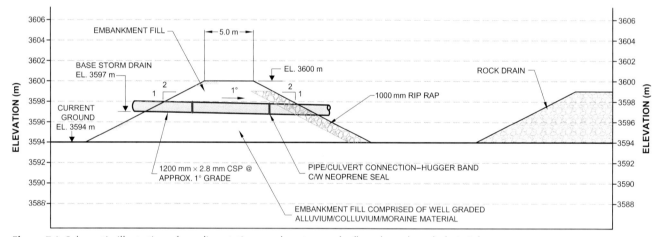

Figure 7.4: Schematic illustration of a sedimentation pond upstream of a flow-through rock drain inlet

Where it is not feasible to construct a sedimentation pond or inlet pond immediately upstream of the inlet, or where there remains a significant concern of sedimentation of the flow-through drain, such as during the construction period when vegetation has yet to become established, it may be feasible to cover the inlet with filter fabric to help prevent the transportation of fines into the rock drain. However, periodic replacement of the filter fabric may be required if it becomes clogged with sediment or damaged. Where expected flow rates are high, the use of filter fabric over the inlet of a rock drain may result in an unacceptable restriction of the flow.

7.3.5 Outlet flow

The most vital component of a rock drain design is the drain outlet. The outlet should be capable of transmitting the design flow without compromising the stability of the final downstream slope. Where the consequences of instability of the dump are severe, the probable maximum flood may be the appropriate design level. Alternatively, if feasible, designed overflow channels could be incorporated into the design.

Design aspects to be considered for the drain outlet are the final geometry of the embankment slope, the height of the seepage face predicted to occur on the slope during the flood event, seepage forces that could destabilise the slope, and scour forces at the outlet that could undermine the toe. In addition, environmental effects downstream of the dump, including potential scouring of downstream facilities such as bridges and piers, should be assessed. A procedure for calculating rock drain outlet flow capacity was presented by Leps (1973) and applied to an actual rock drain by Claridge *et al.* (1986).

7.3.6 Overflow channel

It may be desirable to provide an overflow channel in the event of extreme flood flows, blockage of the inlet of a flow-through rock drain, or as an added safety against drain voids plugging with fines. The function of an overflow channel is as follows:

- to channelise and convey flow that is not directed into the inlet, in a location where it will have the opportunity to seep downward through the dump/stockpile into the underdrainage system
- much like the spillway of a dam, to convey flows in a controlled manner, so that the integrity of the downstream portion of the dump or stockpile is not compromised during the flood event.

7.3.7 Gradation

The ability of a waste dump or stockpile to convey flow is determined by the physical properties of the rock contained in the dump and/or in the underdrainage system. The most important physical property is particle size distribution, which controls the hydraulic conductivity of the material. The particle size distribution is, in turn, determined by the susceptibility of the various rock types to degradation by mechanical and physico-chemical processes. Mechanical degradation refers to the breakdown of rock due to the mining and dumping process. Physico-chemical breakdown applies to weathering processes that may cause further size decay following dumping. Freeze–thaw cycling will affect rock breakdown at the exposed faces of the drain, but is less likely to occur within the dump or stockpile core, where above-freezing temperatures are expected to be maintained throughout the year.

The extent of mechanical degradation attributable to mining is affected by several factors, such as blasting, the amount and type of handling, crushing if transported by conveyor, the amount of rolling along the dump face, the loading of overlying material in the dump, the height of dumping and the overall amount of impact energy that the rock is subject to during placement. The relative susceptibility of different rock types to mechanical degradation can be evaluated by means of the Los Angeles abrasion test, as well as unconfined compressive strength tests and point load index tests. Weathering properties can also be determined by slake durability and freeze–thaw tests. Procedures for these types of tests are discussed in Chapter 5.

The rock making up the drain should consist of coarse-grained, clean, competent, rockfill within a specific gradation range. Depending on the source of the drain rock, the available gradation will most likely vary significantly. Ideally, drain rock should be relatively coarse with a high void ratio. However, the actual gradation will vary depending on the source (i.e. waste rock, quarried rock, fluvial rock) and whether it is screened or unscreened. Where appropriate, the gradation of the drain rock should be chosen with due consideration of the granular filters that may be required.

Where waste is expected to be of poor quality, containing appreciable fines and/or degradable materials, a filter zone should be placed on top of the rock drains as illustrated in Figs. 7.3 and 7.5. The purpose of this filter is to guard against migration or piping of fines from the waste rock into the core of the rock drain, where they could cause plugging. To confirm the compatibility of the filter material with the drain rock, the filter material must satisfy the following criterion:

$$\frac{(D_{15})_{DR}}{(D_{85})_{F}} < 4 \qquad \text{(Eqn 7.4)}$$

where:

$(D_{15})_{DR}$ is the grain size of the drain rock in mm with 15% passing by weight

$(D_{85})_{F}$ is the grain size of the filter in mm with 85% passing by weight.

For example, where $(D_{15})_{DR}$ = 290 mm (i.e. only 15% of the particles of the drain rock are smaller than 290 mm) and $(D_{85})_{F}$ = 90 mm (i.e. only 15% of the particles of the filter material are larger than 90 mm), the ratio calculated from the above values is ~3.2, which satisfies the criterion.

Another potential cause of drain plugging is squeezing of softened, fine-grained foundation materials into the drain voids as a result of loading by equipment or the advancing dump. To help avoid this potential problem, softened foundation materials should be removed and drains founded directly on competent soils. This may be difficult to achieve where the rock drains will be founded on lacustrine soils and, in this case, placement of a sand/gravel filter zone beneath the rock drain may be required.

7.3.8 Geometry

In general, the minimum thickness and width of a rock drain is governed by the maximum expected drain rock particle size. Notwithstanding this, where rock drains are excavated into the foundation, the minimum width may be controlled by the dimensions of the equipment used to create the excavation. If an excavator is used to excavate a channel for the rock drain, the minimum design width of the drain should be no less than the width of the typical bucket used for excavation. Where rock drains are excavated into the foundation, due consideration must also be given to sideslopes, particularly where it may be necessary for personnel to enter the excavation to install

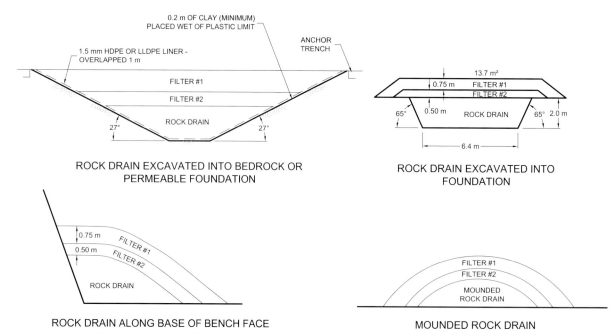

Figure 7.5: Schematic illustration of mounded drains and drains excavated into the foundation

filter fabric on the bottom and sides of the excavation to help prevent fines migration into the rock drain. Excavation sideslopes should be sloped at angles that will remain stable during construction. Where it is not warranted or possible to excavate rock drains into the foundation, drains can be mounded above the foundation surface or constructed along the toe of road cuts or benches (Fig. 7.5).

Where granular filters are required, the thickness of the granular filter should be sufficient to avoid short circuiting of the filter in the event that a portion of the filter material is scoured out by the impact of coarse rock blocks rolling down the face of the dump.

Where there is a concern for loss of seepage from rock drains into the underlying bedrock or foundation, it may be necessary to install an impermeable liner in the rock drain excavation before commencing drain construction, as illustrated in Fig. 7.5.

7.3.9 Long-term performance

7.3.9.1 Degradation of rock drain

Degradation is a particular concern where rock drains are constructed using sedimentary rocks, which may be prone to mechanical or chemical weathering, including slaking, stress relief, thermal expansion and contraction, freeze–thaw and dissolution. Drain rock materials should be hard and tough enough to resist crushing, degradation and disintegration resulting from production, transportation and placement. Drain rock must be resistant to breakdown and disintegration from weathering (wetting/drying and freezing/thawing) or they may break apart and cause

premature failure of the underdrainage system. Methods for assessing the durability of drain rock materials are the same as those described for assessing the durability of general rockfill materials in Chapter 5.

7.3.9.2 Sedimentation

Over time, sedimentation of both inlet ponds and rock drains can occur. Inlet ponds should be sized to provide adequate stilling and detention time to allow the majority of suspended sediments to fall out of suspension. Calculated flow volumes should include multipliers (i.e. factors of safety) to account for potential clogging of the drains over time by sediment and precipitates.

Where good quality material is scarce, there is a risk that poor quality, fine-grained materials could inadvertently be placed in direct contact with a rock drain. Alternatively, runout associated with local instability could result in fine-grained material being deposited directly on top of a drain. In addition, in the absence of a properly sized and graded filter zone, fine-grained material could pipe into the rock drains, causing plugging. Ensuring that only clean, durable, good quality material is placed directly on top of the drain can significantly mitigate the potential for piping and plugging of drains. Segregation by dumping good quality material over a minimum 20 m high repose angle face should result in a well-graded filter zone above the drain that should prevent the migration of fines. Alternatively, consideration could be given to placing a granular filter zone on top of constructed rock drains, including finger drains, as conceptually illustrated in Fig. 7.5 (two-stage filter zone shown). In general, filter fabrics should be avoided in this role as they tend to plug up with fines over time, reducing the efficiency of the underdrainage system.

7.3.10 Precipitates

The precipitates of elemental iron and sulphate, and iron oxide/hydroxide (from acid rock drainage processes), calcite or other minerals, may have the potential to reduce the flow capacity of rock drains. Geochemists should be consulted if there is concern regarding the potential for such precipitates.

7.3.11 Instrumentation

When sediment is deposited in an underdrainage system, the flow capacity will be reduced, resulting in higher than expected water levels. Ongoing measurement of water levels in the underdrainage system should be conducted to monitor drain performance and provide early warning of potential clogging. Water level measurements and climatic conditions should be recorded on a routine basis and during periods of high precipitation and runoff. By developing a database of climatic conditions and associated water levels in the underdrainage system, it should be possible to identify changes in underdrain flow capacity before they become a concern.

One of the simplest methods currently available to monitor the water level in an underdrainage system is using vibrating wire (VW) piezometers. A series of double (redundant) VW piezometers should be installed in the underdrainage system to monitor drain performance. For protection, VW transducers should be installed in perforated steel pipes at the base of each rock drain excavation, and cables should be protected by steel conduits buried in trenches excavated in the dump foundation. Cables should be terminated at a convenient location beyond the toe of the ultimate dump to facilitate long-term monitoring. Typical piezometer installation details are illustrated in Fig. 7.6. Transducer cables should be routed beyond the toe of the dump to a suitable monitoring point. For more detailed information regarding instrumentation used for water level monitoring, refer to Chapter 12.

Where it is necessary to measure water quality from time to time, standpipe piezometers may also be installed in the underdrainage system. Standpipe piezometers have the added benefit of allowing the injection of tracer dyes into the underdrainage system. Data loggers can also be installed in standpipes or connected to VW piezometers to allow continuous monitoring of flow depth. Where rock drains daylight at the toe of a dump, it may be possible to measure flow volume by using a weir.

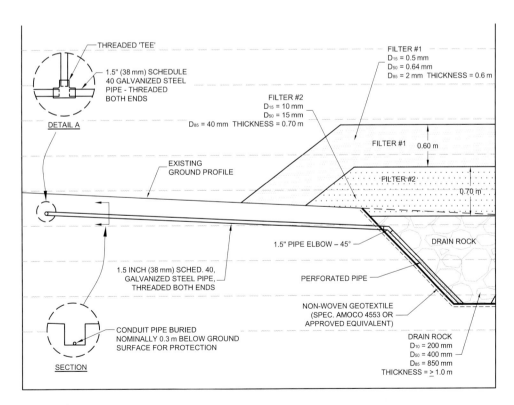

NOTE: WHERE NECESSARY TO TIE A SECOND PIEZOMETER CONDUIT INTO AN EXISTING CONDUIT USE THREADED "TEE "AS SHOWN IN DETAIL A.

Figure 7.6: Piezometer installation details

7.4 Other drainage elements

7.4.1 Drainage blankets

'Underdrainage' refers to the collection and conveyance of excess water (e.g. groundwater, meteoric water) or leachate through a network of interconnected drains or a drainage layer constructed in, or on, the foundation. By preventing the development of a phreatic surface within a waste dump or stockpile, static and seismic slope stability is improved.

Underdrainage systems should be designed with adequate capacity to remove all potential flows and must be constructed of free-draining materials. Gradients typically greater than 2% are recommended to avoid significant pooling in low spots created during construction.

Development of an effective underdrainage blanket may be accomplished through natural segregation of end dumped, good quality rock. As for rock drains, a minimum lift height of 20 m is generally considered necessary to yield adequate segregation. Where the dump is constructed using lift heights of less than 20 m, or from poor-quality material, and a underdrainage system is required, pre-construction of an underdrainage blanket or a series of French drains may be necessary.

7.4.2 French drains

A French drain is typically a closed-ended drain with an outlet but no inlet. The French drain is named after Henry Flagg French, who in 1859 published a book titled *Farm Drainage*, in which he described several drains composed

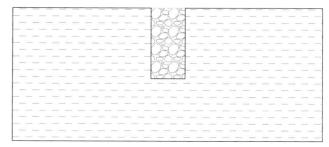

Figure 7.7: Schematic illustration of a French drain composed of a ditch filled with drain rock

Figure 7.8: View of a rock drain under construction showing drain rock and granular filters. Source: J. Hogarth

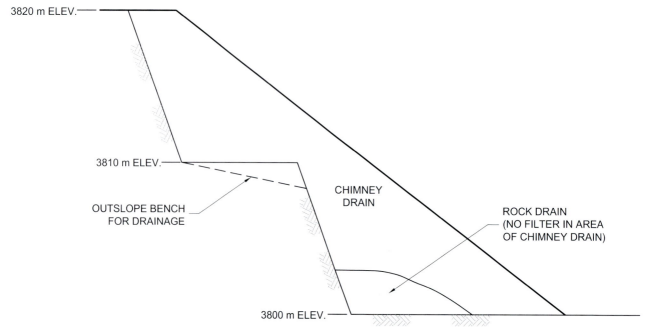

Figure 7.9: Schematic illustration of a chimney drain

3820 m ELEV.

3810 m ELEV.

OUTSLOPE BENCH FOR DRAINAGE

CHIMNEY DRAIN

ROCK DRAIN (NO FILTER IN AREA OF CHIMNEY DRAIN)

3800 m ELEV.

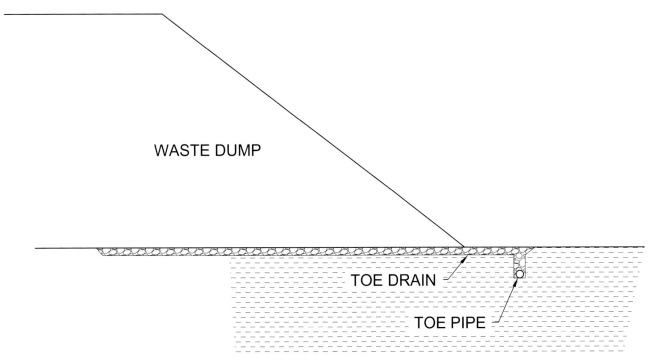

Figure 7.10: Schematic illustration of a toe drain with an integral drain pipe

of a ditch or channel filled with gravel, drain tiles or roofing tiles (Fig. 7.7).

Where the quantity of good quality waste rock is limited and natural segregation cannot be relied upon to yield a suitable rock drain underdrainage layer, pre-construction of a series of French drains may be necessary (Fig. 7.8). Secondary French drains can also be pre-constructed in strategic gullies and natural drainages and along existing roads and ditches, creating a dendritic pattern of subsurface drainage to convey groundwater into the main rock drains.

7.4.3   Chimney drains

Where the topography is too steep to construct rock drains (e.g. steep natural slopes, pit slopes) but continuity of underdrainage is required, it may be necessary to construct a chimney drain. Chimney drains are vertical or steeply inclined forms of rock drains (Fig. 7.9). Chimney drains can be formed from the top down by dumping drain rock in a repose angle slope, or from the bottom up by creating a void at the rear of a platform that can be backfilled with drain rock material as successive lifts are placed. It is important to note that it may be difficult to provide for adequate filtering of chimney drains, particularly where top down construction is employed.

7.4.4   Toe drains

A toe drain consists of a rock drain or French drain located beneath or immediately downstream of the embankment toe to collect seepage flows from the foundation and embankment and convey it to a suitable collection or discharge location.

A toe drain can be a simple, effective method of controlling or collecting seepage and can be used in conjunction with a drainage blanket to capture seepage flows (Fig. 7.10). To increase capacity, a perforated pipe can be added to the design.

8

STABILITY ANALYSIS

Mark Hawley, James Hogarth, John Cunning and Andy Haynes

8.1 Introduction

Analysis of the stability of a waste dump or stockpile is required to objectively demonstrate that the structure meets or exceeds the minimum stability acceptance criteria (Section 8.3) and to optimise the geometry and construction sequence. The level of detail required for the analyses should consider the stage of the project, the potential for instability and the level of risk. Chapter 3 provides detailed guidance on the potential for instability as represented by the waste dump and stockpile stability rating (WSR) and hazard class (WHC). Understanding of the dump or stockpile classification using the waste dump and stockpile stability rating and hazard class system can be helpful when planning the level of stability analysis.

8.2 Factors affecting stability

As detailed in Chapter 3, there are many factors that can affect the stability of mine waste dumps and stockpiles (Fig. 3.1). However, from a geotechnical modelling and stability analysis perspective, these can be reduced to the key input variables outlined below.

8.2.1 Foundation geometry

The geometry of the foundation is a necessary input to all stability analysis models. This information is easily derived from topographic plans or digital terrain models. Most modern stability analysis codes are able to import digital topography directly, which is especially convenient for complex 3D modelling. For 2D analyses, foundation profiles can also be imported directly or digitised from topographic maps or sections. Refer to Chapter 4 for a more detailed discussion of sources of foundation geometry and site physiography information.

8.2.2 Foundation conditions

To be able to develop a geotechnical model of the foundation for stability analysis purposes, an understanding of the stratigraphy of the foundation is required. Based on this information, the foundation may be divided into one or more geotechnical units. Depending on the analysis technique, each geotechnical unit would be assigned unique properties (e.g. unit weights, shear strengths, deformability parameters) derived directly from laboratory or *in situ* testing, indirectly from correlations to index properties, or based on literature values, results of back-analysis or judgement. Strength anisotropy and discrete geological structures may also need to be considered and accounted for either explicitly or implicitly in the model. Methods of characterisation for foundation material properties are presented in Chapter 5.

8.2.3 Waste dump and stockpile geometry and construction sequence

The geotechnical model must also incorporate the geometry of the waste dump or stockpile. Advanced analyses would also consider the development sequence, as the ultimate configuration may not be the most critical in terms of stability. As for topography, most analysis codes are capable of importing complex 3D waste dump and stockpile geometries directly from CAD or common mine modelling software platforms. For 2D analyses, geometries may also be imported directly or digitised from sections cut through the model.

8.2.4 Waste rock and stockpile material characteristics

The distribution and properties of the waste rock and stockpile materials must also be input into the analysis model. While stockpiles tend to be composed of more uniform materials, complex waste dumps may incorporate a wide variety of materials, each with unique properties. In such cases, both the spatial distribution and properties of the materials must be determined. The design may also specify the spatial distribution of the various materials to optimise stability and drainage and minimise environmental impacts, so it is important to understand the quality and release schedule of the materials. Material quality and release schedule are determined by the mine plan, but it may also be possible to modify the plan to meet the objectives of the waste dump designer. In some cases, it may be necessary to quarry materials if the mine cannot produce the required quantity and quality of material when it is needed. A good example of this is the Pierina Mine in Peru, where it was necessary to quarry competent andesite to construct a toe buttress and internal drainage to contain and underdrain large quantities of poor quality, clay altered materials (Hawley *et al.* 2002). Methods for determining key material properties required to support stability analyses (e.g. density/unit weight, strength, durability, chemical stability) are described in Chapter 5.

8.2.5 Surface and groundwater conditions

An understanding of the expected groundwater conditions, both within the waste dump or stockpile and in the underlying foundation, is key to developing a reliable geotechnical model. Pore pressures reduce the effective stress and the available shearing resistance of the materials that compose these structures and their foundations; thus, they have a direct impact on stability. In extreme cases, high pore pressures can result in liquefaction failure with potentially catastrophic consequences. Where the foundation is composed of saturated, fine-grained soils with low hydraulic conductivity, the potential for construction-induced pore pressures and undrained failure also needs to be considered. Methods for characterising groundwater conditions and managing pore pressures in foundations and within the waste dump or stockpile are described in Chapters 4 and 6, respectively. Groundwater instrumentation and monitoring systems are described in Chapter 12.

8.2.6 Seismicity

It is generally accepted within the geotechnical community that the design of certain types of embankments, such as water retention dams and tailings dams, must consider the potential impact of earthquakes. While there is less agreement among practitioners as to the impact of earthquakes on waste rock dumps and stockpiles, regulatory agencies in some jurisdictions require that seismicity be explicitly considered in the design of these facilities. In terms of stability analysis, the input parameters required to account for seismicity vary depending on the analysis technique. Simple pseudo-static analyses require determination of a seismic coefficient, which is estimated as a fraction of the maximum acceleration that the structure is expected to experience over the relevant design window (i.e. during its operating life or over the longer term following closure). Maximum

acceleration information is typically derived from regional or site-specific probabilistic or deterministic seismic hazard studies. Post-earthquake static analyses require an assessment of the impact that the design earthquake may have on the available shear strength of the embankment and foundation materials. In the case of sensitive materials, post-earthquake shear strengths may approach residual values. More complex dynamic analysis may require definition of the expected ground motion profile (response) during the operating or design earthquake event. Expected ground motion profiles may be derived by either scaling or scaling and modifying actual historical earthquake records or by generating entirely synthetic profiles.

8.3 Acceptance criteria

8.3.1 Historical evolution of stability acceptance criteria

Among the first organisations to propose minimum stability acceptance criteria for waste rock dumps was the US Mine Enforcement and Safety Administration (MESA). In its 1975 guidelines on the design of coal refuse dumps, MESA suggested a range of minimum factor of safety (FoS) values depending on the perceived level of hazard and certain key underlying assumptions of the stability analysis (Table 8.1).

In the 1977 *Pit Slope Manual*, the Canadian Centre for Mining and Metallurgy (CANMET) proposed minimum FoS guidelines for waste embankments (which included waste rock dumps) based on the assumed shear strength parameters and consequence of instability (CANMET 1977; Table 8.2).

Table 8.1: 1975 stability acceptance criteria

Assumptions	High hazard	Moderate hazard	Low hazard
Designs based on shear strength parameters measured in the laboratory	1.5	1.4	1.3
Designs that consider the maximum seismic acceleration at the site	1.2	1.1	1.0

Source: After MESA (1975)

Table 8.2: 1977 stability acceptance criteria

Assumptions	High consequence	Low consequence
Peak shear strength parameters	1.5	1.3
Residual shear strength parameters	1.3	1.2
100-year return period earthquake	1.2	1.1

Source: After CANMET (1977)

In 1982, the US Bureau of Mines (USBM) slightly modified the original MESA guidelines, which were intended for coal refuse dumps, expanded their application to include tailings dams and waste rock dumps, and published the recommendations presented in Table 8.3.

The first stability acceptance criteria developed specifically for mine waste rock dumps were published by the BC Mine Waste Rock Pile Research Committee (BCMWRPRC) in 1991 in its *Investigation and Design Manual – Interim Guidelines* (BCMWRPRC 1991a). These

Table 8.3: 1982 stability acceptance criteria

Assumptions	High hazard	Moderate hazard	Low hazard
Shear strength parameters from representative testing	1.5	1.4	1.3
Based on maximum expected seismic acceleration	1.2	1.2	1.1

Source: After USBM (1982)

Table 8.4: 1991 stability acceptance criteria

Stability condition	Suggested minimum design values for factor of safety[1]	
	Case A	Case B
Stability of spoil surface		
Short-term (during construction)	1.0	1.0
Long-term (reclamation – abandonment)	1.2	1.1
Overall stability (deep-seated stability)		
Short-term (static)	1.3–1.5	1.1–1.3
Long-term (static)	1.5	1.3
Pseudo-static (earthquake)[2]	1.1–1.3	1.0

Case A
- Low level of confidence in critical analysis parameters
- Possibly unconservative interpretation of conditions, assumptions
- Severe consequences of failure
- Simplified stability analyses method (chart, simplified method of slices)
- Stability analysis method poorly simulates physical conditions
- Poor understanding of potential failure mechanism(s)

Case B
- High level of confidence in critical analysis parameters
- Conservative interpretation of conditions, assumptions
- Minimal consequences of failure
- Rigorous stability analysis method
- Stability analysis method simulates physical conditions well
- High level of confidence in critical failure mechanism(s)

Source: BCMWRPRC (1991a)
Notes:
1. A range of suggested minimum design values are given to reflect the different levels of confidence in understanding site conditions, material parameters, consequences of instability and other factors.
2. Where pseudo-static analyses based on peak ground accelerations which have a 10% probability of exceedance in 50 years yield a FoS of less than 1.0, dynamic analysis of stress–strain response and comparison of results with stress–strain characteristics of dump materials is recommended.

minimum recommended FoS criteria, reproduced in Table 8.4, included consideration of the potential scale of instability (shallow sliver failures versus overall, deep-seated failure), design basis (short-term [construction/operations] versus long-term [closure]), confidence (in the input parameters and analysis method) and the consequence of instability.

8.3.2 Suggested stability acceptance criteria

Suggested stability acceptance criteria for the design of waste rock dumps and stockpiles are summarised in Table 8.5. These criteria incorporate many of the underlying concepts upon which the historical criteria described above were based. In line with current trends in other areas of practice in geotechnical engineering, and in accordance with the approach to defining stability acceptance criteria adopted in *Guidelines for Open Pit Slope Design* (Read and Stacey 2009), in addition to the classical deterministic criteria that define minimum acceptable values of FoS, complementary probabilistic criteria in the form of maximum allowable probabilities of failure have been introduced for static analyses. Further, in recognition of the unique behaviour of waste dumps and stockpiles under seismic loading, and the emerging application of numerical modelling in addition to more traditional methods of assessing the impact of earthquakes on these structures, new provisional criteria that suggest limits to the maximum allowable deformation of the waste dump under seismic loading and require convergence of numerical models have been introduced. For the purposes of these criteria, the amount of allowable deformation is represented by 'strain', which is defined herein as the cumulative shear displacement along a defined failure path due to a seismic event, divided by the length of the failure path, expressed as a percentage. Definitions of convergence vary depending on the practitioner and numerical modelling technique employed, but convergence is usually characterised by a numerical velocity threshold after a specified number of model time steps. Where acceptance is based on the results of a numerical model, the number of time steps and velocity threshold, or other convergence criteria, should be specified and defended.

Table 8.5 follows a similar format to Table 9.9 in Read and Stacey (2009), but substitutes 'Consequence' for 'Scale' and introduces some additional complexities to allow for consideration of 'Confidence' (or reliability) in the key input parameters and analytical technique used, and for limitations on allowable strain. 'Consequence' includes consideration of both the impact of potential instability, which is often related to the scale or size and mechanism of instability, and the service life of the structure. Inclusion of service life under 'Consequence' is intended to retain the sensitivity of the acceptance criteria to the design basis for the structure (i.e. short-term [construction/operations] versus long-term [closure]), a concept that was introduced in the 1991 *Investigation and Design Manual – Interim Guidelines* (BCMWRPRC 1991a). The notes to Table 8.5 provide guidance on the selection of consequence and confidence levels.

The relationship between FoS and probability of failure (PoF) is complex and depends on many factors, including the reliability and distribution of the analysis input parameters and the analytical technique used to derive the final probability distribution. For the purposes of this publication, PoF is considered to be related to both FoS and consequence according to the simplified relationships illustrated on the semi-log plot in Fig. 8.1. Consequence in this context is defined using the guidelines provided in the notes to Table 8.5 and relies upon the premise that the acceptable FoS should increase, and the tolerable PoF should decrease, as the consequence of instability increases.

There is very little published data on the relationship between FoS and PoF in a mining environment, and most of what is available is related to the stability of open pit slopes (e.g. Tapia *et al.* 2007; Jefferies *et al.* 2008; Read and

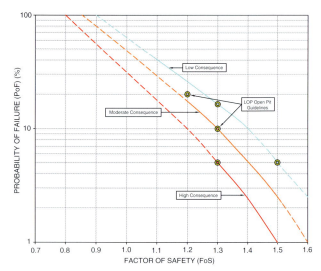

Figure 8.1: Factor of safety versus probability of failure

Stacey 2009). As a result, the upper- and lower-bound envelopes illustrated in Fig. 8.1 are based largely on the guidance provided by Read and Stacey (2009) and should be considered as provisional and subject to change as

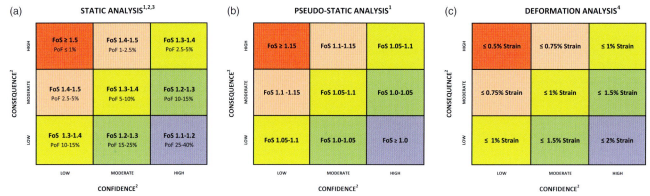

Figure 8.2: Acceptance criteria charts

Notes:

1. The primary stability acceptance criteria for both static and pseudo-static stability analyses are based on a minimum allowable FoS calculated using conventional limit equilibrium methods. Suggested minimum FoS values are illustrated in Figs. 8.2a and 8.2b, and depend on both the consequence (potential impact) of instability and confidence (reliability) in the input parameters and analysis model. In this context, the minimum allowable FoS increases with increasing consequence and decreasing confidence.

2. Guidance for selecting confidence and consequence is given in the notes for Table 8.5. In cases where the guidance conflicts or is unclear, selection of the appropriate level should be based on judgement, and the rationale for the selection should be documented.

3. In the case of static analysis, acceptance may also be characterised in terms of the maximum tolerable PoF. Suggested values for the maximum tolerable PoF for static analyses are also shown in Fig. 8.2a. The maximum tolerable PoF for static analysis depends on both FoS and consequence. The relationship between FoS, consequence and PoF that has been used to estimate the maximum tolerable PoF is illustrated in Fig. 8.1. For a constant FoS, the maximum tolerable PoF decreases with increasing consequence. The primary acceptance criteria should be based on FoS; PoF should be used as a supplementary criterion only.

4. Strain in this context is defined as the ratio of cumulative displacement along a defined failure path due to a seismic event to the length of the failure path, expressed as a percentage. Strain could be manifested by displacement of the crest (scarp development), bulging of the toe, or other signs of deformation. A conservative estimate of the potential displacement or deformation that may occur during a seismic event may be obtained using a variety of methods (e.g. Newmark 1965; Swaisgood 2003). For the purposes of these acceptance criteria, displacement or deformation estimates derived using these methods can be converted to an 'equivalent' strain by dividing the estimated cumulative displacement or deformation by the length of the critical failure path and expressing the result as a percentage. If the length of the critical failure path is unknown or not easily obtainable, the overall slope length or height of the embankment can be used as a conservative estimate for the length of the critical failure path. If numerical techniques are used, then convergence of the numerical model is necessary before the cumulative displacement can be calculated. A non-convergent numerical model would not meet the acceptance criteria.

Table 8.5: Suggested stability acceptance criteria

Consequence[1,3]	Confidence[2,3]	Static analysis		Pseudo-static analysis	Maximum allowable strain[5]
		Minimum FoS	Maximum PoF[4]	Minimum FoS	
Low	Low	1.3–1.4	10–15%	1.05–1.1	≤1%
	Moderate	1.2–1.3	15–25%	1.0–1.05	≤1.5%
	High	1.1–1.2	25–40%	1.0	≤2%
Moderate	Low	1.4–1.5	2.5–5%	1.1–1.15	≤0.75%
	Moderate	1.3–1.4	5–10%	1.05–1.1	≤1%
	High	1.2–1.3	10–15%	1.0–1.05	≤1.5%
High	Low	≥1.5	≤1%	1.15	≤0.5%
	Moderate	1.4–1.5	1–2.5%	1.1–1.15	≤0.75%
	High	1.3–1.4	2.5–5%	1.05–1.1	≤1%

Notes:

1. Consequence

Low – waste dumps and stockpiles with overall fill slopes less than 25° and less than 100 m high and repose angle slopes less than 50 m high. No critical infrastructure or unrestricted access within potential runout shadow. Limited potential for environmental impact. Long-term (more than 5 years) exposure for sites subject to very low to low (less than 350 mm) annual precipitation; medium-term (1–5 years) exposure for sites subject to moderate (350–1000 mm) annual precipitation; short-term (less than 1 year) exposure for sites subject to high (1000–2000 mm) annual precipitation; dry season construction/operation only for sites subject to very high (more than 2000 mm) annual precipitation or intensive rainy season(s).

Moderate – waste dumps with overall fill slopes less than 30° and less than 250 m high or repose angle slopes less than 100 m high. No critical infrastructure or unrestricted access, or robust containment/mitigative measures to protect critical infrastructure and access within potential runout shadow. Potential for moderate environmental impact, but manageable. Long-term (more than 5 years) exposure for sites subject to moderate (350–1000 mm) annual precipitation; medium-term (1–5 years) exposure for sites subject to high (1000–2000 mm) annual precipitation; short-term (less than 1 year) exposure for sites subject to very high (more than 2000 mm) annual precipitation or intensive rainy season(s).

High – waste dumps with overall fill slopes more than 30° and more than 250 m high, or with repose angle slopes more than 200 m high. Critical infrastructure or unrestricted access within potential runout shadow with limited runout mitigation/containment measures. Potential for high environmental impact that would be difficult to manage. Long-term exposure (more than 5 years) for sites subject to high (1000–2000 mm) annual precipitation; medium-term (1–5 years) exposure for sites subject to very high (more than 2000 m) annual precipitation or intensive rainy season(s).

2. Confidence

Low – limited confidence in foundation conditions, waste material properties, piezometric pressures, analysis technique or potential instability mechanism(s). Poorly defined or optimistic input parameters; high data variability. For proposed structures, investigations at the conceptual level with limited supporting data. For existing structures, poorly documented or unknown construction and operational history; lack of monitoring records; unknown or poor historical performance.

Moderate – moderate confidence in foundation conditions, waste material properties, piezometric pressures, analysis technique or potential failure mechanism(s). Input parameters adequately defined; moderate data variability. For proposed structures, investigations at the pre-feasibility study level with adequate supporting data. For existing structures, reasonably complete construction documentation and monitoring records; fair historical performance.

High – high confidence in foundation conditions, waste material properties, piezometric pressures, analysis technique and instability mechanism(s). Well-defined, conservative input parameters; low data variability. For proposed structures, investigations at the feasibility study level with comprehensive supporting data. For existing structures, well-documented construction and monitoring records and good historical performance.

3. In cases where the guidance for consequence or confidence conflicts or is unclear, selection of the appropriate level should be based on judgement, and the rationale for the selection should be documented.

4. For the purposes of these criteria, PoF is assumed to be related to FoS and consequence as illustrated in Fig. 8.1.

5. A simplified and conservative estimate of the potential displacement or deformation that may occur during a seismic event may be obtained using a variety of methods (e.g. Newmark 1965; Swaisgood 2003). For the purposes of these acceptance criteria, displacement or deformation estimates derived using these methods can be converted to an 'equivalent' strain by dividing the estimated cumulative displacement or deformation by the length of the critical failure path and expressing the result as a percentage. If the length of the critical failure path is unknown or not easily obtainable, the overall slope length or height of the embankment can be used as a conservative estimate for the length of the critical failure path. If numerical techniques are used, then convergence of the numerical model is necessary before the cumulative displacement can be calculated. A non-convergent numerical model would not meet the acceptance criteria.

experience with application of these criteria is gained. Furthermore, because PoF can vary widely depending on the reliability and statistical treatment of the input parameters and the analysis technique used, and there is currently no consensus among practitioners on its application, it is recommended that the primary acceptance criteria for static stability continue to be based on FoS, with PoF being used only as a supplementary criterion and to help assess the sensitivity of stability to changes in input parameters or slope geometry. For a more comprehensive discussion on the concepts of FoS and PoF, the reader is referred to Chapter 9 of *Guidelines for Open Pit Slope Design* (Read and Stacey 2009).

The suggested acceptance criteria are also illustrated graphically on the charts in Fig. 8.2. These charts represent the suggested acceptance criteria using a matrix approach that has similar attributes to a qualitative risk assessment. In this context, the product of consequence and confidence may be considered an indirect index of risk. A low consequence rating combined with a high confidence rating equates to a relatively lower overall risk and results in a lower required FoS, higher tolerable PoF and higher allowable strain. Conversely, a high consequence rating combined with a low confidence rating equates to a relatively higher overall risk and results in a higher required FoS, lower tolerable PoF and a lower allowable strain. The concept of risk assessment and risk-based design is discussed in more detail in Chapter 10.

It is important to emphasise that the suggested ranges of FoS summarised in Table 8.5 and illustrated in Fig. 8.2 represent *minimum* acceptable values. Designs that result in calculated FoS values that exceed the proposed minimum ranges would also be acceptable. Likewise, the suggested PoF ranges represent *maximum* tolerable values, and designs that result in lower calculated PoF values would also be acceptable. When coupling FoS with PoF within a defined range of values, the higher FoS value in the range should be associated with the lower PoF value in the range. For example, in the case of a static analysis with a moderate consequence and moderate confidence, the suggested *minimum* FoS acceptance criteria would be in the range 1.3–1.4, and the *maximum* allowable PoF would be in the range 5–10%. In this case, if a design minimum acceptable FoS of 1.3 were chosen, the applicable acceptable maximum PoF would be 10%, and if a design minimum acceptable FoS of 1.4 were chosen, the applicable acceptable maximum PoF would be 5%.

8.3.3 Application of stability acceptance criteria

Figure 8.3 illustrates the suggested sequence for applying the stability acceptance criteria presented in the preceding section. The first step is to conduct static stability analyses and compare the results to the static stability acceptance criteria that have been selected for the structure. If the results do not meet the minimum static acceptance criteria, an iterative step is required wherein the design is modified and additional static analyses are conducted. This iterative cycle repeats until a design that meets the static criteria is achieved. In the case of structures on sites where the seismic risk is low and an earthquake analysis is not considered necessary or mandated by regulators, if the static analyses results meet the minimum acceptance criteria, further analyses may not be needed to support the design.

Designs that pass the static test and must be assessed for earthquake stability are next subjected to either pseudo-static analysis or post-earthquake static analysis, or both. Pseudo-static analysis is conducted by applying a constant horizontal acceleration (and in some cases a constant vertical acceleration) to the waste dump or stockpile mass to represent the transient seismic accelerations that may be induced during the design earthquake. A static analysis that incorporates these accelerations is completed and a corresponding pseudo-static FoS is calculated. A more detailed discussion of pseudo-static analysis is given in Section 8.6.1.

While pseudo-static analysis has been in use for many decades and is the standard that is used by some regulators

to adjudicate waste dump designs, some practitioners prefer to use post-earthquake static analysis as the initial screening tool to assess the potential impact of an earthquake on dump stability. In this technique, the shear strength parameters that are applied to the dump foundation, and in some cases to the dump material, are reduced from their initial values to account for potential strain softening that may occur during the design earthquake. Strain softening may be a very important consideration in the case of sensitive foundation materials such as brittle clays with a high ratio of peak to residual strength. Selection of the appropriate shear strength to apply in a post-earthquake static analysis requires an understanding of the strain history and shear strength behaviour of the foundation materials, and an appropriately experienced geotechnical specialist should be consulted before embarking on this type of analysis.

If results of the pseudo-static and/or post-earthquake static analyses meet or exceed the minimum FoS acceptance criteria, the design is accepted and no further analyses are required. If the results do not meet the criteria, two alternative paths are possible: the design can be modified and subjected to another round of static and pseudo-static/post-earthquake analyses, or a simplified deformation analysis can be conducted to estimate the potential strain that may be induced by the design earthquake. This estimated strain is then compared to the maximum allowable strain criteria. Conservative estimates of the potential displacement or deformation that may occur during a seismic event may be obtained using a variety of simplified methods (e.g. Newmark 1965; Swaisgood 2003). For the purposes of these acceptance criteria, displacement or deformation estimates derived using these methods can be converted to an 'equivalent' strain by dividing the estimated cumulative displacement or deformation by the length of the critical failure path and expressing the result as a percentage. If the length of the critical failure path is unknown or not easily obtainable, the overall slope length or height of the embankment can be used as a conservative estimate for the length of the critical failure path. This 'equivalent' strain would be compared to the maximum allowable strain, and if it is less than the acceptance criteria, the design is accepted and no further analyses are required. Note that the simplified Newmark method calculates an expected displacement based on a horizontal yield acceleration for a FoS of 1.0, but can also be adapted to cases where the minimum required pseudo-static FoS is greater than 1.0 by redefining the horizontal yield acceleration based on the required minimum FoS.

If the estimated strain exceeds the maximum allowable, two alternative paths are possible: the design can be modified and subjected to another round of static, pseudo-static/post-earthquake and, if required, simplified

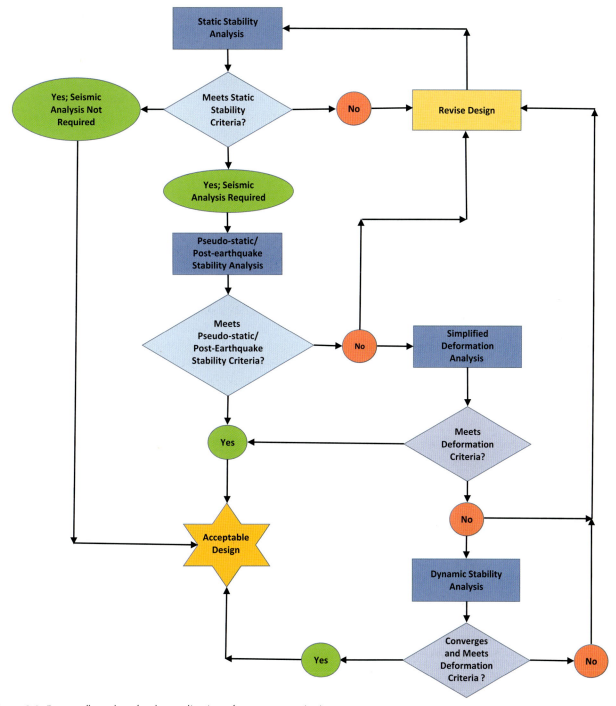

Figure 8.3: Process flow chart for the application of acceptance criteria

deformation analyses, or a dynamic analysis can be conducted to estimate the amount of cumulative deformation/strain using a variety of numerical modelling techniques. Note that convergence of the numerical model is necessary before the cumulative displacement can be calculated. A non-convergent numerical model would not meet the acceptance criteria. If the dynamic analysis converges and the cumulative strain meets the acceptance criteria, the design is accepted and no further analyses are required. If the analysis does not converge, the design must be modified and re-analysed beginning with a static analysis. As discussed in Section 8.3.2, definitions of convergence vary depending on the practitioner and numerical modelling technique employed. Where a numerical model is used, the number of time steps and velocity threshold, or other convergence criteria, should be specified and defended.

There may be special cases where the simplified pseudo-static/post-earthquake analysis approaches are insufficient to validate the design. Such cases may include waste dumps founded on potentially liquefiable soils or on saturated clays with low undrained shear strength. In these cases, dynamic analyses and/or static analyses that include liquefaction assessments and detailed assessment of post-earthquake shear strength behaviour may be required. Such analyses should be conducted by geotechnical specialists with experience in assessment and design of earthfill structures in areas subject to high earthquake risk.

8.4 Failure modes

Knowledge of basic modes of deformation and failure of typical angle of repose waste dumps and stockpiles is necessary to be able to select the appropriate analysis techniques and for designing appropriate dump monitoring programs. Various failure modes that have been described in literature (e.g. Pernichele and Kahle 1971; CANMET 1977; Caldwell and Moss 1981; Blight 1981; Campbell 1981, 1986, 2000; McCarter 1985; CANMET 1992), including key factors contributing to instability, are summarised in the following sections. Failure modes that are applicable to dragline spoils are discussed in Chapter 13.

8.4.1 Waste dump or stockpile material failure modes

8.4.1.1 Shallow

The three modes of shallow instability that are typically observed in waste dumps and stockpiles are edge or crest slumping, plane failure and flow failure (Fig. 8.4). These failures are typically shallow and do not involve the foundation materials. The hazard associated with these types of shallow instability is generally managed through operational procedures including restrictions on material placement zones, monitoring, buffer zones and impact berms that can be placed to contain or limit runout downslope of the dump toe.

Edge slumping, also referred to as crest slumping or sliver failure, is probably the most commonly observed mode of failure in large waste dumps and stockpiles. All repose angle dumps are potentially subject to development of sliver failures. This failure mode involves sliding of a thin wedge of material, usually originating at or near the crest, parallel to the dump or stockpile face. This type of failure generally results from over-steepening of the dump or stockpile face at the crest. Over-steepening may be due to the presence of fines or cohesive waste materials, or due to moisture. Failure commonly occurs when heavy precipitation relieves negative pore pressure in the fines, resulting in a loss of apparent cohesion. Edge slumping may also occur where slaking of dump materials creates a low permeability layer on the dump face, permitting development of high pore pressures at a shallow depth.

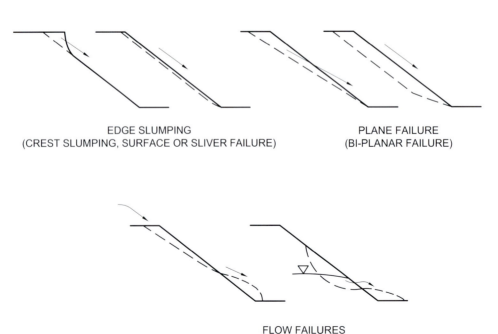

EDGE SLUMPING
(CREST SLUMPING, SURFACE OR SLIVER FAILURE)

PLANE FAILURE
(BI-PLANAR FAILURE)

FLOW FAILURES
(DEBRIS FLOW, MUD FLOW, FLOW SLIDES)

Figure 8.4: Shallow waste dump and stockpile failure modes

Heavy precipitation may also trigger failure in this case. Over-steepening of the dump/stockpile crest may also occur in coarse rockfill slopes. Interlocking of rock blocks may result in overly steep repose angle slopes at low confining stress. Subsequent slope creep, dynamic disturbances or stress changes may result in failure of the interlock, resulting in edge slumping.

Edge slumping commonly results in loss of the dump crest area and is most likely to occur in dumps constructed by end dumping in thick lifts, or where dump or stockpile material contains abundant fines or is subject to ongoing degradation (i.e. slaking). Push dumping, where dozers are used to push material over the crest rather than end dumping directly over the crest, and rapid rates of crest advancement also tend to promote over-steeping of the crest. Analysis of this mechanism is typically not beneficial, as the FoS of angle of repose slopes is unity by definition. These failures typically mobilise relatively small volumes of material and are managed by operational procedures and monitoring.

Plane failure involves sliding along a single plane of weakness within the waste dump and is a common mode of failure of dragline spoils (Rizkalla 1983). Where the plane of weakness does not daylight on the waste dump face or at the toe, some shearing through the dump or stockpile material at the toe may occur in a bi-planar failure mode. Weakness planes may be created parallel to the waste dump face if poor quality or fine materials, such as overburden or altered waste rock, are dumped over the crest and form a zone or layer parallel to the face. Weakness planes may also form where material is dumped over accumulations of ice or snow on the face, or if a zone of susceptible dump materials slakes or degrades due to exposure or shear strain within the dump. High pore pressures within the waste dump may also contribute to plane failure. In the case of a plane of weakness parallel to the waste dump face, plane failure is similar to edge slumping, except that the failure surface tends to be deeper within the waste dump and failure can result in substantially more breakback than typically observed due to edge slumping. Specific analysis methods for simple plane failure (i.e. rigid block sliding on a plane) are applicable. Where failure involves shearing through the toe, more complex analysis methods (i.e. bi-planar analysis) are required.

Flow type failures from the waste dump face can typically be shallow in nature and can be characterised as debris flows, mud flows or flow slides. These failures generally involve shallow slumping and subsequent fluidisation of saturated or partially saturated material. The volume and velocity of the flow may increase downslope as a result of lateral confinement, increasing momentum, and erosion and entrainment of underlying material. Flows may develop in response to saturation of the dump due to high precipitation and infiltration, development of perched water tables, and/or concentration of runoff on the dump. In general, the potential for flow type failures is higher for low density, loose fills composed of fine-grained materials and lower for very dense, consolidated fills with few fines. Infinite slope analysis, with consideration of seepage forces, is the traditional approach for assessment of flow failures.

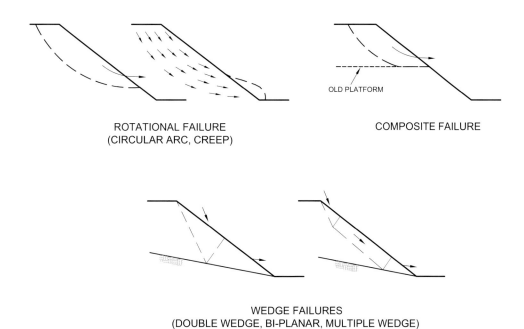

ROTATIONAL FAILURE
(CIRCULAR ARC, CREEP)

COMPOSITE FAILURE

OLD PLATFORM

WEDGE FAILURES
(DOUBLE WEDGE, BI-PLANAR, MULTIPLE WEDGE)

Figure 8.5: Deep-seated waste dump and stockpile failure modes

8.4.1.2 Deep-seated

Deep-seated failure modes that may be observed in waste dumps or stockpiles typically include rotational failures, composite failures and wedge type failures (Fig. 8.5).

Rotational failure of the waste dump involves mass failure along a circular or ellipsoidal/curvilinear failure surface formed wholly within the embankment. Creep failure is a special case of rotational failure, involving widespread rotational shearing through the waste dump, without the movement being focused along a single failure surface. Creep failure commonly manifests itself by long-term, progressive bulging at the toe of the waste dump. Rotational failures are commonly associated with homogeneous waste dumps composed of weak or fine-grained materials. In cohesive materials (e.g. overburden, highly altered rock), they may be precipitated by constructing the overall waste dump, or individual lifts, too high or too steep. Rotational failures may also be triggered by high pore pressures in the waste dump. A wide variety of proven methods are available for analysing rotational failures, including slip circle analysis and various methods of slices.

A composite failure surface is similar to rotational failure, except that a portion of the failure surface is circular and a portion is planar. The planar portion may occur along the interface between the waste dump or stockpile and the foundation (i.e. base failure), or along a plane of weakness within the foundation. This type of failure may also occur along weak zones, such as weakness planes formed by layers of finer-grained material or where material is dumped over accumulations of ice or snow.

Composite failure is analysed using various methods of slices for general failure surfaces.

Shear strength of the waste rock or stockpile material is typically higher than in the foundation materials, and wedge failure may occur when the slope of the foundation at the dump toe is steep (typically more than 25°) or foundation pore pressures are not allowed to dissipate in response to the rate of loading from the dumping. Wedge failure involves a mass failure of the waste dump material through a series of interactive blocks or wedges separated by planar discontinuities. The failure includes an active or driving portion that is in a state of limit equilibrium, and in which the material shear strength is fully mobilised, and a passive or resisting portion in the toe. Wedge failures may occur in several ways, depending on the number and configuration of blocks involved (e.g. sliding wedge, double wedge, bi-planar, multiple wedge). Conditions necessary for the development of wedge failure are similar to those required for composite failure. Wedge failure may be analysed using general methods of slices as well as specific wedge analysis techniques.

8.4.2 Foundation failures

8.4.2.1 Overburden soils

Where a waste dump or stockpile is founded on overburden soils, there is a possibility that a failure surface may extend into the foundation. Where foundation soils are of significant thickness and are weaker than the materials composing the dump, or where high pore pressures exist within the foundation, rotational foundation failures may occur (Fig. 8.6). A wide variety

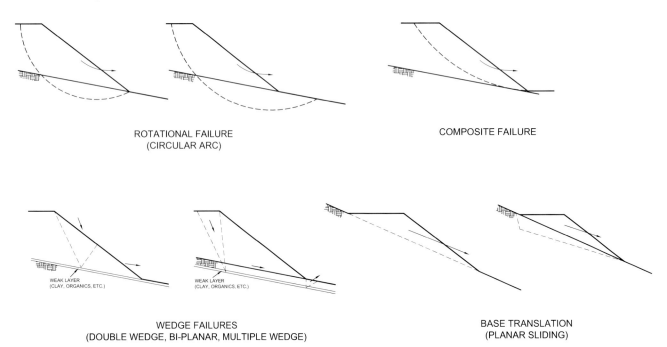

ROTATIONAL FAILURE
(CIRCULAR ARC)

COMPOSITE FAILURE

WEDGE FAILURES
(DOUBLE WEDGE, BI-PLANAR, MULTIPLE WEDGE)

BASE TRANSLATION
(PLANAR SLIDING)

Figure 8.6: Waste dump and stockpile foundation failure modes

of proven methods are available for analysing rotational failures, including free-body slip circle analysis and various methods of slices.

Where foundation soils are relatively thin or the waste dump is founded on a steep slope composed of a thin soil veneer over bedrock, or high pore pressures are present in the foundation, discrete weakness planes may form and composite or wedge type failures may occur (Fig. 8.6). These failure mechanisms have been classified by Caldwell and Moss (1981) as a type of foundation spreading. Similar to failure modes discussed above, failure modes that may propagate through overburden soils in the foundation may be analysed using the general methods of slices and specific wedge analysis techniques. Where waste dumps are founded on sloping terrain underlain by weak foundation soils, mass movement of the entire dump as a rigid block may also occur (Fig. 8.6). This mode of failure is referred to as base translation and is typically assessed using plane failure analysis.

8.4.2.2   Rock mass

Where a waste dump is underlain by a weak, very poor to poor quality rock mass, and shear stresses at the base of the waste dump are high, deep-seated rotational type failure through the rock mass could develop. Accordingly, when analysing deep-seated failure, an assessment of the rock mass strength is usually required. As noted above, slip circle analysis and various methods of slices can be used to assess rotational failures.

8.4.2.3   Structural weakness

Where adversely oriented planes of weakness (e.g. cross-joints, bedding planes, faults, clay layers) exist within the foundation rock mass, base translation failures may occur. Composite or wedge failures associated with structural weakness in the foundation rock mass could also occur. The potential for these types of failures may be exacerbated by high pore pressures in the foundation. An assessment of the bedrock geology and structure beneath any proposed waste dump or stockpile should be conducted to evaluate the orientations and continuity of any faults or other zones of weakness within the bedrock foundation that might adversely affect stability. Failure modes that may exploit structural weakness in the foundation may be analysed using the general methods of slices and wedge analysis techniques.

8.4.2.4   Toe loading

Toe loading failure, also known as toe or foundation spreading, involves localised slumping of the dump toe (Fig. 8.7) due to yielding or failure of foundation soils or loss of confinement and is the most common form of foundation failure (McCarter 1985). Toe failure may occur where local foundation soils are weak, where foundation slopes are locally steep, or where high pore pressures exist in the foundation, including localised liquefaction. Where weak foundation soils are particularly susceptible to strain, toe failure may result in rapid progressive failure of the overall dump. Toe failure may be recognised by bulging of the toe and disruption of foundation soils downslope, beyond the toe. While toe failures may be difficult to assess, recognising their potential to occur and making allowances in the design is the key to success. Similar methods to those used for assessing rotational failure can also be applied in analysing toe failure.

8.4.3   Liquefaction

If liquefaction of foundation soils or a discrete soil stratum in the foundation (or within the waste dump) occurs, the entire dump may be translated *en masse* or progressive failure may occur. For example, seismically induced liquefaction of natural sand lenses has been identified as a contributing factor to the Turnagin Heights landslide near Anchorage, Alaska, which occurred during the 27 March 1964 earthquake with an epicentre located 125 km east of Anchorage (National Academy of Sciences 1968–1973). In the mining industry, liquefaction phenomena are more usually associated with tailings dams. No cases of liquefaction of the foundation for a waste dump, stockpile or non-impounding dump, or dumps related to leaching operations (which are not considered herein) are noted in the literature. However, there may be cases where, given

TOE LOADING FAILURE
(TOE SPREADING)

Figure 8.7: Toe loading/foundation spreading failure modes

the right set of circumstances, liquefaction failure could also occur in a waste dump or stockpile foundation; consequently, the potential for liquefaction failure of a waste dump or stockpile cannot be completely discounted and needs to be considered in the design process.

Liquefaction can occur under both static and dynamic conditions. In both cases, saturation (or near-saturation) and generation of excess pore pressures under undrained conditions are required. Excess pore pressures can be generated by dynamic forces such as earthquake shaking (i.e. dynamic liquefaction) or by collapse of the soil structure due to strain softening (i.e. static liquefaction). Soils that are most susceptible to liquefaction are saturated, loose to medium density, non-cohesive granular soils. Historically, liquefaction was thought to be limited to loose sands and silts, and coarser grained soils, such as gravel, were considered to be too permeable to sustain generated pore pressures long enough for liquefaction to occur. However, recent experience tends to support broadening of the gradation envelope for liquefaction-susceptible soils to include non-plastic silts and clays to coarse gravel. According to CANMET (1992), laboratory triaxial testing results suggest that even coarse, well-graded sandy gravels are susceptible to this type of behaviour at relatively low void ratios.

Liquefaction can occur in a saturated granular material that derives all or most of its shear strength from friction (i.e. non-cohesive soils) if the pore pressure in the voids approaches the confining stress. When this occurs, the effective stress approaches zero and the available shear strength becomes very small. One characteristic of liquefaction-type failures is that, due to the low frictional strength and associated excess pore pressures, the failure mass is typically very mobile and runout distances can be substantial.

Liquefaction susceptibility is influenced by the gradation of the soil. Well-graded soils are generally less susceptible to liquefaction than uniformly graded soils because the voids between the coarse particles in a well-graded soil are more likely to be filled with fines. Well-graded soils are typically subject to less volume change, and hence lower pore pressure generation, when subject to shear loading in undrained conditions. Particle shape also influences susceptibility to liquefaction. Soils with rounded particles densify more easily than soils composed of angular particles; hence, soils with rounded particles are more susceptible to liquefaction. Soils that contain an appreciable component of plastic, clay-sized particles are less susceptible to liquefaction than those that are clean or contain non-plastic fines.

In recent years, several groups have studied the liquefaction phenomenon. Evaluation of liquefaction potential due to dynamically generated pore pressures (i.e. earthquake induced) is typically based on empirical criteria. In 2003 a paper on the relationship between *in situ* density, based on standard penetration testing results, and liquefaction potential was published (Seed *et al*. 2003). More recently, papers relating cone penetration testing results and liquefaction potential have been published (Cetin *et al*. 2004; Moss *et al*. 2006). Assessments can also be carried out using shear-wave velocity for seismic soil liquefaction potential (Kayen *et al*. 2013). Liquefaction potential due to rapid dump loading may be based on an assessment of pore pressure generation and dissipation rates determined through laboratory testing, field testing, field monitoring of test fills, or monitoring performance of trial dumps.

While significant progress has been made recently in understanding the liquefaction phenomenon and in developing procedures for predicting its occurrence, due to the complex nature of the phenomenon, engineering procedures, including analytical modelling and laboratory testing, have not been developed to the point where they are generally agreed upon or can be applied with confidence (Finn *et al*. 2010). If potentially liquefiable materials are thought to be present either in the foundation or within the waste dump, a geotechnical specialist with experience in the evaluation of liquefaction stability should be consulted.

8.5 Static limit equilibrium analysis

Limit equilibrium analysis is the most widely used technique to calculate the FoS of soil and rockfill embankments, including waste rock dumps and ore stockpiles (Eberhardt 2006). Numerous limit equilibrium analysis procedures have been developed based on solving one or more of the three conditions of static equilibrium (i.e. moment equilibrium, vertical force equilibrium and horizontal force equilibrium). Some of these procedures solve all three equilibrium conditions explicitly, while other, less rigorous methods only solve for some of the equilibrium conditions while making assumptions about the other equilibrium conditions to calculate a solution. The principal differences between these procedures are the conditions of static equilibrium they satisfy implicitly and/ or explicitly and the assumptions used by each to obtain a statically determinate solution. However, they all assume that the ratio of the forces resisting instability to the forces driving instability is the same everywhere along the failure surface.

The selection of the most appropriate static limit equilibrium analysis technique depends on several factors. These include the following:

- dump geometry, including overall height, lift height, lift face angle and overall slope face angle
- foundation conditions

- location, orientation and geometry of potential or existing failure surfaces
- distribution of different material types in the dump and the foundation
- drainage conditions in the dump and foundation materials
- design earthquake loading.

It is important to remember that static limit equilibrium tools are based solely on equations of statics and do not incorporate the effects of strain or displacement. To assess stress–strain relationships, a finite element or other numerical analysis is required.

8.5.1 Infinite slope analysis

Infinite slope analysis, also known as plane translational slip analysis, is a limit equilibrium analysis method that assumes a planar failure surface at a shallow depth, parallel to the slope face (Fig. 8.8), such as where a thin layer of overburden soil or weathered bedrock overlies competent bedrock. Infinite slope analysis approximates near-surface stability only, and is strictly valid only where the thickness of the failure is negligible in comparison to the length of

the failure plane. Application is normally restricted to homogeneous slopes.

The FoS is based on vertical and horizontal force equilibrium of a unit width (vertical) slice of the failure mass. Moment equilibrium is implicitly satisfied. Infinite slope analysis is analytically simple, easy to use and amenable to hand calculations. Seepage forces can be accounted for and chart solutions are available (Abramson et al. 2002).

8.5.2 Plane failure analysis

In plane failure analysis, the FoS is based on force equilibrium of an entire block (Fig. 8.9), which is assumed to be rigid, and moment equilibrium is implicitly satisfied. The effects of various piezometric surfaces and tension cracks can also be simulated easily. The analysis is analytically simple and amenable to hand calculations or computer solution (Abramson et al. 2002).

8.5.3 Wedge failure analysis

Wedge failure analysis assumes that the failure mass is divided into two or more interactive blocks or wedges (Fig. 8.10). The waste rock in the upper block or wedge is

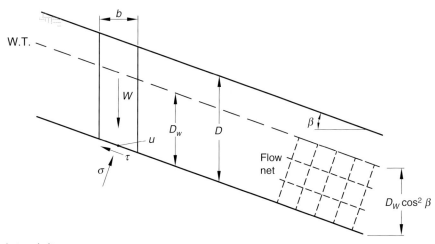

Figure 8.8: Plane translational slip

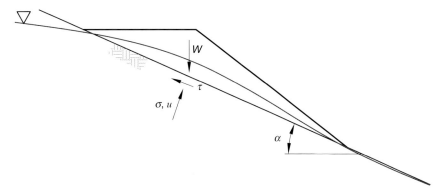

Figure 8.9: Plane failure analysis

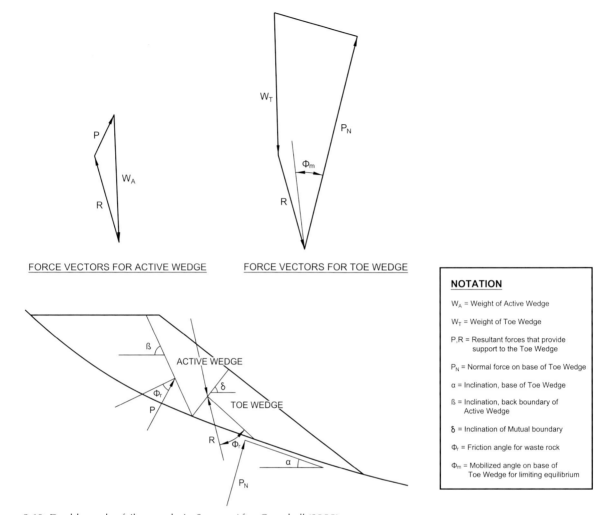

Figure 8.10: Double wedge failure analysis. Source: After Campbell (2000)

considered to be in the active Rankine state of stress, the shear strength of rockfill material at the edges is fully mobilised and the limiting equilibrium FoS of the active wedge remains at unity. Stability of the perimeter of the dump structure is dependent on the ability of the lower block or toe wedge to provide support to the active wedge. As the FoS of the active wedge is unity, the FoS of the toe wedge must remain greater than unity for the dump to remain stable. Figure 8.10 presents a summary of the methods of stability analysis for the double wedge mechanism (Campbell 2000).

The double wedge analysis technique illustrated in Fig. 8.10 (Campbell 2000) considers the stability of the driving wedge or active wedge and that of the passive wedge or toe wedge separately. The active wedge is assumed to be in a state of limiting equilibrium, and the forces R and P are calculated for a trial set of wedge boundaries. The value of R is then applied to the passive wedge, and the value of ϕ_m corresponding to limiting equilibrium of the passive wedge is calculated. Using an iterative process, the

maximum value of ϕ_m corresponding to limiting equilibrium of the passive wedge is determined. The FoS is calculated as the ratio of the tangent of the friction angle available in the foundation soil to the tangent of the friction angle mobilised.

Wedge failure analysis appears to model some types of dump behaviour very well, especially where dumps are located on steep slopes with a soil foundation. The models are analytically simple and amenable to hand calculations or computer methods. Excel spreadsheet solutions are available for some models, making parametric studies relatively easy. In addition, the analysis may be generalised to three dimensions (Golder Associates Ltd 1987). Some wedge analysis formulations consider both force and moment equilibrium, while others only consider force equilibrium in one or two directions (vertical, horizontal).

In common with methods of slices, wedge analyses require some assumptions regarding inter-block forces to render them statically determinate. Also, the FoS calculated by these methods is based on a relatively small

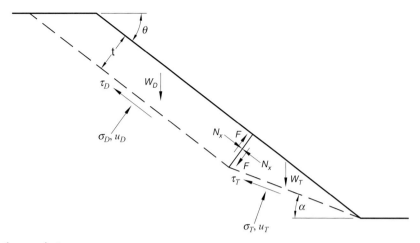

Figure 8.11: Bi-planar failure analysis

portion of the failure surface, unlike most limit equilibrium methods, which base the FoS on the shearing resistance of the overall plane. Consequently, caution must be exercised when interpreting results or comparing them to results of other analysis methods.

8.5.4 Bi-planar failure analysis

Similar to wedge failure analysis, the FoS determined using the bi-planar analysis, also referred to as bi-linear slab analysis, is based on force equilibrium of two blocks (Fig. 8.11), which are assumed to be rigid, and moment equilibrium is implicitly satisfied. The block or wedge forming the toe of the slope supports the upper block. The equations of equilibrium are used to resolve the net driving force imparted to the toe block by the upper active or sliding blocks. The FoS is then based on the resistance of the toe block to sliding along its base. The effects of various piezometric surfaces and tension cracks can also be simulated easily. The analysis is analytically simple and amenable to hand calculations or computer solution (Abramson *et al.* 2002).

As with other limit equilibrium analyses, this model provides no insight into the internal deformations of the

failure mass, which are commonly observed in conjunction with overall failure.

8.5.5 Methods of slices

Methods of slices are the most commonly used limit equilibrium techniques, whereby the failure mass, defined by a general slip surface, is typically divided into several slices (Fig. 8.12). Equilibrium conditions for each slice are considered, allowing for the variation of soil and groundwater conditions throughout the slope. With the exception of the technique developed by Sarma (1973), these methods are restricted to simulating the failure mass using vertical slice boundaries. The equilibrium of each slice is considered individually, and a FoS is generally based on the sum of the shear stresses and available shear strength on the base of each slice. When analysing deep-seated bi-planar or multi-wedge style failure mechanisms, the Sarma method is capable of modelling the observed slice boundary orientations and is able to assign strengths discreetly or automatically, based on materials intersected. Refer to Chapter 13 for a more detailed discussion of the Sarma method in the context of dragline spoil instability.

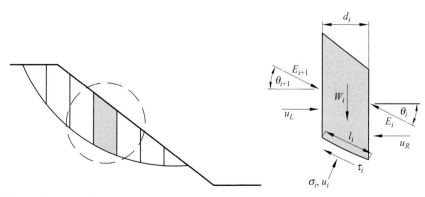

Figure 8.12: Method of slices schematic diagram

Methods of slices can simulate a wide variety of failure surface shapes, dump configurations and non-homogeneous slopes. They are a widely used, understood and accepted approach to engineering analysis of slope stability. Methods of slices also provide a more rational assessment of the normal stresses along the failure surface than more simplistic methods, such as slip circle analysis, and provide some insight into internal deformations within the failure mass. Several of the methods discussed below can also be applied to failure surfaces of almost any shape, allowing them to be applied to a much wider range of problems.

Some assumptions regarding internal stress distributions are required to render the analyses statically determinate. The various procedures differ primarily with respect to the assumptions that are made to yield a statically determinate solution. As a result, some procedures yield more accurate results than others. However, as discussed below, there is a good understanding of the differences among the various procedures and the effect on the FoS values that are obtained by the different procedures.

A variety of easy to use computer programs (e.g. Slide [RocScience 2015], SVSlope [Soilvision 2009], SLOPE/W [Geo-Slope 2012c], Slope [Oasys 2014], FLAC/SLOPE [Itasca 2011] and Galena [Clover Technology 2015]) are commercially available for many of the more common methods of slices. In spite of this, problems with convergence of iterative solutions can sometimes limit their usefulness, and it is very important that the user has knowledge of these potential problems.

For the purposes of this discussion, the methods of slices have been divided into two basic groups: simplified methods and rigorous methods.

8.5.5.1 Simplified methods

Simplified methods, including the ordinary method of slices (Fellenius 1936), Bishop's simplified method (Bishop 1955), Janbu's simplified method (Janbu *et al.* 1956) and various force equilibrium methods, do not satisfy all conditions of static equilibrium. The ordinary method of slices neglects inter-slice forces, allowing the direct calculation of a unique FoS. Bishop's simplified method neglects inter-slice shear forces, and requires iteration to solve for FoS. Force equilibrium methods, such as Janbu's simplified method, neglect moments and also require iteration to reach a result.

In comparison to rigorous methods, simplified methods are relatively quick and easy to use, and can be solved using hand calculations, though these tend to be tedious and not very efficient. Simplified methods are commonly accurate to within 15% of the rigorous methods, which in many cases may be all the accuracy that is needed. Simplified methods are also useful for preliminary assessments of stability and sensitivity, and as

a check on the validity of results from more detailed calculations.

In some cases, such as with the ordinary method of slices, when pore pressures are high or slopes are flat, results tend to be overly conservative. In addition, the ordinary method of slices is suitable only for total stress analysis. Also, convergence problems may occur with Bishop's simplified method where failure surfaces are very steep.

8.5.5.2 Rigorous methods

Rigorous methods, such as Janbu's generalised procedure (Janbu *et al.* 1956), Spencer (1967), Morgenstern and Price (1965) and Sarma (1973), satisfy all conditions of static equilibrium. The general limit equilibrium (GLE) method (Fredlund and Krahn 1977; Fredlund *et al.* 1981) encompasses the key elements of several methods of slices and is based on the solution of two equations, one based on moment equilibrium and one based on horizontal force equilibrium. The GLE method also incorporates a range of inter-slice shear-normal force conditions, making it the most rigorous of all the methods, satisfying both force and moment equilibrium for both circular and non-circular failure surfaces (Krahn 2003). All of these methods require iteration to reach a solution, and FoS values calculated by the various methods are usually within ~5% of each other.

Rigorous methods are more complex and difficult to use than simplified methods. Accordingly, selection of modelling parameters and interpretation of results require a detailed understanding of the analytical basis and assumptions each method employs.

Because they satisfy all conditions of static limit equilibrium, rigorous methods implicitly provide more realistic models of the physical mechanics of failure than do simplified methods. Available computer programs allow detailed parametric studies, and a wide variety of internal stress conditions can be modelled through informed selection of analysis parameters.

8.5.6 Compound or complex failures

The development of a waste dump or stockpile typically involves advancing lifts of varying thickness over variable terrain that may be composed of a range of foundation conditions. In these cases, it is not uncommon for stability analyses to be controlled by a complex failure surface geometry. Steep foundation slopes and/or lack of confinement were considered contributory factors in several of the failures reported in Appendix 1, while poor foundation conditions were cited as the most frequent cause of instability in many of the failures listed in Appendix 1.

As noted previously, 2D limit equilibrium analyses are based on a static model, so the waste dump geometry and the distribution of different material types in the waste dump and foundation does not vary laterally. Accordingly, while it may be possible to develop a model that incorporates a worst case combination of slope geometry and foundation conditions in two dimensions, the analysis results may still not be adequately representative of the actual FoS due to the effects of 3D confinement or lateral variation of materials in the waste dump or the foundation.

Where the waste dump geometry, failure surface geometry, or the distribution of materials in the waste dump and foundation are complex, it may be necessary to consider a 3D analysis. The FoS value determined by 2D limit equilibrium analysis is generally conservative when compared to the results of a 3D analysis. According to Gitrana *et al.* (2008), 3D factors of safety can be as much as 30% higher than the 2D FoS values. Given the size of many waste dumps/stockpiles, the use of a 3D analysis to optimise designs may increase storage capacity in a given area by allowing the use of steeper inter-ramp or overall slope angles. The volume increase resulting from the use of steeper slope angles may also reduce haul distances, increasing the favourable economics of a given design.

Several of the simplified and rigorous analysis techniques discussed in Section 8.5.5, including Bishop, Morgenstern-Price, GLE and Janbu, can be extended to analysis in three dimensions, resulting in what is referred to as the method of columns. Whereas 2D analysis does not consider inter-slice forces perpendicular to the plane of the failure surface, these lateral forces are considered in 3D, yielding a much more realistic model. There are a few commercially available 3D limit equilibrium programs that employ a method of columns approach, the most widely used of these being SVSlope 3D (Soilvision Systems 2009) and Clara-W (O. Hungr Geotechnical Research 2009).

Where waste dump failures may involve more complex mechanisms (e.g. internal deformation, strain hardening or strain softening [elastic-plastic], brittle fracture, progressive failure, liquefaction), limit equilibrium methods may be inadequate to accurately assess the FoS and more sophisticated numerical methods should be used.

8.5.7   Probability of failure

Typically, there is variability or uncertainty in the shear strength data determined from *in situ* or laboratory testing. This variability can be due to sample disturbance, particularly where undisturbed samples are required for testing, or the natural variability within the materials themselves. Where this variability can be described by a probability distribution function and the mean and standard deviation of the shear strength is known or can be reliably estimated, probabilistic analyses can be used to assess this variability and calculate the PoF. The probability density function describes the probability that

the value of a random variable will assume a specific value. For the assessment of waste dumps and stockpiles, the most commonly used probability density function is the normal distribution, although, depending on the dataset, other probability density functions may be more representative (e.g. log normal, bi-normal).

The PoF is defined as the number of analyses with a FoS less than 1.0 divided by the total number of FoS results. Most commercially available software for analysis of slope stability now has the capability to carry out probabilistic analyses. This allows the user to determine the impact of the variability or uncertainty in input parameters on the FoS, and allows the calculation of a PoF and determination of the failure path with the greatest PoF (i.e. critical failure path).

It is important to note that many factors can impact the shear strength of materials, including the following:

- strain-softening behaviour
- scale effects in laboratory testing
- rate of shearing
- *in situ* stress conditions
- anisotropy of natural materials
- simplifying assumptions
- groundwater assumptions.

Accordingly, selection of modelling parameters and interpretation of results requires a detailed understanding of the limitations of the investigation methods employed to obtain samples for testing and the uncertainties that affect each testing method.

8.6   Seismic stability analysis

This section presents seismic stability analyses, which include pseudo-static, pseudo-dynamic and dynamic stability analysis methods. As illustrated in the stability analysis flow chart in Fig. 8.3, seismic stability analysis may be required to support the design of a waste dump or stockpile.

Pseudo-static and pseudo-dynamic methods are considered appropriate for early design stages (conceptual to pre-feasibility) and for design validation of waste dumps or stockpiles with moderate to high overall stability ratings (WSR higher than 40; WHC I, II and III). For high or very high hazard waste dumps or stockpiles (WSR less than 40; WHC IV or V), supplemental assessments of stability under seismic loading, possibly including full dynamic analysis and/or static analyses that include liquefaction assessments and detailed assessment of post-earthquake shear strength behaviour, may be required at more advanced stages of design. Such analyses should be conducted by geotechnical specialists with experience in assessment and design of earthfill structures in areas subject to high earthquake hazards.

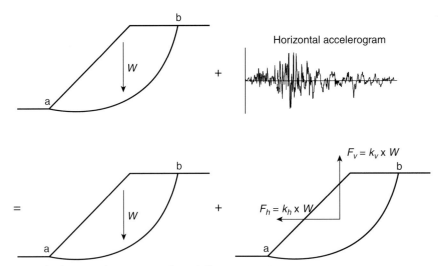

Figure 8.13: Pseudo-static method approach. Source: Melo and Sharma (2004)

8.6.1 Pseudo-static analysis

The pseudo-static stability analysis method is based on the limit equilibrium method and can be used to evaluate the potential effect of seismic loading on waste dump or stockpile slopes. The key assumption is that the horizontal and vertical seismic forces generated by an earthquake can be simulated by applying constant horizontal and vertical accelerations to the slices in the dump slope analysis. The applied seismic forces are proportional to the mass of the slope and are estimated based on seismic coefficient factors (i.e. K_h and K_v). These seismic forces are typically applied either at the centre of gravity of each slice or at the bottom of the slices. Terzaghi (1950) considered applying the seismic forces at the centre of gravity of the entire sliding soil mass. In this sense, Seed (1979) noted that the computed FoS depends on the selected location for seismic forces application.

Seismic forces are included in the limit equilibrium method to obtain the final pseudo-static FoS (Fig. 8.13). Limit equilibrium software, such as SLOPE/W (Geo-Slope 2012c) and Slide (RocScience 2015) among others, has the

Table 8.6: Range of horizontal seismic coefficient used in pseudo-static stability analyses

Horizontal seismic coefficient, k_h	Source and comment	
0.05–0.15	Typical values used in United States	
0.12–0.25	Typical values used in Japan	
0.1	'Severe' earthquakes	Terzaghi (1950)
0.2	'Violent, destructive' earthquakes	
0.5	'Catastrophic' earthquakes	
0.1–0.2	FoS ≥ 1.5	Seed (1979)
0.10	Major earthquake, FoS > 1.0	USACE (1982)
0.20	Great earthquake, FoS > 1.0	
1/2–1/3 of horizontal peak ground acceleration (PGA) in g's	FoS > 1.0	Marcuson and Franklin (1983)
1/2 of the horizontal peak ground acceleration (PGA) in g's	FoS > 1.0	Hynes-Griffin and Franklin (1984)

Source: After Melo and Sharma (2004)

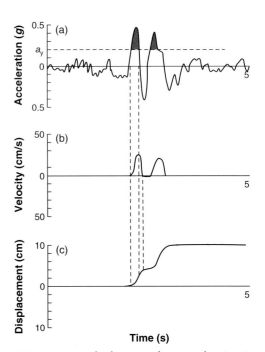

Figure 8.14: Permanent displacement from acceleration-time record. Source: Wilson and Keefer (1985)

option to include these seismic forces in the stability analysis in a user-friendly manner.

In addition, consideration should be given to whether drained or undrained shear strength of the foundation materials should be adopted for the purpose of pseudo-static analysis. Undrained shear strength is typically used for foundation materials that are below the water table and may respond in an undrained manner during seismic loading. Waste dumps and stockpiles are typically unsaturated and include coarse particles; hence, the drained shear strength of the fill material should be used for both static and pseudo-static analyses, unless there are portions of the dump or stockpile that are submerged and capable of behaving in an undrained manner during earthquake loading.

When undrained shear strength is used, proper choice of material properties and slope stability scenarios is critical to obtain reliable results; consequently, a specialist with experience in the analysis of earth embankments subject to seismic loading should be consulted.

The selection of the appropriate seismic coefficient is an important input into the pseudo-static analysis. Table 8.6 presents a range of recommended horizontal coefficients from different sources and countries.

Although this method is recommended by most design manuals and codes, it has some limitations. For example:

- The method does not consider the effects of time, frequency and body waves travelling through the soil during the earthquake (Chakraborty and Choudhury 2013).
- Seismic coefficients used in practice generally correspond to acceleration values well below the predicted peak accelerations (Kramer 1996).
- The method assumes a horizontal force, derived from the seismic coefficient, is constant and acts in one direction (Stark 1997).

Pseudo-static analysis should be used for preliminary screening purpose. Slopes that fail to meet the minimum FoS defined in the design should be subjected to supplemental analyses (as shown in Fig. 8.3) before being considered unsafe for the defined seismic condition.

In cases where an estimation of the permanent displacement of the waste dump or stockpile due to an earthquake is required, displacement methods such as the Newmark method (Newmark 1965) can be used. This method is based on modelling the waste rock as a rigid block on a potential sliding plane (Chowdhury et al. 2009). When the acceleration of the block, $k(t)$, exceeds the yield acceleration, a_y, the block starts to slide along the plane. Yield acceleration is calculated using pseudo-static analysis in which the acceleration that reduces the FoS of the slope to unity is calculated. A strong-motion earthquake record is selected, and those parts of the record that exceed the yield acceleration are integrated twice to obtain the cumulative earthquake displacement of the slope (Fig. 8.14).

8.6.2 Dynamic analysis

In special cases, such as a waste dump or stockpile constructed on a potentially liquefiable or very weak or sensitive foundation, a detailed analysis of stresses and/or deformations may be required to examine the response of the structure to an earthquake. In these cases, a full dynamic analysis utilising a finite element or other numerical analysis methods may be needed.

In these types of analyses, the incremental strains induced in each element by cyclical shaking during a simulated earthquake are accumulated to obtain the permanent deformation of the slope (Kramer 1996). The results of these analyses are sensitive to the input ground motion and engineering properties. Consequently, a prerequisite for using these procedures is a comprehensive seismic hazard analysis and detailed site and material characterisation (Patel and Sanghvi 2012).

8.7 Numerical methods

As indicated above, numerical methods enable the simulation and analysis of complex slopes subject to both static and dynamic (earthquake) loading. Common numerical methods that are sometimes used to analyse both the static and dynamic stability of waste dumps and stockpiles include the finite element method (FEM) and the finite difference method (FDM). Codes that support these methods can be used to simulate slope stability problems in both two and three dimensions.

The main advantages of numerical modelling over limit equilibrium analysis for the assessment of waste dump/stockpile stability include the following:

- No assumptions are needed about the shape or location of the failure surface, or the direction or magnitude of slice side forces (Griffiths and Lane 1999).
- Numerical models are able to simulate the development of progressive shear failure (Griffiths and Lane 1999).
- Numerical models can simulate complex slope configurations in two and three dimensions, and can model virtually all types of deformation and failure mechanisms (Mihai et al. 2008).

Unlike limit equilibrium methods which calculate a FoS along a defined failure path, most numerical methods utilise the shear strength reduction (SSR) approach (Dawson et al. 1999) to calculate an 'equivalent' FoS. This methodology involves systematically reducing the shear strength of the material until failure occurs. The 'equivalent' FoS is then calculated as the ratio of the originally estimated shear strength to the reduced shear strength at failure. In the context of waste dumps and

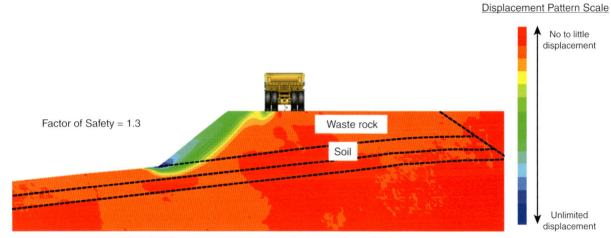

Figure 8.15: FLAC analysis of haul truck on a waste rock dump platform above a fill slope

stockpiles, this approach provides advantages over limit equilibrium methods, including:

- natural development of failure modes, with no need to specify a specific failure surface or range of failure surfaces in advance
- natural development of multiple failure surfaces, if the conditions exist for them to develop.

8.7.1   Finite element codes

There are several commercially available FEM codes. Among the most popular are RS^2 (formerly $Phase^2$), developed by RocScience (2016a), and Plaxis (2014). RS^2 is a 2D numerical modelling program that offers a wide range of support modelling options. This code can be used with either Mohr-Coulomb or Hoek-Brown strength parameters. A 3D version (RS^3) has recently been released. The Plaxis FEM code also includes advanced constitutive models that enable simulation of the non-linear, time-dependent and anisotropic behaviour of both soils and rock in 2D and 3D.

In addition to RS^2 and Plaxis, fine element software packages such as Geostudio (Geo-Slope International Ltd) and SVOffice (Soilvision Systems Ltd), are commercially available and include FEM codes that can be coupled to evaluate slope stability based on a wide range of stress scenarios for both static and dynamic conditions, and in 2D and 3D.

8.7.2   Finite difference codes

Finite difference codes such as FLAC (Fast Lagrangian Analysis of Continua), FLAC3D and FLAC/SLOPE, developed by Itasca (2016, 2012, 2011), are alternatives to FEM codes. These codes iteratively solve equations of motion, including inertia terms, to arrive at a solution. An example of a slope stability analysis using FLAC is illustrated in Fig. 8.15.

The use of finite difference codes for assessment of waste dump stability is becoming more commonplace. However, because codes like FLAC and FLAC3D provide a complete solution of the stress-strain, equilibrium and constitutive equations, more geomechanical input parameters are required for these analysis as compared to limit equilibrium methods. These include the porosity, elastic modulus and Poisson's ratio. As noted by Gómez *et al.* (2002), little of this data is available in the literature for waste rock; however, with respect to elastic parameters, some laboratory testing data associated with rockfill dam materials are available. For a simple analysis model that uses Mohr-Coulomb yield criteria, required input parameters for a FLAC/SLOPE analysis include density, cohesion, tension, friction angle and dilation angle.

9

RUNOUT ANALYSIS

Oldrich Hungr

9.1 Introduction

Mine waste dumps are some of the largest man-made structures on Earth, built over large areas and with heights which can be hundreds of metres from the crest to toe. The previous chapter presented methods of stability analysis for mine waste dumps and stockpiles. In addition to stability analysis, it is common to undertake an evaluation of the consequence of a mine waste dump failure by considering the potential distance that runout debris may travel should failure of the dump occur. This enables the identification of infrastructure that may need to be protected or relocated, and provides a basis for delineated exclusion zones in operation guidelines. Appendix 1 presents a summary of the failure records of mine waste dumps in the province of British Columbia, Canada, for the period 1968 to 2005, and includes records for 209 waste dump failures that reported runout events ranging from 10 to 2800 m in distance from the toe of dump before failure and runout volumes ranging from 600 to 30 000 000 m³.

This chapter presents a summary of the methods for mine waste dump runout analysis.

9.2 Materials

9.2.1 Mine waste

Mine waste is a material with wide-ranging characteristics (see Chapter 5), from angular fragments of strong rock to graded sand and gravel mixtures and silty sands to silt and clay. The mechanical character of mine waste grains is different from that of most natural soils. Transported natural soils have been modified and abraded during transport by water, wind or ice. As a result, their grains are rounded to varying degrees and relatively strong, and their compressive and shear strength depends largely on inter-particle forces. Mine waste is essentially crushed rock material whose angular, freshly formed grains may be susceptible to further breakage. Mine waste is also typically more widely graded (poorly sorted) and heterogeneous than other soils.

Natural residual soils generated by weathering are likely to have more similarities with mine waste, although they also differ, having been subjected to a long maturing process. What is important is that the extensive empirical knowledge of soil mechanics, which has been compiled primarily by testing natural soils, may not always apply to mine waste (e.g. Morgenstern 1992). For these reasons, common 'textbook' empirical rules derived from testing of natural soils may not apply to mine waste.

Dry mine waste in a loose condition can behave like an ideal granular medium, with nearly constant friction angle and ductile post-failure movement. This is evidenced by the large deformations of many metres or tens of metres that commonly take place at dump crests, forming tension cracks and sag features, without brittle acceleration (e.g. Hungr *et al.* 2002). However, the shear failure of some saturated waste can be accompanied by a tendency for volumetric contraction, leading to pore pressure increase and possible liquefaction. Sudden strength loss and brittle, rapid failure and runout can result. Contractive failure behaviour in mine waste may be promoted by grain crushing. As inter-particle stresses cause breakage of asperities or grain crushing, the soil skeleton can contract volumetrically, possibly causing pore pressure increase. The resulting process has been documented in undrained ring shear tests and is called sliding-surface liquefaction (e.g. Sassa 2000). The same process can simultaneously reduce permeability and hinder drainage of shearing waste layers.

Classification of mine waste materials is discussed in Chapters 3 and 5. For the purposes of determination of post-failure behaviour and runout, the following types of mine waste materials can be recognised:

- Strong, coarse waste (Fig. 9.1a) – this is material originating from strong, durable rock, in the form of angular fragments ranging from fine gravel (2 mm) to boulders. In a loose condition, common to waste dumps, this material will fail in a slow, ductile manner, retaining its frictional character. It is free draining, so pore pressure increase during failure is limited. In a dense condition, some strength loss due to dilatancy may be expected.

■ Weak, coarse waste (Fig. 9.1b) – this results from breakage of weak rocks to form fragments of sand to boulder size. Loss of frictional strength, volumetric contraction and permeability reduction take place during failure in a loose state. Grain crushing and sliding surface liquefaction may occur if sufficient moisture is available. Some brittleness due to grain breakage can be expected even in a dense condition.

■ Non-plastic fine waste – graded waste resulting from crushing of non-argillaceous rock may contain fines down to silt size, but without significant clay content and plasticity ('rock flour'). In a loose, saturated condition, this material may form liquefaction flow.

■ Plastic fine waste – graded waste sourced from weak, argillaceous rock or from weathered or geochemically altered rock masses can have substantial clay content and measurable plasticity. The potential for liquefaction is greatest with small amounts of clay and probably reduces at clay content in excess of ~10–20%, although no precise boundary has yet been established.

Generally, the various components are randomly mixed during loading in the pit. Within the waste dump, the operator may achieve some degree of sorting that may be beneficial to stability. For example, clay-bearing loads from weathered pit zones may be taken to different locations than more granular loads from strong rock.

Characteristic sorting of waste dumps results from segregation of material during end dumping. As shown in Fig. 9.2, a dump constructed by end dumping contains several characteristic zones (Nichols 1987):

■ Downslope ('foreset') layers form parallel with the dump face (Fig. 9.3). As a truck load is emptied at the crest of the dump, the fine-grained waste, usually moist and having some apparent cohesion, deposits first and spreads in a thin sheet down the face from the crest. Coarser rock fragments roll down the face of the dump. Like natural talus deposits, the coarse fractions become gradationally sorted downslope, as the dominant grain size gradually increases with distance from the crest. The fines concentration near the crest may cause local over-steepening maintained by apparent cohesion, while the lower slopes deposit at the lower limit of the angle of repose.

■ The top horizontal layer, forming the surface of the dump platform, is compacted by traffic and becomes dense and less pervious.

■ The largest fragments segregate from the rest of the material and roll to the toe of the dump and accumulate there, forming a basal zone, usually coarsely granular and pervious, overlying the foundation.

Concentrations of brittle (or liquefiable) mine waste will most likely be distributed in the upper parts of the fine foreset layers in the stratigraphy. The basal coarse accumulation is not likely to be brittle.

Sheet and gully erosion of the dump surface and landslides often disrupt this idealised stratigraphy and may form concentrations of finer material in the lower parts of the dump. In particular, small planar sliding failures of the over-steepened dump crest can distribute fine mine waste over the front surface of the dump, forming sloping fine-grained layers.

Figure 9.1: (a) A ample of coarse mine waste from strong sedimentary rock, (b) sample of fine mine waste rock from sedimentary rock. The reference frame is 2 m wide. Source: Dawson (1994)

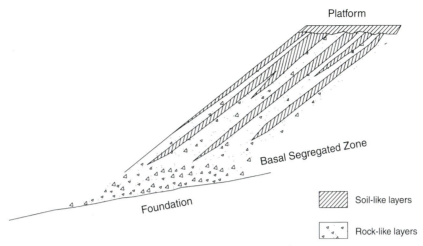

Figure 9.2: Typical stratigraphy of an end-dumped waste dump. Source: Hungr *et al.* (2002)

As discussed in Chapter 6, hydrology of the dump is influenced by the stratigraphy. Infiltration on the surface of the platform may be retarded by the high compaction of the top platform layer. On the other hand, the interior of the dump is drained through the coarse basal segregation zone. Thus, the interior of the dump does not often sustain positive pore pressure. However, fine-grained layers and zones in the stratigraphy may locally retain various degrees of capillary saturation (Dawson *et al.* 1998).

Figure 9.3: An excavated cross-section of a waste dump. Source: O. Hungr

9.2.2 Foundation materials

Foundation characterisation is presented in Chapter 4. Dump foundations may consist of any material forming the surface of the dump site, depending on its geomorphological setting. The foundations inherit any existing propensity of the natural surface towards failure and formation of landslides. Inherently stable foundations on flat or gently sloping ground may be formed in bedrock, residual soil, aeolian sand, glacial till or glacio-fluvial gravel and similar relatively dense granular deposits. Unstable units may include loose saturated alluvial or deltaic sands, marine or lacustrine clays, aeolian silts and organic soils.

On sloping ground, the surficial material is usually of colluvial origin, overlying any of the previously listed materials. The stability of colluvial deposits varies widely. It is nearly always necessary to assume that the top layer of the foundation will be looser and more strongly weathered than the underlying parent soil. Some colluvium (especially debris of planar slides or earth flow deposits) may stand at the point of failure, even before it is loaded by waste. It may also contain surfaces pre-sheared to residual strength by previous instability. Other colluvium may be loose and liquefiable, or may contain weak depositional or weathered horizons. Bedrock, even if completely weathered to form saprolite, can have unfavourable structure of jointing or faults, which may promote foundation failure. Some of these 'relict' structures may be altered by weathering, filled with clayey gouge and possibly pre-sheared to residual strengths.

Surface colluvial layers are commonly more pervious than the underlying substrate and may support a perched water table separate from the seepage regime of the rest of the slope. In tropical climates, laterisation of the top layers in the saprolite (or derived colluvium) may, on the other hand, retard exfiltration of slope water and increase local pore pressures within the foundation.

Permafrost foundations present special conditions. In the summer, the active layer presents a highly saturated layer, loosened by frost heave and solifluction and prevented from draining by the underlying, impervious permafrost table.

In surface mining practice, topsoil or other surficial soil may not be removed (stripped) from the dump footprint, and even vegetation may not be completely removed. Thus, an organic layer may exist at the interface of waste and the foundation. However, stripping of topsoil has become increasingly common in recent years either due to the mine's desire to subsequently reuse the stripped materials in reclamation, or as a result of regulation.

As in the case of any engineering structure, the stability of a mine waste dump depends on the characteristics of the foundation deposits. Potential behaviour such as plastic yielding under stress, structural failure along discontinuities or weak layers, brittle failure through cohesion loss, pore pressure increase due to loading, liquefaction of sand or silt, or remoulding of sensitive clay may influence the stability and potential mode of failure of waste dumps.

Dump foundations also inherit the pre-existing groundwater regime of the site. Of special concern are seepage discharge zones, where groundwater pressures may be increased by the presence and weight of the overlying waste. While the coarse segregated basal zone of the waste dump may act as a drain, the underlying topsoil and surficial soil, compressed under the weight of waste, may act as an aquitard/aquiclude, trapping seepage within the slope. Given the rapid placement rate of some mine waste rock dumps, development of excess pore pressure by undrained loading of the foundation needs to be considered in all but the coarsest foundation materials.

9.3 Landslides resulting from failures of waste dumps

9.3.1 Initial failure mechanisms

Basic failure mechanisms of mine waste dumps have been described in Chapter 8. In the following section, these mechanisms are briefly reviewed from the point of view of potential runout behaviour.

A report from a research project commissioned by the British Columbia Ministry of Energy, Mines and Petroleum Resources in 1992 defined the following failure types in coal mine dumps (BCMWRPRC 1992a):

Yielding toe/toe failure (YT) is caused by yielding of a weak layer in the foundation, at or just beneath the contact with the waste material (Fig. 9.4a). The toe failure may be initiated due to weak foundation soils at the toe, steep foundation slopes in the toe area or high foundation pore pressure. The toe yielding removes support from the body of the dump and is followed by the development of a steep failure surface passing through the waste. The resulting geometry is often approximated using a model, where an active wedge within the waste drives forward a passive wedge based on the failing foundation surface. The upper part of the sliding surface is usually parallel or sub-parallel to the foreset depositional layers within the dump stratigraphy, as illustrated in Fig. 9.2. The resisting stresses derive from the residual strength of the foundation plane, the steep back-scarp surface and the strength of an internal shear zone separating the two wedges. This failure type is the most common among large slides on waste dumps. It may start as being relatively shallow and then enlarge by retrogression, possibly involving the entire dump. Its dynamic behaviour depends primarily on the possibility of full or partial liquefaction occurring in the foundation, in the waste interior or both.

Sliver failure (S) is a shallow failure of the dump crest, over-steepened by rapid construction and the apparent cohesion of fine spoil (Fig. 9.4b). The failure is controlled entirely by the strength of the waste material. Because of the low saturation of near-surface material, sliver failures are usually not susceptible to liquefaction strength loss and are therefore usually not very mobile. This failure mode is very common, displaces relatively low volumes and often runs out on the lower slopes of the dump.

Basal failure (B) is a failure of the entire waste deposit along a shallow weak surface within a sloping foundation (Fig. 9.4c). The kinematics of such failure depends on the geometry of the foundation. Most commonly, it is a planar slide occurring on a steep foundation slope. However, a variety of geometries may be controlled by concave or convex foundation slopes. The important distinction is that the resisting stresses are derived predominantly from the strength of weak zones and associated pore pressures within the surficial layers of the foundation. Liquefaction within the foundation can cause rapid motion.

Rotational embankment failure (Re) is a deep, approximately rotational failure of the mine waste itself, excluding the foundation. It occurs due to the weak, relatively homogeneous nature of the waste material (usually fine grained and containing clay), combined with elevated pore pressure due to undrained loading caused by rapid dump construction, seepage within the waste or discharge from the foundation. To the writer's knowledge, mine waste material that is sufficiently saturated to be liquefiable in bulk is very rare. Most such failures, therefore, have the character of ductile grain flow.

Rotational (or compound) foundation failure (Rf) is a deep failure of weak foundation material. Deeper foundation failures may take the form of rotational or compound slides familiar from natural slopes, the waste acting mainly as surcharge. The common cause is high foundation pore pressures (often due to rapid loading) and weak foundation soils. If the foundation material is brittle, such as liquefiable sand/silt or sensitive clay, the failure may be accompanied by liquefaction and may be retrogressive. Such deep weak foundation deposits usually occur on valley bottoms and flat ground.

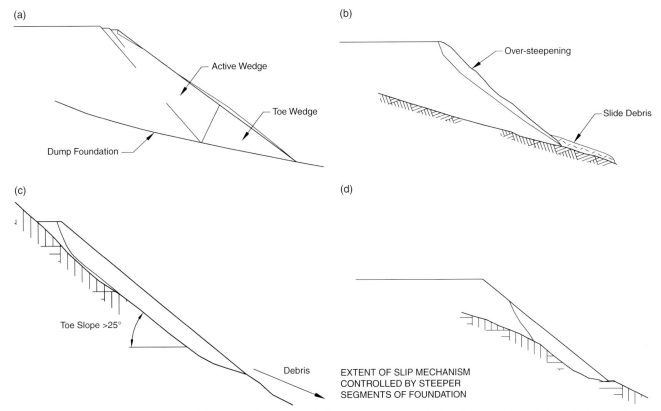

Figure 9.4: Basic failure mechanisms of dumps on sloping foundations: (a) toe failure, (b) sliver failure, (c) basal failure, (d) basal failure. Source: Campbell and Kent (1993). Published with permission of CRC Press/Balkema

Table 9.1: Distribution of failure mechanisms in observed dump failures

Failure type	Number of failures	%	% Total volume	Avg. volume per failure (10⁶ m³)
Yielding Toe (YT)	15	35	67	1.6
Sliver (S)	5	11	2	0.14
Basal (B)	9	20	14	0.56
Rotational, Embankment (Re)	7	16	5	0.26
Rotational, Foundation (Rf)	8	18	12	0.54
Total	44	100	7.6×10^7 M³	0.82

A database of mine waste failures compiled by BCMWRPRC (1992a) had a distribution that is summarised in a simplified form in Table 9.1.

BCMWRPRC (1992a) noted that liquefaction was involved in ~14% of the failures (YT, S and Rf types). A further discussion of liquefaction can be found in Section 9.4.3. The noted causes of failure were poor-quality material (18%), high loading rate (14%) and steep topography (9%), followed by high pre-failure precipitation, unconfined toe and excessive pore pressure generation.

9.3.2 Source volume and failure character

A runout analysis of a waste dump is preceded by a stability analysis (see Chapter 8), which can be used to suggest the most likely shape of the critical sliding surface and the extent of the most probable potential initial failure volumes within the range of failure modes described in the previous section.

For the purposes of runout analysis, we need to define several useful terms (Fig. 9.5). The 'rupture' (sliding) surface is the surface separating the real or potentially unstable mass from the underlying stable soil or rock. The 'source volume' (or 'source') of a potential landslide is the volume between the rupture surface and the original ground surface. The plan extent of a source volume is

referred to as the 'source area'. The 'deposit' is the volume of the landslide following the cessation of movement, and its plan area is the 'deposit area'. The 'path' of the landslide is a strip of land connecting the source and the deposit (provided that the landslide runs far enough to separate the deposit from the source). 'Entrainment' volume is the volume of material picked up by the landslide from the path and from the upper (proximal) part of the deposit area. The total volume of the deposit equals the source volume, expanded by any bulking that may occur during failure, plus the entrainment volume sourced from the path. If entrainment occurs within the deposition area, some of the deposit volume may be of local origin.

Landslide behaviour will depend on mechanisms that may cause strength loss during failure. As shown in Fig. 9.6a, the resultant driving (Fd) and resisting (Fr) forces acting on the source volume must be momentarily in equilibrium at the point of failure, when the factor of safety (FoS) equals 1. Because the gravity driving force Fd cannot change very rapidly, failure acceleration results primarily from the imbalance caused by the decrease in the resisting force, Fr. As shown in Fig. 9.6b, the two forces continue changing as the slope deforms. From the work-energy theorem, the kinetic energy that the landslide acquires during displacement equals its mass multiplied by the shaded area between the two curves. The upper curve depends primarily on the shape of the rupture surface and is usually relatively gradual. The lower curve is determined by strength loss, and its shape is modified by rate-dependent strength changes. The more rapid the strength loss during failure, the greater the velocity. This property of a material to lose strength has been termed 'brittleness' by Bishop (1973).

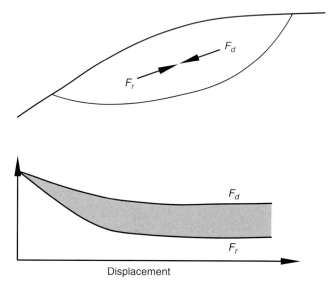

Figure 9.5: Schematic diagram illustrating the definition of key terms for a long-runout landslide

Figure 9.6: Schematic diagram explaining the development of kinetic energy of a landslide, following failure

If sliding occurs completely within non-brittle material, then the landslide runout will very likely be limited and take the form of sliding or granular flow, as described in the following section. Such materials include dry waste, or foundation made of moderately compact granular soil or insensitive clay. These materials do not lose much strength during failure, so acceleration of the landslide following failure will be limited. In such cases, it is appropriate to assume that the source volume will coincide with the volume above the 'critical sliding surface' determined from stability analysis.

Long runout will occur if some part of the source volume suffers significant strength loss, such as loss of cohesion, pore pressure increase, liquefaction or brittle remoulding of sensitive clay. Such processes often do not affect only the precise extent of the initial instability. Because initial failure of brittle material can be rapid, the initial source volume may be quickly emptied and the failure may retrogress upslope and involve additional material. On the other hand, a rapid failure may cover loose material downslope from the initial source and cause enlargement of the unstable zone by rapid undrained loading (Section 9.4.5). In other words, the source volume of a flow slide in brittle material may be substantially larger than the volume above the 'critical' sliding surface as determined by slope stability analysis.

It is necessary in such cases to carry out additional analysis, whether quantitative or based on judgement and experience, to determine the volume that will be involved in brittle sliding, influenced by the strength loss processes described above. Advanced numerical analysis using strain-softening material models can provide guidance, but requires special expertise.

9.3.3 Flowslides

In natural landslides, the term 'flowslide' is applied to an extremely rapid failure resulting from liquefaction of saturated or submerged loose sand (Casagrande 1940) or due to remoulding of sensitive clay (Meyerhof 1957). In such materials, liquefaction may be complete, affecting the bulk of the landslide volume. Thus, a classic flowslide would entail a change of the entire landslide volume from a frictional solid to a viscous fluid by structural collapse of the soil fabric, without any change in total stress or water content (Hungr *et al.* 2014). As applied to landslides in mine waste, the term means that substantial liquefaction has occurred, so a certain proportion of the material in the vicinity of the sliding surface can be assumed to be liquefied. Because waste dumps tend to be incompletely saturated, the rest of the sliding body has normal (or even negative) pore pressures and is deforming as a frictional mass. This concept is well suited to dynamic modelling of a frictional fluid, as described in Section 9.6.1.

9.4 Mechanisms of failure propagation

Strength loss is the most important determinant of landslide mobility. In a geo-material controlled by the Mohr-Coulomb strength law, such loss results from loss of cohesion and/or friction of the material, or from pore pressure increase (particularly, liquefaction).

The mechanisms that control the development of shear strength along the rupture surface of an unstable landslide are discussed in the sections below.

9.4.1 Sliding

During sliding, the rupture ('sliding') surface remains relatively thin and discrete, acting as a displacement discontinuity. This is usually the case in rock, dense granular material or stiff clay. These materials are usually dilatant, so pore pressure increase initially does not occur. What does occur at failure, however, is the rapid loss of cohesion, whether it is true cohesion due to overconsolidation, cementing or rock bridges, or apparent cohesion created by negative pore pressures. There is also a less abrupt loss of friction as a result of asperity crushing and volumetric dilation. Sliding surface liquefaction, described in Section 9.4.3, may take place during initial motion.

As a result of the above phenomena, when sliding motion is analysed using runout models, the dynamic strength of the sliding surface will be somewhat lower than the static strength used in the stability analysis, even in the absence of pore pressure effects. How much lower depends on the type of material. Cohesion will always fall to zero in dynamic analyses. The friction angle of angular, broken rock will most likely be reduced from a peak static value of ~38° to a dynamic value of ~30–32°. A more accurate value could be estimated by back-analysis of small failures and by measuring the slope angles of the dumps, which should tend to approximate the dynamic angle of friction in their lower reaches.

If there are saturated or partly saturated zones within the potential source volume, the possibility of sliding surface liquefaction, which could lead to further significant reduction of shear strength during motion, may need to be considered. Often, the 'bulk friction angle' acting on the slide base is a compound angle resulting from the averaging of dry and saturated zones.

9.4.2 Granular flow

Dry or moist, loose granular material fails in wide shear zones, with distributed strain. The resulting deformation can be described as 'granular flow'. This is a very common process at the free frontal surface of end-dumped waste dumps, where freshly deposited waste in loose and essentially dry condition forms shallow flows, transporting material from the crest to lower reaches of the dump face (an S type failure mechanism). Because there is very little strength loss during dry granular flow, the failure velocity tends to be limited and the travel distance is small. Most granular flows from the dump crest deposit lower on the dump face, forming an 'angle of repose' slope, which is a few degrees lower than the maximum 'static' angle of internal friction that occurs near the crest of a slowly built pile. If a deep-seated failure occurs in dry or moist, loose waste as a result of foundation yielding (a YT type failure mechanism), the shear strength contributed by the sliding surface segment seated within the waste can be assumed to approximate the dynamic friction angle, usually around 32°. For a rapid failure to occur, the required brittleness (strength loss) must occur in the foundation.

9.4.3 Sliding surface liquefaction

Shear displacement along thin shear bands developed in dense or compact waste will cause grain crushing and breakage of asperities. As this occurs, the grains will be able to pack more densely, promoting volumetric contraction of the sheared material. If water is present, this may lead to an increase in pore pressure, a phenomenon called 'sliding surface liquefaction' by Sassa (2000). This is a novel concept that has so far not found its way into standard soil mechanics literature. The reason is that sliding surface liquefaction requires displacement of several metres along the sliding surface, which cannot be obtained in routine laboratory tests. It can be obtained in ring shear tests, but these must be carried out in an undrained manner to be able to sustain and measure the pore pressure increase. Undrained ring shear testing is only available in a few laboratories in the world.

There is a fundamental difference between spontaneous liquefaction (Section 9.4.4) and sliding surface liquefaction. The former occurs instantaneously at the point of failure and requires saturated loose material. The latter requires several metres of displacement caused by some other strength loss mechanism and can involve materials that are relatively dense and even incompletely saturated.

The undrained ring shear laboratory experiments reported by Sassa *et al.* (2004) showed that excess pore pressures reaching almost to the limit of complete liquefaction can be generated by grain crushing. The effect can be observed even in partially saturated materials (down to a degree of saturation equal to ~70%). It is less prominent in waste derived from strong rock such as igneous and metamorphic rocks and calcareous sediments, and more prominent in waste derived from weak, especially argillaceous sediments (including coal and related lithologies). It is also more intense under large normal stress, related to deeper surfaces (and larger waste

dumps). Hence, the effect very likely depends on the volume of the landslide.

The quantitative effect of sliding surface liquefaction is not easy to interpret from laboratory tests because it depends on an unknown thickness of shear bands in the field. As a result, it is suggested that the quantitative amount of excess pore pressure to be used in dynamic analyses should be derived from back-analyses of full-scale case histories. The 'bulk friction angle' concept, described in Section 9.6.5, is the bulk friction angle observed in typical deep-seated slides in coal mine waste, involving weak or moderately strong sedimentary rocks, is often of the order of 20°.

9.4.4 Earthquake and spontaneous liquefaction

Some of the most damaging and unpredictable natural landslides involve liquefaction of loose, saturated soil. Liquefaction is a sudden loss of shear strength caused by the collapse of the soil structure under undrained conditions and complete, or at least major, transfer of stress from inter-grain contacts into pore fluid pressure. Liquefied soil has very low effective stress and, consequently, very low shear strength and rigidity, approaching that of a heavy liquid (Casagrande 1940).

Soil liquefaction may occur due to earthquake shaking, or spontaneously due to overstress. A body of solid material suddenly fully or partially changes into a liquid without appreciable change in water content or mean total stress. Of course, such a process has dramatic consequences on the behaviour of a slope. Geotechnical engineers have long used the term 'flowslide' (Terzaghi and Peck 1967) to describe the extremely rapid, mobile landslides resulting from liquefaction (Hungr *et al.* 2014).

Liquefaction can play a role in any of the waste dump failure mechanisms described in Section 9.3.1 except, perhaps, sliver failures that involve only drained surficial material. In the course of any of the deeper failures, a certain segment of the sliding surface, or a zone within the sliding body, may liquefy. However, it is unlikely for any mine waste failure to liquefy completely because much of the waste volume is unsaturated, or may be too dense to liquefy or too coarse to maintain undrained conditions. Thus, typically, liquefaction will affect only certain saturated zones situated within the sliding body, or in the foundation, while the remainder of the volume will deform by shearing as dry or moist frictional material. Localised liquefaction may occur in parallel with sliding surface liquefaction as described in Section 9.4.3.

The presence of liquefaction in an observed landslide may not always be simply evident. For example, some highly mobile flowslides involve liquefaction of thin discrete layers overlain by an inert mass of non-liquefiable

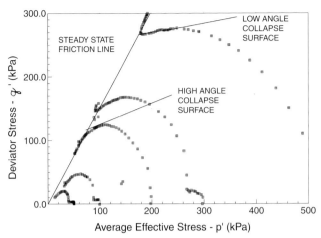

Figure 9.7: Stress paths in undrained triaxial tests on fine-grained coal mine waste, displaying collapse and partial liquefaction behaviour. Source: Dawson *et al.* (1998). © Canadian Science Publishing or its licensors

soil. Several cases of long-runout flowslides from coal waste dumps built in the mountainous terrain of British Columbia display clear evidence of strength loss due to liquefaction (Dawson *et al.* 1998). Dawson *et al.* (1998) also described standard undrained triaxial testing of fine coal waste exhibiting liquefaction (Fig. 9.7).

Remoulding of certain sensitive clay has effects similar to liquefaction, and the resulting extremely rapid landslides have also been referred to as flowslides (e.g. Meyerhof 1957). Sensitive ('quick') clays can sometimes be

found in the foundations of mine waste dumps on marine or lacustrine clay deposits on valley floors and may be the cause of brittle toe yielding, followed by flowsliding.

True liquefaction can occur only in specially pre-conditioned materials. Hunter and Fell (2003) summarised the indices marking liquefaction susceptibility (Fig. 9.8). Laboratory undrained triaxial testing indicates that liquefaction behaviour can occur in silty sands and in sands and sandy gravels with low silt contents, including some coal mine waste and coal (Eckersley 1990; Dawson *et al.* 1994). Clayey or silty sands, such as colluvium derived from residual soils, can also liquefy.

Soil susceptible to liquefaction is usually nearly fully saturated and loose. The maximum relative density criteria, also reviewed by Hunter and Fell (2003), vary within a wide range, depending on gradation and grain properties. Relative density of graded materials, such as mine waste, is extremely difficult to determine. In civil engineering practice, practitioners usually refer to empirical methods based on the standard penetration test (SPT) or cone penetrometer test (CPT) measurements. Normally, SPT values of less than ~8–10 are considered indicative of liquefaction potential. But in poorly sorted waste material, neither the SPT nor the CPT perform well. Again, the estimation of liquefaction potential in mine waste may have to rely on judgement, guided by field experience.

At times in the past, opinions were expressed in the literature that liquefaction is more frequently a problem in coal mining waste than in waste from metal mines. Such

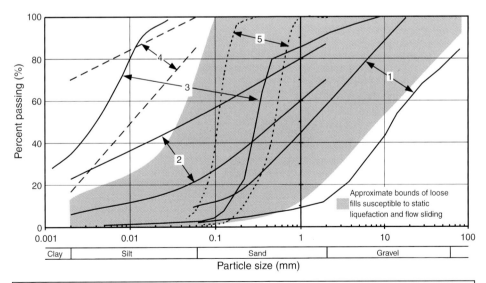

1 - Coarse grained coal mine waste (Dawson et al. 1998; Taylor 1984; Bishop et al. 1969; Hutchinson 1986)
2 - Loose silty sand fills, Hong Kong (upper and lower quartile of pre 1977 fills (HKIE 1998))
3 - Hydraulically placed mine tailings and fills in dam embankments (various published sources)
4 - Sensitive clays (indicative limits from various published sources)
5 - Sub-aqueous slopes, natural and fill slopes (Koppejan et al. 1948; Kramer 1988; Sladen and Hewitt 1989; Cornforth et al. 1974)

Figure 9.8: Range of gradation of soil materials which exhibited liquefaction flowslide behaviour. Source: Hunter and Fell (2003). © Canadian Science Publishing or its licensors

a simplification is not justified, as there are spectacular examples of liquefaction flowslides from gold and copper mines. However, the occurrence of flowslides does seem to be confined to moist climates. Mines in arid regions do not commonly experience this type of failure. Unfortunately, no simple quantitative rule can be given and we must rely on local experience.

Recently, concerns have been expressed about the potential of liquefaction in heap leaching piles, which may possess some of the characteristics of liquefiable waste deposits (Smith 2002).

Sand liquefaction or extra-sensitive clay remoulding within foundation deposits under static or earthquake conditions must be investigated using routine methods of liquefaction susceptibility analysis, as summarised, for example, by Cubrinovski and Ishihara (2000).

9.4.5 Rapid undrained loading

Each of the strength loss processes described in Sections 9.4.1 to 9.4.4 can cause propagation of a landslide to a certain distance. The toe of the landslide mass will displace forward and override soil material covering the path downslope from the source area (i.e. downslope from the toe of the dump). If the soil is saturated, an increase in pore pressure will occur as a result of rapid undrained loading (Fig. 9.9). This is a fundamental process mobilising natural flow-like landslides such as earthflows (Hutchinson and Bhandari 1971) and debris avalanches (Sassa 1985).

The pore pressure response of a saturated soil to rapid loading can be expressed using Bishop's pore pressure coefficient $\bar{B}$:

$$\Delta u = \bar{B}\Delta\sigma_v \qquad \text{(Eqn 9.1)}$$

Where Δu is the pore pressure increase and $\Delta\sigma_v$ is the increase in total vertical stress. When the stress increase is rapid (undrained) and the soil is fully saturated, $\bar{B}$ approaches 1.0. Thus, the bulk of the total stress increase is translated into excess pore pressure and the low effective stress that existed in the soil before loading remains unchanged. Surficial soil was lightly stressed before the arrival of the landslide front, with negligible effective stress. Following loading, the total stress and shear stress have dramatically increased, but the effective stress

remains negligible. The effect of this process is comparable to liquefaction and it results in extremely low shear strength of the saturated soil at the base of the landslide mass. The liquefied soil fails in shear and is entrained by the landslide sliding over it. The slide front advances further downslope, overriding and entraining more soil in an unstable progression.

The distinction between spontaneous liquefaction and undrained loading must be understood. In the former case, the soil skeleton collapses as a result of overstress under static condition. The total stress remains constant, but the effective stress is reduced to a low value. During undrained loading, the total stress increases due to movement, but the effective stress remains at a low value. Either case results in a soil zone that is extremely weak and often fluid-like. The term 'flowslide' should be reserved for a landslide that exhibits spontaneous liquefaction and contains liquefiable material. Undrained loading can occur in any soil with sufficiently high saturation, and is therefore the consequence of the landsliding process rather than material characteristics. Under natural conditions, the process occurs in debris avalanches, earthflows and debris flows (Hungr *et al.* 2014).

The properties of soil liquefied by undrained loading are similar to those of a soil subject to spontaneous liquefaction. In dynamic runout analyses, this can be simulated by replacing frictional sliding parameters with flow parameters at the point where undrained loading begins, usually just downslope of the toe of the source volume, at the beginning of the landslide path, or wherever loose, saturated surficial material is first encountered. The characteristics of the path material must be considered in such analysis and may be the determining factor of long runout. Loose colluvial soils and saturated organic soils are particularly prone to these effects.

9.5 Empirical methods of runout analysis and prediction

9.5.1 Travel angle

Empirical methods attempt to correlate the runout of waste dump landslides with a variety of measurable

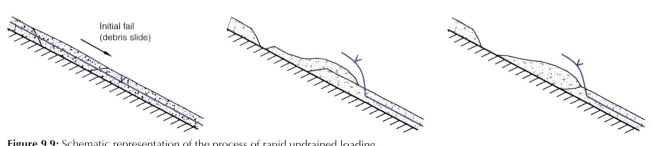

Figure 9.9: Schematic representation of the process of rapid undrained loading

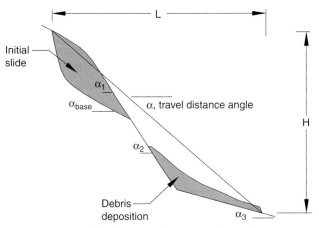

Figure 9.10: Variables describing the runout of a mobile landslide. Source: Hunter and Fell (2003). © Canadian Science Publishing or its licensors

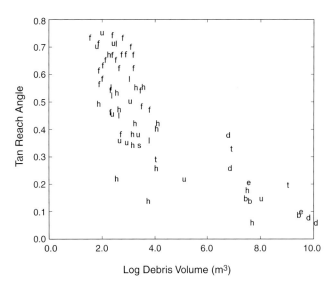

Figure 9.11: Relationship between the tangent of the reach angle (*H/L*) and volume for natural landslides classified as debris avalanches and debris flows. The symbols u and h indicate landslides that have unobstructed runout or are channelised, respectively. The remaining symbols indicate obstructions due to sharp bends, adverse slopes or mature forests. Source: Corominas (1996). © Canadian Science Publishing or its licensors

characteristics. The first correlations of this type published in the literature used data from large rock avalanches in the European Alps (Scheidegger 1973). It was shown that the vertical angle measured from the crown of the landslide source to the toe of the deposit, called the 'travel distance angle' (*fahrböschung* in German) or 'reach angle', decreases systematically with increasing volume of the landslide. This relationship has been used in the literature to make approximate predictions of runout for large natural rockslides. The tangent of the travel angle is the ratio of the overall fall height, *H*, to the horizontal distance between the crown of the source volume and the toe of the deposit, L (Fig. 9.10).

Similar conclusions have been reached by Corominas (1996), who explored the relationship between travel angle and volume for a variety of small and large natural landslides, both in soils and rock. Figure 9.11 shows a plot for landslides classified as debris flows and debris avalanches, sorted according to perceived obstructions to flow. Even though the data are plotted on a log-log scale,

the correlation is very poor, particularly for landslides smaller than 10 000 m³ in volume. It can be observed, however, that the 'unobstructed' (u) and 'channelised' (h) runouts are more mobile. The data contain several true channelised debris flows, which cannot be directly compared to other landslide types (e.g. Hungr *et al.* 2014).

In the 1990s, Golder Associates Ltd completed a comprehensive research project for the British Columbia Mine Waste Rock Pile Research Committee, dealing with the runout behaviour of landslides from coal mine waste dumps in south-eastern British Columbia (BCMWRPRC 1992b, 1995) (Appendix 1). Figure 9.12, compiled from 44 cases of waste dump failures, shows the runout paths of

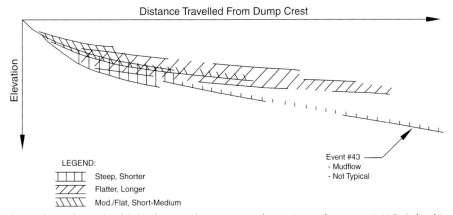

Figure 9.12: Travel path envelopes for 44 landslides from coal mine waste dumps in south-eastern British Columbia. Source: BCMWRPRC (1995). © Province of British Columbia. All rights reserved. Published with permission of the province of British Columbia

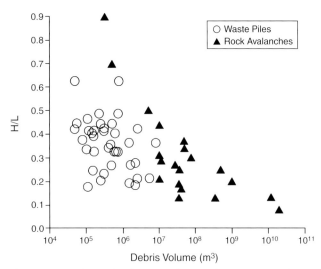

Figure 9.13: Relationship between the travel ratio (*H/L*) and landslide volume for 44 cases of mine waste dump slides and rock avalanche cases from the European Alps. Source: Hungr *et al.* (2002)

landslides, joined at the crest of the source volume as a common point. A relationship between the ratio *H/L* and volume for these cases is plotted in Fig. 9.13 and compared with rock avalanche data used by Scheidegger (1973). The decreasing trend (increasing mobility) is clearly indicated for the rock avalanche data (triangular symbols), ranging over six orders of magnitude in terms of volume. The mine waste data (round symbols) range only between 10 000 and 10 000 000 m³ and show much worse correlation. In fact, as noted by Golder Associates Ltd and Hungr (BCMWRPRC 1995) and Hunter and Fell (2003), there is practically no useful correlation between *H/L* and volume for waste dump landslides. The reason for this is the wide range of mobilisation processes acting in waste dump slides, as described in Section 9.4.

In an effort to improve the correlation, Golder Associates Ltd and Hungr (BCMWRPRC 1995) classified the data by degree of confinement, determined by the topography of the runout. 'Confined' (C) events are those where the landslide follows a gully or valley, the path is substantially narrower than the source area and the deposit area widens into the form of a fan. 'Partly confined' (P) events have a path of approximately equal width as the source area. 'Unconfined' (U) events occur on open slopes or convex hillsides, where both the path and deposit area are wider than the source. As re-plotted by Hunter and Fell (2003) and shown in Fig. 9.14, the classification reduces the scatter of the plot to some degree, with the confined events displaying the major part of the highly mobile data. There are probably several reasons for this. First, confinement of the travel path maintains thicker flow, thus increasing the basal total normal stress and promoting undrained loading. Second, confined valleys or draws are more likely to contain thicker and more highly saturated soil deposits, including organic soils.

The relationship shown in Fig. 9.14 is probably the best of the available *H/L* versus volume correlations for empirical estimation of mine waste landslide runout.

Hunter and Fell (2003) also noted that flowslides, involving spontaneous or earthquake liquefaction of loose saturated sands or remoulding of sensitive clays, produce travel angles that are much smaller than those reviewed so far (Fig. 9.15). Where such materials occur in the foundations of waste dumps, no empirical correlations such as those reviewed here are useful and the resulting estimated runout distances may be underestimated.

9.5.2 Other empirical correlations

A universal correlation exists for most landslides between the plan area of the deposits and the deposited volume. Assuming geometric similarity, the two parameters should

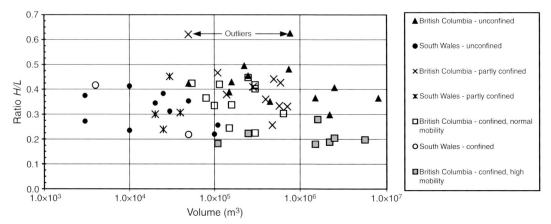

Figure 9.14: Relationship between the tangent of the travel angle (*H/L*) and landslide volume for mine waste dump slides, sorted by degree of confinement. Source: Hunter and Fell (2003), based mostly on data from BCMWRPRC (1995). © Canadian Science Publishing or its licensors

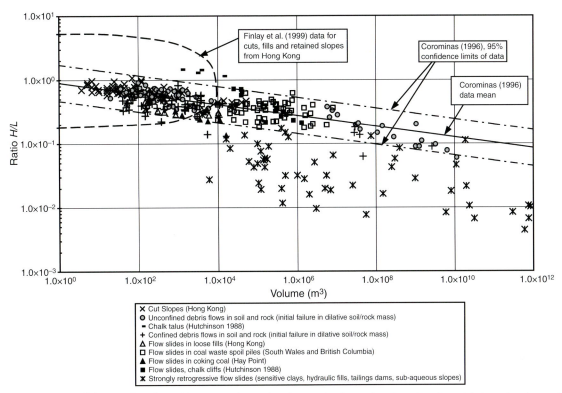

Figure 9.15: A summary of the travel angle (*H/L*) versus volume data from a database of rock avalanches, debris avalanches, mine waste dump landslides and various natural flow slides. Source: Compiled by Hunter and Fell (2003). © Canadian Science Publishing or its licensors

be related by an exponential function, with an exponent equal to 2/3. This is confirmed by a plot of values from the British Columbia database of waste dump landslides, compiled by Golder Associates Ltd and Hungr

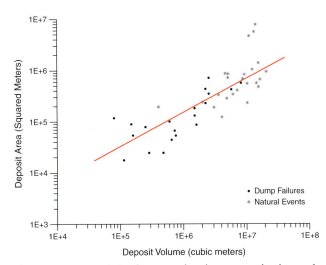

Figure 9.16: A correlation between the plan area and volume of deposits from British Columbia coal mine waste dump slides. Source: BCMWRPRC (1995). © Province of British Columbia. All rights reserved. Published with permission of the province of British Columbia

(BCMWRPRC 1995), shown in Fig. 9.16. This provides an alternative method for estimating runout. With a known source volume, the deposit area can be estimated from Fig. 9.16 and distributed over the expected deposition surface (usually at slopes of less than ~10°). However, the high scatter of the correlation, even on a log-log plot, again makes the application of this method difficult.

Other correlations examined by BCMWRPRC (1995) include relationships between runout distance and dump height and the so-called 'excessive runout distance' (Hsü 1975). However, these do not offer useful trends.

Hunter and Fell (2003) correlated the *H/L* ratio for coal mine waste dump slides with the slope angle of the runout path (defined in Fig. 9.10), as shown in Fig. 9.17 (based mostly on data compiled by Golder Associates Ltd and Hungr [BCMWRPRC 1995]). The correlation is relatively weak. Also, the selection of the angle α_2 is ambiguous in cases where the runout path extends over irregular, convex or concave slope. Hunter and Fell (2003) recommended an iterative procedure, where an initial estimate of *H/L* is based on an assumed value of α_2. Then, an initial estimate of *L* is made and a new average value α_2 is obtained, averaged over this distance. After a few iterations, the final α_2 should correspond to the average value of the slope between the toe of the source volume and the deposit. Again, this correlation is very approximate.

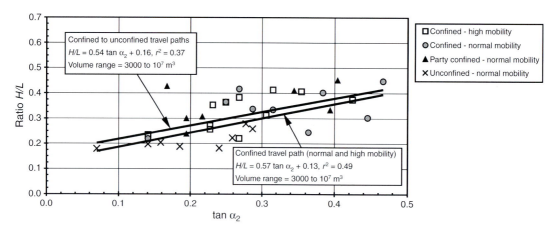

Figure 9.17: Correlation between the travel angle (*H/L*) ratio and the of the runout path, α_2 (see Fig. 9.10) for coal mine waste dump slides. Source: Hunter and Fell (2003). © Canadian Science Publishing or its licensors

In summary, the use of empirical correlations offers only very approximate means of predicting the runout of waste dump landslides. Perhaps the best approach is to seek precedents from data on failures observed in a given region and from physical and geomorphological settings comparable to those of the case being analysed. It is essential to make maximum use of local experience. The form of these site-specific correlations can be based on the relationships reviewed in this section.

9.6 Dynamic runout analysis

The advantage of a dynamic analysis in comparison to empirical methods is that it can take into account the actual geometry and mechanism of a potential slide. Also, it can supply information on flow velocities and depths and the distribution of deposits, parameters that are useful for hazard mapping and design of remedial measures. However, unlike stability analysis, dynamic runout analysis is not yet a routine analytical task. Its methodology is not well developed. Many numerical models exist (e.g. Dynamic Analysis of Landslides DAN/w by Hungr 1995), but their use requires good understanding of the processes involved and their results must be viewed with caution and verified against field observations and experienced judgement.

The usual approach to analysis in engineering is to derive a material constitutive relationship from laboratory tests and apply it in an analytical solution. This is generally not possible in connection with landslide dynamics, because the strength loss processes described in Section 9.4 are highly variable and scale-dependent and often occur only under full-scale field conditions. In practice, an alternative approach called the 'equivalent fluid approach' (Hungr 1995) is used. A flow analysis is carried out using the rheology of a relatively simple, usually non-Newtonian, fluid, selected empirically so that the dynamic behaviour

of the model will approximate the behaviour of the real landslide process in its overall characteristics. The properties of the equivalent fluid cannot be determined by laboratory tests, but must be determined by calibration, based on the back-analysis of real full-scale landslides sufficiently similar to the case being analysed.

9.6.1 Framework of dynamic analysis

Engineering dynamic analyses can be applied to the motion of rigid bodies, or to deforming bodies controlled by stress–strain relationships, or to fluids, characterised by rheology. When considering landslides, there is potentially a role for each of the three branches of mechanics. Initial failure of geomaterial most often involves sliding, during which the sliding body can be assumed to remain rigid (Fig. 9.18a). Unless dealing with an ideal planar failure of a block on a straight sliding surface, subsequent motion over uneven terrain involves straining of the moving mass, which in geomechanics is often simplified by assuming distributed through-going shear failure (so-called 'grain flow' or, in mechanical terms, 'frictional flow'), as shown schematically in Fig. 9.18b. This framework could be called a 'deforming frictional mass' model. It assumes that the internal deformation of the moving mass obeys frictional deformation principles and is controlled by an internal friction angle, while a weak basal zone may be in the process of sliding or flowing facilitated by pore pressure and controlled by a certain rheological law. Only if complete liquefaction of the moving body takes place and internal deformation of the moving body becomes dominant can we apply conventional principles of fluid dynamics (Fig. 9.18c).

Dynamic methods utilising rigid body motion exist and can be applied to the sliding motion of rock blocks (e.g. Hungr *et al.* 2005). These methods have little application to mine waste landslides. Dynamic elasto-plastic stress–strain analyses, which model in detail both

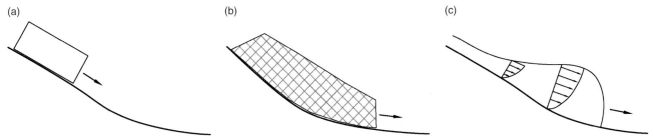

Figure 9.18: Three possible kinematic configurations of landslide motion: (a) rigid sliding block, (b) deforming frictional mass, (c) fluid flow

the internal and boundary deformations of the failing body, are available in modern computer programs but are difficult to apply. Fluid mechanics methods, which require full liquefaction of the entire failure volume, are not applicable to mine waste landslides.

The most common framework used in most recently available solutions is that of a deforming frictional mass, sketched in Fig. 9.18b, where a flexible block of frictional material whose interior is in the state of distributed shear deformation is moving on a weak basal surface. This observation is in agreement with the most common flow mechanisms of waste dump failures, as described in Section 9.3.3. Input data for this type of analysis include rheological parameters of the basal shear zones, which include pore pressure effects, and a constant 'internal friction angle' that is assumed to control internal deformation of the moving mass. The selection of the internal friction angle is somewhat less important in determining runout than that of the basal resistance parameters (Mancarella and Hungr 2010).

If the internal friction angle is zero, the entire mass is considered a fluid and the solution becomes equal to a conventional unsteady flow fluid mechanics solution (Fig. 9.18c). Such an assumption would be appropriate for fully liquefied flowslides (which are not common in connection with waste dump failures).

9.6.2 Two- and three-dimensional differential stress-strain analyses

The spectacular recent growth in computing capacity has brought several sophisticated elasto-plastic stress–strain models for dynamic analysis of solid continua. Most of these problems are solved by finite difference or finite element numerical methods. Their complexity depends mainly on handling boundary conditions and material behaviour, including post-failure strength loss. When used for runout modelling, the finite difference or finite element codes must have the ability to handle large deformations. Usually, this requires re-meshing routines to correct excessive distortion of the reference framework.

An excellent example of a complete dynamic solution simulating both laboratory sand flow tests and a natural rock avalanche can be found in Crosta *et al.* (2009). This

solution determines the full extent of internal deformation within the moving mass, in both two and three dimensions, and also accounts for entrainment of material from the path. At this time, however, differential analysis of runout can only be conducted by specialised teams as a research exercise. It is not a tool that is available for routine practical work.

9.6.3 Depth-integrated unsteady flow models

Solutions of unsteady flow in shallow open channels have been routinely available for flood routing applications for many decades. The standard fluid dynamic theory is based on momentum conservation, integrating stresses in a direction that is either vertical or perpendicular to the flow bed. Dealing with fluids, this approach results in the well known Navier-Stokes equations (e.g. Strelkoff 1970). The depth integration of the equations of motion is facilitated by the assumption of shallowness: if the normal depth of the flow is much less than its lateral extent, it is possible to neglect shear stresses on normal or vertical surfaces and to assume hydrostatic stress distribution with depth.

After integration, a 2D flow problem (like the flow in an open channel of constant width) is reduced to a 1D differential equation, while a fully 3D problem becomes two-dimensional mathematically. Confusingly, many mathematicians and hydraulicians then refer to their 3D models as 2D. In this work, flow that is of constant width where the velocity vectors are always assumed to be in a plane is referred to as 2D, while a general flow over 3D terrain is considered 3D.

While the depth-integrated equations of continuity and momentum conservation are well understood, their application to the movement of landslides is not straightforward. The reason is, first, that the initial conditions of landslide motion are more complex than normal flow hydrographs and the flow paths are steep. Second, the influence of internal strength of flowing landslide material precludes easy application of the laws of hydrostatic pressure.

The first model that recognised the dual nature of landslide motion discussed in the previous paragraphs was derived by Savage and Hutter (1989). The goal of their derivation was to simulate the flow of sand in a smooth-base laboratory flume. The internal deformation of the flowing body of sand was controlled by an internal friction coefficient, ϕ_i, of 29°. On the other hand, the basal sliding occurred on the interface between the sand and the smooth base of the flume, measured as 23°.

Savage and Hutter recognised that hydrostatic stress distribution is not appropriate for the flow of granular materials. The longitudinal stress in a mass of sand that is stretching (e.g. because it is moving over a convex slope path) is related to the bed-normal stress by the active earth pressure coefficient, which is less than 1.0. In a mass of sand under compression, on the other hand, it is related through the passive earth pressure coefficient, often much greater than 1.0. The Savage-Hutter model keeps track of longitudinal strain in the moving mass, so that the internal stress ratio can change between passive and active conditions as the landslide moves over concave or convex path segments. The passive and active earth pressure coefficients are calculated based on the user-defined internal friction coefficient and the friction and slope of the landslide base.

The original model of Savage and Hutter (1989) used frictional strength both for internal shearing and basal flow resistance. Hungr (1995) extended the model by allowing the basal shear layer to have a variety of rheological properties: frictional, plastic, viscous, turbulent or a combination of these rheologies. While the rheology of the basal shear zone can be described by a variety of rheologies, the internal deformation of the moving mass remains frictional as in the original Savage-Hutter model. This is consistent with the concept of a deforming frictional mass, introduced in Section 9.6.1. Several research groups around the world have now developed their own codes for this type of analysis, in both 2D and 3D. However, most codes found in publications are of a research nature and only a few are available commercially.

Hungr (1995) also introduced the concept of a pseudo-3D analysis. The mathematical solution of the equations of motion is the same as a 2D depth-integrated solution (i.e. the controlling differential equation is 1D). However, the continuity equation, which is used to determine the flow depth after each time step, is adjusted for the changing width of the flow path, specified by the user. Thus, deepening of the flow that corresponds to increasing path confinement can be accounted for, although momentum changes due to lateral spreading are neglected. Experiments have shown this latter inaccuracy not to be significant (e.g. Hungr and McDougall 2009).

9.6.4 Boundary conditions for flow analysis

The initial condition for most flow models is the geometry of the slope subject to failure (i.e. the waste dump), its foundation and the surface of the ground downslope. The models then require input of the shape and volume of the landslide source. Figure 9.19 shows a hypothetical case of a waste dump situated on a slope. The original ground surface, which will form the foundation of the dump, is shown in Fig. 9.19a. The waste dump is added in Fig. 9.19b. Figure 9.19c shows a potential rupture (sliding) surface of an ellipsoidal shape, passing through the waste and through a part of the foundation. This surface has been determined from stability analysis, combined with considerations of likely retrogression of sliding following initial failure and the extent of material susceptible to strength loss. The shape and size of the landslide source volume is obtained by subtracting the digital elevation model (DEM) of the rupture surface from the DEM of the ground surface of the waste dump (b) and multiplying the resulting volume by 1.1 to allow for expected bulking of the moving mass by 10% during failure. The resulting source volume of 1.46 million m^3 is shown in Fig. 9.19d.

Input into a 3D dynamic analysis requires the DEM of the rupture surface and flow path (Fig. 9.19c) and the 'thickness file' describing the source volume (Fig. 9.19d). An example 3D analysis in progress of motion is shown in Fig. 9.19e. The DEM of the flow path should not be excessively detailed and should be smoothed to a certain degree. Typical grid spacing for a model is of the order of 1/20 to 1/50 of the estimated runout distance. An excessively rough flow surface DEM could introduce unrealistic additional apparent friction into the dynamic model.

If a 2D analysis is carried out, the geometrical input data consist of a pre-failure cross-section of the slope, waste dump and expected rupture surface, as shown in Fig. 9.20a. A central cross-section of the source volume can again be obtained by subtracting the rupture surface cross-section from that of the pre-failure ground surface. However, except in the case of thin, shallow slides, the central cross-section represents the maximum depth rather than the average depth of the landslide. 2D solutions work with the 'hydraulic depth' (i.e. the average depth obtained by dividing the flow cross-section area by the surface width). If the central cross-section shown in Fig. 9.20a were multiplied by the width of the source area as shown in Fig. 9.20c, a volume of more than 2 million m^3 would be obtained. It is therefore necessary to enter the source volume cross-section as an average depth, as shown by the blue line in Fig. 9.20b (including a 10% bulking increase). This will result in the correct source volume. Often, the best way to determine the source thickness is to use trial and error, so as to obtain a correct estimate of the source volume, which is an important factor in determining the mobility of the landslide. Use of a reasonably smooth path cross-section is again recommended. The recommended number of cross-section input points is 20 to 50. Some software incorporates path smoothing functions.

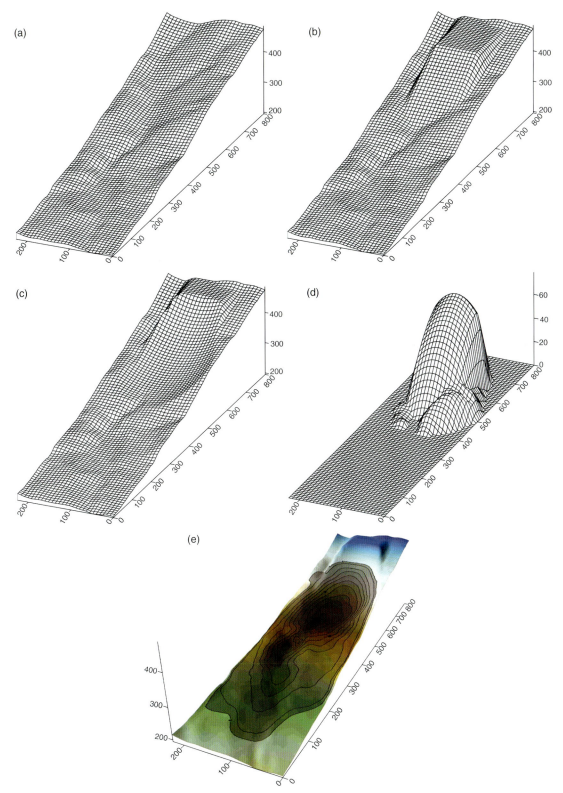

Figure 9.19: Initial conditions for a hypothetical model of a waste dump landslide: (a) original ground surface, (b) waste dump, before failure, (c) rupture surface (sliding surface); the unstable 'source volume' has been removed, (d) source volume, obtained by subtracting the DEM in (c) from the one in (b) and multiplying by the 'bulking coefficient' of 1.1, (e) 3D model in progress

In pseudo-3D analyses, the user is responsible for outlining the width of the flow path downhill from the source, as shown in Fig. 9.20c. The path width is sometimes constrained by strong channelling. At other times, the likely lateral spreading of the moving mass must be estimated by judgement. The path width function, shown in Fig. 9.20d, becomes the last element of input. Generally, the path width function does not influence the analysis too strongly. Figure 9.20e shows a pseudo-3D model in progress.

The source thickness diagram should be enlarged by any bulking (loosening) that occurs during failure. This depends on the nature of the source material. If the source was bedrock (a rockslide), a bulking of some 20–25% should be allowed to estimate the volume of the flow. During slides from waste dumps, the only bulking involves transition from a moderately dense in-place condition of the soil to a loose flow condition and may amount to ~5–10%.

9.6.5 Rheological relationships for basal flow resistance

The ability to select a variety of basal rheological relationships for flow models opens an avenue for

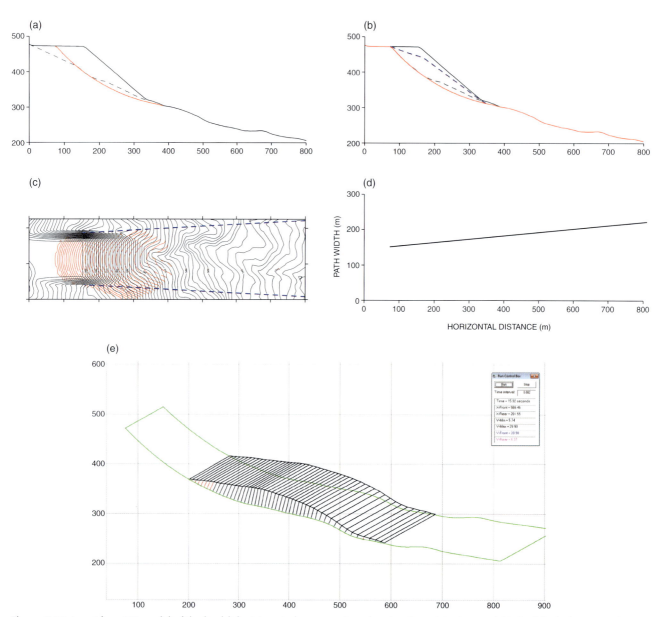

Figure 9.20: Input for a 2D model of the landslide: (a) central cross-section showing the original ground (dashed line), the waste dump and the rupture surface (red), (b) the source volume, represented as an average cross-section (hydraulic depth), determined so as to obtain the requisite total failure volume (1.46 million m³, after 10% bulking), (c) the anticipated widening of the path shown on a topographic map, (d) the corresponding path width function, (e) the 2D ('pseudo-3D') model in progress

calibration within the above-described concept of equivalent fluid. The basal shear stress resisting the flow, τ (in kPa), can be calculated by an infinite number of alternative expressions, depending on the flow depth, velocity and the characteristics of the mobile shear zone forming the base of the flowing mass. Some of the simplest and most frequently used relationships are listed below (e.g. Hungr 1995; Hungr et al. 2005).

9.6.5.1 Frictional flow

The flow resistance during the motion of granular soil is determined by the frictional strength:

$$\tau = (\sigma - u)\tan\phi \qquad \text{(Eqn 9.2)}$$

where:
- σ is the total normal stress acting on the base of the flow
- τ is the dynamic friction angle of the material
- u is the pore pressure acting within the shearing basal layer.

Pore pressure within a moving, fully or partly liquefied layer is practically impossible to measure. The simplest assumption that can be made is that the pore pressure is proportional to the total normal stress:

$$\tan\phi_b = (1 - r_u)\tan\phi \qquad \text{(Eqn 9.3)}$$

where r_u is the ratio between the pore pressure and the total normal stress. This is analogous to the pore pressure ratio, and is commonly used during limit equilibrium slope stability analysis in cases where pore pressure is poorly defined. The 'bulk friction coefficient', ϕ_b, now represents friction reduced by pore pressure effects, which can be substantially less than the dry friction of granular soil. In fact, for a fully liquefied material, ϕ_b may approach zero as the flowing material becomes similar to a fluid.

Note that the frictional strength is rate-independent and this may lead to unrealistic results at high velocity flow. As the pore pressure rises towards the geostatic value and the frictional strength decreases, the viscous or turbulent resistance characteristic of fluids gain in importance, requiring the use of one of the velocity-dependent relationships listed below.

The advantage of assuming a constant r_u is that the resistance term remains frictional, similar in form to the resistance of dry material (although it may be much smaller):

$$\tau = \sigma\tan\phi_b \qquad \text{(Eqn 9.4)}$$

On theoretical grounds, during fully or partly liquefied flow, the assumption of constant r_u is not well justified. In fact, rapid changes in depth, typical in a highly unsteady flow, as well as overriding and entrainment of saturated material from the flow path under undrained conditions,

would be more likely to produce something like a constant undrained strength, unaffected by the level of total stress. Current research aimed at clarifying the role of pore pressure in liquefied flows is under way, but results are not yet available for practical use. The assumption of a constant bulk friction angle, as expressed by Eqn 9.3, is probably reasonably justifiable in case of a flow that is not fully saturated, where a part of the resisting strength derives from shearing through relatively dry soil. Experience has shown that models based on the frictional rheology expressed by Eqn 9.4 perform well for flowslides that do not contain a large proportion of fully liquefied material on the basal surface, including typical mine waste flowslides that do not entrain large quantities of saturated material from the path (BCMWRPRC 1995).

9.6.5.2 Turbulent flow

The turbulent flow of a fluid, where τ is a function of mean velocity squared, V^2, the unit weight of the material, γ, flow depth, H, and the Manning roughness coefficient, n, is given by:

$$\tau = \frac{\gamma V^2 n^2}{H^{1/3}} \qquad \text{(Eqn 9.5)}$$

Turbulent flow would apply to watery flood flows of small depth and small solids concentration–usually not the conditions of mine waste flowslides.

9.6.5.3 Voellmy fluid

This rheology, combining a frictional and turbulent term, was proposed for snow avalanches by Voellmy (1955):

$$\tau = f\sigma + \gamma\frac{V^2}{\xi} \qquad \text{(Eqn 9.6)}$$

Here, the friction coefficient, f, is equivalent to $\tan\phi_b$, as defined in Eqn 9.4. ξ is a turbulence parameter equivalent to the Chézy coefficient of hydraulics, with dimensions of L/T2. Note that ξ can be related to Manning's n as:

$$\xi = \frac{1}{nH^{1/3}} \qquad \text{(Eqn 9.7)}$$

However, n, being an empirical coefficient determined for water flow, cannot meaningfully be used to derive ξ for landslides.

Voellmy (1955) intended the turbulent term to account for all velocity-dependent resistances acting on snow avalanches, including surface air drag. For landslides, the term represents turbulent energy losses within the pore fluid and its interaction with solid grains and, possibly, velocity-dependent changes in pore pressure. Dry, cohesionless, granular materials appear always to be frictional with little rate dependence (e.g. Hungr and Morgenstern 1984). Therefore, the turbulent term is needed only for saturated materials. The frictional model should be considered as a member of a series, where the material is essentially dry and the velocity dependence is negligible (i.e. ξ approaches infinity). Under such conditions, Eqn 9.6 becomes equivalent to Eqn 9.4. As more saturation appears, the frictional resistance decreases and the turbulent term becomes more important.

9.6.5.4 Plastic flow

The shear resistance of fine-grained, cohesive material moving in an undrained manner at moderate velocities can be assumed to be constant, controlled by a yield shear strength c:

$$\tau = c \qquad \text{(Eqn 9.8)}$$

The assumption of purely plastic strength in dynamic analyses can lead to unrealistic results. In the absence of rate dependence, the model can easily attain high velocities. At flow velocities in excess of ~1 m/s, it is necessary to account for a velocity-dependent stress increase (viscous or turbulent) by using one of the rate-dependent models.

9.6.5.5 Bingham flow

Bingham flow is the flow where the basal shear resistance is velocity dependent and consists of a constant yield shear strength, similar to c above, flow depth, velocity and Bingham viscosity. An explicit expression for the Bingham resisting stress, similar in form to the previous equations, is not available and the dynamic model requires the solution of a third order equation. But, with Bingham viscosity approaching zero, the model reduces to that of Eqn 9.8.

The Bingham model was one of the first used in numerical analyses of landslides (e.g. Sousa and Voight 1991). It was also found useful for modelling some flowslide failures of tailings impoundments (e.g. Bryant et al. 1983). Some limited modelling work shows that the model may be suitable for liquefied flows that contain large fractions of fine matrix, silt and clay. However, there has so far not been sufficient testing of the Bingham model in the context of mine waste flow slides.

9.6.5.6 Other rheologies

Various researchers have reported simulations of natural flowslides and debris flows/debris avalanches with a variety of other rheologies, usually more complex than those reviewed above. The Herschel-Bulkley rheology, being a generalisation of the Bingham model, has been used by several research groups for fine-grained mudflows (e.g. Laigle and Coussot 1997; Imran et al. 2001). Other theories range from coupled sliding-consolidation models (Pastor et al. 2014) to complicated two-phase theories that

attempt a simultaneous analysis of the interaction of solid and liquid phase in mixtures (Iverson and Denlinger 2001). A commercial model suitable for analysis of debris foods on colluvial fans uses a three-parameter model, combining frictional, viscous and turbulent terms (O'Brien et al. 1993). While many successful back-analyses of various landslide types have been reported, none of these models has been sufficiently calibrated for cases of waste dump flowslides. It is to be noted that more sophisticated rheologies require a larger number of input parameters, which must be constrained by calibration. The basic rheologies reviewed in this work rely on one or two parameters and are therefore easier to constrain, though they may not be as accurate in simulating the flow process.

9.6.6 Material entrainment

Once large displacement of the front of a waste dump occurs, the landslide will deposit on the ground surface downslope. If loose, saturated soil is present, rapid loading will occur as described in Section 9.4.5 and the over-ridden soil may be entrained by the moving mass. The process has two consequences: (1) the volume of moving material is increased and (2) the undrained shear strength and rheology of the entrained material will determine the shear resistance at the slide base.

Some theoretical research that attempts to derive the erosion depth and amount from first principles, where entrainment estimates depend on simple properties of the basal material, has been published. However, the natural process is very complex and cannot be reliably simulated. In most cases, the amount of entrainment is limited by the thickness and characteristics of loose, saturated soil that is available for entrainment (so called supply-limited condition). The entrainment thickness, therefore, must be determined by the analyst as an input parameter, based on detailed knowledge of the stratigraphy, characteristics and saturation of the soils at the surface of the path, downslope from the toe of the landslide rupture surface and on observation of previous events. In practice, the user must determine an 'entrainment zone' within the potential path of the landslide, which is described by the thickness of material considered available for entrainment. The rate at which material is picked up depends on the type of entrainment algorithm used in the model.

In the model described by McDougall and Hungr (2005), for example, the entrainment is proportional to the thickness of material overriding a given location and its displacement amount. This results in a natural exponential growth of the landslide with displacement. Although empirical, the method has a physical basis, as the intensity of undrained loading leading to failure within the path material depends on the flow depth. The proportionality constant, termed 'entrainment rate', is sometimes set within the code to ensure that the full specified entrainment depth is reached when a given point within the entrainment zone is passed by the entire current volume of the landslide (Hungr 1995). Alternatively, the entrainment rate must be specified by the user to fulfil a similar condition. McDougall and Hungr (2005) give an equation that approximates a suitable value for this factor. The entrainment rate must be set carefully. If it is too large, the leading front of the landslide may grow excessively fast, distorting the moving mass. If too low, the prescribed entrainment amount may not be reached. Trial and error adjustments are sometimes required.

Another input item required of the user of these models is the change of rheology that may accompany entrainment of saturated soil. The initial slide may involve only partial liquefaction. However, once undrained loading of loose saturated soil occurs within the entrainment zone, the basal layer of the moving mass may become fully liquefied. The user may then have to specify a rheology change at that point, increasing the pore pressure ratio and possibly adding a rate effect such as turbulence (e.g. the Voellmy model).

Several existing dynamic models are capable of simulating entrainment (e.g. Hungr 1995; McDougall and Hungr 2005; Crosta *et al.* 2009; Pastor *et al.* 2014). However, calibration of these models is limited.

9.6.7 Calibration and forecasting

The authors of dynamic models should provide evidence of verification against controlled observations and alternative solutions to ensure that the solutions are correct. Nevertheless, for practical use, any model must also be calibrated against real-life landslide cases. The ideal process for using a dynamic model to forecast the runout of a potential landslide is as follows:

1. Define and describe the type and mechanism of the potential landslide.
2. Find real cases of similar landslides, ideally in the area being considered, or elsewhere if necessary. The calibration cases should be similar in terms of geological setting, climatic conditions, the characteristics of the waste material and the foundation. They should also be similar in terms of size (volume).
3. Back-analyse the calibration cases using the chosen dynamic model. Choose the most appropriate rheological model and select the best fitting input parameters. The fit of the back-analysis should be judged by the total runout distance, the areal distribution of deposits and the flow velocities and durations.
4. Analyse the potential landslide using the best fit parameters (or a range or distribution of parameters as described below).

The selection of the best rheology and parameters during back-analysis requires experience. In general,

frictional rheology is appropriate for relatively dry granular soils (waste and foundation). Pore pressure effects that may act over a part of the sliding surface can be simulated by specifying an average pore pressure ratio. The frictional analysis will generally predict high flow velocities and a deposit that is thin at the front and thickens in the proximal region. If a greater amount of saturated material is involved in a slide consisting primarily of granular soil, the introduction of rate dependence (e.g. through the use of the Voellmy rheology) may be required. For a given extent of runout, the velocities will be smaller with increasing turbulence or viscosity and the deposits will be thicker at the front. If the material contains large proportion of silt and clay, it may be appropriate to experiment with the Bingham model. This tends to produce high velocities and a fairly uniform thickness of deposits, even on steeper slopes.

Where entrainment occurs, it may be necessary to use two rheologies, one describing the movement of the initial slide (usually frictional) and a second that applies from the upslope margin of the entrainment zone.

Flow velocity or duration estimates for evaluation of back-analysis results are the most difficult item of evidence to obtain. In some cases, eyewitness reports or video footage may be available. In others, velocities of calibration cases can be determined by analysing superelevation of the flow in curving paths, or run-up against adverse slopes or obstacles. Such evidence should be sought wherever possible.

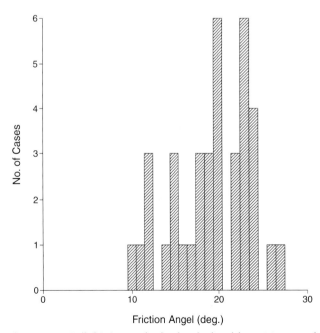

Figure 9.21: Bulk friction angles back-calculated from 44 cases of waste dump landslides from the coal areas in south-western British Columbia. Source: BCMWRPRC (1995). © Province of British Columbia. All rights reserved. Published with permission of the province of British Columbia

If several back-analyses are completed, it may be possible to determine an approximate distribution of runout parameters, or at least a set that applies to the most likely situation and that is considered extreme.

Unfortunately, only a few systematic calibration programs can be found in the literature. The most relevant to this chapter is the back-analysis of 44 landslides from the waste dumps of several mountain coal mines in south-eastern British Columbia, Canada, completed by Golder Associates Ltd and Hungr (BCMWRPRC 1995; see also a summary in Hungr *et al.* 2002).

For each landslide in the database, a bulk friction angle was found using the rheology described by Eqn 9.4, so as to produce the required displacement along the slide path. The distribution of the resulting values of the angle is given in Fig. 9.21. Some 70% of the cases were successfully simulated with angles ranging between 18 and 24°, averaging 21.2°. Estimating 32° as a typical effective dynamic friction angle of the waste material, the mean pore pressure ratio can be estimated from Eqn 9.3

as 0.37 (37% of full liquefaction). These cases are referred to as 'normal' runout cases. The remaining cases were more mobile and could not be well simulated by the frictional model. In these cases, it was necessary to implement the Voellmy model with a friction coefficient of 0.05–0.1 (i.e. ϕ_b of 3–6°) and a turbulence coefficient of 200 to 500 m/s^2.

All of these 'mobile' runouts occurred on paths that followed confined valleys or gullies containing loose colluvium or organic soils with high water tables. An example is shown in Fig. 9.22. In this case, a waste dump failure of 700 000 m^3, caused by the instability of a weak bedrock and colluvium foundation, ran into the head of a gully, lined with colluvial and organic soils. A flowslide was generated, flowing at speeds of up to 28 m/s for 2 km as determined by superelevation analysis (Hungr *et al.* 2002). The distal deposit of the flowslide, shown in Fig. 9.22, consists primarily of grey-coloured waste rock. However, brown organic soil from the substrate, liquefied by rapid undrained loading, is seen extruded to the surface

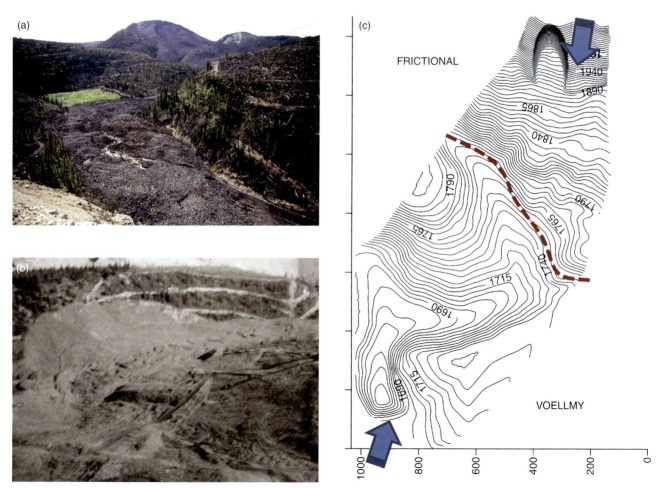

Figure 9.22: A flowslide of 700 000 m^3 of coal mine waste in south-eastern British Columbia. The topographic map in (c) was used to set up an analysis using the DAN3D model, with two rheologies. The arrows in (c) indicate the locations of two photos of the landslide path. Source: O. Hungr

in several places. This liquid material was also found lining the base of the flowslide in test pits.

Similar flow characteristics and properties were obtained by back-analyses of flowslides in waste from a gold mine in Indonesia (unpublished).

9.7 Hazard and risk mapping

Output from empirical runout determination will consist of an outline of the area likely to be impacted by the landslide ('hazard area'). Output from forward dynamic analysis of waste dump landslides can include a map of the area that may be impacted by the movement and maps showing the potential distribution of the deposits and the distribution of maximum flow depths and velocities. These last two parameters combine to produce a map of landslide hazard intensity. Again, all these output parameters can be presented as average (most likely) and extreme (worst case scenario), or in stochastic terms (i.e. distribution of intensity parameters in various sectors of the hazard area).

As discussed in Chapter 10, risk equals the probability of hazard times consequences. Under the ideal scenario, a complete description of the landslide hazard will be obtained by constructing the above mentioned hazard intensity maps from the runout analysis. Consequences of the hazard can be determined by combining the hazard intensity values with the vulnerability of people, structures and other elements at risk situated within the hazard zone.

Hazard probability is a complex concept. A detailed risk analysis requires establishment of several alternative hazard scenarios (e.g. small, medium or large failure). Each scenario is assigned an annual probability of occurrence, and a runout analysis is carried out to determine the corresponding hazard intensity map. After estimating the vulnerability of all the elements at risk to the various hazard intensities, a specific risk for each element in each scenario can be calculated. The total risk for the project is obtained by adding all the specific risks.

This detailed procedure is unlikely to be followed in connection with typical mining projects, and various degrees of simplification will probably be required. However, the above discussion provides an idealised framework for quantitative runout risk analysis for waste dump facilities.

9.8 Protective measures

If the risks determined following the procedure mentioned in the previous section are found unacceptable, various types of mitigation of the possible damage caused by a waste dump landslide must be contemplated.

The first possible line of defence includes measures to prevent large-scale waste dump failures from occurring. This may include:

- controlling the quality of the waste placed in critical locations to reduce failure and liquefaction susceptibility
- controlling placement direction to reduce exposure of steep frontal dump slopes on steep foundations
- treatment of dump foundations (removal of topsoil or colluvium)
- resloping or buttressing of critical sectors of the dumps.

If it is not possible to positively prevent the potential for rapid failures, measures within the runout areas may be required:

- Personnel and facilities can be excluded from the identified hazard areas.
- Diversion berms may be constructed in such a way as to deflect potential mobile failures away from critical objects or areas. There are no published accounts of such protection structures, but they have been used successfully in mines to divert flowslides from facilities and equipment. The dimensioning of such structures can be aided by dynamic analyses. For example, several groups participating in the 2007 Hong Kong benchmarking exercise were able to simulate the deflection of a model sand flow by an oblique diversion dyke with their dynamic models (Hungr *et al*. 2007; see also McDougall and Hungr 2004).
- Terminal berms may be constructed in a direction perpendicular to the expected movement of the landslide, to limit its runout. These berms can be dimensioned using suitable run-up formulas or a dynamic model (Mancarella and Hungr 2010).
- Rigid and flexible barriers consisting of concrete walls or cable fences can be used to limit the impacts of potential landslides on sensitive structures. Various authorities in Asia and Europe have recently proposed standards for design of such structures (e.g. Lo 2000).
- Structures can be reinforced to resist potential landslide impacts, determined from the intensity maps mentioned in Section 9.7.

In the context of waste dump failures, the construction of terminal berms to help contain failure debris and limit runout distance below a dump or stockpile is probably the most cost effective measure of those listed above. Dimensioning of terminal berms should include an allowance for freeboard as a contingency or factor of safety against potential variability in the size of any potential failure, or to provide adequate containment volume where multiple failures are anticipated. Actual freeboard allowances should be based on the designer's experience with similar failures and the confidence in the calibration

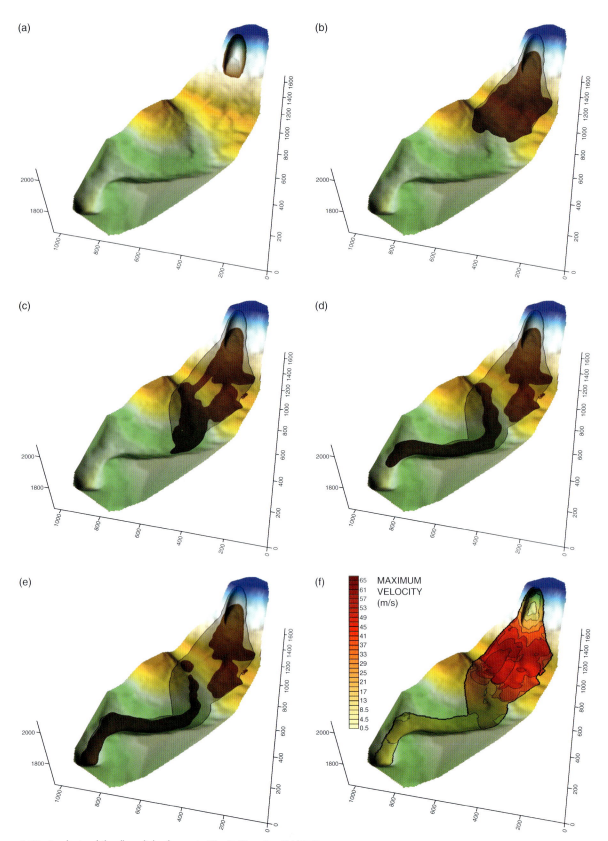

Figure 9.23: Analysis of the flowslide shown in Fig. 9.22, using DAN3D

of the model/methodology being used to determine the potential run-up on the upstream face of the terminal berm.

9.9   An example runout analysis

Hungr *et al.* (2002) described a highly mobile flowslide that occurred in a waste dump of one of the coal mines in south-eastern British Columbia in the mid-1990s. The failure is shown in Figs. 12 and 14 of that paper and in Fig. 9.22 of this chapter. The waste dump was built on a steeply sloping (~30°) foundation consisting of shallow colluvium underlain by sedimentary bedrock. A toe failure developed due to the weakness of the colluvial veneer, aggravated by shallow instability of bedrock layers under the toe of the waste dump. The failure was manifested by extensive yielding of the dump crest, forming series of head scarps with a cumulative displacement of some tens of metres. A few months after the cessation of spoiling, 700 000 m³ of waste suddenly failed, sliding out on the foundation at the toe, with the head of the source volume rising steeply to the level of the platform. The failure scar is visible at the top of the frontal view in Fig. 9.22. The sliding surface is visible as a niche in the contour map shown in the figure.

The sliding mass progressed beyond the toe of the dump and down the foundation slope. At the base of the slope is an ephemeral drainage gully, curving away from the source scar. The base of this gully contains an accumulation of colluvium, derived from glacial till and a thin deposit of organic soil. At the time of the failure, there was thin snowpack on the ground and the soil was saturated, The sliding waste over-rode the gully deposits and flowed down the gully, following a curving path. In the first bend of the path, the waste stream superelevated. An analysis of the superelevation, produced an estimate of velocity of 16 m/s. Because of the low thickness of the surficial layers, the volume entrainment was neglected. Hungr *et al.* (2002), in their Fig. 22, show a back-analysis of this landslide using the 2D model DAN-W (Hungr 1995). To recognise the importance of entrainment of saturated loose and organic soil, the distal part of the path, following the gully, was simulated using the Voellmy rheology (Eqn 9.6), with an $f = 0.05$ to 0.1 and a ξ of 200–500 m/s². In the proximal part of the path, including the base of the sliding surface and the steep foundation slope, the flow resistance was assumed to be frictional (Eqn 9.4), with a basal 'bulk' friction angle of 21.4°. This corresponds to a dry friction angle of 32°, combined with a pore-pressure ratio of 0.37.

The same analysis was repeated recently using the 3D program DAN3D (McDougall and Hungr 2004) and a topography shown in Fig. 9.22c. The assumed dividing line between the frictional and Voellmy flow regions is shown on the topographic map. The results are summarised in Fig. 9.23. The distribution of the debris deposits and the maximum flow velocities correspond well with the results obtained with DAN-W. The back-calculated best fit parameters are 21.4° for the bulk friction angle, $f = 0.06$ and $\xi = 680$ m/s². These parameters indicate that the presence of saturated organics in the path made the flowslide particularly mobile.

The resulting back-calculated parameters could be used to make forward predictions of landslides in similar circumstances.

10

RISK ASSESSMENT

Brian Griffin

10.1 Introduction

This chapter provides an overview of risk assessment and its application to the design of mine waste dumps, dragline spoils and stockpiles. Topics include a broad introduction to risk terminology in the context of the risk management process, a description of typical types of risk receptors, highlights of different risk analysis methods and a summary of the risk mitigation step in the risk management process.

10.2 Definition of risk

The general risk management process discussed in this chapter is presented in Fig. 10.1 after similar figures from internationally recognised references such as *Risk Management– Principles and Guidelines* (AS/NZS ISO 31000:2009) from the International Organisation for

Standardisation and based on the *Risk Management Standard* from Standards Australia and Standards New Zealand (AS/NZS 4360:2004). Figure 10.1 provides a framework for risk assessment in the risk management process.

The terminology used to describe risk is important because risk is all around us and we tend to describe risk in many ways according to our experience in life and work. Risk terms and definitions have evolved from many different guidelines to a common international guideline, which has helped promote a common understanding of risk.

According to AS/NZS ISO 31000:2009, risk is defined as the 'effect of uncertainty on objectives'. Risk is often expressed as a combination of the consequences from an event and their associated likelihoods of occurrence:

- The **consequence** is the outcome of an event affecting objectives.
- The **event** is the occurrence or change of a particular set of circumstances (can be uncertain).
- The **likelihood** is a general description of probability or frequency (chance of something happening).
- The **hazard** is a potential source of harm.

A more systematic approach to managing risk allows decision-makers to increase the likelihood of achieving their objectives and making informed decisions. In addition to managing threats, risk management also provides opportunities to improve performance. The main process steps presented in Fig. 10.1 are:

- **communication** – internal and external stakeholders, two-way, throughout the process, facilitating understanding
- **context** – internal and external parameters, risk management scope, risk criteria
- **risk identification** – what can happen, causes and potential consequences, defining when, where, how and why risk occurs

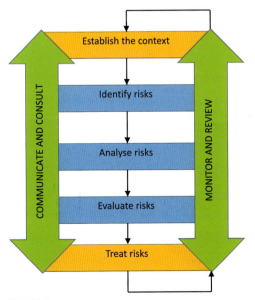

Figure 10.1: Risk management process

- **risk analysis** – methodology, existing controls, positive and negative consequences, likelihood of those consequences occurring
- **risk evaluation** – comparing risk estimate with criteria, ranking and prioritising actions
- **risk treatment or mitigation** – options for risk mitigation, cost effectiveness
- **monitoring** – effectiveness of all steps, continuous improvement.

10.3   Types of risk receptors

After defining the context (objectives and scope) of a risk assessment, the hazard and risk identification step is used to identify both the hazards and the hazard scenarios. This is a critical step because if a hazard is not identified, it will not be assessed. To illustrate the components of a hazard scenario, a schematic is presented in Fig. 10.2.

The hazard scenario in Fig. 10.2 begins on the left with a source of risk or hazard (e.g. an earthquake). A series of causes may lead to an event or accident (such as a mine waste dump failure). Following an event, there may be several consequence factors that determine the effects being analysed. There may be one or more different types of risk receptors being assessed (such as mine employees,

Figure 10.2: Anatomy of a hazard scenario

operational downtime), and the severity of effects for each type (such as injury or fatality, amount of downtime) is assessed. The 'bow-tie' diagram illustrated in Fig. 10.2 is discussed further in Section 10.5 to illustrate the objectives of different types of risk mitigation measures to block the hazard scenario (shown as a red line in the figure) from progressing left to right.

Effects on the far right of the Fig. 10.2 schematic generally include impacts to three categories of receptors as illustrated in the different coloured sections of Fig. 10.3, representing people, environment and financial aspects. At the centre of Fig. 10.3 are the risk sources, and the risk receptors radiate outwards.

Typical receptors and indicators for each of these three categories are shown in Fig. 10.3. People receptors include

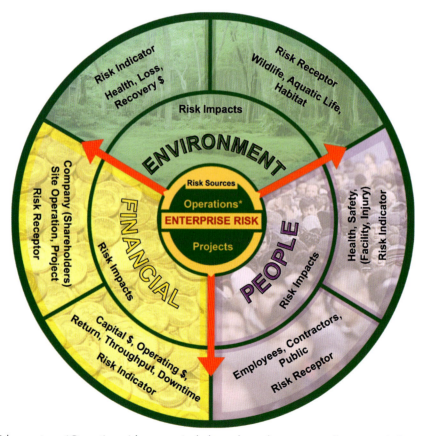

Figure 10.3: Types of risk receptors. *Operations risk sources include unplanned events as well as normal planned events.

the mine employees, contractors and the public that may be potentially exposed. Indicators of the effects on people include health, safety and social expectations. Environment receptors include wildlife, aquatic life, habitat and the ecosystem. Indicators of the effects on environment include health, loss, and recovery costs. Financial receptors include the company (shareholders), the mine operations or a particular project, and public infrastructure (e.g. a public road). Indicators of the effects on financial aspects include loss of facility, throughput or reputation and increased capital or operating costs. The temporal and spatial aspects of the consequences (when and where) may also define the hazard scenario risk. The hazard identification step is completed when material hazard scenarios that represent the risk are identified in sufficient detail to achieve the analysis objectives.

10.4   Types of risk assessment

A risk assessment includes the analytical steps of risk identification and estimation in addition to the evaluation of risks (as shown in the three blue boxes in Fig. 10.1). The objective of a risk analysis is to understand the risk and its nature and then measure it. A risk evaluation helps determine which risks may require mitigation and will help establish the priorities for risk mitigation. Following the risk assessment, if risks have been prioritised for mitigation (or treatment), then the next step involves selecting and implementing the best option for mitigating the risks.

10.4.1   Qualitative to quantitative

A wide variety of risk analysis methods have been developed and adapted within many industries, economic sectors and organisations. Different methods may address one or more of the risk analysis steps described in Section 10.1. One classification of methods follows the requirements ranging from a general understanding of risk to the most detailed understanding of risk.

In such a classification, risk analysis will involve qualitative, semi-quantitative or quantitative analyses with increasing levels of analytical detail. Qualitative methods involve descriptive assessments of risk and include multidisciplinary group evaluation, expert judgements, interviews, questionnaires, what-if analyses and checklists. Semi-quantitative methods involve numerical index descriptions of risk, and may include failure modes and effects analysis and more specialised index-based analysis. Quantitative methods involve more rigorous numerical descriptions of risk and include actuarial analysis of historical data, consequence and probability modelling, and logic trees including fault and event tree analyses. One class of methods is not superior to another, but the appropriate method should be selected to achieve the objectives. Detailed descriptions of risk analysis methods

may be found in references such as *Risk Management – Principles and Guidelines* AS/NZS ISO 31000:2009 (Standards Australia and Standards New Zealand 2009), the *Risk Management Standard* AS/NZS 4360:2004 (Standards Australia and Standards New Zealand 2004) and *Guidelines for Chemical Process Quantitative Risk Analysis* (American Institute of Chemical Engineers 1989).

The selected risk analysis method should match the risk management objectives and scope. Uncertainty is inherent in risk analysis and therefore needs to be evaluated and communicated with the results as it can affect decision-making. Finally, sensitivity analyses may be carried out to test the effect of uncertainty in assumptions or data and the effect of alternative risk mitigation measures. A few of the risk analysis methods commonly used by the mining industry and applicable to the geotechnical guidelines in this book are summarised in the remainder of this chapter.

10.4.2   Failure modes and effects analysis

The failure modes and effects analysis method was originally developed to assess the detailed risk associated with parts and components of equipment. Adaptations to this method include studying large systems rather than small components, identifying existing risk mitigation measures, and estimating and ranking the risk using the risk matrix approach. This broader systems failure modes, effects and criticality analysis (Systems FMECA) method covers all of the standard risk assessment steps previously described.

The Systems FMECA method can be used to evaluate systems such as waste dumps and stockpiles by identifying and analysing potential failure modes and estimating the risks. In Chapter 3, a waste dump and stockpile stability rating and hazard classification system was introduced. A new system was presented incorporating 22 key factors or attributes arranged in seven groups that are thought to affect stability. These groups, shown in Fig. 3.1, are Regional Setting, Foundation Conditions, Material Quality, Geometry and Mass, Stability Analysis, Construction and Performance. An index rating system is combined for each factor to classify structures according to an overall stability rating. These key factors serve as a checklist for things to consider when evaluating stability issues and design options for such structures. The stability rating system references the potential failure modes described in Chapter 8 and their contributing factors, and can also provide the failure mode input for a Systems FMECA. However, the stability rating system is not a risk assessment method as consequences of failure and risk (effects and criticality) are not explicitly addressed.

The Systems FMECA approach first identifies material failure modes and their associated consequences using an assessment protocol and the knowledge base of the risk assessment team. Existing treatments and controls to mitigate risk are also identified. The safety, environmental or financial risks (as defined in the objectives) are then estimated for each failure mode and associated consequence (hazard scenario) using a risk matrix approach.

The Systems FMECA approach is implemented by a team of experienced personnel assessing risks in a systematic workshop process. A team may include operations, engineering, environmental and management personnel and is led by an experienced risk assessment facilitator. The experience of team members provides the knowledge base, and the workshop format provides a method to build synergies among the different types of experience. Other stakeholders, such as the public or government, may also participate when appropriate to promote communication and understanding in a transparent process. In addition to team knowledge, background information, including the operational or industry history, engineering and design information, operational and monitoring procedures and surrounding environmental exposures, is collated before the workshop. Specific background information requirements depend on the objectives and scope of the risk assessment.

A Systems FMECA is a comprehensive process designed to identify potential significant failure modes associated with the system being assessed (e.g. a waste dump). The failure mode describes how a system may fail and includes possible causes ranging from natural hazards, such as earthquakes, to equipment failures, operator errors and management system deficiencies. Potential consequences and safety, environmental or financial effects (as defined in the study objectives) are also identified for each failure mode. For example, environmental effects may be measured in terms of environmental damage from the runout of a failed waste dump caused by a shallow embankment failure. A series of events usually needs to occur before a failure mode results in an effect, and therefore the complete series of events is

assessed. Following the identification of this series of events, the risk or 'criticality' is estimated using a risk matrix approach. For each of the significant failure modes and corresponding consequences (hazard scenarios – see Fig. 10.2) identified in the Systems FMECA, a measure of the associated risk is estimated using a risk matrix. A risk matrix is composed of one index representing the measure of likelihood and another index representing the measure of consequence severity. When a 'failure mode and consequence' scenario is identified, the associated risk is estimated by locating it within the risk matrix. The risk matrix may be used to present risk estimates derived from any risk analysis method, including more detailed quantitative analysis.

In the example risk matrix format shown in Fig. 10.4, the likelihood (defined here by frequency) index on the left of the matrix ranges from a 'rare event' to a 'probable event' and is more formally defined in terms of frequency with an events per year value. Events could also be defined in terms of probability (with a scale from 0 to 1). The example index is divided into orders of magnitude with the expectation that the knowledge base of the team and the historical industry performance record will be sufficient to estimate the level of risk to this accuracy. The likelihood index could also be divided into a lesser or greater number of divisions with different definitions depending on objectives.

In the example presented in Fig. 10.4, five categories of consequences are being assessed:

- people impacts measured according to:
  → health and safety
  → community concerns
- environment impacts measured according to:
  → magnitude of impacts on the biological or physical environment
- financial impacts measured according to:
  → operational downtime
  → total costs.

		CONSEQUENCE SEVERITY				
	Category	(A) Very Low	(B) Low	(C) Moderate	(D) High	(E) Very High
	I Health & Safety					
	II Community					
	III Environment					
	IV Operations					
	V Cost					
LIKELIHOOD						
Index	Events/year					
5 Probable	>1	Moderate	High	Critical	Critical	Critical
4 Likely	1 to 1/10	Moderate	High	High	Critical	Critical
3 Possible	1/10 to 1/100	Low	Moderate	Moderate	Critical	Critical
2 Unlikely	1/100 to 1/1000	Low	Moderate	Moderate	High	Critical
1 Rare	1/1000 to 1/10,000	Low	Moderate	Moderate	High	High

Increasing risk

Figure 10.4: Risk matrix format

There may be many other categories of consequences and definitions of the severity of consequence levels depending on the risk assessment objectives. The consequence severity index could also be divided into more or fewer levels. In general, all categories may be addressed as people, environmental or financial impacts (see Fig. 10.3). Any of the categories may be relevant for a particular risk assessment, and all of these risks can be estimated using the risk matrix method.

The severity of effects for each category of consequence is defined by an index ranging from very low to very high. Although these indices are blank in the figure, they would be defined as appropriate for the risk management program. In total, there are five separate risk matrices shown in Fig. 10.4 (one for each consequence category), and each hazard scenario would be located in one or more of these five matrices as appropriate. Further analysis may be required following the Systems FMECA to refine either the likelihood or consequence severity estimates.

Following the identification and measurement of risks, they can be evaluated and then managed appropriately. The evaluation of risk requires determining the priority of risk as defined through the different locations (or risk values) within the risk matrix developed for the risk assessment. The criteria for evaluating risks are developed for the risk management program and are useful for comparing risks, such as among different operations, or for prioritising risks. The risk matrix shown in Fig. 10.4 is divided into four groups representing the criteria for managing risks. These groups are colour-coded from green (lowest risk) to yellow to orange to red (highest risk). Examples of the associated criteria for managing risks are shown in Fig. 10.5.

Most often, the grouping to define risk priority reflects the corporate risk management program. If the grouping is defined through a mathematical function of the two indices, care must be taken not to estimate risk incorrectly. For example, if each index is arbitrarily given a value of 1 to 5, then arbitrary values from 1 to 25 can be calculated for each location of the risk matrix by multiplying the respective index values. The appropriate criteria should be

carefully developed to match the risk assessment objectives.

Risk mitigation measures are described in Section 10.5. One benefit of the risk matrix is to clearly illustrate how risk has been reduced by decreasing likelihood, consequence severity or a combination of the two. Different cost implications are often associated with the selection of measures that address these different approaches to reduce risk.

Results from the Systems FMECA are documented in a risk register that includes the following information:

- system, unit description
- failure mode
- causes
- consequences (one or more for each failure mode)
- existing controls (safeguards)
- residual risk estimate according to likelihood and consequence location in each risk matrix
- notes (documenting uncertainty).

There are many variations of the risk register, and an example format is presented in Fig. 10.6. The risk estimate may also be shown before and after the successful implementation of risk mitigation measures.

The risk matrix method and risk register approach may be used to estimate and/or present risk results derived using other risk analysis methods and are not limited to the Systems FMECA approach.

10.4.3 Logic trees – fault and event trees

Logic trees include fault trees, event trees and their combination in cause–consequence trees (sometimes referred to as bow-tie diagrams). These trees provide graphical presentations of the sequence of events describing a hazard scenario as depicted in Fig. 10.2. Looking at this figure, a fault tree could be developed to show the sequence of events (and the logic connecting them) for the left part of the hazard scenario leading to the top event, a hazardous accident. Then an event tree could be developed to show the different pathways leading from the hazardous accident to different consequences.

Risk level (identified in the risk matrix)		Risk management action
	Critical	Action required. More detailed risk analysis may be carried out
	High	Assess risk mitigation options and prioritise resources to manage these risks before Moderate or Low ranked risks
	Moderate	Assess risk mitigation options and manage these risks
	Low	Monitor risks

Figure 10.5: Risk evaluation

Study facility: Expanded waster dump
Objectives: The project will increase the capacity of the existing waste dump to xx million m³, and its area to xx ha.

Failure mode	Cause	Consequences	Planned risk mitigation measures	Type of control	Frequency (Risk matrix)	Severity (Risk matrix)					Notes
						Health & Safety	Community	Environment	Operation	Cost	
1 Dump movement	1.1 Earthquake	1.1.1 ARD-generation materials from the waste dump release into offsite environment and local rivers.	Geotechnical design; construction quality assurance; geotechnical investigations and monitoring;	Engineering Administrative Recovery	1	–	–	D	C	D	xx m height limit and xxH:xxV slope
		1.1.2									
	1.2 Inadequate design										
	1.3										

Figure 10.6: Risk register format

A combination of the fault and event trees results in a cause–consequence diagram. This application to the hazard scenario in Fig. 10.2 is just an example, and both types of logic trees could be applied to different events in the hazard scenario as needed.

A fault tree begins with a top event, usually a failure of interest, and then causes are deduced through a series of events that lead to the top event. Events are connected through logic gates that can be described using the example in Fig. 10.7. The top event is Dump Failure, and causal event sequences were developed that lead to this failure. The first logic gate is an 'OR gate' connecting three causes: Other Causes, Inadequate Design or Inadequate Construction QA/QC. The OR gate logic says that if any one or more of these three causes occur, then the Dump Failure occurs. The other principal logic gate is an 'AND gate', which is shown leading to the Inadequate Construction QA/QC cause. The two events connecting through this AND gate are Rapid Construction and Poor Foundation Preparation. The AND gate logic says that only if both of these two causes occur does the Inadequate Construction QA/QC cause occur. There are many variations to these two basic logic gates, but all of them can be created from these two gates. Two additional OR gate sequences are also shown in Fig. 10.7, and the triangles simply indicate that the associated branch continues elsewhere.

Fault trees can be complex models, but simple versions (using a few independent events) have also proven to be effective tools for risk assessments with the following benefits:

- **Identification of all factors** – the creative method to develop a sequence of events leading to the failure

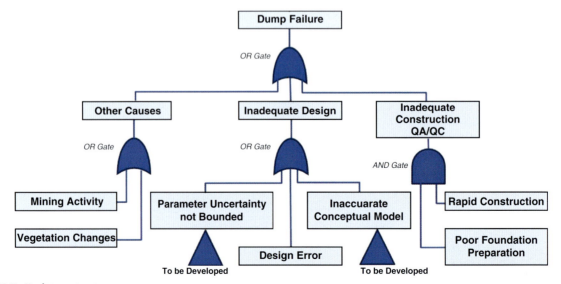

Figure 10.7: Fault tree structure

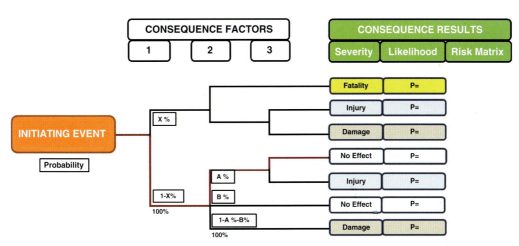

Figure 10.8: Event tree structure

provides a systematic framework for members of a team to apply their different types of expertise to identify hazard scenarios. Events can range from natural events, such as earthquakes, to equipment failures, operator errors and management system deficiencies.

- **Focus on one system failure** – only one top event is defined and this keeps the assessment on track.
- **Provides a picture of failure** – the graphical presentation of the issues helps communicate the results among members of the assessment team and to other interested parties, such as management or stakeholders.
- **Qualitative or quantitative** – developing a fault tree can provide valuable insights for risk management, and its structure can be used to develop risk mitigation strategies. Should a quantitative approach be required, the fault tree model provides the mathematical framework for combining probabilities (point values or

distributions) of lower events through the logic gates to estimate the probability of the top event.

- **Risk estimate, sensitivity** – an estimate for the top event likelihood can be easily calculated for simple fault trees and sensitivity analysis can be carried out to determine the more significant events and where best to look for risk mitigation measures. Event probabilities are multiplied through an AND gate (resulting in a lower probability for the gate output) and added (if small) through an OR gate (resulting in a higher probability for the gate output). The more complex a fault tree, the more complex the analysis becomes.
- **Risk mitigation** – the risk impact of implementing various risk mitigation measures can be estimated both qualitatively and quantitatively.

Event trees are also logic diagrams, but start with an initiating event such as a failure and show the pathways

Figure 10.9: Event tree example

leading from this failure to the different possible consequences. The event tree structure can be described using the example in Fig. 10.8. The Initiating Event could be a dump failure, and the consequence sequence could involve as Factor 1 the potential for slide material to run in two directions (upper or lower branch of the tree). Factor 2 could be the volume of slide material, which for the upper tree has two options and for the lower tree has three options. Factor 3 is applicable to only two of the pathways shown. Each pathway ends with a consequence severity, which in this example is colour-coded according to four possibilities: Fatality, Injury, Damage and No Effect. The likelihood is calculated by multiplying the Initiating Event probability by the percentage associated with each node of the pathway (the total of each branch probability from a node is 100%). One of the seven pathways is highlighted in red as an example leading to No Effect. Total probabilities for a specific consequence severity (e.g. Injury) can be added together (in this example there are two injury pathways). Finally, the results can be presented in a risk matrix format as discussed in the previous section.

An event tree application to assess risks of a contaminant leaking through a stockpile and into the environment is presented in Fig. 10.9. This example follows possible pathways for a contaminant solution to escape through a liner, then through an underdrain and into the groundwater, and finally be transported in a particular groundwater flow direction. Flow rates and concentrations are also calculated along the flow pathways, and likelihoods are estimated based on engineering information such as groundwater modelling.

10.5   Risk mitigation and management

The risk management process was introduced in Fig. 10.1. As shown, following the assessment of risk described in Section 10.3, if risks have been prioritised for treatment, the next step involves selecting and implementing the best option for mitigating the risks. Where risks do not meet the established acceptance criteria or tolerance, potential risk mitigation options are developed and cost-effective solutions are analysed to mitigate risks to the necessary criteria level (or possibly eliminate them). Several measures may be applied to treat risks depending on the evaluation. Not all treatments may be financially based, depending on the criteria and management requirements. Another iteration of the risk analysis may be appropriate to confirm the estimated risk benefit for implementing the selected mitigation option.

Usually, there are several controls in place, such as monitoring or detection systems, engineering analyses or emergency response plans that will prevent the

development or limit the consequence severity of a hazard scenario. Such controls or risk mitigation measures need to be successfully implemented in order to manage the risks to expectations. The hazard scenario schematic presented in Fig. 10.2 was adapted to illustrate four different classes of risk mitigation measures and where they are applied along the hazard scenario pathway progressing to a consequence effect. Typical risk mitigation measures are shown in these four classes as grey barriers to the progression of a hazard scenario in Fig. 10.10.

Each class of mitigation measures, described as Failure Prevention, Safeguard, Impact Limitation, and Receptor Protection in Fig. 10.10, provide an additional layer of protection. The first two classes can prevent an accident from occurring, while the last two can mitigate the severity of consequences from an accident. Any number of measures may be implemented for a given hazard scenario, and if the associated risk levels are still above criteria, then additional measures may be added.

The successful implementation of controls and mitigation measures is critical to estimating risk, and the risk management program will include tasks to assess and monitor the efficiency of all such measures (or re-evaluate the risk). These controls can be organised in an ordered hierarchy of classes according to their potential effectiveness from most to least effective:

- **Elimination** – remove the hazard and eliminate the risk (most effective)
- **Substitution** – change or replace a hazard source with something that will not create the hazard
- **Engineered controls** – redesign to control the exposure; hazards are not eliminated but receptors are less exposed
- **Administrative controls** – change how people work
- **Personal protective equipment** – protect people (least effective).

There are many industry variations of this hierarchy to include issues such as receptors other than people and recovery procedures. Risk assessment methods such as the Systems FMECA described in Section 10.3 often include consideration of the hierarchy of controls, and the example risk register in Fig. 10.6 shows a column to record the type of control.

Once implemented, the effectiveness of risk controls and mitigation measures can be assessed to ensure that they are operating as expected and that risks are not underestimated. This may be part of a risk assessment method such as the Systems FMECA or it may be a separate analysis. Effectiveness can be estimated according to a checklist addressing availability, reliability and effectiveness when working. These attributes can also be analysed though models and monitoring programs. The risk analysis techniques that were previously discussed,

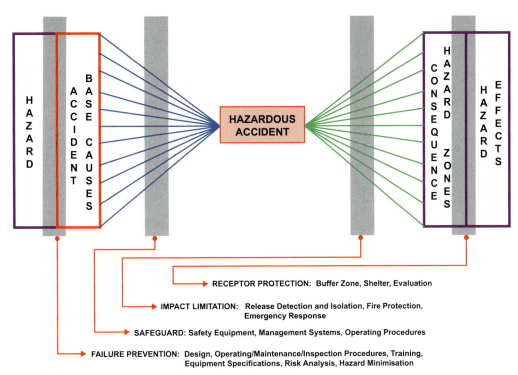

Figure 10.10: Typical risk mitigation measures applied to a hazard scenario

including the use of 'bow-tie' diagrams such as illustrated in Figs. 10.2 and 10.10, can be applied specifically to assess the effectiveness of multiple risk controls. For example, if appropriate training is a risk mitigation measure, then the program for training, the time between training sessions, competency monitoring and continuous updating are issues that can impact the effectiveness of that mitigation (and therefore the estimated risk).

Risk management includes the program to assess, implement and monitor the risk mitigation measures identified during a risk assessment. The risk register illustrated in Fig. 10.6 can be extended with further columns or a companion document to identify an owner for the risk mitigation measure and management data such as what the options are, who is evaluating them, what the schedule is, when the measures are implemented and how future effectiveness is monitored.

Finally, risk mitigation and management are part of a continuous process, as illustrated in Fig. 10.1, that applies throughout the life cycle of all facilities.

11

OPERATION

Andy Haynes and Geoff Beale

11.1 Dump and stockpile management plan

Construction of the waste dump or stockpile is generally the responsibility of the mine's operations department and involves several activities:

- transport of waste and stockpile materials from the mine or mill to the dump or stockpile
- placement of the materials in the dump or stockpile in accordance with the planned development and operating plans, including lift height and location
- access road construction and maintenance
- clearing of new areas for dumping or stockpile construction, foundation preparation and drain construction as required
- maintenance, upgrade and expansion of surface water management facilities
- environmental monitoring of conditions at the dump or stockpile, including seepage water, surface water and groundwater quantity and quality
- dump and stockpile performance monitoring and documentation, including stability, erosion, consolidation and creep.

Environmental monitoring is normally the responsibility of the mine environmental group and may include surface water, sediment, seepage water, groundwater levels and quality, dust and air quality, and geochemistry sampling.

11.1.1 Operational guidelines and standard operating procedures

The operation of a waste dump or stockpile must be consistent with the design basis and assumptions. Operational guidelines or standard operating procedures should be developed using the design basis. Regular contact between those involved with design and those involved with operation is essential to ensure the continuity of information, especially in the event of staff turnover.

11.1.2 Roles and responsibilities

The mine manager is ultimately responsible for all aspects of mine operation. However, it is common practice to delegate responsibility to operations managers and supervisory personnel. The roles and responsibilities of the various staff must be clearly identified and understood by all relevant parties. The tasks necessary for safe and effective operation will vary for each site and will depend on many factors, including the size and configuration of the facility, the mining rate, the location of the active dumping areas, and the climatic setting.

11.1.3 Monitoring protocols and trigger action response plans

Monitoring of the performance of dumps and stockpiles is essential to confirm that performance is consistent with design expectations and to allow adjustments to be made to the dump plans as necessary to accommodate changes to the mine plan. It is common for operations to constantly adjust the active dumping and stockpiling areas based on visual observations and measured deformation rates and to appropriately accommodate materials of varying quality. The monitoring program must also consider potential failure mechanisms.

The operational guidelines or standard operating procedures should identify the monitoring protocols be used as the basis for developing a trigger action response plan (TARP) that identifies the appropriate actions for various operating scenarios. The type of measurements and threshold levels must be established for each area of the facility and must reflect the type of dumping operation (e.g. end dumping or push dumping), the material type and quality, the size of the dump or stockpile and the consequence of failure. Instrumentation and monitoring systems and TARPs are discussed in more detail in Chapter 12.

The operational guidelines or standard operating procedures should clearly identify status categories (e.g. open, standby, closed) and the applicable constraints around general access and dumping activities. A robust communication/signage system should be in place to ensure that all relevant parties are aware of the status and the appropriate actions.

11.2 Foundation preparation

The requirements for foundation preparation must be identified at the design stage. The requirements will depend on several factors:

- strength of surficial materials
- presence of marketable trees
- presence of other vegetation
- requirement for drainage
- corporate or permit requirements for topsoil salvage for reclamation.

Foundation preparation must proceed far enough in advance of the toe area to avoid potential hazards created by the ongoing dump operations. Where it is necessary to work within the runout zone of the dump, it may be necessary to modify or suspend dumping and verify the stability of the dump before foundation preparation can proceed. Alternatively, it may be possible to utilise impact berms or deflection berms between the dump toe and the foundation preparation work to protect personnel and equipment.

11.3 Climatic conditions

The prevailing climatic conditions are an important factor for waste dump operations. The operating procedures must account for rainfall and surface water runoff, snowfall, temperature and wind conditions that may impact the integrity of the dump and the safety and efficiency of the dumping operations.

11.3.1 Surface water management

Typically, there are three goals for surface water management on the waste dump or stockpile facility:

- **Minimising infiltration** – the operation may need to consider rapid shedding of water from the dump surfaces to minimise infiltration, particularly where there is the potential to cause a build-up of pore pressure or to create poor-quality seepage water.

The operating facility should normally minimise the area of any flat surfaces and hollows.

- **Minimising erosion** – it is often necessary to use surface water collection channels to minimise uncontrolled runoff that may lead to erosion. The surface water channels may need to include drop structures or other best management practices to minimise erosion or control sediment. The operating plan may also include the placement of crest berms to prevent uncontrolled runoff down the outer faces.
- **Minimising the development of poor-quality runoff** – the operational surface water management plan may need to minimise the contact of water with zones of more reactive waste materials and/or to minimise the transportation of contact (reactive) sediment.

Good design practice for the facility drainage system will typically include upstream diversions to prevent run-on to the facility and to maximise the proportion of non-contact water that is available for downstream mixing. Ongoing construction and extension of upgradient and lateral diversions are often required as the footprint area of the dump/stockpile increases. The use of flow-through rock drains may be considered as an alternative to convey upgradient flows beneath the dump, particularly where diversion of the flow is not feasible.

Operating practices for the facility will often include the following:

- surface water collection channels routed along the side of access ramps
- surface water collection channels along permanent access benches between lifts (these channels often need to be lined)
- sloping of working areas towards collection channels and haulroads
- minimisation of flat areas and hollows to reduce the potential for ponding water to accumulate in topographic low points on the dump surface
- prevention of uncontrolled runoff down external faces by placement of crest berms
- avoidance of steep channel profiles that may increase the potential for erosion, or the incorporation of drop structures into the channel design
- incorporation of sediment traps within the drainage design, as needed.

Dump surfaces that include a high proportion of fine-grained materials with a low proportion of hummocky or blocky areas will have an increased potential for runoff. Peak runoff flows and cumulative volumes will usually be greater when active drainage measures are installed.

Although the curve number approach can be used for assessing the magnitude of runoff from the dump, the above factors must be carefully considered, and caution should be used when selecting the curve number value (Chapter 6) because of the complex geometry and variable surfaces of most operating dumps. The dynamic nature of a working dump means that, for any given rainfall intensity and duration, the proportion of the rainfall that contributes to runoff can vary with time as the dump evolves. Consequently, the development of simple runoff coefficients (with an appropriate envelope for uncertainty) is often appropriate.

In some (unusual) cases, it may be advantageous to encourage surface water infiltration into the surface of the waste dump, and therefore minimise runoff and erosion potential. This may be applicable in situations where one or more of the following conditions apply:

- The waste rock materials are inert and there are no concerns for seepage water quality.
- The waste rock (and possibly foundation) materials are permeable and there are no concerns regarding pore pressure.
- Mean annual precipitation is low and there is little potential to saturate the waste rock materials even if storm events are allowed to infiltrate.
- The runoff may create erosion and generate a large amount of sediment, so it is easier to manage controlled seepage from the facility than uncontrolled surface water runoff.

In wetter climates, surface water management is an integral part of waste dump operations. Routine inspection and maintenance of surface water control systems are therefore important. Geotechnical staff must work with the mine operations group to implement the required repair and maintenance work. Regular cleaning of control structures is often required immediately before, and regularly throughout, the wet season.

The surface water management plan may need to consider the potential for ongoing settlement of the materials and, in particular, how settlement may create hollows in the dump surface and the need to actively manage these (usually by placement of additional fill material).

11.3.1.1 Case study from Botswana

At a mine site in Botswana, a significant rainfall event (120 mm in 48 h) led to the development of a washout gully on the side of a large waste dump (Fig. 11.1). The surface area of the dump was ~1000 m by 800 m, and materials within the dump were mostly old tailings that were relatively coarse and free draining. The materials were layered because of stacker alignments, but no saturation was noted in the walls of the gully (Fig. 11.2). No seepage was evident in any of the toe area, and it was concluded that the gully was caused solely by surface water erosion.

The surface of the dump was flat and included a significant proportion of fine-grained materials (partially wind derived). Significant ponding of surface water had previously occurred at low points on the dump surface (as was evidenced by recent, thin, fine-grained sediment deposits). Based on the rainfall records, it was calculated that the *average* year-round infiltration rate to the dump surface was less than 10^{-8} m/s which, based on site observations, was lower than any potential unsaturated or saturated vertical permeability of the materials, suggesting that the presence of any saturated layers within the dump was unlikely.

Figure 11.1: Botswana case study: looking from the toe area to the erosion gully. Source: G Beale

Figure 11.2: Botswana case study: layering of the exposed materials in the sides of the erosion gully. Source: G Beale

It was also calculated that the peak rainfall event may have produced $\sim 10^{-3}$ m/s short-term (less than 30 min) water accumulation potential and $\sim 2 \times 10^{-6}$ m/s daily water accumulation potential. Both of these values were likely to have been higher than the infiltration capacity of the fine-grained surface materials. Therefore, it was evident that a considerable amount of surface water runoff occurred during and immediately following the rainfall event. The field inspection indicated that, because of the flat upper surface of the dump, a significant portion of the uncontrolled runoff had flowed over the crest, thus creating the gully. Surface water erosion on the dump surface above the crest of the gully was also evident, with smaller runoff channels leading towards the main gully.

Because of the large, flat surface area of the dump, it would have been impractical (and very costly) to carry out enough regrading to implement a program of surface water collection and runoff control on the dump surface. Therefore, the following alternative controls were implemented:

- Regrading was done to stabilise the area of the gully before the next rainy season.
- Test drill holes and piezometers were installed to confirm there was no saturation or pore pressure contributing to the instability.
- Control berms were installed on the dump surface to prevent future runoff from reaching the crest areas. Prior to construction of these, a coarse topographic survey of the top of the dump was carried out to help determine the design and sizing of the berms and to identify future areas where water may pond. The size and width of the berms was increased around the

(marginally) lower points on the crest, and particularly above any infrastructure that was present below the dump toe areas.

- Runout control berms were installed below toe areas where there was a need to protect infrastructure.

11.3.2 Groundwater management

Groundwater management is often part of the dump design, but is usually less of a concern for operations, except in cases where one or more of the following points apply:

- Ongoing construction/extension of rock drains is required as the footprint area of the dump/stockpile increases.
- Ongoing extension of the underdrainage system is required as the footprint area increases.
- There is a need to construct management facilities to collect and control seepage in the downgradient toe areas.

There is also a need to work around and protect monitoring facilities. For example, the cables from vibrating wire piezometers or time domain reflectometry systems may need to be extended (with adequate protection) to allow dumping of materials above the original installation if the original drill hole collar becomes buried by subsequent dump lifts.

11.3.3 Snow and avalanche management

Large accumulations of snow can present significant operational and geotechnical issues for waste dumps. Wind-blown snow on the dump surfaces may become mixed and entrained within the waste rock. If the depth of snow is sufficient to maintain a discrete and through-going layer or lens in the dump, it may have the potential to impact stability. This situation usually becomes more important when the entrained snow begins to melt. An issue may arise where active end dumping is occurring on leeward faces where snow is concurrently accumulating. This may lead to an increased concentration of moisture in the lower toes, which may, in turn, increase the potential for seepage.

Snow and avalanche management plans may need to be developed if snowfall regularly accumulates on high leeward slopes. Where the avalanche hazard is severe, it may be necessary to seek the advice of a specialist. The plans will vary depending on the location and magnitude of typical snow accumulation, but will typically include the following:

- seasonal adjustments to the dumping plan or to access locations
- active removal of snow from areas where accumulations become significant
- the use of designated snow dumps, where possible.

If there are no snow-free dumping areas available, the following options may need to be considered:

- dumping on lower sections of the dump where hazards are lower (but considering the avalanche potential from the slopes above)
- dumping in areas that afford greater stability through topographic diversity or confinement (e.g. gullies that are not sheltered and prone to snow accumulation)
- dumping at the end of a side hill dump, extending the dump parallel to the valley; thus, the snow-covered surface will possibly be oriented such as to lessen the impact on stability
- dumping on the horizontal surface of the dump platform at the back of the dump (i.e. away from the face) to form a lift
- dumping on exposed platforms where the potential for accumulating snow is reduced because of the wind.

11.4 Concurrent reclamation

Guidelines on waste dump closure and reclamation are presented in Chapter 16. Closure and reclamation activities can be performed at the end of operations and/or during

Figure 11.3: Example of concurrent dump reclamation carried out as part of operations with the goal of stabilising the dump faces and optimising surface water runoff chemistry (Mogalakwena Mine, South Africa). Source: G Beale

the operational phase. If reclamation works are to be performed during the operational phase, the work must be integrated into the plans for the active areas of dumping.

The goals of a concurrent reclamation program are typically to improve surface water management and to reduce the burden of work following full mine closure. In many cases, the dump faces can be reclaimed without the need for regrading (Fig. 11.3). However, concurrent reclamation can have the added benefit of increased stability where dump slopes are regraded to a flatter angle.

11.5 Material quality control

The quality of waste material obtained from open pits is typically variable. Some waste dumps are designed such that they can accommodate all materials without differentiation. However, such designs may be unnecessarily conservative. Other dumps are designed such that some differentiation of waste materials is required. Differentiation may be based on strength, particle size, geochemical properties or geochemistry. Practices for differentiation may include the following:

- placing materials of lower strength in portions of the dump less critical to stability
- placing materials of lower strength in dumps for which the risk and/or consequences of failure are lower
- using coarse, clean, high permeability materials for rock drains
- using high strength and/or permeable materials on downstream faces
- segregating potentially acid generating and/or reactive materials (see Chapter 14).

If differentiation is necessary to achieve the desired operational performance, a quality control program will be required to ensure that the plans for differentiation are correctly implemented.

Where possible, overburden soils should be stockpiled for reclamation use. Where it is necessary to place weak and/or fine-grained overburden soils in the main waste dumps, the scheduling should be such as to blend the lower strength materials with higher quality waste, or to consign the lower strength materials to areas of the dump that are not susceptible to overall instability, and where they will not adversely affect internal or subsurface drainage.

11.6 Dumping operations

11.6.1 Crest berms

Safety berms must be maintained on all dump crests and must be of sufficient height to prevent the largest mine

equipment from inadvertently driving over the crest. The height may be defined by local regulatory authorities, but is typically no less than half the height of the largest haul truck tyre.

Safety berms are not meant to be used as a wheel stop when backing up and their use as such should not be permitted. Where there is any doubt, the operating practice would be to dump short of the crest and to push the dumped materials using a dozer. It is also important that some of the dumped material is retained on the crest of the dump to provide the dozer with sufficient material for ongoing crest berm construction. There may not be adequate materials for berm construction if the entire loads are dumped over the crest.

11.6.2 Dump platform

The condition of the dump platform must be monitored visually for any signs of instability. As a dozer is typically used on the crest of an active dump, it is generally the responsibility of the dozer operator to ensure that the surface of the dump is maintained in good condition and to spot the trucks as they back up. However, in some operations, a dump person (i.e. spotter) is used to direct trucks to appropriate dump locations on the dump platform. In these operations, the dump person, who is typically on foot, may be in a better position to assess the condition of the platform and the need for maintenance.

The dump platform should be maintained with an uphill grade to the crest. Because of the ongoing settlement of an active dump, this will require periodic work. A grade of not less than 2% should be maintained to facilitate surface water drainage away from the crest and so that trucks do not have to back downhill.

11.6.3 Signs

Adequate signage should be maintained on dump platforms and access roads to ensure smooth traffic flow. Signs will be required to clearly mark active and closed dumping areas, direction of travel (i.e. left-hand drive or right-hand drive), crossovers and other traffic directions or instructions.

11.6.4 Dump lighting

Most of the mines included in the 2013 Large Open Pit waste dump survey (Appendix 2) maintain lighting on the dumps when operating at night. Lighting plants should be adequate to effectively light the active dumping area and particularly the crest berm. The lights should be placed in such a way as to avoid shining directly into truck operators' eyes during the approach or dumping procedure.

The American National Standard Practice for Industrial Lighting (ANSI/IES 1984) provides suitable guidelines for minimum illumination of active dump platforms. The ANSI specification for quarries (the most relevant category) recommends 50 lux (or 5 foot-candles).

11.6.5   Safety slings

Safety slings, or cables, are used for emergency towing of equipment. Their use on the dump platform would be required in the event of a haul truck becoming disabled on unstable ground near the dump crest. A dozer operator would connect the cable to a suitable attachment point on the truck and proceed to tow the disabled vehicle to stable ground as quickly as possible. Safety slings should be kept in a highly visible and accessible location close to active dumping areas. Supervisors, dump persons, dozer operators and truck drivers should be familiar with locations and use of the equipment.

11.6.6   Roads

Maintenance of road conditions is beneficial for safety and vehicle maintenance. Care should be taken to remove any loose rocks from roadways. Visual inspection of roadways near dump crests should reveal any problems due to subsidence or cracking. Regular inspections of the stability of any dump faces above active roadways should be carried out, with the frequency of inspection dependant on the operating plans and rainfall patterns. Inspection of surface water drains should also be carried out, particularly during the wet season or at times of the year with increased frequency or potential for high runoff events. Haul truck operators, dump persons, dozer operators and any others who routinely visit the dumps should be made aware of the inspection and maintenance programs, and should be trained in the recognition of hazards and reporting procedures. Normally, any identified hazards should be reported immediately to the dump supervisor.

11.6.7   Rock roll-out

As material is dumped or pushed down the face of a dump, a potential hazard from rolling rocks is introduced. The bulk of the waste material stops on the face or at the toe, but some larger rocks have the potential to roll and bounce well beyond the toe area. The hazard can typically be managed by either controlling access to the roll-out zone or by installing impact berms or tyre berms downslope of the toe to attenuate rolling rocks.

11.6.8   Runout zones

Runout zones/distances must be clearly marked in the field via berms and/or signage when activities will occur beneath an active part of the dump, or when access by the public is possible. Runout is described in Chapter 9.

11.6.9   Dumping sequence

The dumping sequence should consider the following factors:

- **Haulroad configuration** – typically, dumping will commence as close as possible to the open pit; however, some dumps may require that materials be placed preferentially in particular areas to achieve adequate stability in subsequent lifts. In this case, additional haulroads may be required to facilitate access to different dump areas at the same time.
- **Foundation conditions** – where steeper foundation slopes occur, it is often preferable to advance the dump along the contours rather than down the slope. Where lower strength materials exist in the foundation, the designer would identify whether the materials must be removed or can remain in the foundation. In cases where they can remain without causing instability, it is often necessary to sequence the dump development to confine such materials and avoid local interim stability issues.

11.7   Advance rate

As described in Chapter 8, dump stability is influenced by many factors, including dump height, dump materials, dump geometry, climatic conditions, foundation materials, foundation geometry and surface and groundwater conditions, making it difficult to develop specific loading rate guidelines. However, it is clear from studying case histories of mine dump failures that there is a relationship between dump stability, dump height and crest advance rates.

A compilation of the rates of crest advance was made using data obtained from the 1991 British Columbia waste dump survey and the 2013 Large Open Pit waste dump survey (see Chapter 1 and Appendix 2). Crest loading rates reported by survey respondents were given in terms of volume dumped per linear metre of crest per day ($m^3/m/d$) or in metres per day (m/d) of crest advance. These data are presented in Fig. 11.4.

A curve, labelled 'lower bound of most failures' in Fig. 11.4, was drawn such that most failed dumps fall above and most stable dumps fall below the line. As a guideline for new mines or dumps with no performance history upon which to base dumping rate limits, it is suggested that the initial relationship of dumping rates and dump height be governed by the second curve, labelled 'guideline' in Fig. 11.4. This curve has been offset from the lower bound curve to offer a degree of extra safety. As site-specific data become available, it may be possible to

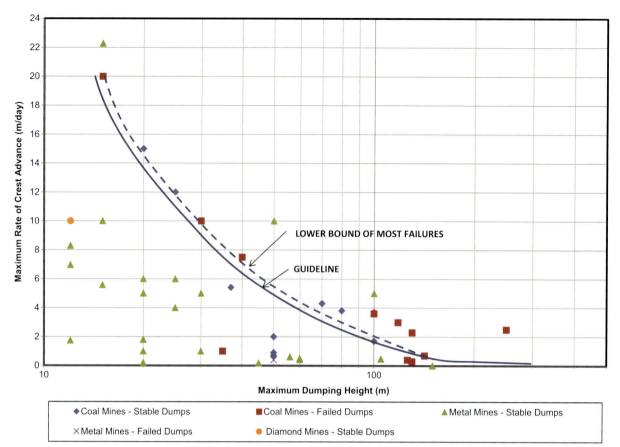

Figure 11.4: Plot of dump crest advance rate versus dump height (1991 and 2013 survey results)

revise the criteria for a particular dump based on its proven performance. The fact that some stable dumps plot above this guideline illustrates this point. However, had the dumps which did fail been restricted by this guideline, it is likely that some, if not most of the failures, could have been avoided.

It is valuable to develop a site-specific relationship similar to the one above, which may potentially be refined to reflect some of the other variables and assist selection of appropriate crest advance rates. This can include specific relationships based on waste material quality and foundation conditions. An example of such relationships is illustrated in Fig. 11.5.

By keeping detailed records of loading rates (i.e. crest advance rate), crest displacement/deformation rate, foundation conditions (i.e. slope, material, groundwater levels), material quality, weather conditions, construction method (i.e. end dump or push dump), dump height and time of year, it should be possible to develop site-specific crest advance rate relationships.

It is important to distribute the waste material relatively evenly along the available active dump crest length. Equal load distribution across the entire active dump crest reduces the risks of instability by reducing the pressure applied to the foundation materials and the resulting potential pore pressure increases. The management procedures should include methods for monitoring and managing the distribution of waste materials.

(a)

Material quality – foundation condition	Maximum toe–crest height (m)						
	10	25	50	100	150	200	300
	Maximum daily crest advance (m)						
Very good waste – very good foundation	20.00	10.00	5.00	2.50	1.50	1.25	1.00
Very good waste – fair foundation	15.00	7.50	3.75	2.00	1.25	1.00	0.75
Very good waste – very poor foundation	5.00	2.50	1.00	0.50	0.30	0.15	0.05
Fair waste – very good foundation	12.00	6.00	3.25	1.75	1.00	0.75	0.50
Fair waste – fair foundation	10.00	5.00	2.75	1.25	0.50	0.40	0.30
Fair waste – very poor foundation	5.00	2.50	1.00	0.50	0.30	0.15	0.05
Very poor waste – very good foundation	8.00	4.00	2.00	1.00	0.40	0.25	0.15
Very poor waste – fair foundation	6.00	3.00	1.50	0.75	0.35	0.20	0.10
Very poor waste – very poor foundation	5.00	2.50	1.00	0.50	0.30	0.15	0.05

(b)

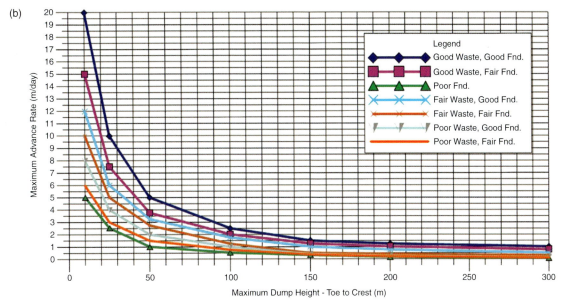

Figure 11.5: Example of relationship between maximum allowable crest advance rate and dump height for different foundation conditions and material quality

12

INSTRUMENTATION AND MONITORING

James Hogarth, Mark Hawley and Geoff Beale

12.1   Introduction

Instability of waste dumps and stockpiles is usually preceded by warning signs, such as increased rates of deformation, cracking and/or settlement of the platform; bulging of the face; bulging of the toe or bulging or heaving of the foundation in front of the toe (i.e. toe spreading); seepage on the face; increased pore pressures in the foundation or embankment; and, in some cases, increases in the ambient noise level within the embankment. Many of these warning signs can be observed visually, and the importance of routine, periodic visual inspections to identify signs of developing instability early cannot be understated.

The purpose of waste dump and stockpile monitoring is as follows:

- Maintain safe operating conditions.
- Provide advanced warning of developing instability so that mitigative measures can be implemented to avoid instability or lessen its impact.
- Provide information about the instability mechanism.
- Quantify displacements and rates.
- Establish and maintain a record of facility performance.

As noted by Clayton *et al.* (1995), instrumentation utilised to monitor waste dump and stockpile conditions should be as simple, reliable and robust as possible, and capable of measuring the required parameters to an acceptable level of accuracy, with an acceptable speed of response to changes that might occur in the parameter being measured.

12.2   Visual inspections

Visual inspection is critical to ongoing safe operation of a waste dump or stockpile, and is the easiest, most common

and most practical method of monitoring (BCMWRPRC 1991b). Visual inspection is a key component of an effective waste dump and stockpile monitoring program. At many mines, operations staff, equipment operators, surveyors and other personnel that regularly visit the waste dumps and stockpiles are trained to recognise potential signs of developing instability, including:

- excessive or abnormal cracking
- excessive crest deformation
- excessive over-steepening of the crest
- abnormal platform tilting
- seepage breakout on the face
- bulging of the face
- toe spreading.

Observations of any of these indicators should be brought to the attention of the geotechnical or mine engineering department and evaluated in the context of available instrumentation monitoring data to help identify potential instabilities that may be developing.

Inspections of the crest and active and inactive platforms should look for evidence of slumping; development of large, arcuate tension cracks; normal or obsequent scarps; and/or tilting of the platform that might be indicative of developing deep-seated instability. Inspections should also identify over-steepening of the crest beyond the normal angle of repose. Over-steepening can result from an accumulation of wet, fine-grained material at the crest and lead to rapid onset of sliver failures. Excessive over-steepening can be addressed by cutting down the crest and refilling the cut-down area with coarse material. Tension cracks should be graded over to help prevent infiltration of runoff and maintain a smooth platform for optimum trafficability.

Inspections of the slope or face of the waste dump or stockpile should look for signs of bulging or seepage breakout. Bulging in the upper portion of the face is typically related to over-steepening caused by poor segregation, placement of fine-grained materials, excessive advance rate or saturated material. Bulging at the toe can be an indication of developing deep-seated instability. Where toe bulging is observed, a review of monitoring records should be carried out to help determine the cause. In many cases, minor bulging of the face does not lead to instability, and routine advancement of the crest and face will restore the normal angle of repose. However, if the bulging persists or expands, appropriate slope monitoring methods should be employed to confirm that movement rates are within acceptable limits and not increasing. If necessary, material placement should be discontinued or slowed until movement rates reduce to within acceptable limits.

Many waste dump and stockpile failures can be linked, at least in part, to foundation failure. Visual indicators of foundation failure include heaving/deformation of the foundation in front of the toe. Adverse foundation conditions are typically managed by limiting dump height and/or advance rate or by removal of weak foundation materials in advance so the toe is founded on more competent materials.

Any visual monitoring program should establish a systematic approach to the inspection and recording of observations. Where possible, routine visual inspections should be carried out by the same people. This continuity may help to identify subtle changes in the facility that may not be evident to someone who is less familiar with the facility. However, while continuity is important, periodic reviews by external geotechnical specialists should also be conducted to help ensure that inspections do not become too routine, and to add the perspective of a different set of eyes.

12.3 Displacement monitoring systems

Monitoring of waste dumps and stockpiles by means other than visual observation is employed at most mines where these structures are subject to one or more of the following conditions:

- high face height (≥ 100 m)
- steep overall slope angles (≥ 25°)
- steep, adverse foundation slopes (≥ 15°)
- lack of topographical confinement
- poor-quality foundation materials
- poor-quality waste or stockpile materials
- aggressive or unfavourable construction methodology
- unfavourable piezometric or climatic conditions
- high advancement rates (≥ 50 m²/d [crest advance in metres per day × face height] – see Chapter 3)
- poor historical performance.

As discussed in Chapter 3, these attributes are important factors in the waste dump and stockpile stability rating and hazard classification system, and any one factor, or combination of factors, can have a significant impact on stability.

The mechanics of the various modes of instability are well understood (see Chapter 8). Different modes of failure tend to manifest themselves in different ways. While ongoing consolidation settlement routinely induces deformation and cracking in large waste dumps and stockpiles, the rate of deformation is generally linear or decreases over time. Failures, however, are typically preceded by increasing/accelerating rates of deformation. Where a failure mechanism is deep-seated, evidence of deformations well behind the crest may be observed. Understanding the likely failure mechanism responsible for observed deformation patterns and rates is important when designing a monitoring program, including the type of monitoring system best suited to assessing that failure mechanism.

Several of the most common types of monitoring systems are discussed below. The reader is also referred to Chapter 12 in *Guidelines for Open Pit Slope Design* (Read and Stacey 2009) for a comprehensive discussion of deformation monitoring systems.

12.3.1 Prisms

Prism monitoring is an integral component of the monitoring program at many mines and can be used to quantify vertical, horizontal, total vector, and slope distance displacement and displacement rates. In its simplest (manual) form, the use of prisms to monitor crest displacement requires mobilisation of a survey crew and processing and interpretation of the survey data. This can be a time-consuming process and can result in unacceptable delays when a real time response is needed. In cases where the rate of deformation is increasing relatively quickly, the use of simple surveying techniques may be inadequate to provide sufficient warning of incipient failure, due to the time required to obtain and process data. Similarly, the use of prisms to monitor an active crest is not recommended due to the need to continually reposition the prisms. The use of prisms is better suited to the monitoring of inactive waste dumps or stockpiles, where prisms can be installed without being at risk of damage by mining equipment and where constant repositioning is not required.

The use of a dedicated robotic total station (RTS), which combines a theodolite, electronic distance measurement (EDM) system, servo motors, automatic target recognition and a controller, can help reduce the time required to obtain survey data (Fig. 12.1). The controller is typically a database program that tells

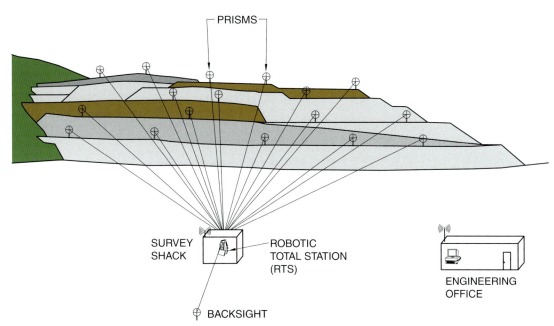

Figure 12.1: Components of robotic total station system

the RTS in the remote survey hut to turn on, take a backsight reading and then cycle through the prisms. The raw survey data are transmitted back to the controller for storage in the geotechnical database. Typically, the controller is located in the mine engineering office, but it may also be located in a convenient location that is accessible by personnel responsible for processing and reviewing the survey data. Once the data have been loaded into the database, another computer program can automatically query the database, calculate the movement rates for the prisms and compare them to preset movement rate thresholds.

The RTS should be located on stable ground and requires at least one backsight (also on stable ground) located a similar distance from the RTS as the prisms being monitored. To help reduce the impact of atmospheric conditions on the RTS, it should be located in a climate-controlled survey hut. In general, the absolute position of the prism is of less interest than the relative position of the prism and how that position changes between readings.

The three fundamental types of readings required to calculate the position of a prism in three dimensions are slope distance, vertical angle and horizontal angle. Based on these three readings, the location of a prism can be calculated with reference to the backsight. If the position of the backsight is known, the location of the prism can then be calculated.

It is important to consider sources of error inherent to prism surveying. There are numerous sources of survey error, but they can generally be grouped into three main

categories: human error, equipment error and environmental error.

Human error can be minimised by having detailed work procedures for specific surveying tasks. Having the same person conduct the surveying can also help to reduce human error. Using a remotely controlled RTS system can help to eliminate human error, assuming it is correctly installed.

Equipment errors may arise from poor equipment calibration and equipment limitations, and typically define the potential accuracy of the survey system. This error can have a significant impact on velocity calculations. For example, if the accuracy of the survey equipment is ± 5 mm, the potential error between two consecutive readings could be as much as 10 mm. Where prisms are being monitored and are subject to a specific movement rate threshold, the time required between two successive readings to ensure that the apparent velocity can be attributed to the accuracy of the survey method can be large. As can be seen in Table 12.1, for a survey accuracy of ± 5 mm, a survey interval of less than 1.3 days could yield a velocity error of greater than the 7.5 mm/d threshold. Similarly, a reading interval of less than 0.39 days could yield a velocity error of greater than 25.4 mm/d.

Environmental errors are typically related to variations in climatic conditions. These include variations in angular measurement caused by refraction of light as it passes through air of different densities, and errors in the distance measurement caused by differences in the speed of light as it passes through air of different densities. In a normal open pit environment, these errors cannot

Table 12.1: Example comparison of monitoring thresholds to minimum time between readings such that apparent movement rate is within the threshold

Threshold		Minimum interval (days) for ± 5 mm error
From (mm/day)	To (mm/day)	
0	7.5	> 1.3
7.5	12.5	1.3–0.8
12.5	25.4	0.8–0.4
25.4	> 25.4	< 0.4

practically be avoided or eliminated. However, it is possible to reduce the effects of environmental errors by using several data filtering techniques. For example, by using the average of several readings taken over the period of a day or several days, diurnal effects can be negated to some extent. Similarly, employing a running average data analysis approach can help minimise the effect of errors. Where feasible, using readings taken nearest to the same time of day will help reduce errors related to atmospheric factors, assuming weather conditions do not vary significantly from day to day.

By incorporating global positioning system (GPS) technology into a RTS monitoring system, individual survey accuracy can be further enhanced. By integrating a GPS receiver into the RTS survey hut, GPS survey data can be recorded at the same time the RTS is surveying individual prisms. In doing so, any change in position of the RTS can be calculated. This information can then be used to more accurately calculate the position of each prism surveyed.

12.3.2 Wireline extensometers

Wireline extensometers measure the change in distance between two points using a thin cable or wire and are the most commonly used instrument for monitoring waste dump and stockpile crest deformation. Extensometers can be simple, manually read devices or more sophisticated electronic devices capable of continually measuring movement and transmitting the readings to a receiver connected to a computer/database where the data are readily accessible for processing and interpretation.

In their simplest form, extensometers consist of a tripod and pulley, a weight, a measurement scale or measurement reference point, a wireline and an anchor point. The wireline is generally anchored at or near the crest and the tripod is positioned on stable ground behind the most distal tension crack or zone of deformation (Fig. 12.2).

By the nature of the design, only changes in the distance between the anchor point and the tripod pulley are measured. Accordingly, depending on the position of the tripod and the anchor, it is possible that the amount of movement measured can range anywhere from almost 0–100% of the actual total movement. The measured movement is a function of the angle of the wireline with respect to the true direction of movement. Adding an intermediate support between the anchor and the tripod can help increase the percentage of the total movement that is captured, as illustrated in Fig. 12.3.

Note that where monitoring thresholds and trigger action response plans (TARPs) are based on deformation rates measured without the use of intermediate supports, adding an intermediate support can result in significantly

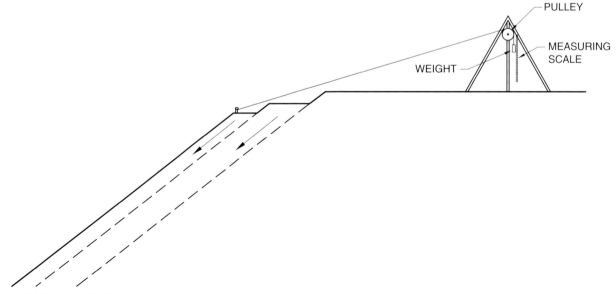

Figure 12.2: Schematic illustration of a manual extensometer

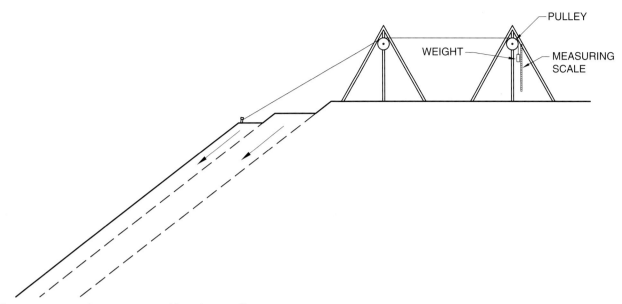

Figure 12.3: Manual extensometer with an intermediate support

higher apparent rates of deformation, and threshold limits may need to be adjusted.

Where intermediate supports are used inside the zone of deformation (Fig. 12.4), the amount of measured deformation may be reduced significantly. This is a significant limitation of the wireline extensometer that needs to be considered when interpreting the data and setting thresholds.

Another limitation of wireline extensometers is that, where crest advance rates are relatively high, extensometers will need to be relocated frequently to provide continuous coverage.

In addition to the simple wireline extensometer illustrated in Figs. 12.2 and 12.3, several other varieties of wireline extensometers are commonly used at mine sites globally. These include buried wire extensometers of various types and automated wireline extensometers. Automated wireline extensometers measure movement and communicate the information to a central computer via a telephone line, or wirelessly via radio signal, cellular network, or satellite transmission. Figure 12.5 shows the SlideMinder® automated wireline extensometer system with near real-time data telemetry, developed by Call & Nicholas Instruments, and Fig. 12.6 shows a portable

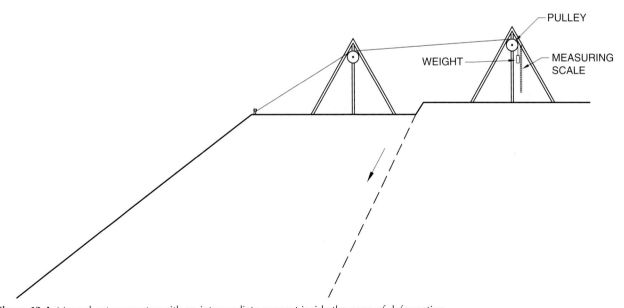

Figure 12.4: Manual extensometer with an intermediate support inside the zone of deformation

Figure 12.5: SlideMinder automated wireline extensometer system developed by Call & Nicholas Instruments, Inc. Source: Call & Nicholas Instruments, Inc

Figure 12.6: Trailer-mounted digital wireline and laser extensometer systems equipped with telemetry and portable power supply. Source: M Hawley

extensometer system that combines a SlideMinder with a laser distance measuring system developed by RST Instruments Ltd (see Section 12.3.12). The portability of this trailer-mounted system makes it ideal for rapid deployment and for monitoring situations that require frequent relocation, for example, to accommodate rapid crest advancement.

For waste dumps or stockpiles that are subject to a moderate or higher hazard of instability (i.e. a waste dump and stockpile hazard class [WHC] of III, IV or V [Chapter 3]), additional instrumentation should be considered to provide a record of total deformation. This could include periodic surveys of the facility using prisms or GPS, or the use of light detection and ranging (LiDAR), interferometric synthetic aperture radar (InSAR) or laser imaging/scanning surveys.

12.3.3 Global positioning systems

Most mines today utilise some form of GPS for routine surveying and design layout. GPS surveying uses signals from orbiting satellites to determine the location of a point on Earth. Using triangulation, a GPS receiver can determine its position in three dimensions by receiving signals from at least four satellites.

Among the key advantages of GPS based survey systems are that they can be used day or night and in inclement weather conditions. However, as noted by Read and Stacey (2009), there are some limitations that affect the application of GPS to routine slope monitoring, including the following:

- GPS surveys require personnel to access monitoring points unless receivers are permanently mounted on the point being monitored.
- Surveys may be slower than using a total station or RTS system if travel to each survey point is required.
- GPS receivers require direct line of sight to the satellites.

Several sources of error affect the accuracy of locations measured by GPS receivers, including errors due to atmospheric effects, clock drift, multipath, dilution of precision, selective availability, and anti-spoofing. The effect of atmospheric conditions on satellite signals can be significantly reduced by using dual-frequency GPS receivers. While some personal GPS units are single frequency, most commercially available, survey grade GPS units currently employ dual frequency receivers. Clock errors due to drift in the atomic clocks on the satellites can occur. This drift is continually monitored by the US Department of Defence and the on-board clocks are routinely adjusted to account for this drift. Multipath errors occur when the receiver is located close to a large reflective surface, such as a large building, lake or open pit slope, and GPS signals do not travel directly to the receiver from the satellite. Multipath errors can be reduced by using a specialised choke ring antenna that has several concentric rings around the antenna to trap unwanted reflected signals. The three error types discussed above affect the calculated distance between the receiver and the satellite.

Where the readings are based on satellites that are poorly distributed in the sky, dilution of precision errors can occur due to overlapping error zones; dilution of precision can affect the accuracy of vertical, horizontal and 3D position readings. These types of errors can be minimised by using as many satellites as possible for determining position. Selective availability and anti-spoofing are errors introduced by the US military forces to reduce the accuracy of the signals received by civilian GPS receivers.

Many of the errors affecting GPS measurements can be significantly reduced or eliminated by using differential measurement techniques. By employing two GPS receivers, a stationary base station and a mobile surveying unit, very accurate surveys are possible by using the differential GPS survey method. Using this technique, the relative positions of the base station and mobile unit can be measured to an accuracy of less than ~20 mm over a few kilometres. By establishing the base station at a known position, the theoretical distance to the various satellites can be calculated and the error between the theoretical distance and the distance calculated using the satellite signals can be determined and broadcast to the mobile surveying unit to correct the readings in real time.

Figure 12.7: Remote global positioning system monitor deployed at crest of inactive waste dump. Source: J Hogarth, published with permission of Teck Coal Limited

Like wireline extensometers, where remote GPS systems are used to monitor waste dumps and stockpiles that are under construction, relocation and resetting are required as the crest and face advance. The amount of time between resets is dependent on the rate of advancement of the crest. Where remote GPS monitors are located near the crest, they are also at risk due to sliver failures. Accordingly, remote GPS systems may be better suited to the monitoring of inactive facilities where crest deformations and consolidation settlements related to advance rate are minimal and where deep-seated failure modes and settlement are more of a concern than shallow sliver failures (Fig. 12.7).

12.3.4 Slope inclinometers

Inclinometers are widely used to monitor the stability of natural and mined slopes and constructed embankments.

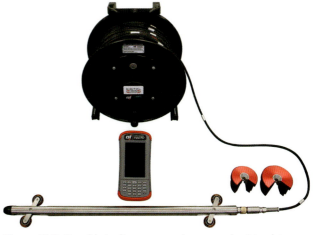

Figure 12.8: Portable inclinometer probe, control cable, data logger and casing adaptors. Source: Courtesy of RST Instruments Ltd

They are suitable for measuring both distributed creep deformations and shear movements along discreet discontinuities or zones of deformation within a waste dump or stockpile and the underlying foundation. Two types of inclinometers are typically used to monitor waste dump and stockpile stability. One of these uses a

cylindrical probe with spring-mounted wheels that fit into grooves machined into a plastic or metal casing that is inserted into a borehole (Fig. 12.8). These are referred to as portable inclinometer probes (also known as torpedos) because they must be inserted into the casing and lowered and raised every time a reading is required. The other

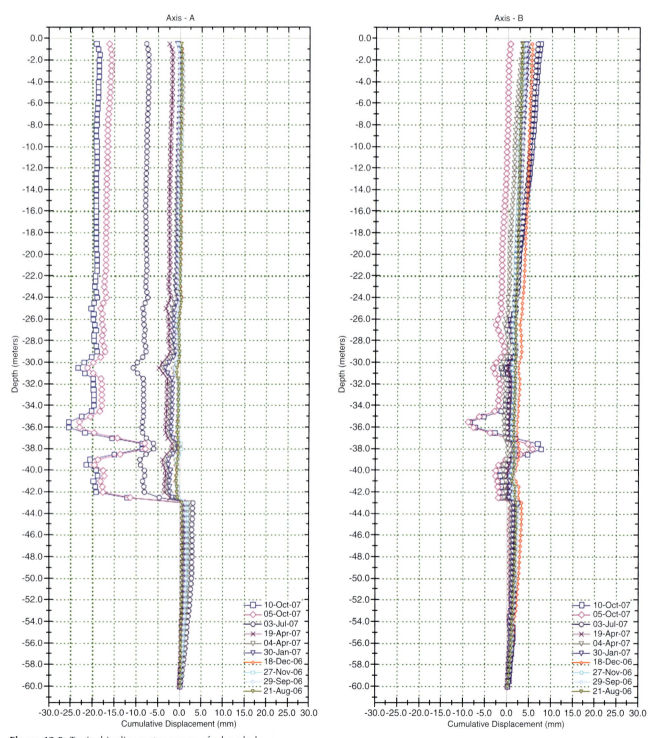

Figure 12.9: Typical inclinometer survey of a borehole

Inclinometer casing controls
orientation of sensors

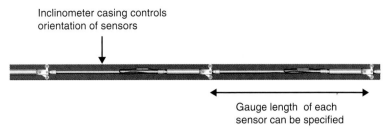

Gauge length of each
sensor can be specified

Figure 12.10: Typical in-place inclinometer installation. Source: Courtesy of Durham Geo Slope Indicator

type, in-place inclinometers, can also be lowered into a grooved casing, but are designed to be positioned at one location, measuring movement at that location. Multiple in-place inclinometer units can be installed at discreet depths in the same casing, or connected together in a linear array.

Of these two types of inclinometers, the portable inclinometer is the most commonly used type for waste dump and stockpile monitoring. The inclinometer probe is connected to a cable that is used to lower the probe to the bottom of the casing. The probe is then raised in increments, typically equal to the spacing between the two sets of guide wheels. At each increment, a reading is recorded at the surface by a readout box connected to the upper end of the cable or wirelessly by a handheld readout unit, depending on the inclinometer system being used. The casing is typically installed in a vertical to steeply inclined borehole, with one set of internal grooves oriented parallel to the direction of expected movement (referred to as the 'A+' and 'A–' grooves) and one set (the 'B+' and 'B–' grooves) oriented perpendicular to the direction of expected movement. Most current slope inclinometer probes use biaxial sensors, requiring a single survey of the borehole with the wheels of the probe fitted into the A+ and A– grooves. Some older uniaxial inclinometer probes require that the probe be lowered and raised in the A+/A– grooves, then turned 90° and lowered and raised in the B+/B– grooves.

It is generally recommended that inclinometer boreholes be extended into stable ground so that the bottom of the hole can serve as a stationary reference point for all subsequent profiles. As well, to obtain the most accurate readings possible, the casing should initially be measured for spiral deflection, which can be used to adjust initial baseline readings and all subsequent readings.

After the initial survey of the hole, subsequent surveys are compared to the original to determine the magnitude and direction of any movement that may be occurring with respect to the bottom of the hole (Fig. 12.9).

The tilt of the probe is measured using tilt sensors that employ either accelerometers or microelectromechanical systems (MEMS). Microelectromechanical systems convert a measured mechanical signal (e.g. gravity) into a voltage signal. As a result of recent advances in microfabrication

techniques, MEMS devices can be very small, measuring a few millimetres to less than a micron, making them well suited for use in geotechnical instrumentation. Based on technical specifications published by several inclinometer probe suppliers, MEMS devices are also more capable of withstanding shock loading and are more accurate than typical force-balanced accelerometer devices.

In-place inclinometers are typically installed as a string of individual sensors, at fixed positions, and allow monitoring of remote sites without the need for a physical presence of personnel (Fig. 12.10). By using a data logger and a data acquisition system, readings can be obtained, profiles plotted and trend plots prepared within minutes of readings being obtained. Because sensor lengths can be customised, it is feasible to configure a string of sensors with shorter sensors in the zone of interest and longer sensors above and below to help minimise the number of sensors. Using a system that employs a single signal cable that interconnects all the sensors in series eases installation, reduces cable costs, eliminates the need for a multiplexer and simplifies the data connection for a data logger. It is also possible to recover sensors for reuse, assuming that deformation has not damaged the casing to the extent that the sensor units cannot be removed from the casing. In-place inclinometer systems are available from several suppliers.

12.3.5   Time domain reflectometry

Time domain reflectometry (TDR) is a remote sensing technique that was originally developed in the 1950s to help locate breaks in power transmission and communications cables. In the 1970s, TDR technology was adapted for use in geotechnical applications. Geotechnical TDR systems typically use a coaxial cable, similar to the coaxial cable used for cable television, grouted into a borehole. A voltage pulse is produced by a signal generator at surface and transmitted down the coaxial cable. The distance to any deformations in the cable can be calculated if the propagation velocity of the signal in the (undeformed) cable is known. As illustrated in Fig. 12.11, ground movements deform the coaxial cable, changing the impedance at the point of deformation. This causes the voltage pulse to be reflected, and the travel time of the reflected pulse determines the location of the deformation.

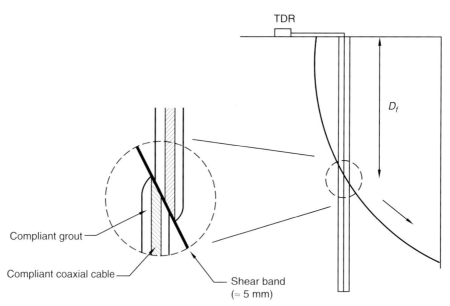

Figure 12.11: Schematic illustration of a grouted time domain reflectometry (TDR) sensor installation showing shearing mechanism. Source: Dowding *et al.* 2003

The amplitude of the reflected signal is proportional to the amount of deformation in the coaxial cable.

A coaxial cable consists of an inner conductor core and a cylindrical outer conductor, separated by a dielectric insulator. This dielectric insulator is typically a foam or solid polymer. When a voltage pulse is triggered by a signal generator, it creates a current between the two conductors

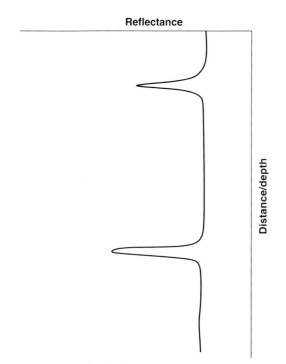

Figure 12.12: Example plot of reflectance versus distance for a time domain reflectometry (TDR) cable

that travels along the cable. The rate and characteristics of propagation are functions of the dielectric material and distance between the inner and outer conductors. If the cable remains undeformed, the impedance remains unchanged and a pulse will not be reflected. However, if the cable is deformed, some of the pulse energy is reflected back up the cable. The ratio of the transmitted voltage to the reflected voltage is referred to as the 'reflectance'. Reflectance is typically plotted versus distance for interpretation purposes (Fig. 12.12).

When installing a TDR in a waste dump or stockpile, the compressive strength of the grout must be well matched to the waste dump/stockpile material. In general, the grout should be of sufficient strength to shear the coaxial cable, but not so strong that it will not be deformed or sheared if the waste dump or stockpile shears or deforms. The grout stiffness needs to be carefully calibrated with the shear strength of the embankment material. Dowding *et al.* (2003) suggested a shear capacity for the grout of 5–10 times the shear strength of the embankment material. Dowding *et al.* also suggested that, for installations greater than 50 m in length, long lead cables should be of the 75 ohm F11 (low loss) type, due to potential attenuation and noise problems. For installation lengths less than 50 m, 50 ohm RG/U type cable should be sufficient.

12.3.6 Slope stability radar

The use of ground-based interferometric radar, commonly referred to as slope stability radar (SSR), to monitor open pit slopes is now relatively commonplace. Monitoring of waste dumps and stockpiles with SSR is less common, but

Figure 12.13: Examples of (a) real aperture and (b) synthetic aperture radar systems. Source: Courtesy of GroundProbe

is growing in popularity. Slope stability radar units can monitor large slope areas, detecting movements with millimetre and, in some cases, sub-millimetre accuracy, even in some of the harshest climatic conditions. Operating ranges between 10 and 4000 m are possible with currently available units. Accuracy is dependent on the distance between the slope being monitored and the radar system.

Slope stability radar systems work by emitting signal pulses of known frequency and amplitude towards the slope being monitored. The signals reflected by the slope are measured by the radar unit and, using the interferometry technique, the signals are processed and the magnitude of the deformation between successive scans can be calculated.

Two general types of ground-based interferometric radar systems are currently used to monitor slope movements: synthetic aperture and physical or real aperture systems. Synthetic aperture radar systems continually scan the entire slope with a relatively wide beam, utilising a small antenna that moves along a fixed guide to 'simulate' a larger antenna. Physical aperture radar systems typically use an oscillating dish that can scan a much greater field of view than synthetic aperture radar systems. Both types of radar have strengths and weaknesses that need to be considered before a specific system is chosen. Examples of real aperture and synthetic aperture type radar systems are shown in Fig. 12.13.

Slope stability radar units can be deployed rapidly, with set-up times of less than an hour possible. Target acquisition times can also be relatively low, with the minimum time required to complete a scan of a slope being of the order of a few minutes, depending on the equipment manufacturer and area being scanned. Scan resolution is determined by pixel size and is largely dependent on the distance between the slope and the radar unit. Pixel size is typically of the order of a few square metres at a distance of 1 km.

The data recorded by the radar unit are typically transmitted to a central computer where they are reduced and plotted for viewing. By comparing the signals received from successive radar scans, differential movement between individual scans in a direction parallel to the line of sight between the slope and the SSR can be determined and plotted. Displacement plots for specified periods of time can also be generated. The software that is responsible for the operation of the radar system and reduction of the acquired data can also initiate alarms if movement rates exceed predetermined thresholds and issue email notifications to responsible individuals. In some instances, data can be reviewed in the field at the radar unit itself.

One advantage of SSR systems is that personnel are not required to access the slope being monitored, unlike some of the other methods discussed above. However, because the movements recorded by the radar system are one dimensional, it has generally not been possible to determine the direction of movement. Accordingly, radar systems have typically been used in combination with other survey systems to help define the direction and nature of movement. Recent advancements in radar system design, application and analysis techniques, and associated software, suggest that the next generation of SSRs will be able to generate fully geo-referenced scans, and to reliably and accurately resolve movement directions as well as magnitudes.

Radar systems are typically deployed where real-time monitoring of slope deformations is critical to the safety of mining operations located immediately below unstable or potentially unstable slopes. The ability to rapidly deploy, and the capability to monitor large areas, makes radar systems well suited for this task. Slope stability radar systems are typically best suited for monitoring stationary slopes, such as pit slopes or inactive portions of waste dumps or stockpiles. In the case of active waste dumps or stockpiles, deformations may be masked by activities associated with construction, such as end dumping or dozer pushing along the crest, or sliding and roll-out of material dumped or pushed over the crest. However, there are some situations in which SSR monitoring of active waste dumps or stockpiles may be useful. For example, where waste dumps or stockpiles are advancing over poor foundations, it may be possible to use an SSR system to monitor for foundation deformations in front of the toe of the advancing embankment. In some cases, SSR systems may also be able to filter out normal operational background 'noise' and movements associated with advancement of the face and crest, and detect local bulging or anomalous deformations or 'hot spots' on the advancing face, and possibly even over-steepening of the crest.

12.3.7 Laser imaging/scanning

Terrestrial laser scanning is commonly used for topographic mapping, but its use in slope monitoring is becoming more commonplace. While laser scanning technology is similar to radar scanning, laser scanners use LiDAR technology to measure distances. Due to the differences between the wavelengths of the signals used, laser scanning is typically less accurate than radar, but accuracy can be increased to ± 1 cm by calibrating scan results with reference targets surveyed using first-order survey techniques. Operating ranges up to 6000 m are possible with currently available units; however, accuracy is dependent on the distance between the slope being monitored and the laser scanner location.

Several manufacturers produce laser scanning systems. A typical laser scanning unit is shown in Fig. 12.14.

Due to their accuracy limitations, the use of laser scanners for real-time monitoring is currently limited, but

Figure 12.14: Laser scanning unit. Source: Courtesy of Maptek

the use of these systems to conduct rapid topographical surveys of inaccessible slopes is becoming quite common. As with radar systems, laser scanners are usually better suited for monitoring inactive dumps and stockpiles than active facilities.

12.3.8 Acoustic monitoring

It has been observed that failures of waste dumps composed of coarse waste rock can be preceded by acoustic emissions for some time before a failure (McCarter 1976). This noise is interpreted to be due to the fracturing of waste rock blocks caused by movement along the failure plane. By recording acoustic emissions using an array of at least three geophones installed on and around a waste dump or stockpile, it may be possible to determine the location of a specific acoustic event. Acoustic emissions tend to attenuate rapidly in unconsolidated fill materials, so it is important to place the sensor/geophone array as close as possible to the expected source of the emissions. This can be accomplished by installing the sensors in the foundation and/or within the embankment itself. Another limitation is that it is not possible to determine the magnitude of movement associated with the emissions. However, when used in conjunction with other instrumentation systems, such as slope inclinometers or TDRs, acoustic monitoring may help to develop a three-dimensional picture of a failure surface.

12.3.9 Tiltmeters

Tilting of waste dump and stockpile platforms can be indicative of rotational failure modes or differential settlement related to internal deformation. Surface mounted tiltmeters measure angular orientation relative to a vertical gravitational vector and are compatible with data loggers and data telemetry systems. Tiltmeters use an accelerometer coupled with a transducer that generates a specific output voltage based on the amount of tilting experienced. Biaxial systems facilitate tilt measurement in two orthogonal directions simultaneously.

The most commonly used tiltmeters employ servo-accelerometers, MEMS, vibrating wire (VW) sensors and electrolytic cells to measure tilting. Measurement accuracy reported by manufacturers is up to ± 0.0125% of full scale or 0.03 mm/m, with a resolution of up to 0.01 mm/m (2 arc seconds). Tiltmeters have the advantage of being light weight, durable, easy to install, relatively stable over the long-term (i.e. subject to limited creep) and low cost (Darling 2011). A typical tiltmeter unit is shown in Fig. 12.15.

12.3.10 Crack monitoring

One of the simplest forms of displacement monitoring is to observe and measure the rate of extension across surface tension cracks that often develop on waste dump and

Figure 12.15: Uniaxial tiltmeter unit. Source: Courtesy of RST Instruments Ltd

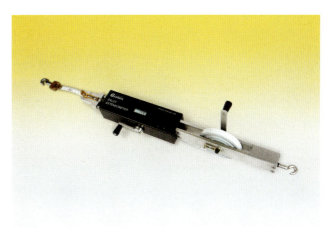

Figure 12.16: Tape extensometer unit. Source: Courtesy of Geokon, Inc

stockpile platforms in response to settlement and deformation. Several simple, easy to install, easy to monitor measuring systems are applicable to the measurement of displacement across surface cracks. These can be as simple as anchoring two stakes or pins on either side of a tension crack and periodically measuring and recording the distance between the stakes or pins with a tape measure. By recording the date and time of the readings, the rate of movement can be calculated.

More sophisticated and accurate crack monitor systems that utilise tape extensometers or electronic measuring systems are also available. A tape extensometer consists of an electronic measuring unit with an integral measuring tape. The device is attached to pins installed on either side of a surface crack and the tension in the tape adjusted to a preset value. Once the preset tension value is reached, the distance between the anchor points is displayed and recorded. By comparing successive readings, the rate of movement can be calculated. Typical measurement accuracy reported by several manufacturers is of the order of ± 0.1 mm, with tape lengths up to 50 m available. A typical tape extensometer unit is shown in Fig. 12.16.

12.3.11 Tell-tales

Tell-tales are simple devices that are placed to span a surface crack or series of cracks. In their simplest form, tell-tales are composed of two components which are anchored on opposite sides of a crack and connected via an indicator or pointer. Movement across the crack causes a relative movement of the two pieces that can be used to determine both the magnitude and rate of movement. Examples of simple tell-tale units are shown in Fig. 12.17.

Evidence of movement can also be indicated by the dislocation of a paint or picket line placed across a surface crack or zone of deformation. Even small amounts of

Figure 12.17: Example of simple tell-tale devices. Source: J Simmons

Figure 12.18: Pickets installed across a tension crack. Offset of pickets due to movement is clearly evident and can be measured using a tape measure. Source: J Hogarth

dislocation of the paint or picket line can be identified visually, as illustrated in Fig. 12.18.

12.3.12   Laser distance measuring systems

Laser distance measuring (LDM) systems use a fixed, tripod-mounted laser to measure the distance between the laser and a remote target. The target is typically a circular, rectangular or triangular flat surface made of metal sheeting or plywood with a smooth, painted surface, and is mounted on a stake or tripod located at or near the crest of the waste dump or stockpile. Similar to RTS systems, laser systems can be configured to take repeated measurements at a specific frequency and transmit the data using a telemetry system to a central computer for storage and processing. Linear movement rates can be calculated based on the change in distance and time between measurements and displayed in near real-time. Laser systems are much less expensive than RTS systems and provide linear (1D) extensional movement data similar to the type of data provided by wireline extensometers. However, unlike wireline extensometers, laser systems do not use wirelines that can restrict equipment movement and hamper dumping platform operations. While this system works well in cases where the movement direction is predominantly along the line of sight of the laser (i.e. extensional movement across tension cracks), it does not work well in cases where there is a significant component of settlement. In these cases, depending on the size and position of the target, frequent manual retargeting of the laser may be required, which may limit practicality. Examples of LDM systems are shown in Figs. 12.6 and 12.19.

Figure 12.19: Example of a laser distance meter system. Source: Courtesy of RST Instruments Ltd

12.3.13   Autonomous wirelessly networked sensors

While still in the research and development stage, in the future, an individual sensor or an array or series of autonomous sensors (pore pressure, tilt, acceleration) encased within a capsule and buried in a waste dump or stockpile could transmit real-time data back to a central monitoring point without the need for a physical connection between sensors. Figure 12.20 shows a prototype of an autonomous sensor capsule equipped with accelerometer and pore pressure sensors, capable of measuring and transmitting orientation and piezometric pressure data wirelessly to a surface receiver, allowing real-time measurement of deformation and pore pressure. The main limitation of this system is that the effective transmission distance through waste rock is currently only a few metres. However, the telemetry system within the capsule is able to communicate and transmit data to nearby capsules, allowing the possibility to develop a wireless array within the embankment using a series of closely spaced capsules, transmitting data from one capsule to another using a relay system.

Figure 12.20: Autonomous sensor capsule. Source: Courtesy of Elexon Electronics

12.4 Surface water and groundwater monitoring

12.4.1 Introduction

The amount and type of surface water and groundwater monitoring data required will depend on the site setting and the types of materials consigned to the waste dump or stockpile. Hydrogeological monitoring can be minimal where facilities are located in arid climates and where the waste dump and stockpile materials are strong and not subject to degradation over time. In this case, monitoring may be limited to upgradient and downgradient groundwater chemistry for permit compliance. Conversely, an extensive array of instrumentation may be required for facilities located in wet or seasonally wet climates, where the waste dump or stockpile materials and/or foundations are weak, or where the topography is steep.

12.4.2 Types of surface water and groundwater monitoring data

The following types of surface water and groundwater monitoring may be required:

12.4.2.1 Climatological monitoring

Types of climatological data that may be required for the design and operation of a waste dump or stockpile are described in Section 4.3.4. Climatological monitoring typically forms part of the site-wide monitoring plan and is not usually specific to individual waste dump or stockpile locations. However, in wet areas, and/or regions of steep topography, it may be necessary to operate a rainfall station close to the area of a particular facility.

12.4.2.2 Monitoring of upgradient conditions

Upgradient groundwater water level and chemistry monitoring (and potentially surface water monitoring) will normally be required throughout the operating life of the facility and beyond. The instrumentation and procedures for obtaining these data are described in detail in Chapter 6. Monitoring requirements specific to the operational stage are discussed below.

12.4.2.3 Monitoring within the facility

The following operational monitoring may be required for the waste dump/stockpile hydrogeology program:

- the nature and distribution of the materials being placed (e.g. material types, tonnages/volumes and spatial distribution)
- water content of the placed materials
- the geochemical nature of the waste dump and stockpile materials, typically with a quarterly sampling program that involves acid–base accounting and static

testing and aims to confirm the nature and variability of the materials
- VW piezometers and/or soil moisture probes installed in the foundations or in the basal layer of the waste dump or stockpile
- thermistors to assess freeze-back in the base of the facility or the outer (downgradient) toes
- VW piezometers installed in the initial lifts
- VW piezometers or standpipe piezometers installed from benches between lifts; the use of standpipe piezometers enables water quality sampling of the pore water chemistry
- regular visual inspection and monitoring of any seepage zones (flow and chemistry)
- regular visual inspection and monitoring of the surface water management system to identify any required maintenance or remediation; in some cases, the installation of surface water monitoring stations for key runoff channels may be beneficial for understanding the site water balance and/or potential downgradient changes in the surface and groundwater system.

12.4.2.4 Monitoring within the near-surface materials

If it is necessary to accurately assess recharge to the surface of the facility, the following types of monitoring may be required:

- VW piezometers and/or soil moisture probes installed in the near-surface materials and/or cover layer
- suction lysimeters to allow pore water sampling of the near-surface materials and/or cover layer
- mapping of snow banks, including measurements of the depth of accumulated snow, and the water content of the snow
- thermistors to assess the surface freeze–thaw layer and freeze-back
- visual inspection of any vegetation programs for the cover.

12.4.2.5 Monitoring of the foundations

Hydrogeological monitoring requirements for the foundation may include:

- pore pressure data obtained from VW piezometers installed in the foundation materials before facility construction (see Section 12.4.3)
- changes in moisture content using soil moisture probes and/or lysimeters installed during construction
- monitoring of pore pressures in toe-area piezometers to ensure that foundation design criteria are being met (see Section 12.4.3).

12.4.2.6 Monitoring of downgradient conditions

Downgradient monitoring will usually include:

- groundwater levels and water quality monitoring in environmental wells installed within the underflow system in the toe area and downgradient of the facility – some type of downgradient groundwater monitoring is required for virtually all major waste dumps and stockpiles
- surface water monitoring stations for regular or continuous flow monitoring, sediment monitoring and water quality sampling – downstream surface water and sediment monitoring is usually required for all waste dumps and stockpiles where surface water has been diverted and/or for facilities that create any downgradient surface water flows. Where flows are continuous, frequent monitoring may be required, or continuous recording stations may need to be established.

One or more downstream compliance points will need to be established for both surface water and groundwater. The optimum location of these compliance points will depend on the site setting and the overall layout of the mine site water management system. Ideally, the compliance points for the waste dump and stockpile facilities can be located downstream from the entire mine site so that they can benefit from mixing with other waters to optimise the chemistry. Corporate environmental staff and regulators often require 'trigger' monitoring points for any runoff and seepage immediately below the waste dump or stockpile, so that the water quality performance of the facility can be assessed on an ongoing basis.

12.4.3   Pore pressure monitoring

As discussed in Chapters 4, 6 and 8, depending on potential failure modes, pore pressures in the facility or its foundation can be one of the most critical factors affecting stability. Where high pore pressures within the facility or its foundation pose a concern for stability, monitoring of pore pressure during construction and operations will be important, particularly where the materials within the facility or its foundation are fine grained and subject to pore pressure generation due to loading. The potential for pore pressure generation and dissipation rates must be evaluated and the results incorporated into the analysis and design.

Measuring of groundwater levels and pore water pressures is typically accomplished using piezometers installed in the facility and/or its foundation. Foundation piezometers can be installed in boreholes drilled through the facility into the foundation, or in the foundation before construction. Some piezometer types can also be installed on existing platforms before the start of construction, or on subsequent, overlying lifts.

Several piezometer types applicable to the monitoring of waste dumps and stockpiles are available. These include

pneumatic, VW, strain gauge (i.e. electrical resistance), standpipe (slotted or porous media) and multiport or multilevel piezometers. Brief descriptions of piezometer types are provided below, along with a discussion of their pros and cons. For more detailed descriptions and theory regarding piezometers and piezometric monitoring, the reader is referred to Beale and Read (2013).

12.4.4   Pneumatic piezometer

A pneumatic piezometer operates by balancing compressed gas pressure and water pressure across a diaphragm within the instrument tip. Pneumatic piezometers have the advantage of being relatively inexpensive, simple to install and operate, and applicable to a wide variety of operational environments. Depending on hole diameter, multiple piezometers may be installed at various depths in a drill hole. Recent technological advances also enable the measurement of negative pressures with specialised installations. These types of instruments are available with 100% non-corrodible construction. One disadvantage of pneumatic piezometers is the need for a compressed gas supply to be connected for each reading. Over the long-term, the use of compressed air can introduce moisture into the supply and return lines, which can lead to freezing of tubes or the diaphragm in cold climates. Where this is a potential concern, the use of nitrogen is recommended. As pneumatic piezometers are typically installed in a sand pack, a considerable volume of water is required to fill the sand pack and cause a change in reading. This limits their usefulness in tight formations where pressure changes may be associated with very little change in the volume of pore water, resulting in slow response of the piezometer. Care must also be taken during installation to avoid pinching or crushing the tubing connecting the tip to the surface. This type of piezometer is not easily adapted to a data logger for continuous monitoring. Examples of typical pneumatic piezometers and readout box are provided in Fig. 12.21.

12.4.5   Vibrating wire piezometer

Vibrating wire piezometers are available from several suppliers and have become very popular over the last decade primarily due to their relatively low cost and ease of installation. This instrument uses a tensioned wire connected to a diaphragm at one end and a stationary anchor block at the other. Changes in the pressure on the diaphragm cause the tension in the wire to change, and hence, the frequency of vibration of the wire. When a reading is taken, the tensioned wire is plucked electronically by a coil and magnet located close to the tensioned wire, causing it to vibrate at its resonant frequency, the frequency of the vibration being proportional to the pressure acting on the diaphragm. This vibration then creates an alternating current in the

Figure 12.21: Pneumatic piezometer with readout box. Source: Courtesy of RST Instruments Ltd

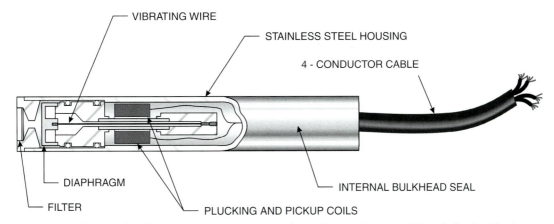

Figure 12.22: Schematic drawing of a vibrating wire piezometer transducer. Source: Courtesy of Campbell Scientific, Inc

Figure 12.23: View of vibrating wire piezometer cables in a steel conduit buried in a trench excavated in the foundation beneath a waste dump. Source: J Hogarth

adjacent coil that is measured by the readout box. This reading can then be converted to an equivalent pressure. A schematic of a VW piezometer is shown in Fig. 12.22.

In boreholes, VW piezometers may be installed using conventional methods with sand pockets around the VW transducer tip and cement–bentonite seals placed above and below. However, because they are very accurate and require very little water displacement to develop a hydraulic connection within the adjacent materials and register pressure changes across the diaphragm, in most materials, the VW transducers may be fully encapsulated in a cement–bentonite grout, eliminating the need for sand pockets. This makes installation in boreholes very easy and allows multiple transducers to be installed in a single borehole.

Vibrating wire piezometers are also relatively stable over the long-term, and are easily adapted to a data logger. In addition, long lead lengths do not affect accuracy; consequently, VW piezometers can be easily installed in foundations before facility construction, and on lifts that will be buried by subsequent lifts as the facility is developed. Where VW piezometers are installed in foundations or on lifts that will be subsequently buried,

the leads should be protected against damage and routed to a suitable monitoring location beyond the toe of the facility. This can be accomplished by placing leads in a steel pipe conduit buried in a shallow trench, as illustrated in Fig. 12.23.

12.4.6 Strain gauge piezometer

Strain gauge piezometers are available from several sources and allow the installation of multiple transducers in a single drill hole. This instrument, which looks very similar to a VW piezometer, uses a conventional strain gauge attached to a diaphragm to directly measure the pressure acting on the diaphragm. One possible advantage of this type of piezometer is that it is capable of continuous measurements, whereas a VW piezometer needs a few seconds to obtain a reading. However, strain gauge piezometers require a more sophisticated type of readout that is more expensive than the type used for VW piezometers. Strain gauge piezometers are also typically more expensive than VW piezometers and less reliable over the long term. As with VW piezometers, strain gauge piezometers are easily adapted for continuous data collection using a data logger, and the data can be telemetered and assessed in near real-time.

12.4.7 Standpipe piezometer

A standpipe, sometimes referred to as a Casagrande piezometer, is the simplest device for measuring the pore pressure or water level in a borehole. It typically consists of a pipe installed into a borehole, perforated or slotted over the area of interest (usually an interval at the bottom of the hole). Some manufacturers offer porous media piezometer tips that can be attached to the bottom of the standpipe instead of a perforated or slotted section. The diameter of the pipe is most commonly 25 or 50 mm, but may be larger if pumping or sampling is required. Small-diameter standpipe piezometers (i.e. 15–20 mm diameter) may also be used in low-permeability materials in which the response to pore pressure changes may be slow.

The annulus between the pipe and the borehole may be filled with sand or other porous media. The installation can be open to atmospheric pressure (open standpipe piezometer) or sealed with low-permeability material (Fig. 12.24). Open standpipe piezometers are suitable for measuring the depth to water or water table elevation in unconfined, surface aquifers. Sealed standpipe piezometers are required to measure pore pressures within discrete layers in confined aquifers. Sealed piezometers typically have a pocket of sand or fine gravel placed around the perforated section of the pipe, with a low-permeability seal placed above, and in some cases below, the sand/gravel pocket to isolate the perforated interval. Seals are commonly composed of bentonite and/or a cement–bentonite slurry that is several orders of magnitude lower

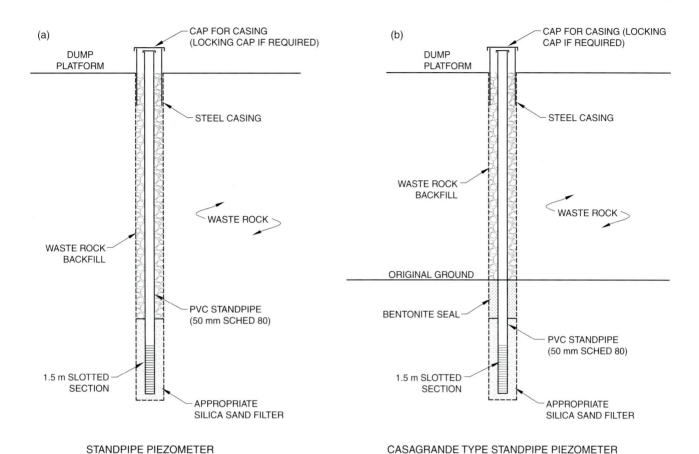

Figure 12.24: Schematic illustration of standpipe piezometers: (a) open standpipe piezometer and (b) sealed standpipe piezometer

in permeability than the surrounding formation. Once this slurry sets, the perforated interval or Casagrande tip is hydraulically isolated and the water level within the piezometer pipe is representative of the head or pore pressure within the formation at the tip or adjacent to the perforated zone.

Shallow standpipe piezometers are simple, inexpensive and relatively easy to install, although installation of sand/fine gravel zones becomes more difficult as the depth of installation increases. Open standpipe piezometers can also be installed in test pits and trenches, which are subsequently backfilled.

Water levels in standpipe piezometers can be measured either using a water level sounding device or by installing a pressure transducer and data logger in the standpipe. Standpipes also have the advantage of enabling *in situ* hydraulic conductivity testing (i.e. falling or rising head testing) of the formation or layer in which the standpipe is completed. Sampling of water from the standpipe may also possible using a bailer or small electric or hand operated pump.

12.4.8 Multiport or multilevel piezometer

Installation of multiple standpipe piezometers in a single, large diameter borehole enables the measurement of pore pressures at multiple depth intervals in the hole. Where multiple standpipes are installed in the same borehole, the interval between completion zones is typically filled with bentonite clay and/or a low-permeability cement-bentonite grout. Figure 12.25a is schematic illustration of a multilevel sealed standpipe borehole installation.

As indicated above, VW piezometers are well suited for installing multiple instruments at different depths in a single borehole. Schematic illustrations of two different methods for installation of multiple VW piezometers in boreholes, one with the transducers encased in sand completion zones and one with the transducers encased in a continuous column of cement–bentonite grout, are given in Figs. 12.25b and 12.25c, respectively. For additional information on the theory of VW piezometers and practical installation guidelines, the reader is referred to McKenna (1995), Mikkelsen and Slope Indicator (2000) and Beale and Read (2013).

12.5 Monitoring guidelines and trigger action response plans

As with open pits, monitoring of waste dumps and stockpiles should be an integral part of mine operations.

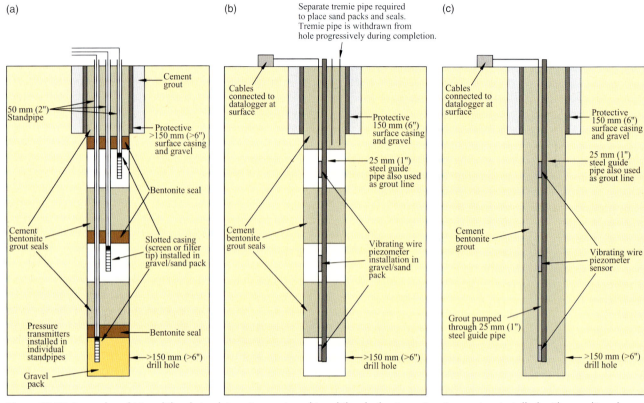

Figure 12.25: Examples of (a) multilevel standpipe piezometers, (b) multilevel vibrating wire piezometers installed with gravel/sand completion zones and (c) fully grouted multilevel vibrating wire piezometers. Source: Beale and Read (2013)

The monitoring methods and requirements depend on the type and scale of the waste dump and stockpile, as well as the possible failure modes and overall hazard level of the facility.

12.5.1 Monitoring program

It is important to initiate an appropriate monitoring program as early as possible in the waste dump/stockpile development cycle to establish an initial baseline for assessing the performance of the structure. Without this reference baseline, it is difficult to evaluate the significance of monitoring results, or to be able to identify and interpret signs of developing instability.

The monitoring program should be developed in collaboration with geotechnical specialists experienced in the design, construction, operation and monitoring of waste dumps and stockpiles. It should be designed to recognise deformations that may be precursors to possible instability modes that were identified during the investigation and design process. The required scope and intensity of the monitoring program will vary depending on the level of hazard as summarised in Table 3.12 in Chapter 3. A flow chart outlining the basics steps involved in developing a waste dump or stockpile monitoring program, including references to specific chapters in this

book, is provided in Fig. 12.26. A summary of suggested initial monitoring, inspection and reporting guidelines for each WHC (see Chapter 3) is provided in Table 12.2. These guidelines are for stable dumps ('Normal' alert level – see Section 12.5.4) and should be considered as an initial starting point. They should be modified to suit the individual facility as site-specific experience with the performance of the facility is gained.

12.5.2 Data acquisition and telemetry

The need for a data acquisition and telemetry system is site-dependent and not necessarily related to the type of instrumentation. Similarly, the need for automated processing of monitoring data can be site-dependent, but may also change over time. The past few decades have seen major advances in the way instrumentation monitoring data are captured and processed. The frequency of readings and data processing requirements depend, in part, on the type of instrument being monitored and, in part, on the speed at which results are required. Some forms of monitoring lend themselves to automated data acquisition and telemetry more readily than others.

Data acquisition systems can range from the manual reading and recording of instruments in the field by suitably trained personnel to the use of data loggers and

Table 12.2: Suggested initial monitoring and inspection guidelines

Waste dump and stockpile hazard class (WHC)	Waste dump and stockpile stability rating (WSR)	Monitoring methods	Inspection requirements[1,2]	Frequency[3]
I	80–100	• Visual	• Operations	• Daily visual inspection
			• Geotechical	• Monthly visual inspection
II	60–80	• Visual • Basic instrumentation (manual wireline extensometers, crack monitors, tell-tales, manually monitored prisms (monitoring frequency weekly–bimonthly) • Piezometers as required	• Operations	• Visual inspection and monitoring at the start of each shift (every 12 h)
			• Geotechnical	• Weekly visual inspection; monthly data review
			• Independent reviewer	• Every 12–24 months
III	40–60	• Visual • Basic instrumentation (manual wireline extensometers, crack monitors, tell-tales, manually monitored prisms (monitoring frequency daily–monthly)) • Redundancy (more than one system and/or multiple instruments) • Piezometers as required	• Operations	• Visual inspection and monitoring every 6 h (twice per shift)
			• Geotechnical	• Daily visual inspection; weekly data review
			• Independent reviewer	• Annual
IV	20–40	• Visual • Automatic, continuous (telemetered) wireline extensometers, LDM, RTS, SSR • Redundancy (multiple systems) • Specialised instrumentation (tilt sensors, LDM, TDRs, inclinometers, acoustic emissions) as required • Piezometers as required	• Operations	• Visual inspection and monitoring every 4 h
			• Geotechnical	• Visual inspections every 12 h; continuous (24/7) remote surveillance; biweekly data review
			• Independent reviewer	• Semi-annual
V	0–20	• Visual • Automatic, continuous (telemetered) wireline extensometers, LDM, RTS, SSR • Redundancy (multiple systems) • Specialised instrumentation (tilt sensors, LDM, TDRs, inclinometers, acoustic emissions) as required • Piezometers as required	• Operations	• Visual inspection and monitoring every 2 h
			• Geotechnical	• Visual inspections every 8 h; continuous (24/7) remote surveillance; daily data review
			• Independent reviewer	• Quarterly

Notes:
1. All inspections should be summarised in a suitable format, circulated as appropriate and filed for future reference.
2. Geotechnical department staff and independent reviewer conducting inspections should be geotechnical specialists with appropriate experience in the design, construction and operation of waste dumps.
3. Frequency of inspections subject to local regulatory requirements.

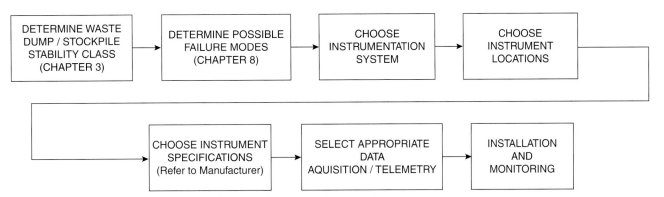

Figure 12.26: Flow chart for development of a waste dump or stockpile monitoring program

remote transmission unit systems that query sensors at a preset frequency, record the data and transmit the data to a central computer for processing and assessment. Data transmission is typically achieved using radio, telephone/cellular or satellite transmissions. The most suitable method of transmission depends on site conditions and existing systems. Where a radio network with suitable bandwidth is already in place, radio transmission may be viable. However, where a cellular network is in place on site, the use of a cellular telemetry system may be more appealing, particularly if there is limited radio frequency bandwidth available. Where site infrastructure allows and there is no cellular network, it may also be possible to use telephone hard line connections. Satellite transmission of data may be more costly than the other options, but is advantageous where sites are remote and it is necessary to transmit the data off site without access to other methods of communication.

When choosing the right telemetry system for a particular site, the following criteria must be considered:

- The chosen system must be compatible with the expected temperature and moisture extremes at site.
- Where possible, telemetry systems should interface directly, without the need for signal pre-conditioning (i.e. the telemetry system should be able to transmit raw data directly). It may, however, be necessary to first convert analogue signals to digital signals before transmission.
- Systems requiring running of long communication wires, which are prone to physical damage or vandalism, or which may be susceptible to lightning strikes, should be avoided or protected. Note that where analogue signals are being transmitted, wire length can affect signal strength.
- Regardless of the telemetry system used, it is advisable to allow for the instruments to be read manually, and for data logger files to be downloaded manually, in the event the telemetry system fails.

12.5.3   Data assessment and reporting

Data processing requirements and presentation formats depends on the specific instrumentation. As noted by Read and Stacey (2009), monitoring programs have failed because the data acquired were not used or interpreted properly, or were never used. The monitoring program must be developed with a clear sense of purpose, which will dictate how the accumulated data are interpreted and reported. The goal of processing and properly presenting data obtained from the monitoring systems is to be able to rapidly assess data and detect changes that require action.

Interpretation of data from monitoring systems typically involves looking for changes from the expected norms. For this reason, as discussed above, early

implementation of a monitoring program is essential to establishing a baseline and determining the accuracy of the specific monitoring systems that have been deployed. Baseline data may also be needed to establish variations in the data due to diurnal or seasonal variations. For example, as discussed in Section 12.3.1, diurnal variations in atmospheric conditions can have a significant impact on EDM and RTS measurements, particularly where the differences in daytime and night-time temperatures are significant. Similarly, phreatic levels within a waste dump/stockpile or its foundation can fluctuate seasonally. These variations must be recognised and considered as part of the interpretation process. With respect to an EDM or RTS system, it may be possible to minimise the potential for diurnal affects by taking measurements at the same time each day. However, if instability develops and it becomes necessary to monitor prisms more often than daily, it will be necessary to filter out the atmospheric effects.

Reporting formats should be tailored based on site requirements. Where monitoring systems are large or complex, it may be worth considering the use of an integrated geotechnical monitoring system developed specifically for mining applications. These systems are designed to allow real-time data acquisition and processing of multiple instrumentation systems, and typically include the ability to issue alerts and alarms. They also allow monitoring data to be stored in a central location that can be queried by authorised users from almost anywhere in the world. They help increase monitoring program reliability while reducing data acquisition and processing costs. Examples of integrated geotechnical monitoring systems that are currently available include GeoExplorer from SiteMonitor Systems, Geoscope from Soldata Inc., Vista Data Vision from RST Instruments Ltd and M2M from SysEng (S) Pte Ltd. Some mines also develop site-specific integrated geotechnical monitoring data platforms that are customised to the suite of instruments in use at the mine, although the cost of developing such systems can be prohibitive.

12.5.4   Trigger action response plans

A trigger action response plan (TARP) should be established for the key monitoring method(s) used to assess the short-term performance and safety of the waste dump or stockpile. To be effective, the TARP should be based on changes in readings in excess of the monitoring method's accuracy or, in the case of piezometers, a predetermined piezometric pressure or head that may be associated with a specific factor of safety or probability of failure. A generic example of a TARP for monitoring of a dump using wireline extensometers is given in Table 12.3.

The generic TARP in Table 12.3 does not provide specific movement thresholds as these will depend on many factors, including the monitoring method, the size

Table 12.3: Example trigger action response plan (TARP) for a waste dump monitored using wireline extensometers

Alert	Green	Yellow	Orange	Red
Platform condition	No new tension cracks; no recent movement on old tension cracks	Existing cracks opening and new tension cracks developing close to the crest	Cracking extends well behind the crest; open cracks; obvious differential settlement and/or development of steps or grabens on the platform	Rapidly coalescing, arcuate cracks that extend well behind the crest; deep, open tension cracks and/or grabens; platform not trafficable
Crest, face and toe condition	Normal toe, face and crest geometry; no toe spreading or heaving, face bulging, crest over-steepening or excessive ravelling	Minor changes in dump geometry (minor toe spreading/heaving, and/or over-steepening); minor face and crest ravelling	Pronounced toe spreading/heaving, and/or face bulging; substantial over-steepening of crest; small silver failures; active face and crest ravelling/breakback	Extensive toe spreading/heaving and/or face bulging; active face and crest slumping/ravelling, frequent silver failures; active instability
Deformation rate	$< X$ mm/h, extensometer monitoring frequency every A h; dump advance rate $< D$ m/d	Rate of movement $> X$ mm/h and $< Y$ mm/h, rate increasing; extensometer monitoring frequency every B h; dump advance rate $< E$ m/d	Rate of movement $> Y$ mm/h and $< Z$ mm/h, rate increasing; extensometer monitoring frequency every C h	Rate of movement $> Z$ mm/h, rate increasing
Responses				
Operations control	Normal operational controls	Contact mine geotechnical engineer; notify all affected personnel of yellow alarm level	Contact mine geotechnical engineer; notify all affected personnel of orange alarm level	Contact emergency response committee and mine geotechnical engineer; notify all affected personnel of red alarm level
Mine workers	Continue with normal duties	Become familiar with location and potential change in waste dump condition during shift; report significant changes in conditions to shift supervisor	Elevate level of awareness; visually observe and monitor waste dump and platform conditions; report significant changes in conditions to supervisor	Comply with emergency evacuation procedures and remain outside waste dump and potential runout area
Geotechnical engineer	Routine inspections and monitoring	Visually inspect area; determine frequency of inspections, monitoring and need to reduce loading rate; notify management of any recommended changes; communicate with mine workers and advise on the location, nature and expected conditions associated with a potential failure	Evaluate the monitoring data and provide recommendation for TARP level advance; determine frequency of inspections, increased monitoring, and mitigative measures; notify management of any recommended actions	Inspect, investigate and formulate recovery plan (risk assessment required); report findings to management
Shift supervisor	Continue with normal operations	Monitor conditions throughout shift; report any noticeable change in conditions to the geotechnical engineer; consult with geotechnical engineer on need to reduce loading rate; report any change of conditions or change in TARP level to next shift	Close dump, idle trucks or redirect to alternative dumps; communicate with workforce that orange level has been reached; closely monitor conditions throughout shift; report any noticeable change in conditions to geotechnical engineer; report any change of conditions or change in TARP level to next shift	Communicate with workforce that a red level has been reached and withdraw personnel and equipment to a safe location; secure to prevent entry; inspect area from outside the instability zone and report to mine superintendent/mine manager; implement recovery plan once formulated (risk assessment required)

(Continued)

Table 12.3: (Continued)

Alert	Green	Yellow	Orange	Red
Mine superintendent	Monitor production	Monitor production; communicate with geotechnical engineer	Liaise with shift supervisor; assess situation and inspect as required; communicate with geotechnical engineer; notify mine manager of situation as appropriate	Inspect area from outside the instability zone and report to mine manager; implement recovery plan once formulated (risk assessment required)
Mine manager	Normal operations	Monitor situation as required	Prepare to evacuate affected areas; monitor situation as required	Evacuate affected areas; approve recovery plan; notify corporate, mine inspectors, emergency services; monitor situation as required

Source: Adapted from Read and Stacey (2009)

Table 12.4: Suggested simple wireline movement thresholds for waste dumps and stockpiles with no performance history

Waste dump and stockpile hazard class (WHC)	Normal		Caution		Warning		Close	
	Monitoring frequency (h)	Movement threshold cm/h (cm/d)	Monitoring frequency (h)	Movement threshold cm/h (cm/d)	Monitoring frequency (h)	Movement threshold cm/h (cm/d)	Monitoring frequency (h)	Movement threshold cm/h (cm/d)
I	Daily visual monitoring only		12	3.0 (72)	6	4.0 (96)	4	5.0 (120)
II	12	1.5 (36)	6	2.0 (48)	4	3.0 (72)	2	4.0 (96)
III	6	1.0 (24)	4	1.5 (36)	2	2.0 (48)	2	3.0 (72)
IV	4	0.75 (18)	2	1.0 (24)	2	1.5 (36)	1	2.0 (48)
V	2	0.5 (12)	2	0.75 (18)	2	1.0 (24)	1	1.5 (36)

Notes:
1. Movement rates should be above the threshold on two consecutive readings before advancing to a higher alert level, and below the threshold on four consecutive readings before dropping to a lower alert level.
2. Closed dumps should not be reopened until the rate has dropped to the Caution level for at least four consecutive readings.
3. Thresholds are for simple wireline extensometer monitoring (manual or automatic/telemetered systems) without intermediate supports and with the wireline positioned in the middle of the portion of the crest that is being advanced.
4. Thresholds are intended as a starting point for facilities that have no performance history, and should be adapted and modified to suit site conditions and as experience is gained.

and geometry of the facility, foundation conditions, material types, the method and speed of construction, and, most importantly, the performance history of the facility. For waste dumps and stockpiles that have no performance history, suggested initial movement thresholds are provided in Table 12.4. Suggested thresholds in Table 12.4 vary depending on the potential hazard of instability as represented by the WHC described in Chapter 3. Lower thresholds are suggested for facilities with a higher hazard class. Note that the WHC does not specifically address the potential consequences of instability or risk (see Chapter 10). Hence, even lower thresholds may be appropriate for cases where the consequences of instability and risk are high. Note also that the suggested thresholds in Table 12.4 are intended for simple wireline monitoring systems (without intermediate supports – see Section 12.3.2). These guidelines are intended as an initial starting point only. They should be adapted and modified to suit site conditions as experience is gained.

13

DRAGLINE SPOILS

John Simmons and Robert Yarkosky

13.1   Draglines

A dragline is the single largest unit of equipment at an open cut mining operation, with initial capital costs exceeding US$100 million for a new large-class machine. The high capital cost and limited mobility of draglines are offset by their capability to move large volumes of material quickly and at minimum cost. Draglines operate by lifting and swinging large bucket loads of material to the closest possible dumping point. This means that the most cost-efficient use of a dragline will result in the steepest and highest dump slope profile that the machine can construct with acceptable stability for the geometric and material constraints of the site.

The operating characteristics and terminology for draglines are relatively simple (Fig. 13.1). The design of modern 'walking' draglines dates to the first three decades of the 20th century and consists of a long boom to carry the bucket, with two sets of ropes for dragging and lifting. The boom mounting arrangement and rope winders are located within a machine house, which is mounted on a large circular support structure known as a tub. The bucket filling and dumping cycle occurs within the fore–aft vertical plane of the machine house. Lateral movements of the bucket are controlled by reversible motors, also located within the machine house, that swing the orientation of the fore–aft vertical plane using a ring bearing mounted on the (stationary) tub. The fore–aft load cycle requires the boom and suspended load to be offset by a counterweight in the aft of the machine house. Currently, the majority of draglines used for mining purposes fall into two size classes: ~3000 tonnes and 6000 tonnes. Boom tip operating radii range from 75 to 110 m, tub diameters range from 17–26 m and suspended loads may range from 130 to 275 tonnes, of which ~50% comprises the material payload.

Draglines 'walk' by means of large shoes connected to bearings on each side of the machine housing. The shoe bearings are linked to a propulsion mechanism that drives the bearings through an elliptical path within vertical planes parallel to the fore–aft machine axis. 'Walking' is achieved when the shoes take up as much as 80% of the machine weight as the bearings approach the bottom of their path while lifting the aft part of the tub. As the tub is lifted, the propulsion mechanism pulls the tub towards the aft, effectively sliding the machine tub aft-wards while the shoes remain in static contact with the operating bench.

Currently, the majority of draglines used for mining purposes exert average tub–ground contact pressures of between 110 kPa and 160 kPa and average shoe-ground contact pressures of between 250 kPa and 400 kPa. Tub and shoe dimensions can be modified to suit the bearing capacity and deformability characteristics of the ground materials. Relatively large ground deformations can be tolerated under tub and shoe load applications and are amenable to analysis using the geotechnical principles of initial and repeated loading of surface foundations.

The dragline load cycle consists of filling (dragging the bucket through the dig profile towards the machine house) followed by lifting. As the filled bucket leaves the ground, the superstructure above the tub commences swinging towards the dump point. The bucket load is lifted towards the boom point during swinging, and the swing motors are reversed ('plugging') to minimise rotational velocity by the time the bucket reaches the dumping point. Dumping is achieved by releasing the drag ropes, after which the superstructure is swung back to the filling point while the bucket is lowered to the filling position.

Swinging and plugging exert large torsion reactions at the tub–bench contact, while filling and dumping cause large fluctuations in the shear reaction force and eccentricity of the normal reaction force at the tub–bench contact. During a complete filling and dumping cycle, the tub and working bench reactions can cause significant deformation of the bench surface.

(a)

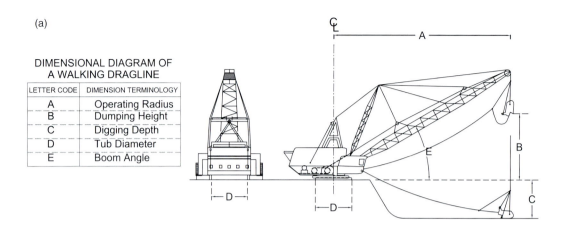

DIMENSIONAL DIAGRAM OF
A WALKING DRAGLINE

LETTER CODE	DIMENSION TERMINOLOGY
A	Operating Radius
B	Dumping Height
C	Digging Depth
D	Tub Diameter
E	Boom Angle

Figure 13.1: Operating characteristics and terminology for draglines. Source: J Simmons

13.2 Dragline operating methods

Draglines work most efficiently by uncovering target resources in strips, with the width of each strip related to the geometry of the machine and the mining plan. Greatest efficiency involves the machine working from a bench that is designed to exploit the dig depth and dump height capabilities of the machine's boom and rope configuration. The stripping process is inherently three-dimensional due to the swing-dump part of the operating cycle, but on completion a dragline dump slope may be treated as a 2D profile consisting of up to four components: the bench with its undercut batter (bench face), the dump

with its rilled batter (repose angle slope face), the profiles of any previous strip slopes and the uncovered material at the base of the bench (Fig. 13.2).

Draglines are ideally suited to stripping deposits that consist of relatively flat-lying target resource seams and interburdens, and are most widely deployed for coal mining. For a single-seam deposit, the optimum positioning and operating sequence are relatively simple. For multiple-seam deposits, the relative thicknesses of interburdens and seams result in a more complex operating sequence where dragline cuts may be supplemented by other stripping methods. Where the

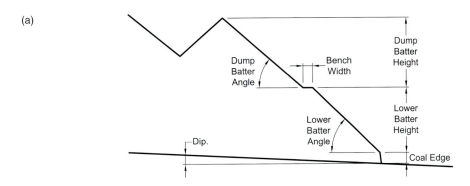

(a)

Figure 13.2: Typical dragline in-pit bench operation showing characteristics of dump slope profile. Source: R Yarkosky

thickness of overburden material exceeds a machine- and deposit-specific threshold, pre-stripping above the dragline horizon has to be undertaken either by other methods or by additional rehandling where the dragline executes a pullback operation to provide spoil room for its resource-uncovering cycle.

Many different dragline operating methods have been developed to most efficiently utilise a range of machine configurations for wide-ranging combinations of geological and geotechnical conditions. All methods have some common features, and some have particular significance for management of risks associated with potential geotechnical instability hazards involving spoil dump profiles, dragline operating benches, or both. All methods require a dozer to maximise the efficiency of

dragline operation. Mirabediny and Baafi (1998) described the following methods that have been developed for Australian coal mines:

- simple sidecast
- extended bench
- split bench
- chop-cut in-pit bench
- extended key cut
- a variety of multi-pass techniques.

Sidecasting (Fig. 13.3) can be used where there is only a minimal distance between the filling and dumping locations. Sidecasting results in the initially excavated materials being placed sequentially at the lowest locations within the spoil profile, and this may have

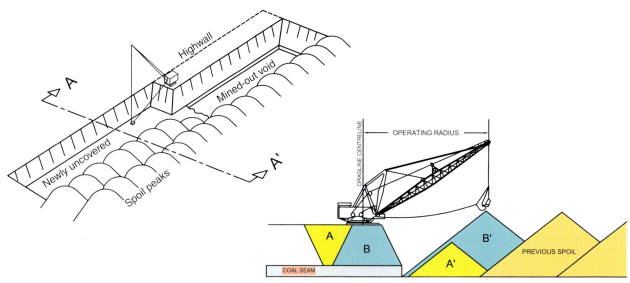

Figure 13.3: Dragline simple sidecast operating method

significant implications for the geotechnical stability of the dump profile. All other operating methods involve the dragline working from a bench and swinging material from the excavation side (highwall) to the dump side (low wall).

Machine loading of a bench has the potential to cause involuntary tub slip unless the working surface is suitably prepared, but may also cause instability of the bench resulting in uncontrolled movement of the machine. Different operating methods result in different exposures of the machine and operating bench to instability hazards. Exposure to bench instability hazards is minimised by the key-bridge method (Fig. 13.4), which involves preparation of a bridge between highwall and low wall, followed by

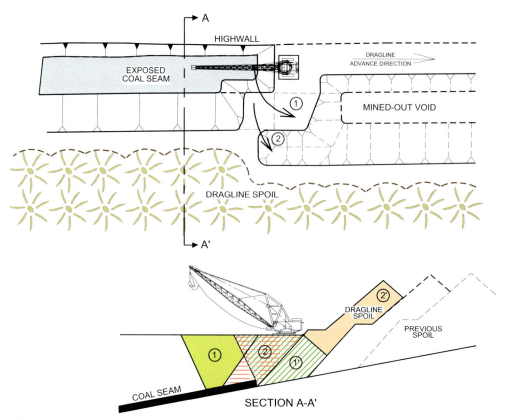

Figure 13.4: Dragline key-bridge operating method

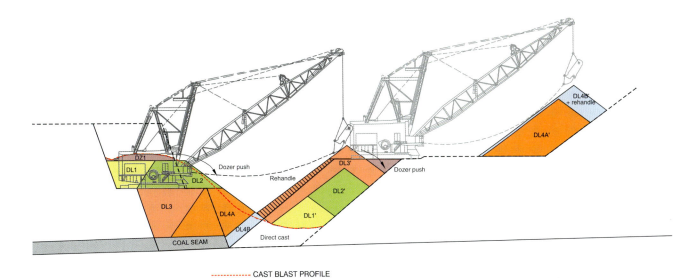

Figure 13.5: Dragline split bench, extended key cut, extended bench, multi-pass method

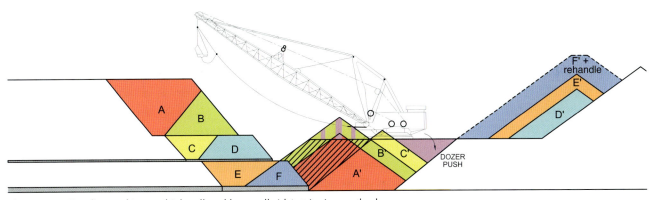

Figure 13.6: Dragline multi-pass (highwall and low wall side) stripping method

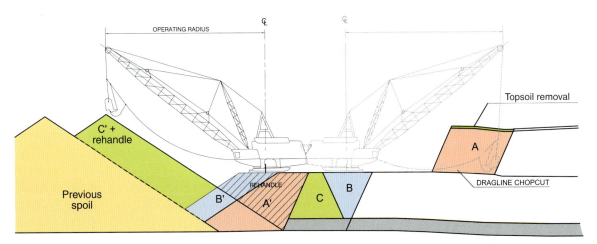

Figure 13.7: Dragline chop-cut in-pit bench method

sequential removal of material from a key on the highwall side and then from the block left on the low wall side. All of the other techniques (Figs. 13.5 to 13.7) offer greater efficiencies in exploiting the machine's capabilities for particular geological settings, but may also result in greater exposure to potential bench instability under the combined effects of machine loading and minimal lateral support of the bench profile.

(a)

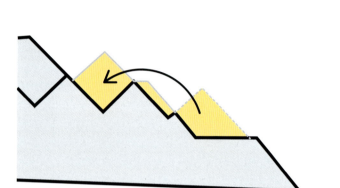

(b)

Figure 13.8: Dragline pullback with machine positioned on elevated bench with doubled cut batters. Source: J Simmons

Figure 13.8 shows an example of a dragline pullback operation. The machine is established on an elevated bench position where it usually sits near its previous dump peak elevation and rehandles previously dumped material to a position further back from the void. The dragline load on a pullback elevated bench involves a much higher cut slope face and exposure of previously dumped materials, and may require non-routine stability assessment.

13.3 Dragline tub slip

Involuntary tub slip may result in a dragline sliding over the edge of a bench with potentially catastrophic consequences (Fig. 13.9). Typical dragline tub and machine designs result in a required interface friction coefficient of at least 0.3 ($\phi' = 16.7°$), which is readily achieved if the contact between the tub and ground is relatively dry and of a consistency and strength that provides adequate bearing capacity for tub and shoe loadings.

Standard dragline operating practices include levelling of the working bench and dozer trimming to create a suitable working surface. In particular, rainfall or water runoff can cause softening of the working surface and may create a thin surface layer of wet, high fines content material that has relatively low short-term or undrained soil shear strength response during the working load cycle. Positioning the tub on an untreated wetted surface may reduce the equivalent frictional coefficient to 0.2 ($\phi' = 11.3°$) or less, which is very likely to result in uncontrolled slip, particularly during the plugging stage of the working cycle.

Operator-induced and controlled slippage at the tub–bench contact zone occurs normally during walking and for spinning of the tub to achieve the desired alignment of the trailing cable. It is equally a matter of training, competent operation and compliance with standard procedures to effectively minimise the likelihood of involuntary tub slip.

(a)

(b)

Figure 13.9: Catastrophic consequences of involuntary dragline tub slip. Source: Annonymous

13.4 Dragline operating bench stability

13.4.1 Machine load action effects

Potential instability of a dragline operating bench may develop from the interaction of the machine loading action with a range of geotechnical hazards. The machine loading action is a sequence of different combinations of normal force distribution, lateral shear force distribution and torque distribution. In principle, these action effects can be broken down into components that change during the complete filling–dumping cycle and that can be analysed using appropriate principles of applied mechanics. In practice, each of the action effects is a distributed variable and the combined action effects become too complex for routine practical analysis.

The simplest approximation to the normal load action effect is to average the gross loaded normal force over the gross tub area. This ignores the variation in resultant normal force eccentricity during the swing cycle, with the eccentricity being towards the low wall side during both the filling and dumping extremes of the working cycle. While the loaded area is circular, most practical analyses of loaded bench stability are undertaken using 2D limiting equilibrium analysis methods. If the normal tub pressure is applied as a 2D distributed load, the actual 3D effect is over-represented as a strip loading. Adoption of the average normal tub pressure as a 2D strip load therefore biases the normal load action effect towards the high end of the working cycle range. Comparisons of 2D and 3D limit equilibrium or stress-deformation analysis outcomes for dragline loadings on benches can be used to show that the 2D average tub pressure approach is both reasonable and reliable for practical design purposes.

Directional shear and torsional load effects due to filling and swinging/plugging are generally ignored for bench stability assessment. This approach may be rationalised in terms of the torsion effects being essentially self-equilibrating in terms of net lateral force, and the sideways reaction to filling being well within the frictional resistance capabilities provided that the bench has been adequately prepared.

The resultant normal force at the tub–bench contact is in reality a 3D laterally variable distributed pressure that is further distributed with depth below the tub contact. There are many geotechnical techniques for analysing such pressure–depth distributions. Elastic theory may be applied using closed-form solutions (Poulos and Davis 1974), by computer-based stress distribution models (e.g. Settle3D (RocScience Inc. 2016b)), or by other methods of stress-deformation analysis. A simple and effective approximation, given the already uncertain variation of applied pressure during the operating cycle, is to consider the total normal force to be uniformly applied to a subsurface area that increases with depth as a simple linear lateral variation that approximates the zone of pressure from an elastic model. In practice, a suitable lateral variation can be obtained by projecting the boundaries of the surface contact area downward at 60° (measured from horizontal) or, for simplicity and convenience, at 2 (vertical): 1 (horizontal), which is equivalent to 63.4° from horizontal. The concept of a lateral distribution line for tub loading (Fig. 13.10) has significant implications for managing bench instability risks, particularly when low strength materials are present within the bench or the bench foundation.

Dragline benches may also be subjected to dynamic loads resulting from blasting or from earthquakes. With respect to blasting, a dragline will normally be located

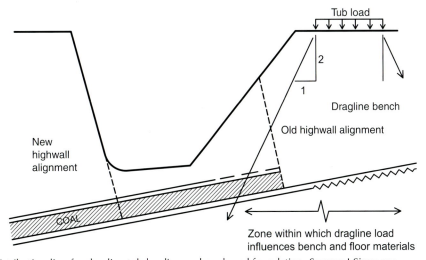

Figure 13.10: Lateral distribution line for dragline tub loading on bench and foundation. Source: J Simmons

outside of a blast exclusion zone to reduce the likelihood of blast-related damage to an acceptable level. Earthquake loading effects on a dragline bench are therefore likely to be much more significant than blasting effects for dragline bench stability. Stability assessment for dynamic blast or earthquake is discussed further in Section 13.5.7.

13.4.2 Geotechnical hazards for dragline bench loadings

Dragline benches will be composed of some combination of *in situ* rock, blasted rock and spoil that has been placed by dragline dumping and dozing. Coal deposits in particular may include weak bedding-parallel layers or surfaces. Blasting may or may not result in disruption of such layers, depending upon the geological location of the layers and also on the blasting method. Weak layers located as shown in Fig. 13.11 are known to have contributed to bench instability under dragline loadings.

In addition, weak, mud-like material or excessive water left within the previous pit void is typically not displaced or stabilised by being covered by some combination of cast blasting or spoil deposition, resulting in a soft and low-strength zone under the bench profile. There are instances where dragline benches that are fully buttressed to the highwall side have nevertheless collapsed both sideways into key excavations and backwards towards the low wall side.

Management of bench stability for any of the above hazard scenarios requires appropriate analysis, most commonly using 2D limit equilibrium methods. Such analyses require reliable information regarding bench geometry, ground conditions within the bench profile, dragline loading and the influence (if any) of groundwater. In principle, it is possible to measure or interpret the effective shear strength of weak bedding-parallel layers, provided that the existence and location of such layers has been identified from prior exploration. In most situations, the operational shear strength of the blasted materials cannot be measured by testing and has to be assumed using experience and judgement, or inferred from back-analysis. The latter option can be technically reliable but should be used with caution and care. Shear strength parameters for spoil materials may be assessed through a justifiable process of assumption, testing and back-analysis, and this is discussed in further detail in Section 13.5 below. The shear strength of soft and weak mud or saturated and remoulded spoil at the base of the bench may be measured by appropriate testing. Given the operational constraints, it is generally not practicable to measure groundwater conditions within the bench.

For the above reasons, any geotechnical model of a dragline-loaded bench may require many assumptions and the exercise of sound engineering judgement. The outcomes of analyses may therefore be reliable only to the extent that reliable judgements were involved in development of the model. The model parameters are likely to be a self-consistent, rather than uniquely correct, set. Under such circumstances, it is preferable to develop a sound working knowledge of actual dragline bench stability conditions by back-analysis of stable profiles, appropriate materials testing, empirical correlations and sensitivity testing of the geotechnical model.

When a weak bedding-parallel bench instability hazard is identified, either through design assessment or as a contingency, some form of mitigation treatment is required. The most appropriate mitigation will depend on a range of site-specific and time-dependent factors, but will generally involve provision of a supporting buttress, either by leaving material in place or by placement of material by dumping or dozing. Buttress material has to be removed subsequently, either by means of dragline stripping during a mining pass or by some form of post-stripping operation by other equipment. Buttressing of a dragline bench will inevitably result in some additional stripping costs.

A weak zone within the base of a bench profile also has implications for the stability of the final dump profile. To stabilise the bench and/or the final low wall, the dragline

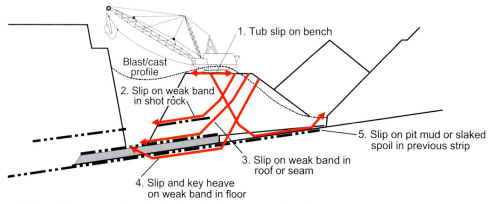

Figure 13.11: Potential instability mechanisms for dragline loading on in-pit benches. Source: J Simmons

may have to use a modified key-bridge method to remove the weak zone while remaining effectively buttressed. This technique is often called a mud-pass, and is difficult to analyse realistically by 2D methods. Empirically derived operating rules, based on arching principles, are usually invoked because the use of 3D analytical methods can be hard to justify due to the large number of unknown parameters involved. To place spoil obtained from a mud-pass, it is necessary to create retention basins called mud-dams within the dump profile (Section 13.5.1). Creation of stable mud-dams, and capping with stronger spoil, requires significant skill on the part of operators. The locations and approximate geometry of such mud-dam zones should be surveyed and recorded so that hazards related to subsequent loading or reshaping of the dragline dump profile can be identified and the associated risks appropriately mitigated.

Dragline pullback operations are a special case of bench loading for which geotechnical stability assessments should be carried out. While adverse groundwater conditions are not normally associated with pullbacks, the overall slope height is much greater than for a normal dragline stripping scenario and stability should be proven rather than assumed.

13.5 Dragline dump profile stability

The most significant aspect of the overall dump profile is that it is the product of sequential dumping and excavation, and therefore must recognise the practicalities of the dragline work cycle as well as the requirements for slope stability. Factors that contribute to stability are the profile geometry, the dumped material characteristics, dump foundation, groundwater regime, surcharge and dynamic loadings. Determinaton of the deformation responses and identification and assessment of dump instability mechanisms require modelling of these factors using appropriate methods.

13.5.1 Dragline dump profile geometry

The discussion of bench loadings included references to several factors that are equally relevant to the final dump profile created by dragline spoiling. Apart from simple sidecasting, all dragline stripping methods result in two dump components that may be significantly different in many respects: the dump profile and the bench profile (Fig. 13.2).

Dump profiles are formed by rilling (placement of material incrementally on an angle-of-repose slope), which leads to localised compartments that are stratified at the rill (repose) angle and where larger particles tend to segregate downward towards the bench level. Rock traps or drains are often dug into the bench surfaces by dragline operators to trap rolling rocks that might otherwise continue rolling and strike either the machine or its trailing power cable. The rill angle is related more to the particle size range and the dynamics of the dumping and rilling processes than to the particle mineralogy. Typical rill angles range from 34 to 38° and may be usefully approximated as 37°, which is equivalent to 3 (vertical) : 4 (horizontal).

In practice, the dumped rill is also formed by swinging to form a lateral sequence of arcuate ridges generated by the positioning of the machine during each block of spoil placement. It is impractical, and unnecessary for design purposes, to capture the fine detail of the dump geometry. An approximation that is suitable for both 2D and 3D analyses is to consider the dump to have a simple peak with linear sides formed at the rill angle, which is either an average measurement from survey or a reasonable assumption.

The spoil bench profile is formed by excavating a planar face inclined at the design batter angle, with each dig in the swing cycle followed by a dump to form the rill slope angle. Materials making up the bench are a combination of cast-blasted overburden (if any) and other materials placed by dozing or dragline dumping. Most excavated spoil materials are effectively unsaturated particulate masses with fines. Stable excavated faces can, in many circumstances, be dug at angles significantly steeper than the rill angle, or at the friction angle determined from shear strength testing. Stability of such faces is dependent on some combination of particle interlock and unsaturated suction effect (apparent cohesion). Some materials cannot be excavated at batters steeper than the rill angle, due to a lack of effective particle interlock or suction.

A remnant bench may or may not be left within the slope profile at the completion of a dragline stripping operation. This is a matter or practice (e.g. to allow access or for management of rock rolling hazards from the dump) or of design (e.g. to provide for adequate profile stability).

13.5.2 Characterisation of dumped waste materials

Dumped waste or spoil material may be described in soil mechanics terms as an essentially normally consolidated granular mass or rockfill with fines that is deposited incrementally. Deposition gives rise to internal inclined surfaces with zones of preferred particle orientations and with the larger particle sizes segregated towards the toe of the spoil because of the dynamics of the placement process. Some heterogeneity may be present, giving rise to potentially anisotropic properties, but characterisation of dumped materials at such a level of detail is impractical.

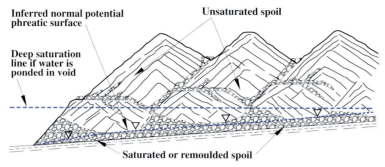

Figure 13.12: Schematic section of dragline spoil dump fabric showing potential phreatic surface and typical material zonation. Source: After Simmons and McManus (2004)

In many dragline operations, the dump profile is constructed over a base of cast-blasted overburden with different particle sizes and distribution geometries. Typically, the mass of dumped waste is unsaturated. Even if the original *in situ* materials were effectively saturated, the processes of fragmentation and placement create voids between particles that will only become saturated under specific boundary conditions and over sufficient time frames. If the dragline dump is constructed by infilling of a void containing water and/or mud, the basal zone of the spoil will be wetted and subjected to rapid breakdown and remoulding if the spoil materials are susceptible to slaking. Cast-blasting into water and/or mud will, more often than not, form a trapped basal layer of softened or remoulded waste. Figure 13.12 is a simplified sketch of a typical dragline dump profile showing internal segregation and layering, together with zonation where unsaturated and saturated/remoulded conditions may apply.

Engineering characterisation of dragline waste is therefore an empirical or at best semi-empirical process. Even the largest strength testing configurations can only

include a limited particle top-size substantially less than 0.5 m, whereas particles larger than 3.0 m are relatively common in many dragline dumps. Most spoil material strength testing is, for practical reasons, undertaken on sub-samples from which particles larger than a few tens of millimetres have been excluded. The effects of particle size scalping on the relationship between shear strength at test-scale and at full-scale are a matter of judgement, although some insight may be gained from laboratory testing (Seif El Dine *et al.* 2010). Back-analysis of full-scale instabilities can provide a connection between these scales, but opportunities for back-analysis are extremely limited and the back-analysed model and strength parameters are more likely to be one of potentially many solutions than a rigorously unique outcome.

Shear strength of rockfill has been extensively studied in the context of rockfill dams and embankments. Where the particles are primarily strong and durable, a view has emerged that the shear strength can be described by a curvilinear envelope (Leps 1970; Barton and Kjaernsli 1981) with reduced tangent friction angle at higher normal

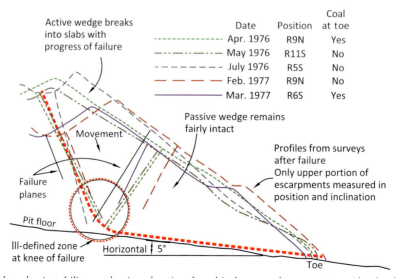

Figure 13.13: Deep-seated wedge instability mechanism showing deep bi-planar wedge geometry with a heel zone of ill-defined complex shear deformation. Source: After Richards *et al.* (1981)

stress levels to represent the loss of interlock due to destruction of asperities. For particles formed of water-sensitive materials, the presence of free water is also likely to reduce the shear strength, but specific testing is required to evaluate such effects.

Large-scale dragline low wall dump instability at Goonyella Mine in the Bowen Basin region of central Queensland, Australia, was the subject of extensive research in the 1970s. Based on external physical mapping and internal measurements of shear deformation locations, a rational model of dragline dump instability emerged (Gonano 1980) in terms of translational sliding on a weak basal surface combined with steeply inclined backscarps and internal midscarps forming a two-wedge mechanism (Fig. 13.13). The weak basal surface was typically associated with saturated waste near the contact with the pit floor, or with weak bedding-parallel strata in the immediate subfloor. The Goonyella experiences led to systematic study of dragline spoil dump stability at all of the (then) BHP Coal open cut mines, including correlations between laboratory shear strength tests and back-analysed shear strengths from full-scale instabilities.

The BHP Coal studies led to the development and implementation of a framework process for assessing dragline dump stability in terms of three material strength states (unsaturated, saturated and remoulded) and appropriate two-wedge noncircular methods for stability analysis (Simmons and McManus 2004). More recent investigations have confirmed the applicability of the two-wedge mechanism (Duran 2012) and provided more insights into basal strength mobilisation (Duran 2013) and laboratory testing for strength parameters (Kho *et al.* 2013; Bradfield *et al.* 2015).

Recent large-scale direct shear testing (Simmons and Fityus 2016) has identified that certain types of routinely encountered Australian coal mine spoil materials do not have typical rockfill curvilinear strength envelope characteristics. For these materials, the Simmons and McManus (2004) category strengths are adequate and possibly conservative at stresses corresponding to dump heights greater than 200 m. However, the categorisation process has significant limitations in terms of adequacy and reliability for spoils formed from low-strength rocks with pronounced slaking and dispersion characteristics. To comply with the category strengths, spoil materials should have the following:

- Source rock materials should have a minimum unconfined compressive strength of ~5 MPa.
- Fines produced by the fragmentation processes should be of low plasticity (liquid limit less than 35%).
- Particles should have low slake and swell potentials and only low or moderate dispersion potential based on a simple slaking test methodology.

The shear strengths for spoil materials that do not comply with the above criteria cannot be reliably predicted from the Simmons and McManus (2004) category process, and specific testing may be required to identify shear strengths for dump design purposes.

13.5.3 Characterisation of foundations

If adequate foundation characterisation is not undertaken before finalisation of the dump design, then geotechnical instability and mining productivity risks will not be understood or controlled. While the consequences of instability may be manageable, impaired productivity may not be recoverable. Foundations may consist of natural ground, previously dumped spoil in the case of pullback operations, or the exposed floor of the mined-out void in the case of in-pit dumping. The key consideration for all foundations is adequate resistance to a combination of downward bearing and lateral sliding deformations. In all scenarios, the preferred method of resistance assessment is to use limit equilibrium or limit state methods of analysis together with an adequate geotechnical model for the foundation conditions.

13.5.3.1 Dumps founded on natural ground

The foundation geotechnical model for the natural ground will include a geological model of all layers or zones together with representative strengths, stiffnesses and other parameters likely to affect deformation and shear resistance. Under some conditions, the bearing deformation of a soft or weak layer may not be of concern if buttressing support can be provided to control lateral deformations and prevent development of slope instability. However, some foundation materials are prone to liquefaction triggered by rapid loading or by dynamic responses to earthquake or blasting.

Geotechnical design principles applicable to embankment engineering should be adopted to evaluate and, if necessary, control risks related to bearing and sliding instability or liquefaction.

13.5.3.2 Dragline pullback foundations

The objective of a dragline pullback operation is to create more spoil room at a lower level by selectively excavating previously dumped spoil and dumping to create a much higher slope. The pullback slope geometry is nominally dictated by the digging and dumping capabilities of the machine, but the mechanisms of potential instability are more complex due to the range of possible foundation materials and deformation scenarios at and below the dragline dig level.

Foundation deformation under pullback loading is not well understood, at least in part because draglines are deformation-tolerant. Instances of instability appear to be rare and have not been discussed in the geotechnical literature. Assessment of foundation response due to pullback should preferably be undertaken by stress-deformation modelling, requiring representative geotechnical strength and stiffness parameters as well as detailed simulation of the loading sequence. Such assessments are well suited to the numerical modelling strength reduction factor method for pullback slope stability assessment.

13.5.3.3 In-pit dump foundations

Draglines are ideally suited to stripping of overburden to uncover gently dipping resources. Under the usual conditions where the dip is inclined in the direction of strip advance (Figs. 13.10 and 13.11), the pit floor inclination has a direct influence on the stability of the dragline dump slope. Provided the shear strength of the floor materials is sufficiently greater than the shear strength of the spoil materials, then the limiting inclination of the foundation interface will be determined by the dump slope profile, the relevant groundwater conditions, the shear strength of the waste materials and the shear strength of the waste-rock interface.

Adequate definition of the geotechnical model for pit floor conditions is essential to minimise the likelihood of instability related to foundation weakness. In most circumstances, the foundation rock mass will be significantly stronger than the waste mass. When the strength contrast is large, the stiffness contrast may also be large and consideration should be given to the potential for the interface shear resistance to be less than the shear resistance of the waste material.

Lateral instability of the foundation can develop when there are weak bedding-parallel surfaces or bands below the floor level but sufficiently close to the floor to form a breakout mechanism resulting in floor heaving and large-scale slope profile slumping (Fig. 13.14). Exploration drilling or in-pit trenching investigations are necessary to identify weak bands and enable the operational shear

Figure 13.14: Dragline dump slope instability with floor heave caused by weak subfloor band, stabilised by buttressing in left foreground (Newlands Mine, Queensland, 2007). Source: J Simmons

strength for the weak surface or material to be determined reliably. An alternative determination may be made from back-analysis of an actual floor breakout instability mechanism, provided that as many of the geological, geometric, shear strength and groundwater pressure parameters as possible can be reliably defined.

Assessment of lateral stability of the foundation requires definition of a representative breakout strength under operational conditions. Observations of breakout show a combination of separation on joints with tensile splitting and inclined shearing, suggesting that breakout is a complex process involving brittle buckling of thin discrete layers together with ductile shear yielding of rock material. The floor breakout zone may be relatively thin and have insufficient defects to enable reliable determination of an equivalent rock mass shear strength. Determination of operational breakout strength from back-analysis can be reliable provided that it is understood to form a consistent and probably non-unique set in combination with the adopted shear strengths of the other materials involved, particularly the weak floor band.

13.5.4 Groundwater conditions within dragline dump profiles

Dragline waste dumps are formed by repeated stripping processes, resulting in a pattern of particle deposition that has predictable features and a scale-dependent segregation of particle sizes that is partly randomised in location. The waste material particles may or may not be water-saturated, and the void spaces within the waste will be variably saturated, depending on the climatic conditions at and following the time of placement. A proportion of precipitation may infiltrate the dump, with the remainder being dissipated by runoff and evaporation.

Surprisingly little is known about the fate of water within the void space of a dragline dump. Observational experience from the Bowen Basin of central Queensland, Australia, which has a relatively hot and dry climate, suggests that very little, if any, free water flows out of dragline dumps except at the lowest point near the contact with the foundation. This implies that most of the free interstitial water will flow downward relatively easily through preferential, inclined, coarser-grained pathways under the influence of gravity. It also implies that some interstitial water may be effectively held within the waste mass either by suction or by slaking processes.

Under the stress levels within a dragline dump, the interstitial void spaces remain relatively large and open and are unlikely to become saturated with free water until a flow barrier is encountered. The most obvious flow barrier is the contact with foundation rock or other low-permeability foundation zones. A basal wetted zone would then be expected to build up to an extent, with the

profile controlled by free-flow diffusion processes and boundary conditions. A model for the wetted zone was developed (Simmons and McManus 2004) from limited direct observations and an extensive body of operational experience and back-analyses that adopt a wetted zone of maximum 5 m thickness capped by a phreatic surface for routine in-pit dragline dumping. Recently, the model has been generalised to include additional observations (Simmons *et al.* 2015), but the basic premise of a saturated basal zone should always be considered. Other scenarios for wetted-zone depth and phreatic surface development can be derived, for example, when a significant depth of water accumulates within a void over a sufficient period for infiltration to develop into a steady-state wetted condition.

Groundwater pressure and flow conditions within dragline waste must be considered carefully in the context of existing knowledge applicable to rockfill and earthfill dams (Fell *et al.* 2005; Fell and Fry 2007; Beale and Read 2013). As in any mixed granular medium, seepage exit gradients and flow velocities may combine with relatively loose packing structures and low confining stresses to cause internal movements of smaller particles (suffusion) that may lead to the formation of preferential water flow channels or 'pipes'. Development of internal piping does not necessarily lead to unlimited or uncontrolled erosion. Many mines operate with multiple dragline pits where some voids may be used on a short-term or long-term basis for water storage. It is common in such situations for significant lateral flows of water to develop through waste, at or near the contact with the foundation rock, and to form concentrated discharges from preferential flow channels (Figs. 13.15, 13.16). It is important to regularly monitor such outflows for evidence of ongoing piping erosion, and to prepare contingency plans for filtering and buttressing to control erosion-induced instability. Monitoring of outflow rates and measurements of wetted zone dimensions are essential components of site-specific models for groundwater in waste.

It is now common for dragline dumps to be surcharged by additional spoil dumping processes. Surcharging increases the stress levels towards the base of the dump, causing compression that reduces the interstitial void space. Depending on the water content and compressibility condition of the dragline waste profile, surcharging may progress to the extent that the more highly stressed basal component becomes mechanically saturated and could lead to static liquefaction failure of the dump. Such a condition warrants specialised assessment because of its implications for the thickness of the saturated zone and the location of the phreatic surface, both of which may be critical factors that control dump stability.

For routine design purposes, the conceptual waste dump groundwater model utilising a 5 m basal saturated

Figure 13.15: Stable concentrated seepage outflow channel at base of 80 m dragline waste dump. Note crusted salts indicating a basal wetted zone thickness of 1.5 m or greater (Oaky Creek Mine, Queensland, 2009). Source: J Simmons

zone (Simmons and McManus 2004) may be adopted, but only after giving due consideration to the expected particle size range within the waste, the prevailing hydrological conditions and relevant local experience regarding groundwater within waste dumps. The basal 5 m

Figure 13.16: Concentrated seepage outflow channel from base of 90 m dragline waste dump. Water head loss of 40 m from adjacent void, length of shortest flow path 160 m. Outflow clear but monitored for evidence of ongoing suffusion of smaller particles. Source: J Simmons

conceptual model is admittedly simplistic, and if there are any uncertainties regarding its applicability, then a more rigorous assessment is required to manage design-level instability risks.

13.5.5 Infiltration and drawdown of water ponded in mining voids

The nature of dragline strip mining results in large and deep voids with catchments large enough to accumulate significant volumes of ponded water. Void flooding may occur as a result of a heavy rainfall event. In other circumstances, voids may be used for water storage or transfer. Ponded water will generate a transient wetting front that progresses into the waste mass, eventually reaching a steady-state condition where the phreatic surface gradient equilibrates to the external pond level. Simultaneous with the pond wetting front, other infiltration mechanisms may also be acting to transfer interstitial water flows into the basal zone. Based on reasonable estimates of the relative permeabilities of the floor rock, basal wetted zone and overlying unsaturated zone with preferential flow paths, the time frame for equilibration of the wetting front to the pond water level is likely to be measured in days or weeks rather than months.

Development of a substantial thickness of inundated spoil will cause some reduction of shear strength and is also likely to cause immediate settlement due to softening of particle contacts and localised collapse of inter-particle structure, which in some cases could lead to localised static liquefaction. In most cases there is some reduction of slope profile stability, and in some materials there will be cracking or thin-skinned face slippage due to the softening effects. Large-scale slope profile instability due to the combined effects of water ponding is rare. Nevertheless, a stability check for the effects of void water ponding should form part of the slope design process.

Dewatering of the void following ponding creates a drawdown condition where interstitial water flows towards the void and creates a hydraulic head gradient that acts to reduce the stability of the slope profile. The rate of drawdown is directly related to the magnitude of the head gradient. In most dragline strip conditions, the rate of drawdown due to realistic levels of pump capacity is likely to be low enough to create minimal adverse head gradients. Drawdown conditions warrant the deployment of piezometers in the dump slope to demonstrate that the intended level of stability is achieved.

13.5.6 Surcharge loadings

Dragline waste dump profiles will be surcharged by additional layers of waste whenever the quantity and depth of overburden exceeds the volumetric and geometric capacities of the machine. Waste surcharges may be placed by dragline pullbacks, or by other methods such as truck dumping or spreaders. The peaks and troughs of typical dragline dump profiles provide a ready means for controlling rockfall risks associated with surcharge placement, while covering by surcharge dumping provides a means for limiting the infiltration of runoff water into the trough sections of dumps.

A surcharge of truck-dumped waste over a dragline waste profile will normally result in greater stability for the surcharged profile than for the dragline profile alone, provided the following are true:

- The toe of the surcharge is retained behind the second spoil peak from the void.
- The overall slope angle for the surcharge profile satisfies the operational tip-head stability requirements for the surcharge waste materials.

While these principles (Fig. 13.17) are usually satisfactory in practice, formal stability assessment for surcharging of dragline waste dumps should be undertaken to demonstrate compliance with overall waste stability requirements.

The differential and absolute settlements associated with surcharging may be quite significant, and sufficiently large to cause differential settlement cracking and post-dumping surface ponding. Ponded water can infiltrate into dumps by two mechanisms: differential cracking and piping erosion. Both may lead to significant increases in the volume of groundwater that eventually reports to the base of the dragline dump. This may give rise to an unplanned increase in the thickness of the wetted zone and the location of the phreatic surface within the dump. In addition to overburden waste, other surcharge loads may be imposed by water storages, stockpiles and large-scale structures. The consequences of absolute and differential settlements, and the likelihood of leakage through settlement-related cracking, should be taken into account when assessing the stability of dumps under such circumstances. Appropriate consideration should be given to the location of settlement-sensitive facilities on surcharged waste dumps.

13.5.7 Dynamic loadings: blasting and earthquake

Draglines are normally located at a sufficient distance from a blast to avoid damage from flyrock and to limit the overpressure to a level that will not cause damage to componentry. Under these circumstances, the machine and its operating bench will still be subjected to ground vibrations, and consideration must be given to the likelihood of slippage between the tub and the bench and of instability of the bench involving the machine. The nature of the ground vibrations transmitted from a blast will depend on many factors, including the bench geometry and pit configuration between the machine and the blasted ground, nature of the blast in terms of total mass and type of explosive, blast firing pattern and stiffness, and strength of the bench materials. Blast-induced ground vibrations experienced by a dragline are typically limited to a duration of a few seconds and

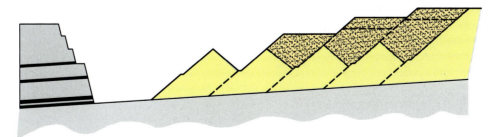

Figure 13.17: Dragline spoil surcharge limit to maximise overall stability

comprise a mixture of wave types, frequencies and amplitudes. In most circumstances, the materials forming the bench will cause relatively high attenuation of the waveform magnitudes and the higher frequencies.

Blast-related dynamic loading of spoil slopes is a normal and routine aspect of dragline strip mining, and, in particular, cast-blasting will also include direct impact of transported particles. Operating experiences indicate that blast-related dynamic loads have negligible consequences for spoil stability, and imagery of blasts typically shows only localised and minimal displacements of finer-grained materials, presumably related more to air movement than to ground shaking. Formal consideration of blast-related slope stability may be considered as a special loading case based on the same principles that are used for earthquake loading.

Ground vibrations from earthquakes will vary greatly with the nature, magnitude and epicentral distance of the event, as well as the topographic characteristics of the pit and the nature of the bench materials. In most circumstances, the earthquake loading duration will be significantly longer, and the frequencies of the vibrations will be significantly lower, than for a blast that produces similar levels of maximum ground movement at the machine.

The principles of earthquake resistant design apply equally to blast vibration and earthquake loadings. With respect to blast vibrations, design considerations may be based on observational experience provided that there is appropriate documentation that includes, where judged necessary, vibration monitoring data at the machine location. For earthquake loadings, design requires an assessment of seismicity at the mine location followed by adoption of design parameters that are appropriate to the severity of the earthquake hazard.

The simplest approach for assessing the effects of dynamic loadings is to apply a pseudo-static screening method to 2D limit equilibrium analysis (Jibson 2011). A widely used screening method (Hynes-Griffin and Franklin 1984) adopts 50% of the design peak ground acceleration as an additional static horizontal body force, resisted by 80% of the static shear strength parameters. These principles are consistent with the load and resistance factor approach to ultimate limit state design, but were developed to provide guidance on the serviceability limit state in terms of slope deformations. Depending on the magnitude of the resulting calculated factor of safety (FoS), the indicated outcomes range from acceptable (deformations insignificant, and therefore the likelihood of instability is extremely low) to unacceptable (significant deformations possible, and more detailed dynamic assessment is required to quantify the magnitudes of movements and the likelihood that the dynamically induced movements will be unacceptable or unbounded).

Many alternative dynamic loading assessment methods are discussed by Jibson (2011) and are not discussed further here, except to add that considerable resources are required for more detailed assessments to account for the many known complexities involved in dynamic loading and deformation responses of soil and rock materials. For this reason, the recommended approach to design is progressive, commencing where possible with observational experience, escalating to the pseudo-static method and then, if justified, to more sophisticated dynamic stress-deformation assessment methods.

13.5.8 Potential instability mechanisms and stability assessment methods

General approaches to stability assessment for mine dumps and stockpiles, including dragline spoils, are described in Chapter 8. Discussed here are particular considerations about potential instability mechanisms and appropriate methods of stability analysis, arising from the depositional structure and construction sequence for dragline spoils and any subsequent surcharge fill loadings.

Shallow-seated and superficial instability mechanisms may arise for many reasons, and are triggered by loss of toe support (undercutting or strength/stiffness reduction on wetting) or heavy rainfall (limited direct infiltration and softening). Shallow-seated instability is more likely to occur on slopes constructed by rilling because deposition-induced size-segregation effects result in banded internal structure inclined at the angle of repose. The angle of repose is a function of many factors relating to particle size and shape, initial velocity of dumping, and the effective friction angle of the particles. In the case of clayey fill materials, the angle of repose will also depend on the degree of saturation of clumps of particles and the short-term shear strength of the points of contact between clumps. It is common in dragline-constructed slopes for the undercut batter angle to be steeper than the angle of repose, and significantly higher than the representative friction angle for the fill material. Analysis of stability for shallow-seated slopes therefore requires careful determination of the operational shear strength to be applied to the observed mechanism. Back-analysis may be the most effective method for determining appropriate operational shear strength parameters, provided that the analysed mechanism matches observations.

Deep-seated instability of dragline constructed slopes is inevitably linked to contrasts in shear strength and stiffness for different parts of the fill mass. In most circumstances, the greatest contrast will develop at the basal spoil saturated zone. An exception arises where a 'mud-dam' is constructed within the dumped part of the profile to retain soft mud excavated from the pit floor (Fig. 13.18), but the basal zone will still be soft and

Figure 13.18: Dragline mud dam within upper spoils creating a potentially weak zone and requiring specific stability assessment. Source: J Simmons

saturated. The key consideration for deep-seated instability is therefore the softened and weakened basal zone, which becomes the locus of lateral spreading movement. If spreading cannot be adequately resisted, lateral movement will progress to form a distinct wedge-shaped mechanism characterised by bulging at the slope toe, development of cracking at or just behind the slope crest, and formation of a reverse-dipping mid scarp to create a lower 'passive' block and an upper 'active' block. Formation of this deep-seated wedge instability mechanism has been observed and modelled in dragline-constructed fills for some time (Gonano 1980; Richards *et al.* 1981) and is equally representative for any fill where a basal saturated zone develops (Campbell 2000).

Stability analysis for the deep-seated wedge mechanism should account for the observed non-vertical wedge boundaries between slices. Sarma's method (Sarma 1973) allows explicit representation of non-vertical slice boundaries and surface-specific shear strength

parameters while fully satisfying equilibrium requirements in 2D limit equilibrium analysis. Several methods based on vertical slice boundaries are also available for 2D limit equilibrium analysis and are discussed in further detail in Chapter 8. With respect to dragline spoils, the choice of potential instability mechanism may have a significant influence on the outcome of calculations (Simmons and McManus 2004; Duran 2012). Analyses should only be carried out for potential mechanisms that are consistent with observations. When undertaking back-analyses of actual instabilities, it is particularly important to note that the analysed parameters form a consistent, but not unique, set that includes the mechanism analysed.

Outcomes of 2D limit equilibrium stability analyses for deep-seated wedge mechanisms acting on dragline slope profiles are shown in Figs. 13.19 to 13.22. These figures illustrate an example of repeated strip development of a slope comprising a 40 m high bench cut batter at 40° and

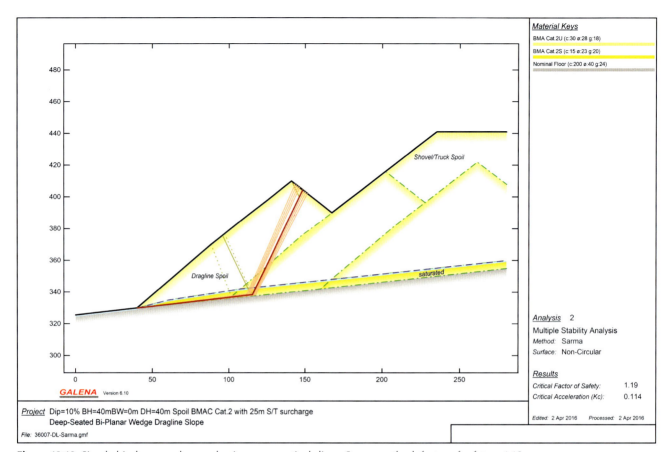

Figure 13.19: Simple bi-planar wedge mechanism, non-vertical slices, Sarma method, factor of safety = 1.19

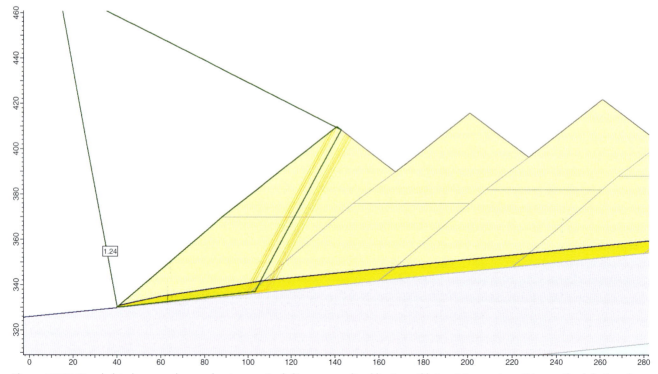

Figure 13.20: Simple bi-planar wedge mechanism, vertical slices, generalised limit equilibrium/Morgenstern-Price method, factor of safety = 1.24

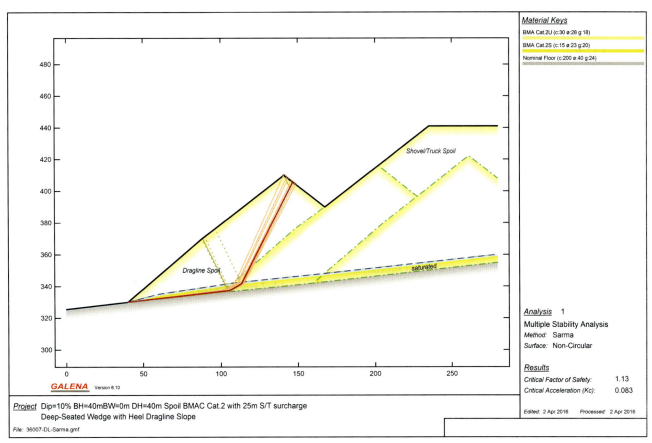

Figure 13.21: Complex bi-planar wedge mechanism, non-vertical slices, Sarma method, factor of safety = 1.13

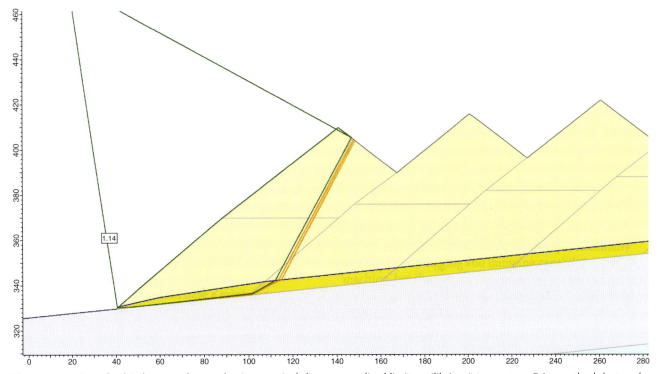

Figure 13.22: Complex bi-planar wedge mechanism, vertical slices, generalised limit equilibrium/Morgenstern-Price method, factor of safety = 1.14

a 40 m high dump batter at 37° with a remnant bench width of 0 m and a uniform floor dip of 10% (5.7°). Shear strength parameters for BHP Billiton Mitsubishi Coal Category 2 spoil were applied, with a basal saturated zone thickness of 5 m drawn down to floor level at the slope toe. Figure 13.19 shows the critical basic bi-planar wedge mechanism with a minimum computed FoS of 1.19 using the Sarma method as implemented in the Galena code (Clover Technology 2013). Figure 13.20 shows the critical basic bi-planar wedge mechanism with a minimum computed FoS of 1.24 using the generalised limit equilibrium (GLE)/Morgenstern-Price method as implemented in the Slide code (RocScience Inc. 2016). Differences in the FoS outcomes can be attributed to the differences in internal slice orientations. The intersection point of the bi-planar wedge implies very high localised internal distortions, but it must be borne in mind that 2D limit equilibrium methods ignore the kinematic admissibility of the analysed geometry. The introduction of a heel to form an intersection zone has a significant influence on the calculated outcomes, with the Sarma and GLE/Morgenstern-Price methods resulting in minimum computed FoS values of 1.13 and 1.14 (Figs. 13.21 and 13.22 respectively).

Figures 13.23 to 13.26 show the outcomes of analyses, including the effect of a surcharge fill 25 m above the second dragline peak. Again, and for illustrative purposes, the minimum computed FoS outcomes are dependent on the chosen method of analysis. Because the mechanism involving the surcharge slope has a more complex geometry than for the dragline slope alone, the effects of localised internal distortions are more subdued and the differences between the FoS values may be minimal. For example, the FoS outcomes shown in Figs. 13.23 and 13.24 are identical to two significant figures.

Application of back-analysed parameters to different mechanisms such as circular arcs may introduce errors into subsequent stability assessments. As discussed in Simmons and McManus (2004), the choice of an arcuate mechanism is inappropriate for assessing dragline spoil stability because such a mechanism is not observed in reality and has a significantly different distribution of internal stresses between slices. Figure 13.27 (Sarma in Galena) and Fig. 13.28 (GLE in Slide) shows outcomes obtained for the example dragline slope profile using the Spencer method as implemented in two different codes. The minimum computed FoS values are in good agreement between code implementations but are

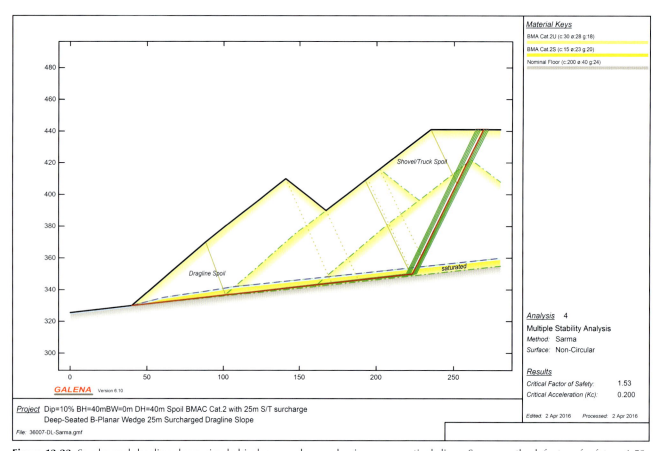

Figure 13.23: Surcharged dragline slope, simple bi-planar wedge mechanism, non-vertical slices, Sarma method, factor of safety = 1.53

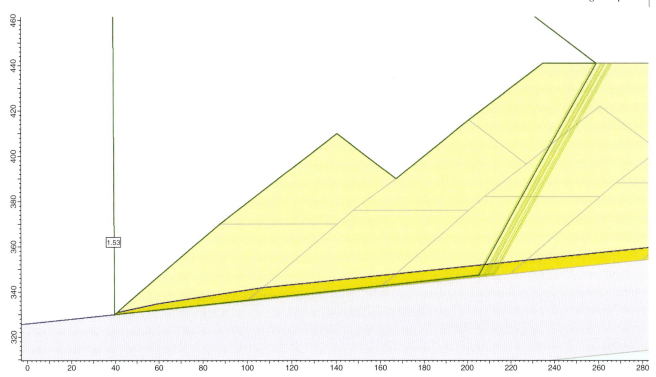

Figure 13.24: Surcharged dragline slope, simple bi-planar wedge mechanism, vertical slices, generalised limit equilibrium/Morgenstern-Price method, factor of safety = 1.53

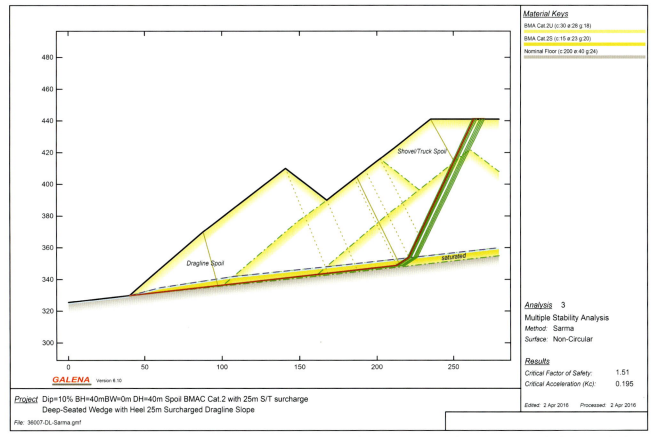

Figure 13.25: Surcharged dragline slope, complex bi-planar wedge mechanism, non-vertical slices, Sarma method, factor of safety = 1.51

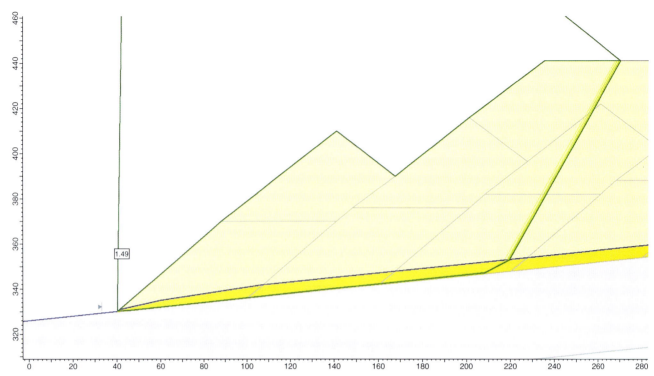

Figure 13.26: Surcharged dragline slope, simple bi-planar wedge mechanism, vertical slices, generalised limit equilibrium/Morgenstern-Price method, factor of safety = 1.49

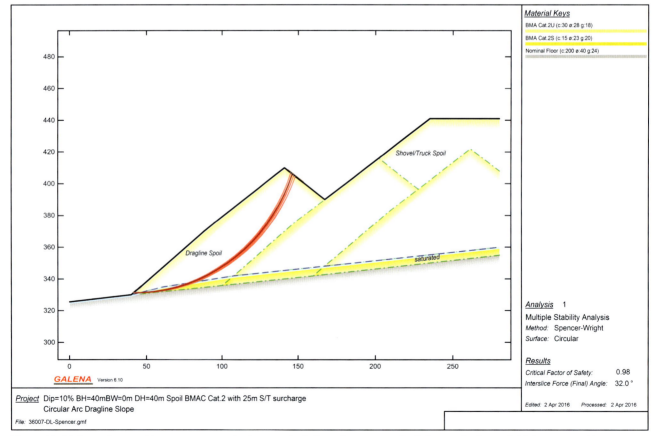

Figure 13.27: Example use of incorrect arc mechanism, vertical slices, Spencer method, factor of safety = 0.98 (compare with acceptable wedge mechanisms, factor of safety = 1.19–1.13, Figs. 13.19 and 13.21, respectively)

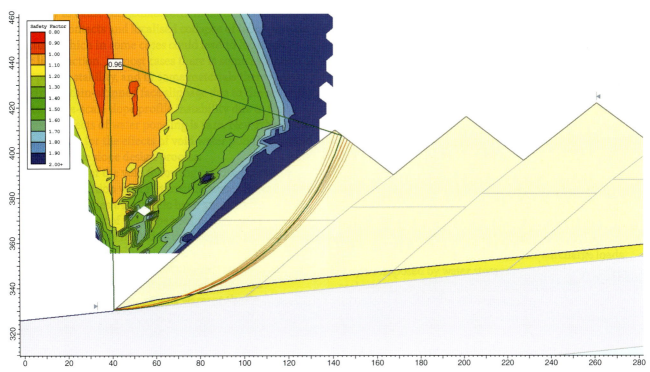

Figure 13.28: Example use of incorrect arc mechanism, vertical slices, Spencer method, factor of safety = 0.96 (compare with acceptable wedge mechanisms, factor of safety = 1.24–1.14, Figs. 13.20 and 13.22, respectively)

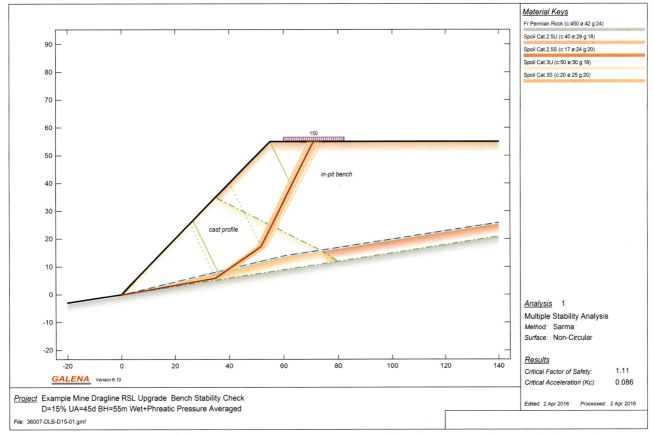

Figure 13.29: Example 2D limit equilibrium stability analysis using Sarma's method with non-vertical slices and dragline loading represented by simplified 2D uniform pressure

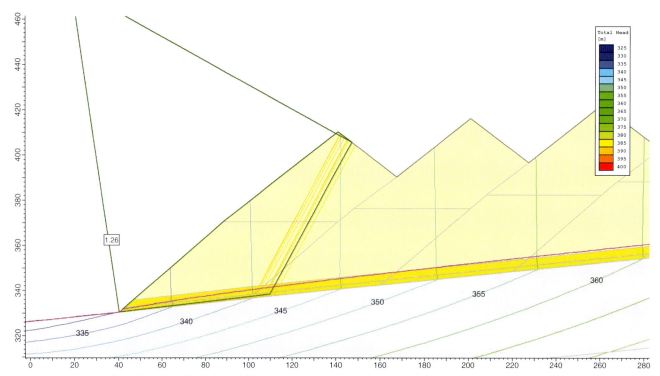

Figure 13.30: Example 2D limit equilibrium model based on Fig. 13.19 with groundwater pressures determined by finite element analysis, vertical slices, generalised limit equilibrium/Morgenstern-Price method, factor of safety = 1.26

dramatically different from the results obtained using deep-seated wedge methods as illustrated in Figs. 13.19 to 13.22. Factor of safety outcomes obtained by analysing arcuate mechanisms for dragline slopes are not appropriate and should not be used in engineering practice.

The examples in Figs. 13.19 to 13.22, 13.23 to 13.26, and 13.27 to 13.28 are reminders that the choice of method of analysis must be expected to have some influence on the calculation result. They reinforce the requirement to regard a calculated FoS value as the output of a process involving assumptions and choices, rather than a property of the slope profile.

Stability analyses for machine loadings on dragline benches involve strictly 3D geometric effects as discussed above. Approximate 2D analysis methods may be used provided that allowances are made for the effects of 3D loading geometry. It is usually not realistic to analyse the effects of pressure distribution changes during the dragline work cycle, and the destabilising effects of dynamic load amplification are also somewhat hypothetical and difficult to justify in the absence of reliable measurements. A robust simplifying assumption that is often made in practice is to consider the gross weight of the dragline as a uniform 3D pressure on the circular tub contact area, and to apply this pressure as an equivalent 2D strip pressure in a 2D analysis. Figure 13.29 shows an example of a 2D stability analysis for a dragline

bench loading scenario where the differences in calculated FoS outcomes for different pressure distributions (not shown) were of the order of 0.005. In this particular example, a basal zone of saturation was included and reduced the calculated FoS to well below the acceptability criterion.

13.5.9 Stress-deformation modelling considerations for dragline spoils

Stress-path dependency of inelastic response is a recurring theme for stress-deformation modelling of mining scenarios, where deformations and localised or unstable yielding may be tolerated provided that the consequences can be managed without unplanned costs. Stability of embankment dams has been studied since the earliest era of geotechnical finite element modelling, when it quickly became apparent that realistic simulation of deformations could only be achieved by allowances for path-dependent stiffness and strength. Dragline spoils create modelling challenges because the incremental construction of a typical dragline slope profile can be achieved by multiple sequences.

Groundwater regimes within dragline spoil masses also present complex challenges. It is far easier to model the groundwater regime within a spoil profile than it is to obtain convincing evidence of appropriate permeability parameters and boundary conditions for flow modelling. It is reasonable to assume that permeability is directly

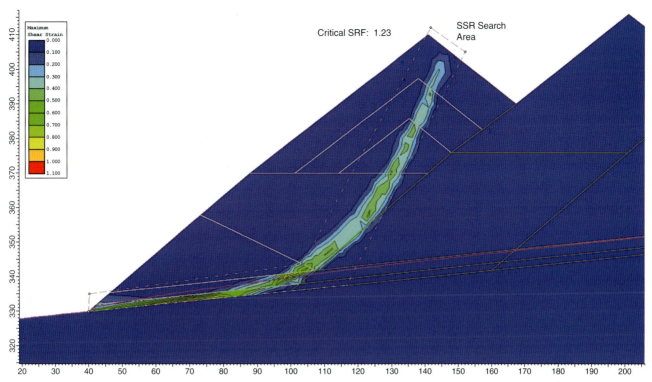

Figure 13.31: Example 2D finite element stress-deformation and groundwater model based on Fig. 13.19 and the critical mechanism using constrained shear strength reduction method, strength reduction factor = 1.23

related to the apertures of the interstitial voids within the spoil mass, implying highly nonlinear variations as a function of degree of saturation and stress level. For this reason, it is strongly recommended that spoil groundwater flow modelling only be carried out in situations where there is a corresponding degree of knowledge based on some combination of observation, testing and instrumentation, to be able to verify that the outcomes of modelling have some basis in reality. Figure 13.30 shows the same dragline slope profile example from Figs. 13.19 and 13.20 with a groundwater flow field modelled using finite element analysis. Correspondingly different outcomes would be obtained by introducing the 'heel' of Figs. 13.21 to 13.22 into the modelled mechanism. The permeabilities and model boundary conditions were based on assumptions from previous modelling experience and adjusted so that the phreatic surface was constrained within a saturated 5 m basal zone. The finite element model included a section of unmined ground, previous strip spoil profiles, and sufficient floor strata to be confident that the lateral and lower boundary locations did not have a significant influence on the pressure distributions within the region of interest for stability assessment.

Stress-deformation modelling can be extended to determine a strength reduction factor (SRF) as an alternative to the calculation of a FoS value using limit equilibrium methods. In principle, this approach takes into account the deformation response of the materials. The SRF is obtained by iterative analyses using successive reductions of the shear strength parameters by what is termed the shear strength reduction method. The shear strength reduction method applies to stress distributions that can be calculated by either elastic or elastoplastic methods, and for which the incremental stiffnesses during the shear strength reduction processes are restricted by the stress-deformation formulation. Because the range of available stress-deformation-strength models is limited to the relatively simplistic versions that have been implemented in existing codes, the outcomes of the shear strength reduction modelling process must be evaluated with great care, and always by comparison with outcomes determined by different techniques and/or observed field behaviour of slopes.

To illustrate this requirement for caution, Fig. 13.31 and Fig. 13.32 show two aspects of a shear strength reduction analysis carried out for the example dragline slope discussed above. Figure 13.31 illustrates the critical mechanism identified by shear strain contours, with a critical SRF = 1.23, but only after constraining the shear strength reduction search area to the approximate location of the critical mechanism obtained by limit equilibrium analysis. Without such constraint, it was possible to obtain SRF values ranging from 1.10 to 1.30 without changing any

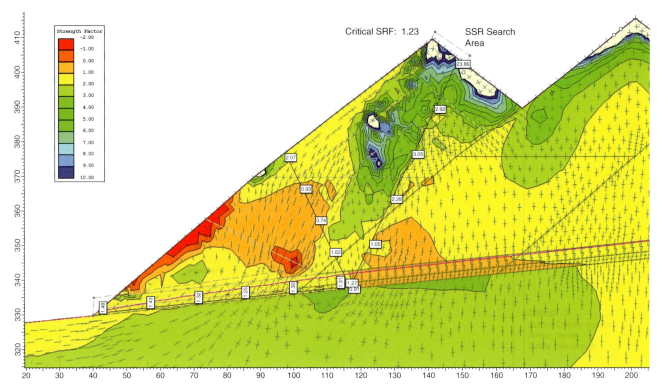

Figure 13.32: Example 2D finite element stress-deformation and groundwater model based on Fig. 13.19, shear strength reduction = 1.23, showing mobilised strength factors along critical surfaces from Sarma's method

of the stress-deformation-strength parameters. Although the critical mechanism has an SRF similar to the simple bi-planar wedge limit equilibrium FoS outcomes, it does not correspond to the typical bi-planar wedge mechanism that is repeatedly observed from monitoring data and field observations. A strength factor can also be calculated from stress-deformation modelling, and Fig. 13.32 shows selected values of the strength ratio along the orientations of the critical sliding surfaces for the critical bi-planar wedge of the example dragline slope. As indicated in Fig. 13.32, the mobilised shear strength should be expected to vary as an instability mechanism forms. Critically, the shear strength reduction technique based on maximum shear strain may not necessarily identify the critical mechanism that is observed in practice, and therefore also may not identify the SRF (or FoS) that applies in practice.

14

MANAGEMENT OF ACID ROCK DRAINAGE

Ward Wilson

14.1   Introduction

Acid rock drainage (ARD) is a phenomenon that has been occurring since the earliest times in mining, dating back thousands of years to Phoenician times in the Iberian Pyrite Belt. Even though ARD has been occurring for millennia, it has only risen to international prominence over the last three or four decades. Acid rock drainage has become a paramount issue to the mining industry, and the total cost to mitigate the current global environmental liability associated with ARD has been estimated to be as high as US$100 billion. Preventing and controlling ARD and metal leaching (ML) from potentially acid forming (PAF) waste dumps is still a largely unsolved environmental problem, with no proven universal solution at hand.

14.2   Principles of acid rock drainage and metal leaching

The mechanisms that drive ARD and ML are a complex, coupled system of biogeochemical reactions that are controlled by the geochemistry and environmental boundary conditions imposed at any given mine site. The principal driver for ARD in waste dumps is the presence of sulphide minerals and rock types associated with the ore bodies. The following sections describe basic geochemical weathering processes together with the influence of climate, waste dump structure and hydrology, as well as oxygen and water transport processes. Special handling techniques for PAF rock types are also presented in a subsequent section.

14.2.1   Drivers of acid rock drainage

Many waste dumps have the potential for the generation of acid through reactions of sulphide minerals contained in waste materials with oxygen and water. This process is referred to as acid rock drainage (ARD). Acid rock drainage is often associated with ML, since all acidic drainages have high metal concentrations. The primary driver responsible for the generation of ARD is weathering of sulphide minerals in the waste material, in particular pyrite, marcasite and pyrrhotite. Acid rock drainage is not just associated with mining, but also occurs naturally. Historical examples of naturally occurring ARD include watercourses such as the Rio Tinto in Spain and Iron Creek in Colorado, USA. Acid rock drainage can also occur where humans accelerate the weathering process, such as through earth moving operations (e.g. in road cuts). In all cases, drainage with a pH below 4.5 is defined as ARD.

High sulphide contents in waste rock (i.e. sulphide-bearing minerals) are often associated with coal mines and ore bodies containing base and precious metals (e.g. copper, lead, zinc, gold and silver). Some of the most well known sources of ARD (previously referred to as acid mine drainage) can be found in abandoned coal spoil piles in the eastern United States. Typical ore deposit types for metal mines most commonly associated with ARD include (1) volcanogenic massive sulphide deposits, (2) high sulphidation epithermal deposits, (3) porphyry copper deposits and (4) skarn deposits. Some of the most dominant factors that control ARD potential and drive reaction rates include (1) the percentage and form of sulphur and iron present in the waste rock, (2) surface area (i.e. rock blocks versus crushed rock versus tailings), (3) presence of water and oxygen and (4) overall geochemistry of the rock matrix and mixture of rock types.

In the most general context, the oxidation of reactive sulphide-bearing minerals in waste dumps is fuelled by the supply of atmospheric oxygen with meteoric water.

14.2.2   Geochemical weathering processes

Complex processes governed by a combination of physical, geochemical and biological factors control the weathering process. The rate of the oxidation for sulphide minerals exposed to weathering is determined by grain size, water content and degree of saturation, oxygen concentration, presence of other oxidising agents such as ferric iron, current acidity, temperature and *Thiobacillus ferrooxidans* (TF) bacteria.

The most common sulphide mineral is iron sulphide, also known as pyrite (FeS_2). A typical reaction sequence of pyrite to generate acid is shown stochiometrically below:

$$2FeS_2 + 2H_2O + 7O_2 = 4H^+ + 4SO_4^{2-} + 2Fe^{2+} \quad \text{(Eqn 14.1)}$$

$$4Fe^{2+} + 10H_2O + O_2 = 4Fe(OH)_3 + 8H^+ \quad \text{(Eqn 14.2)}$$

$$4Fe^{2+} + O_2 + 4H^+ + TF = 4Fe^{3+} + 2H_2O \quad \text{(Eqn 14.3)}$$

$$FeS_2 + 14Fe^{3+} + 8H_2O = 15Fe^{2+} + 2SO_4^{2+} + 16H^+ \quad \text{(Eqn 14.4)}$$

Equation 14.1 is a simple chemical oxidation reaction that occurs under neutral to alkaline conditions and where atmospheric oxygen acts as the oxidant. Note the resulting release of acidity (H^+) and ferrous iron (Fe^{2+}). The reaction shown in Eqn 14.2 can remove some of the ferrous iron under slightly acidic to alkaline conditions, producing relatively insoluble iron hydroxide ($Fe(OH)_3$), but also releases much more acidity. The third reaction (Eqn 14.3) occurs once the pH drops below 4.5 (more acidic). The reaction consumes some acidity but oxidises the ferrous iron (Fe^{2+}) into ferric iron (Fe^{3+}). Once the pH falls to ~3.5 and lower, the TF bacteria dominate the system. Also note that the TF bacteria that catalyse this reaction require a nominal amount of oxygen. The presence of iron-oxidising bacteria (TF) can speed up this reaction by five or six orders of magnitude. In the reaction shown in Eqn 14.4, the abundant dissolved ferric iron results in sufficiently acidic conditions to directly oxidise pyrite. This reaction generates substantially more acidity and the cycle continues.

Acid generation and drainage has undesirable effects on receiving streams. Very low pH waters (less than pH 1) that sterilise downstream watercourses and land tracts receiving seepage can be released. The low pH also promotes metal dissolution from within rock, potentially rendering the water toxic to aquatic life with high concentrations of heavy metals.

Other neutralising reactions can occur that consume acid and release alkalinity. Most carbonate minerals are capable of dissolving rapidly, making them effective acid consumers. A simple example of this reaction is as follows:

$$MeCO_3 + H^+ = Me^{2+} + HCO^{3-} \quad \text{(Eqn 14.5)}$$

where Me is a divalent cation, such as calcium (Ca^{2+}) or manganese (Mn^{2+}), but not iron or magnesium as these release acidity in subsequent reactions.

The alkalinity balances the pH during early phases of oxidation (i.e. Eqn 14.1) and is consumed over time,

preventing a drop in pH. This early phase of oxidation is commonly referred to as the 'lag phase' and can last for years. Many operators are often deceived during this 'lag phase' into believing they do not have an ARD issue, because they see no evidence of it. This may result in preventative measures not being adopted until it is too late, and control measures that are difficult and expensive will be needed. Once the pH falls below 4.5, the microbial reactions dominate and result in a rapid acceleration of acid generation.

14.2.3 Characterising acid rock drainage potential

Test procedures for the characterisation and evaluation of acid generation potential fall under three general categories known as laboratory static testing, laboratory kinetic methods and field testing.

1. **Laboratory static and short-term methods**:
   (a) Acid–base accounting measures acid potential (AP) through independent determination of acid generating and neutralising content. Acid potential and neutralising potential (NP) are combined algebraically to designate whether a sample has a stoichiometric balance that favours net acidity or net alkalinity. Acid potential can be expressed in terms of equivalent carbonate as kilograms of calcium carbonate per tonne or as maximum potential acidity in terms of kilograms of sulphuric acid produced per tonne of rock. With both AP and NP expressed as kilograms of calcium carbonate per tonne, the net potential ratio (NPR) is calculated as:

$$NPR = NP/AP \quad \text{(Eqn 14.6)}$$

   (b) The net acid generation (NAG) test and paste pH tests generate a single value that can be used to indicate the likelihood of acid generation or release of stored acidity. The NAG test involves the reaction of a sample with hydrogen peroxide (H_2O_2) to rapidly oxidise sulphide minerals. Both acid generation and neutralisation occur simultaneously, so the final measure of pH represents a direct measure of the amount of acid generated. A NAG pH of less than 4.5 designates the sample as net acid generating. Acid–base accounting and NAG testing are routinely used together to improve prediction confidence, identify uncertain samples and better define cut-off criteria for material classification as illustrated in Fig. 14.1, taken from the *Global Acid Rock Drainage Guide* (INAP 2014), which illustrates a typical dataset for a given site.

2. **Laboratory kinetic methods** – kinetic testing is more complex and time consuming than static testing, and

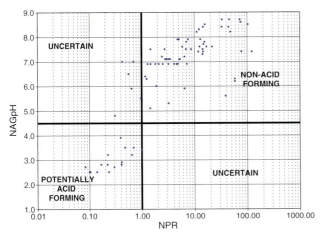

Figure 14.1: Acid rock drainage rock type classification plot based on acid–base accounting and the net acid generation test. Source: INAP (2014)

requires operator experience and skill for consistent results. Samples are subjected to periodic leaching, and the leachate is collected for analysis. The method is intended to accelerate the natural weathering process, and the results can be used to validate static methods, estimate long-term weathering rates and determine the potential for mine wastes and geological materials to release discharges that may impact the environment. Further, it can be used to evaluate both acid generation and metal leaching. Kinetic testing includes both humidity cell and column tests:

(a) Humidity cell tests are a standardised test conducted under fully oxygenated conditions with periodic flushing of reaction products (ASTM D5744; ASTM 2013e) and are generally intended to generate information on weathering rates of primary minerals (e.g. sulphides only).

(b) Column tests – there are no ASTM standards available for column testing. However, the method

Figure 14.2: Field-scale experiment being carried out at Antamina Mine with five 8 m high constructed test piles shown on the left and five 15 m by 15 m lysimeters for testing cover systems over waste rock on the right. Source: W Wilson

can simulate different infiltration rates, temperatures, degrees of saturation and oxygen-deficient conditions. The sample size is typically larger scale than humidity cell tests. In theory, column tests provide information on combined weathering rates of primary and secondary minerals, and may be more suited to evaluation of mitigation measures such as covers and amended waste sites.

3. **Field methods** – on-site test methods range from rapid very small-scale tests to monitoring of full-scale mines for extended periods. Materials are tested on site in ambient conditions, which can include seasonal and discrete event (e.g. rain storm) evaluations. The test should be more representative of site conditions, with a greater amount of material used in the test. Disadvantages include the time required to generate field reaction rates, challenges with respect to adequate control and protection of test installations, challenges related to comprehensive geochemical characterisation and inability to test a large number of different material types, and finally costs.

Larger scale field methods, such as pilot cells, test piles, test plots or test pads, are typically constructed for long-term monitoring with relatively large quantities of material. Large-scale field columns (field lysimeters) that operate under natural precipitation and weathering conditions are perhaps the best method for geochemical characterisation of drainages from waste dumps and stockpiles. Field-scale testing uses more representative samples and minimises effects of conditions that can impact small-scale testing (e.g. sample heterogeneity, reduced grain size and boundary effects). Longer

monitoring durations are generally required (orders of magnitude longer) because of the reduced reactivity of field cell tests relative to the finer grained materials commonly tested in the laboratory. In general, this approach is considered most suitable to operate throughout the complete mine life cycle for establishing accurate long-term releases. Figure 14.2 shows the field experiments constructed at the Antamina Mine by the Universities of British Columbia and Alberta, Beckie *et al.* (2011), Urrutia *et al.* (2011) and Urrutia (2012).

14.2.4 Climate

Conventional methods for the construction of waste dumps and stockpiles do not vary dramatically across the different regions of the planet. For example, the use of heavy haul trucks is universally found on every continent. Furthermore, the geology of ore bodies and rock types tend to be repeated worldwide. What does vary most widely is climate. The mining industry operates in every climate region on Earth, ranging from super humid high rainfall equatorial jungles, to hot super arid deserts, to cold polar regions with permafrost. It is climate that ultimately controls all weathering of rock, both physical and geochemical, as well as how the waste dump impacts the surrounding receiving environment. An understanding of the climate is also paramount for the design of methods used to prevent, control and mitigate ARD and ML.

Figure 14.3 illustrates a useful classification system known as the Holdridge life zones system (Holdridge 1971). This system can be regarded as a global scheme for bioclimatic classification of land-based regions. The life zones are characterised on a tri-linear plot with principal

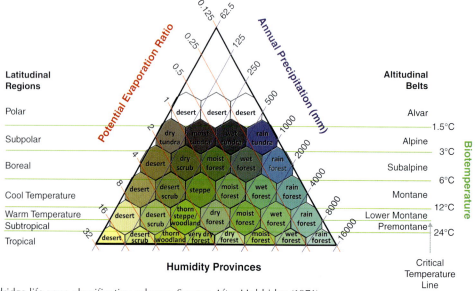

Figure 14.3: Holdridge life zone classification scheme. Source: After Holdridge (1971)

axes of annual precipitation and potential evapotranspiration ratio that infer a third axis defining humidity provinces. Biotemperature (mean annual temperature of the growing season) is correlated with latitudinal regions and altitudinal belts.

The Holdridge life zones tri-linear plot can be divided into four principal regions for selecting techniques to prevent and control ARD and ML in waste dumps, as well as the associated methods for implementing long-term reclamation and closure designs. The first principal division corresponds to a potential evapotranspiration ratio that separates the humidity provinces into either arid or humid, while a second partitioning occurs between the extremes of polar and subpolar versus tropical and subtropical. The application of the Holdridge system will be described further in a following section.

14.2.5 Waste dump structure and hydrology

Waste rock structure and hydrology has been investigated at several mine sites over the past two decades. Wilson (2000) described the results of a 15 million tonne waste rock excavation program that was completed at Golden Sunlight Mine in Montana, USA. Fines *et al.* (2003) and Tran *et al.* (2003) reported the results of a large research program sponsored by the International Network for Acid Prevention to characterise the structure, hydrology and geochemistry in two waste dumps that were excavated for redeposition into open pits. McLemore *et al.* (2009) reported the results for a similar but much larger excavation project completed for the Goathill North waste dump at Questa Mine in New Mexico, USA.

Figure 14.4 shows the waste dump excavation at Golden Sunlight Mine. The upper three lifts, each with an approximate height of 20 m, are shown in the left photograph, while the photographs on the right provide close up views of the dump structure within a single bench and the rock rubble zone found at the toe of each bench. This structure within the dump is caused by end dumping (or pushing) run of mine waste rock with haul trucks as each lift advances. The resulting structure consists of a boulder or rubble zone that ravels to the base of the lift, overlain by a segregated, interbedded formation of coarse and fine rock blocks or fragments dipping at the angle of repose. This structure observed in the dump excavation at Golden Sunlight Mine is a common feature found in waste dumps and is always found in dumps constructed using haul trucks with tips greater than ~6 m. The dump materials shown in Fig. 14.4 had been in place for a period of 8 years or less, but it can be seen that all of the material is oxidised to the full depth of the excavation, and thus can be expected to produce ARD if sufficient infiltration is permitted.

Herasymiuk (1996) proposed the conceptual model for the hydrology of piles shown in Fig. 14.5. Infiltration into the surface of the waste dump is equal to precipitation less runoff and evapotranspiration. The quantity of annual infiltration varies dramatically with climate depending on the annual precipitation rates and evapotranspiration ratios as shown in the Holdridge system (Fig. 14.3). Andrina (2009) observed that ~90% of precipitation infiltrated into trial dumps constructed at the Grasberg Mine in Indonesia. This extremely high rate of infiltration corresponded to an annual average precipitation of ~4000 mm per year with a potential

Figure 14.4: Waste rock excavation at Golden Sunlight Mine. Source: W Wilson

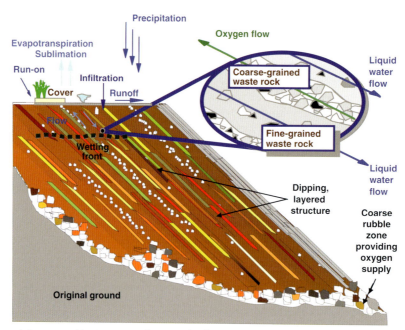

Figure 14.5: Conceptual model proposed by Herasymiuk (1996) for the waste dump structure at Golden Sunlight Mine

evapotranspiration ratio of the order of 10% (i.e. 0.1). In contrast, the infiltration rates at the Questa Mine were estimated to vary between 30 and 40% of precipitation for an annual average precipitation of ~400 mm with an evapotranspiration ratio of ~3. Experience shows that an infiltration rate of the order of 50% of total annual precipitation can be expected for most dumps constructed in sub-humid temperate climate regimes.

14.2.6 Oxygen and water transport

Waste dumps are normally positive topographic structures constructed above the pre-mining land surface and the regional water table. With a few specific exceptions, waste dumps are fully drained and unsaturated. The method of construction produces a particle size distribution that increases along the depth of each tip face; thus hydraulic conductivity increases with depth in each lift, producing optimum conditions for drainage. While this hydraulic characteristic is ideal for constructing stable waste dumps, it also creates a structure that promotes the flow of oxygen into the profile of the dump. The rock rubble zone at the base of each bench allows the free advection of atmospheric oxygen deep into the lateral profiles of the dump as shown in Fig. 14.6c.

Andrina (2009) measured atmospheric oxygen concentrations of 20% in the rubble zone at a distance of ~70 m from the toe of the 20 m high trial dump constructed at the Grasberg Mine. This free advection of oxygen deep into the profile of the constructed waste

dump resulted in the development of rapid oxidation following deposition. Elevated temperatures above ambient conditions were measured within several weeks after construction, both at the base of the pile and within the waste profile above, with temperatures eventually reaching more than 60°C in the interior of the constructed dump profile.

Approximately 4 years after construction was completed, trenches were excavated on the surface of the test piles, parallel to the direction of advance of the tip face. The photograph in Fig. 14.6a shows a typical example of the waste dump materials exposed along the wall of the trench. The dipping beds corresponding to the different material types are clearly evident. The steam that can be seen at the top of the photograph was caused by the exposure of a layer of limestone that had formed a vent or chimney in the profile of the constructed test pile. Figure 14.6c illustrates a conceptual model for the advection and flow of oxygen in waste dumps. This model is based on the observations of Herasymiuk (1996) along with many other excavations in waste dumps such as those completed at the Grasberg Mine by Andrina (2009). In general, the coarse rock rubble zone at the base of each bench or lift of waste forms a free pathway for atmospheric oxygen to flow almost unrestricted through the entire footprint of each lift. This supply of oxygen is drawn by the buoyant forces caused by heating in the reactive materials. The coarse layers form chimneys in the dump that feed the reactive sulphide-bearing rock with fresh oxygen. In summary, the

Advective Pathways Dominate Oxygen Entry

Figure 14.6: Oxygen and water transport in waste dumps with (a) venting and (b) preferential flow shown in the upper photographs, and (c) advective pathways for oxygen in a waste dump. Source: W Wilson

structure of waste dumps created by tip construction with haul trucks is ideal for allowing reactive waste materials to oxidise.

The rock rubble layer and coarse interbedded layers within the waste dump that promote free oxygen convection are fully drained and virtually unsaturated. These materials have no capillary capacity to retain water under negative water pressures (i.e. matric suction). Hence, the permeability with respect to flow in the air phase is at a maximum and these zones become the primary source for oxygenation within the dumps. Conversely, these coarse materials do not transmit significant flow of liquid water. Instead, the primary pathways for liquid water flow are confined to the finer layers within the interbedded structure of the dump profile. It is the fine layers that provide capillary capacity to retain water under negative water pressures, and thus these water retaining layers serve as the principal pathways for liquid water flow. The photograph in Fig. 14.6b shows an example of a saturated fine layer in an excavated dump profile. Andrina (2009) examined the flow in a meso-scale profile of dipping coarse and fine layers of waste materials obtained from the Grasberg Mine. The results of the experiment confirmed that the fine layers of dump materials form a significant pathway for liquid water flow, and that the amount of flow

in individual layers is strongly influenced by the infiltration rate applied at the surface of the dumps.

14.3 Prevention and control of acid rock drainage through special handling techniques

Special handling techniques can be implemented during construction for the prevention, control and mitigation of ARD. Some of the most commonly discussed techniques include segregation, blending, encapsulation, the construction of barriers and seals and, finally, subaqueous disposal. The additional costs that may be incurred with the implementation of such methods usually provide tremendous economic advantages with significant reductions in environmental liabilities associated with closure of the waste dump. Each of these methods will be described briefly in the following sections.

14.3.1 Segregation

Separation of PAF materials from non-acid forming (NAF) materials before deposition is referred to as segregation. In general, the PAF materials are placed in specific engineered configurations that prevent or minimise the potential for oxidation. The most aggressive application of segregation is to separate all PAF material and dispose of it in a completely water-saturated configuration below a permanent water table that isolates all sulphide minerals from contact with atmospheric oxygen. An example of this form of segregation is to place all PAF rock into the tailings impoundment, where it will be inundated or submerged within a saturated tailings profile. Alternatively, PAF rock materials may be placed in engineered cells constructed with compacted low-permeability materials that will inhibit oxygen ingress and minimise water infiltration. In other cases, PAF materials are encapsulated within lifts of NAF materials for the purpose of reducing or neutralising ARD that may occur following deposition, as discussed in Section 14.3.3.

The primary challenges for the successful implementation of segregation programs are to develop a comprehensive block model for the pit materials that identifies the materials that need to be segregated, and to properly identify these materials during construction for separation, handling, transport and proper deposition. Miller *et al.* (2012) provided an excellent example and description for the implementation of a successful segregation program.

14.3.2 Blending

Blending of PAF materials with NAF materials in the dump to create a net neutralising potential within the dump profile has been implemented at many mine sites. The blending ratio for PAF and NAF materials depends on individual sites and individual material types; however, blend ratios for NAF and PAF to give a net neutralising potential ratio between 2 and 3 is frequently implemented. Experience with blending has shown that it can be successful; however, the degree of mixing is paramount. Homogeneous and thorough mixing is generally required to achieve maximum benefit (MEND 2001a, 2001b).

Evidence from some field trials indicates little success with mixing of waste types using haul trucks (Andrina 2009; Andrina *et al.* 2006). For example, Fig. 14.6a shows a dump profile that was theoretically blended to a net neutralising potential ratio of ~3. Nevertheless, it can be seen that the method of deposition produced individual PAF and NAF layers that did not mix, nor did they mitigate ARD from the acid layers, since seepage from the PAF rock flowed parallel to the coarse limestone NAF layers. More successful experiments with blending NAF and PAF materials are described in Chapter 15.

14.3.3 Encapsulation

Encapsulation may involve the construction of internal cells that prevent oxygen ingress to PAF materials that have been segregated for disposal. The encapsulation practice to achieve low oxygen ingress is to construct cells using a thick layer of low-permeability material that maintains high water saturation. The effective diffusion coefficient for oxygen is a function of the degree of saturation as shown in Fig. 14.7. In general, the material used to form the encapsulation cell must maintain a degree of saturation greater than ~90%, which reduces the diffusion coefficient for oxygen by three or four orders of magnitude. The thickness of the encapsulating layer must generally be greater than 1 m, and practical constructability issues may demand even greater thicknesses. The use of high-saturation water-retaining encapsulation layers also depends on an adequate supply of water to maintain high saturation within the encapsulating layer.

Encapsulation layers may also be constructed with NAF waste dump materials for reclaiming PAF waste dumps. An excellent example for the use of NAF materials for the reclamation of an existing PAF waste dump can be found at the Kidston Gold Mine in northern Queensland, Australia. In this case, a competent NAF outer 60 m batter was constructed around the entire perimeter of the existing PAF dump. In addition, a 2 m thick profile of NAF material was also used to construct a cover system over the entire PAF dump. The use of NAF materials for reclamation often requires careful mine planning to ensure that an adequate quantity of non-reactive waste material is available for closure of the dump. A complete description of the reclamation work completed at the Kidston Gold Mine is provided by Williams *et al.* (2003).

14.3.4 Barriers and seals

Barriers and seals may be used to provide liners at the base of the dump or cover systems that limit water infiltration and oxygen entry to the underlying dump materials. Liners at the base of dumps are purpose-built barriers that prevent contaminants flowing from the overlying dump to the receiving environment. In general, they are expensive to build and require a high level of quality assurance and

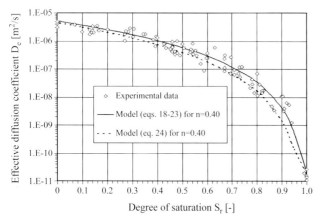

Figure 14.7: Effective oxygen diffusion coefficient as a function of degree of saturation. Source: Aubertin (2005)

quality control, and are generally reserved for specific applications that demand special protection for either highly reactive waste materials or sensitive receiving environments.

Engineered barriers include cover systems. The selection of an appropriate cover system for a given dump is entirely controlled by the climate regime at the mine site and to a lesser exent by the materials available for its construction. Cover systems can generally be classified into four distinct and specific categories as shown in Fig. 14.8.

Low-permeability oxygen barriers are only considered practical in humid climates where annual precipitation exceeds evapotranspiration. The primary reason for this criterion is that oxygen barriers must maintain water saturation in the barrier layer of greater than 90% to function as an effective oxygen barrier. Clearly this is not possible, for example, in an arid climate where potential evapotranspiration greatly exceeds annual precipitation. Compacted clay covers designed as either oxygen barriers or, more commonly, as barriers to infiltration have been documented to fail miserably in arid climates due to desiccation and cracking (Suter *et al.* 1993; Albrecht and Benson 2001).

Store and release covers do not restrict oxygen entry to reactive materials within the stock pile. Therefore, oxidation will continue at some rate since water is always available to drive sulphide oxidation. However, store and release covers have been found to be successful in greatly reducing infiltration rates (i.e. through high evapotranspiration), and thus the volume of ARD generated can be significantly reduced by the application of a well placed cover.

Figure 14.8 also shows a transition zone between the humid and arid climate zones where evapotranspiration may not be sufficiently high to reduce infiltration; nonetheless, covers designed to shed water during storm events can be quite effective at reducing infiltration rates. The fourth class of cover shown at the apex of Fig. 14.8 consists of permafrost covers. Permafrost covers are a special class of cover and are designed to promote freezing instead of limiting infiltration for oxygen entry. These types of covers are by far the most complex cover system to establish.

Further discussion of the design and construction of cover systems can be found in the *Global Acid Rock Drainage Guide* (INAP 2014). The construction of successful barrier-type covers over waste dump systems is proving to be more difficult than initially expected. Conversely, the performance of store and release cover systems on dumps has been found to be more successful.

14.3.5 Subaqueous disposal

Subaqueous disposal is the direct application of a water cover to prevent the exposure of PAF material to atmospheric oxygen. In general, the depth of water cover

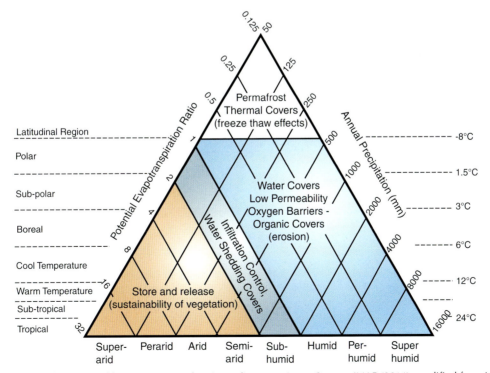

Figure 14.8: Application of covers and barriers systems for given climate regimes. Source: INAP (2014), modified from Holdridge *et al.* (1971) by Wickland and Wilson

required to prevent the oxidation of PAF materials is only ~1 m. While practical construction considerations limit the number of sites at which subaqueous disposal is feasible, as a design option it is attractive due to the potential to completely and securely eliminate oxidation and ARD.

At some mine sites, construction sequencing or other considerations have permitted the subaqueous disposal of waste materials into an existing open pit. Where possible, this is an ideal repository for some or all of the waste materials. Such covers are most likely to be feasible in humid climates (Fig. 14.8), or where regional groundwater regimes establish the formation of a permanent pit lake that would maintain all PAF materials below the water line.

An excellent example where subaqueous disposal of the highly reactive waste rock was successfully implemented is at the Equity Silver Mine in British Columbia, Canada, as described by Aziz and Ferguson (1997). Other examples in the literature include the Flambeau Mine, located near Ladysmith, Wisconsin, USA, where the open pit was backfilled with waste rock mixed with lime, and natural groundwater levels reporting to the pit provided a water cover. A similar approach was undertaken at Owl Creek, near Timmins, Ontario, Canada. There are also numerous examples of subaqueous disposal having been applied for tailings.

Finally, while water covers are extremely attractive for their ability to prevent ARD, it should be noted that, depending on site-specific conditions, there may be the possibility of the release of some metals into the covering water and for the concentration of contaminants of concern in the water to exceed guideline values for discharge (Aube *et al.* 1995).

14.4   Conclusion

The principal processes and drivers of ARD associated with geochemistry, waste dump structure, climate, and oxygen and water transport have been discussed. Various methods for the prevention and control that include segregation, blending, encapsulation, barriers/seals and subaqueous disposal have been summarised. In general, the current technologies available to control ARD are highly developed and well understood. However, problems with ARD from waste deposits still persist, largely because operators continue to use methods for constructing waste dumps that create structures that promote oxygen transport and sulphide. This subject is discussed further in Chapter 15.

15

EMERGING TECHNOLOGIES

Ward Wilson

15.1   Introduction

All acid rock drainage (ARD) and/or metal leaching (ML) problems found in existing waste dumps and stockpiles were created at the time of deposition. Furthermore, all future ARD/ML problems encountered in new dumps to be constructed will be determined at the time of deposition based on the design principles and methods used for construction and closure. Miners and processors generate mine waste that others manage. They do not design the material properties of the mine waste deposits, but rather the design paradigm is to optimise recovery rates and minimise material handling costs. Mine waste specialists, including geochemists and geotechnical engineers, are assigned the task of handling the waste streams they are given. They typically do not specify or design the physical and/or geochemical properties of the waste dump materials or tailings. The purpose of this chapter is to introduce new design methods to improve the physical and geochemical properties of waste dumps for the prevention and control of ARD/ML.

In the case of most mining operations, overburden rock is blasted, excavated and trucked to a waste dump, where it undergoes segregation by gravity. As described in Chapter 14, the coarse large boulders and cobbles ravel out to the base of the dump face, while finer materials are retained up slope. Waste dumps constructed with faces greater than 4–6 m high develop an internal structure associated with segregation, producing coarse and fine layers with a high permeability rubble layer at the base as shown in Fig. 14.4. The combination of the interbedded structure and unsaturated flow conditions creates ideal conditions for weathering of the waste materials through geochemical and biological pathways. Sulphide oxidation and acid generation often begin during the construction of dumps and result in long-term acid drainage.

We will now consider methods for mitigation and control of ARD/ML in dumps that may be implemented for closure in light of the inherent problems associated with the methods of construction as outlined above. Cover systems for the control of sulphide oxidation, infiltration and ARD from dumps have become the common method of choice for many sites. Covers have grown in popularity since the early 1990s, next to the direct collection and treatment of ARD. Indeed, if the construction of a waste dump has been completed or is nearing completion and ARD has already begun, the application of a cover system is the only option available to reduce ARD, short of excavation and removal of the dump materials for subaqueous disposal in an open pit or suitable impoundment.

Recent field observations and practical experience with covers that have been constructed over the past two decades is showing that we need to re-evaluate the use of cover systems for the control of ARD/ML for dumps. Cover systems have generally been most successful for the reclamation of tailings; achieving good performance with soil covers constructed on waste dumps has proven to be much more difficult. Perhaps the most significant feature of waste dumps is that they are positive topographic features that rise high above the surrounding terrain. This creates dump profiles that are free draining and completely unsaturated, resulting in ideal conditions for a constant and everlasting potential for oxidation and ARD.

A typical plot for lime quantities required to treat ARD from a reclaimed waste dump located in a humid mid-latitude climate is shown in Fig. 15.1. The plot shows that lime load decreased following the progressive construction of a cover system that began in 1991 and was completed by 1994. The cover profile was ~1 m thick with a compacted clay till barrier layer. While it can be seen that there is a downward trend in lime treatment requirements after

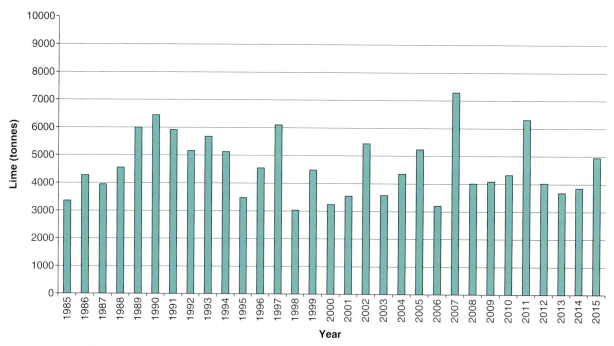

Figure 15.1: Measured lime treatment loads for a waste dump located in a humid mid-latitude climate before and after construction of a barrier cover system. Source: After Wilson (2011)

1994, high peaks continue to occur during subsequent wet years, with the highest load ever occurring almost a decade after completion of the cover system. The constructed cover system is known to be of excellent quality, and field measurements following construction showed the compacted till to have a low hydraulic conductivity in the range of 1×10^{-7}–1×10^{-8} m/s (Wilson 2011). Furthermore, field testing has shown that the cover should be functioning as a suitable oxygen barrier. Nevertheless, a quick examination of Fig. 15.1 should precipitate the immediate question, 'Why is the cover system not providing a substantially greater reduction in lime requirements for the treatment of the ARD'?

One possible mechanism that may help explain why lime treatment loadings do not decrease as expected with the installation of a properly designed and constructed

cover system is shown in Fig. 15.2. Oxidation of reactive potentially acid forming (PAF) material commences at the time of construction, after placement of the first lift of the dump. Figure 15.2 illustrates a typical acid generation curve. Assuming little or no lag, it can be seen that oxidation products become available for leaching as soon as infiltration produces seepage and drainage through the profile of the dump. Progressive oxidation of each subsequent lift of material continues in the same way, while oxidation continues to progress within the previously deposited lifts. Breakthrough seepage eventually occurs, and leachate with oxidation products produces ARD.

The amount of ARD released from the dump profile is less than the rate of oxidation. This occurs since the dump profile is unsaturated and the amount of seepage produced by surface infiltration is orders of magnitude less than the hydraulic conductivity of the waste at full saturation. For example, an annual infiltration rate of 1000 mm/year under a unit hydraulic gradient corresponds to a mobilised hydraulic conductivity of 3×10^{-8} m/s, while a saturated hydraulic conductivity of 1×10^{-1} m/s (i.e. typical waste rock) would produce a seepage flux of 3×10^{6} m/year. Tran *et al.* (2003) demonstrated that the amount of leaching is actually controlled by both flow rate and texture of the waste, with only a very small fraction of the oxidising surface area of the unsaturated dump available for leachate production.

The rate of acid production is shown as the acid generation curve in Fig. 15.2, with the quantity of stored

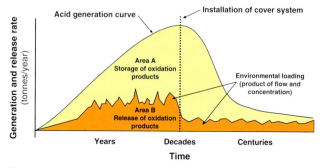

Figure 15.2: Acid generation and release rates for waste dumps. Source: After Wilson (2008)

oxidation products corresponding to Area A. The amount of oxidation products in the dump increases with time as each lift advances along the ARD evolution curve combined with the continuously increasing volume of waste that is progressively placed in the dump. However, the rate of leaching (or release) of the oxidation products corresponds to a substantially lesser quantity represented as Area B. The net result is a continuous increase in the amount of oxidation products that are being stored within the porous matrix of the dump. The total quantity of stored oxidation products can be conceptually visualised as the area under the acid generation curve less the area under the release curve (i.e. Area A minus Area B). In summary, the total quantity of oxidation products stored in the dump continues to increase with time during construction of the dump.

Seepage fluxes will eventually decline as the dump profile drains down after a barrier cover system is installed. The decrease in drainage rate may take several years or even two or three decades to reach equilibrium with the reduced net infiltration delivered by the final cover system. A reduction in the environmental loading can be expected to eventually occur, assuming a constant concentration of oxidation products in the leachate. Nevertheless, the reduction may not be a significant benefit over the long-term. For example, even if the cover provides oxygen limiting effects that reduce the acid generation rate, it must be remembered that all of the oxidation products that were generated and stored within the dump before installation of the cover are still available for release at the reduced net infiltration flux. Furthermore, despite the fact that the release rate is reduced, the initial mass of oxidation products that can be leached remains constant regardless of what new net flux rate is applied. Lowering the net infiltration flux with the installation of a cover simply prolongs time for the release of ARD.

The discussion presented above suggests that the reduction in ARD after the installation of an oxygen-barrier type cover may not significantly reduce mass loadings delivered to the environment or a treatment system, since an abundant quantity of oxidation products have already been stored in the dump profile. Furthermore, while reducing the seepage fluxes may reduce the release rate, the release of ARD at this lower rate may simply result in a prolonged period of treatment, given that the total mass of oxidation products to eventually be leached has not changed.

15.2   Co-disposal techniques

New techniques are needed to prevent ARD from occurring in the first place. This may mean depositing PAF waste in such a way that it remains isolated from atmospheric oxygen.

Various co-disposal techniques have been developed in recent years using mine tailings, which have a relatively low permeability and high moisture retaining capacity, to serve as oxygen seals for PAF waste rock. Wickland *et al.* (2006) described several types of co-disposal for mine tailings and waste rock, depending on the degree of mixing. Co-deposition of waste rock and tailings into the same repository, such as an open pit, is the simplest and easiest form of co-disposal. It has been implemented at numerous mines, such as the Kidston Gold Mine in Australia (Gowan *et al.* 2010). In the case of the Kidston Mine, waste rock was deposited from one side of an open pit while thickened tailings were deposited from the opposite side of the pit. This method of co-disposal is highly effective, as all PAF materials can be placed below a permanent water table in the pit. This permanently isolates the sulphide minerals from atmospheric oxygen. However, implementing this type of co-disposal requires an open void (such as a pit) that is available for backfilling.

Table 15.1 illustrates different forms of co-disposal based on the degree of mixing, beginning with the disposal of waste rock and tailings in the same topographic depression, as described above. The forms of co-disposal identified in the table are organised in a progression upwards based on the degree of mixing between the tailings and the waste rock. The progression starts with the addition of tailings to a dump, then continues on through to the addition of waste rock to tailings, layered co-mingling of waste rock and tailings, pumped co-disposal, and finally paste rock, a material generated by the blending of waste rock and tailings. Paste rock is the most homogeneous form of co-disposal.

Examples of tailings and waste rock co-disposal are presented in the following sections, including waste rock disposal in tailings dams, tailings disposal in waste rock, layered co-mingling of waste rock and tailings, and paste rock created with homogeneous mixtures of dump

Table 15.1: Forms of co-disposal

Co-disposal type	
Homogeneous mixtures – waste rock and tailings are blended to form a homogeneous mass: 'paste rock'.	Increasing degree of mixing
Pumped co-disposal – coarse and fine materials are pumped to impoundments for disposal (segregation occurs on deposition).	
Layered co-mingling – layers of waste rock and tailings are alternated.	↑
Waste rock is added to a tailings impoundment.	
Tailings are added to a waste rock pile.	
Waste rock and tailings are disposed in the same topographic depression.	

Source: After Wickland *et al.* (2006). © Canadian Science Publishing or its licensors

materials and tailings. Finally, brief discussions of the application of blending different types of waste materials and of progressive sealing of dump lifts during construction are outlined in the last sections of this chapter.

15.2.1 Waste rock disposal in tailings storage facilities

Preventing and controlling oxidation and subsequent ARD from high sulphide-reactive dumps constructed above grade may not be possible in some cases. Andrina (2009)

Figure 15.3: (a) Waste rock deposition in the tailings storage facility, (b) pushing waste rock lifts into the tailing pond. Source: W Wilson

observed aggressive oxidation along with an increase in temperature in less than 6 months after placement of reactive waste in a single 20 m lift. The placement of this type of highly reactive waste in a stockpile can lead to long-term ARD that is very difficult or potentially impossible to mitigate with a cover system, leaving long-term collection and treatment as the only viable method of preventing ARD discharge to the environment.

One option is to place the waste material within a tailings storage facility. Properly managed, this can provide subaqueous long-term containment of the waste that will prevent the formation of ARD. One such example of this is at the Phu Kham Copper Gold Operation in Laos. In 2008, Phu Bia Mining Ltd implemented an ARD management plan with all highly reactive PAF waste with total sulphur content greater than 4% deposited in the tailings storage facility as shown in Fig. 15.3. The alkalinity of the tailings liquor discharged into the facility is carefully monitored, and additional lime is added at the mill to ensure that there is adequate alkalinity to buffer any acid that may develop while the reactive material is exposed during construction. The rate of rise of tailings in the impoundment is designed to inundate the reactive waste before significant ARD occurs, leaving all reactive material permanently underwater and thus preventing long-term oxidation and ARD.

In addition to the placement of highly reactive waste in the tailings storage facility at the Phu Kham Mine, less reactive PAF rock is also being placed in the downstream section of the tailings dam. Potentially acid forming cells are being constructed and encapsulated within the dam. Each cell is constructed in the core of the dam and sealed with a compacted encapsulation layer that is 6 m thick on the external face of the embankment. Field monitoring including pore-gas instruments installed within the encapsulation layers has demonstrated that oxygen levels are approximately zero and that oxidation and ARD generation is not occurring. Miller *et al.* (2012) provided a detailed and comprehensive overview of the integrated ARD management plan at the Phu Kham Copper Gold Operation.

15.2.2 Tailings disposal in waste rock

Tailings deposition into waste rock has been attempted at several sites. An example of this is shown in Fig. 15.4. Successful applications require that the tailings be non-segregating (thickened) while at the same time sufficiently fluid to permit flow into the void spaces within the waste rock. The primary advantage of this method is that PAF material within the profile of the waste rock can be sealed with saturated tailings that fill the voids of the waste rock, limiting oxygen transport and potential acid generation. An additional advantage is that a significant storage capacity is available within the waste rock voids. Filling

Figure 15.4: Thickened tailings deposition into waste rock. Source: W Wilson

these voids with tailings permits significant reductions in the total waste volume and footprint required for the storage of both waste rock and tailings. The primary disadvantage of this method is that some form of containment is generally required to prevent fugitive tailings discharge.

15.2.3 Layered co-mingling of waste rock

Layered co-mingling of tailings with waste rock is similar to thickened tailings deposition into waste rock, except that in this case the consistency of the thickened tailings does not permit flow and entry of the tailings into the matrix of the waste rock profile. Lamontagne *et al.* (2000) described the construction of a waste dump profile with

Figure 15.5: Layered co-mingling of tailings and waste rock. Source: W Wilson

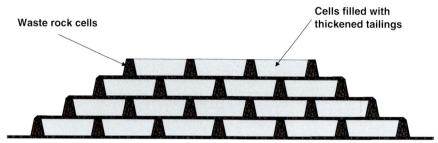

Figure 15.6: Cell construction with thickened tailings and waste rock berms

horizontal layers of thickened tailings, 1–2 m thick, between vertically sequenced lifts of waste rock. The purpose of the tailings layers is to provide barriers to oxygen diffusion and/or advection through the waste rock. The tailings layers maintain high saturation levels within the waste material due to the capillary barrier effect that develops between the moisture-retaining tailings and unsaturated waste rock.

Figure 15.5 illustrates an example of a thickened tailings layer that is ~1 m thick overlying a layer of waste rock. Waste rock berms can be constructed at almost any scale required to contain layers of saturated tailings. A subsequent waste rock layer can be placed over the thickened tailings layer once the tailings layer has dewatered sufficiently to permit adequate bearing capacity for the waste rock. Conversely, the cells can be staggered as shown in Fig. 15.6 to provide internal ribs with the waste rock berms that may improve physical stability as well as allow the rate of rise to increase due to the drainage provided by the more permeable waste rock inclusions.

15.2.4 Paste rock and homogenous mixtures of tailings and waste rock

In the previously described methods of co-disposal, there remains a significant degree of separation between the waste rock and the tailings. In the case of paste rock, the two waste streams are blended in set proportions following a specific design mix to generate an engineered material that is disposed of in a single repository.

Two field trials have been carried out to demonstrate the properties of paste rock blends. The first field study was completed at the Porgera Mine in Papua New Guinea, while the second was conducted at the Copper Cliff Mine in Sudbury, Canada. The Porgera study used four meso-scale columns with designated mixtures of carbon-in-pulp gold tailings and waste rock. At Copper Cliff, five field-scale lysimeters were constructed to test the physical characteristics of blended waste rock, slag and tailings. In the Copper Cliff experiments, the paste rock blends were being evaluated as barrier cover systems that would prevent oxygen entry and reduce infiltration into PAF mine tailings.

At the Porgera Mine, a meso-scale field experiment was conducted using four 1 m diameter by 6 m high columns filled with mixtures of carbon-in-pulp tailings and waste rock. One column served as a control with unmixed waste rock, while the remaining three columns held paste rock blends, with mixture ratios of 5:1, 6:1 and 7:1 (based on dry weight of waste rock to tailings). Figure 15.7 shows a paste rock mixture blended at 5:1. A transit mixer was used to blend carbon-in-pulp tailings with the waste rock. The tailings had been thickened to a solids content of 42%. Waste rock particle sizes greater than 150 mm were removed to ensure oversize rock blocks did not influence measurements in the 1 m diameter columns. Pore water pressures were measured during filling of the columns with electric piezometers, and tensiometers were used to monitor matric suction. Magnetic sensors to measure settlement were also installed along the profile of each column during filling. Flow rates for discharges from the base of each column were measured using a sand drain, with the water table level fixed at the base of each column. After 18 months of experimental operation, the column with the 5:1 paste rock blend was deconstructed by hand excavation from the top to the bottom, allowing direct observation and sampling. Wickland (2006) provides a complete description of the experiment.

Figure 15.7: Placing paste rock at Porgera Mine. Source: W Wilson

Falling head tests on the columns showed that hydraulic conductivity values of the blends were comparable to consolidated paste tailings (i.e. 10^{-7}–10^{-8} m/s). Furthermore, the material was found to have high moisture retention characteristics (i.e. air entry values in the range of 100 kPa), while the settlement measurements showed volume change characteristics similar to waste rock. The paste rock was found to consolidate relatively quickly after deposition with settlement characteristics similar to unmixed waste rock.

The results of the paste rock trials completed at Porgera Mine indicated that it should be possible to construct paste rock structures in landforms that are similar to waste dumps, as the strength and consolidation characteristics appear to be controlled by the waste rock matrix. In addition, the long-term drainage tests demonstrated the capacity of the paste rock to maintain saturation under negative water pressures of up to 5.5 m above a fixed water table, showing that the material will prevent oxygen entry and limit potential oxidation of sulphide-bearing waste rock. The high capacity of paste rock to maintain saturation, combined with its low hydraulic conductivity, suggests the material will be highly resistant to ARD and ML following deposition.

Following the paste rock trials at the Porgera Mine, a study to evaluate the blending of tailings, waste rock and slag to produce a suitable paste rock material for use as a barrier cover system was completed at the Copper Cliff Mine. Several blending ratios of waste rock, slag and tailings were trialled, and a blend of one part rock to one part slag to two parts tailings was found to produce a well-graded material, ideal for the desired application. Five lysimeters, shown in Fig. 15.8, were built to evaluate the performance of the paste rock cover systems. Each lysimeter measured 15 m by 15 m by 2.5 m deep. A central collection sump to measure vertical seepage due to infiltration was installed at the base of each lysimeter. The ability of the paste rock covers to maintain high saturation for minimising oxygen diffusion was also evaluated.

Results for the field-scale lysimeter measurements at the Copper Cliff Mine showed that infiltration and drainage rates can be reduced from 500 mm/year for unreclaimed tailings to values less than 10 mm/year when compacted paste rock is used to construct cover systems. This finding supports the results presented previously for the column experiment completed at the Porgera Mine.

The results from the experiments at both Porgera and Copper Cliff demonstrate that waste rock and tailings can be blended to produce an engineered material with excellent physical and hydraulic properties for the constructing and reclaiming of waste dumps that prevent or control ARD. Paste rock also has been shown to have a density much higher than either conventional tailings or waste rock. This creates further opportunities to reduce surface area requirements for dumps and associated impoundments.

15.2.5 Blending potentially acid forming and non-acid forming waste rock

The principles described for blending waste rock with tailings can also be employed for blending PAF and non-acid forming (NAF) waste to form a dump that is fully net neutralising and will not produce ARD. Andrina *et al.* (2012) and Andrina (2009) presented the results for two constructed waste rock test piles using blended PAF waste rock and limestone at the Grasberg Mine in Indonesia. Figure 15.9 illustrates two excavations in the test piles. Both test piles were constructed using the same blend ratio of PAF and NAF waste rock. Figure 15.9b shows a trench excavated into the trial dump where the materials were

Figure 15.8: (a) Field lysimeters at the Copper Cliff Mine with placement of paste rock cover over tailing and (b) final grassed covers. Source: W Wilson

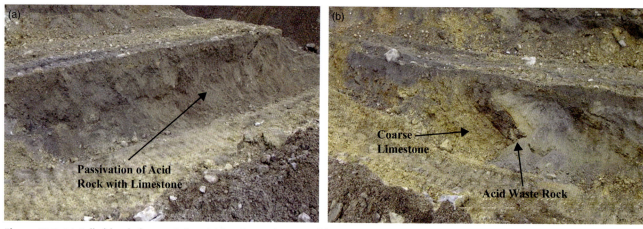

Figure 15.9: (a) Fully blended potentially acid forming and non-acid forming waste rock versus (b) non-blended potentially acid forming and non-acid forming waste rock. Source: W Wilson

simply deposited using haul trucks to dump the material. It can be seen that little mixing occurred and that there are layers of acid generating waste rock adjacent to the layers of coarse limestone. Figure 15.9a shows an exposed cut into a dump profile that had been placed using a stacker following crushing and mixing of the PAF and NAF materials (i.e. using the same mass blend ratio as the profile shown in Fig. 15.9b). It can be clearly seen that the blended rock in the profile placed by the stacker is uniform and unoxidised. This profile retained a pH of 7 and the onset of ARD was completely arrested, suggesting that the blending waste rock has great potential for geochemical control.

15.2.6 Progressive sealing of waste lifts during construction

The co-disposal methods described previously for the purpose of preventing or controlling ARD require considerable effort with respect to design, materials handling, mixing, control and deposition. They are mainly intended to secure the disposal of high PAF reactive waste rock materials. In the case of less reactive waste rock, where there may be significant lag time before the onset of ARD (due to alkalinity within the rock), it may be possible

to implement progressive sealing of each dump lift during construction. The purpose of such progressive sealing is to restrict airflow and prevent the development of advection in the dump profile.

Figure 15.10 illustrates an example where the completed face of the waste dump lift is graded to form an oxygen barrier at the toe of the lift. The high airflow capacity of the permeable rock rubble zone is cut off and buried with fine-grained material that has been pushed down from the crest of the lift. In most cases, the fine-textured rock near the crest of the lift has a sufficiently low air permeability that air advection is severely restricted. This type of construction can potentially prevent early oxidation and the onset of ARD when the alkalinity of the waste rock matrix is sufficient to provide the required lag time.

A method of construction where an airflow barrier is placed at the terminus of each lift in the constructed dump profile is illustrated in Fig. 15.11. This is accomplished by placing a truck-dumped berm at the toe of the lift. Fine-grained material is subsequently dumped from the crest of the lift and allowed to flow to the bottom of the lift, covering the rock rubble zone that exists at the bottom of each lift. This method of sealing the airflow zone that

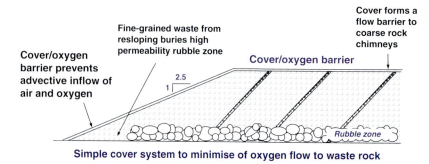

Simple cover system to minimise of oxygen flow to waste rock

Figure 15.10: Rock pile bench graded to restrict oxygen flow into the basal rock rubble zone

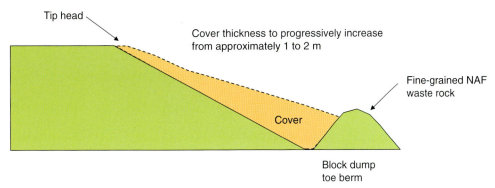

Figure 15.11: Sketch showing the batter and fill method to prevent airflow to basal rock rubble zone

drives early oxidation is relatively simple and inexpensive to implement. While it is believed that construction practice would have benefits reducing and controlling the development of ARD in reactive waste rock materials, no data are currently available to verify such benefits.

15.3 Conclusions

Acid rock drainage and metal leaching problems that occur from waste dumps are created at the time of deposition based on design and the method of construction. Experience has shown that it is very difficult to arrest ARD/ML from dumps after acid drainage has begun. The purpose of this chapter has been to introduce new methods of design that improve the physical and geochemical properties of dumps to prevent and control both ARD and ML. Various methods of co-disposal of waste rock with tailings have been discussed. Co-disposal of reactive waste rock in a tailings pond where the waste rock is rapidly inundated by saturated tailings, as described for the Phu Kham Copper and Gold Mine, has proven to be most effective. Alternatively, research and field testing has shown that emerging technologies based on mixtures of tailings and waste to produce an engineered material known as paste rock may prove to be an effective method of preventing ARD/ML from waste rock piles.

16

CLOSURE AND RECLAMATION

Björn Weeks and Eduardo Salfate

16.1 Introduction

The closure of waste rock dumps presents a unique set of engineering challenges that must be addressed in the context of an overall site closure plan. Typical dump construction practices result in waste rock dumps that contain segregated materials stacked well above the original topography, with slopes commonly at angle of repose. As discussed in Chapter 14, depending on site geology, the waste rock materials in the dumps may be chemically reactive, with the potential to affect the composition and quality of contact waters. Many active waste rock dumps approach the end of their operating life without meeting requirements for the control of long-term risks associated with chemical or physical stability, resulting in the need for costly closure measures to address these risks. In some cases, waste dumps have been abandoned without consideration of the long-term impacts and risks to the environment.

Closure should be considered as an integral component of the life cycle of a waste rock dumps, with planning for the ultimate closure of the facility being an active and iterative process that is modified and adjusted as operations and construction of the dump progress. Progressive closure measures are implemented during the life of the mine to provide reduction of final closure liabilities, as well as a proving ground for evaluating the closure measures. Careful planning from the earliest stages of design ensures that appropriate siting, material segregation (if needed) and dump shaping are undertaken to meet closure goals and provide appropriate risk control. Intelligent cost trade-off analyses are conducted early to appropriately balance factors that affect operating costs (such as haul distances) with the costs for implementation of closure measures such as dump reshaping, cover placement, water diversion and long-term water treatment.

The importance of early planning does not obviate the need to deal appropriately with older, abandoned or imprudently designed mine waste facilities. It is always possible to find the best design for the situation that is at hand, taking into account the site-specific conditions. However, early planning for closure is particularly relevant for waste rock facilities where the possible corrective actions for key closure issues, such as acid rock drainage (ARD) or metal leaching (ML) control, become more limited as the project develops and design flexibility decreases. Where early planning is not undertaken, and major issues are only identified close to the end of operating life, costs tend to increase and the range of effective solutions available decreases.

Much of the attention that has been focused on cover design for waste rock dumps through the 1990s and 2000s was driven by the need to address waste rock dumps where oxidation and the generation of acidic discharges were fully underway, and soil covers were seen as an economic solution that could be evaluated and applied in the last years of operation. While soil covers remain an important design option for waste rock dump closure, learnings of the past 20 years have tempered our expectations for what can be achieved by a skin of relatively fine-grained materials placed over tens or hundreds of metres of waste rock, especially when the impacts of erosion over hundreds of years are considered. Growing awareness of the importance of erosion processes has led directly to the increased discussion in the closure literature of design approaches variously termed 'the geomorphic approach' or 'landform engineering'.

A great deal of research attention and practical design experience has been focused on dealing with the issues related to ARD and ML from waste rock piles. At sites where issues related to ARD and ML are present, their control tends to become the focus of closure and reclamation activities. Even at sites where ARD and ML are not expected, the scale and importance of such issues to mining on a worldwide basis generate a need to conduct the studies to demonstrate their absence. An overview of the range of techniques to control ARD is presented in chapters 14 and 15 of this book. This chapter includes a discussion on how these issues fit into an overall framework for closure and reclamation.

This chapter also provides an overview of some of the key issues associated with the closure of waste rock dumps, along with possible measures that could be implemented during design, operations and closure phases to reduce long-term risks and liabilities. It should be noted that this chapter does not address the social implications of planning for closure and issues such as the input of stakeholders or rights-holders. While these issues are of critical importance to the overall closure, the social dimension does not normally have a specific impact on the closure of the waste rock facilities separate from the closure of other mine facilities.

16.2 Approach to closure and reclamation planning

Various schools of thought exist for closure and reclamation planning, both for waste rock dumps and for mining operations in general. These include the objectives-based approach and the risk-based approach. In the objectives-based approach, closure objectives are defined and criteria established. Closure measures are then developed to meet these criteria. In the risk-based approach, the risks posed by the facility are first identified. Measures to control the risks are then identified and evaluated in terms of their effectiveness and risk reduction potential. These two approaches are not mutually exclusive and can be combined. An example and clear description of the objectives-based approach can be found in the *Guidelines for the Closure and Reclamation of Advanced Mineral Exploration and Mine Sites in the Northwest Territories* (MVLWB/AANDC 2013), while the risk-based approach is well introduced in the International Council on Mining & Metals Toolkit (ICMM 2008), which forms part of a practical framework for closure planning throughout the mining life cycle. The objectives-based approach is also illustrated in the ANZMEC and MAC (2000) *Strategic Framework for Mine Closure*, which incorporates risk-based considerations as part of a planning stage. The interested reader is directed to these documents (which are freely available on the internet) for a more in-depth discussion, including examples of the types of objectives that are typically desired for a wide range of mining installations.

Regardless of the framework adopted, the closure and rehabilitation of a waste rock dump normally have the following three goals, as a minimum:

- protect public health and safety
- prevent or reduce environmental degradation
- allow for productive future land use (either its original use or an acceptable alternative).

Variations on these concepts are enshrined in regulations in most mining jurisdictions around the globe. From an engineering point of view, closure design is also shaped by a desire to attain the preceding objectives in the most cost-effective way possible, considering both the capital expenditures needed to attain the closure and the operating expenditures needed (if any) in the period after implementation of the principal closure works. As the closed site will exist in some form for the foreseeable future, the financial burden (and potential liability) implied by any needed operating expenditures can greatly exceed the capital expenditures. The desire to control or eliminate this cost and associated liability is often expressly stated as the desire for 'walk-away' closure solutions.

Finally, the closure of any waste rock facility may also be defined by local regulations.

16.2.1 Conceptual models for closure

Once a general understanding of the design objectives is obtained, developing a conceptual model of the waste rock facility is a practical approach to developing and evaluating a closure and reclamation plan. The conceptual model allows the designer to focus on the key issues that will affect the closure measures and on the variables that need to be evaluated in the design process. The conceptual model can be developed rapidly, based on available information, and refined as more information becomes available. The first questions to be answered in developing the model are as follows:

- Is the waste rock prone to generating ARD or ML?
- Where is the waste rock located? (For an existing site, the answer is set – but when closure planning is considered early in the design stage there may be an opportunity to incorporate closure considerations in siting.)
- What is the underlying hydrogeology? This includes depth to the groundwater table, direction of groundwater flow and hydraulic conductivity of the material under the dump and in the aquifer matrix.
- What is the surface water setting? This includes the size and topography of the watershed in which the facility is located, any channels, streams or diversion structures in the watershed and any water bodies downstream. Ideally, this will include a

characterisation of surface water flows and a range of design floods.

- What is the downstream use of groundwater and surface water? This includes both human use and sensitive environments.
- What is the climate? At a minimum, this should include average annual and monthly precipitation and evaporation numbers, and how many months per year of sub-zero temperature. Ideally, historical rainfall data should be analysed to develop estimates of storm events for various return periods. In permafrost regions, characterisation of the current and projected future depths of the active layer may be relevant to the closure design.

The above questions all relate to water, the management of which is typically key in the closure design. For physical stability, relevant questions for the conceptual model include the following:

- What is the seismicity of the site?
- What is the geometry of the waste rock facility (for closure of existing facilities), including overall slopes, benches and general configuration?
- What are the strength properties of the waste rock (friction angle, cohesion)? Is the material prone to rapid degradation and change in the material properties?
- What are the foundation conditions under the existing or proposed waste rock dump?

Many of the above considerations are design variables, and the process of evaluating these questions may be iterative as the design evolves.

For rehabilitation considerations, the following questions are relevant:

- What is the natural vegetation in the surrounding areas? Density and type of vegetation? Thickness and quality of soil?
- Is soil available in the surrounding areas for use in revegetation, or can it be scalped and stockpiled for future use?
- What is the surrounding land use: agricultural, residential, traditional hunting grounds, other?
- What is the surrounding topography and how does it compare to the shape of the waste rock facility?

With these basic questions answered, one should be able to develop a good overall conceptual model of the waste rock facility for closure and/or an understanding of the information gaps that need to be addressed before advancing with the closure plan.

The conceptual model can be probed with questions such as these:

- Is there a potential for adverse impacts on water quality (geochemical stability)?

- Is there a risk for physical instability (slope failures with significant consequences, erosion)?
- What would be a final land form or land use compatible with the surrounding conditions?

As mentioned previously, working through these questions may generate design modifications and initiate an iterative process of refining the design. Experience has shown that the answers to these three questions tend to define the majority of the closure and rehabilitation measures needed. Depending on site-specific conditions, it may be necessary to delve into some issues in greater depth, while others can be treated at a more superficial level.

Properly applied, the conceptual model approach allows a rapid evaluation of the waste rock facility (existing or proposed) and identification of the key design challenges that will be faced at a specific facility. This allows a high-level consideration of different closure approaches and their interactions, and different approaches can be considered, as well as the impacts of more radical or out-of-the-box approaches. A more methodical approach to working through individual risk issues and their treatment with individual control measures can be applied at a later stage to ensure that important risks have not been overlooked in the design and to provide documentation of design rational.

16.2.2 Closure criteria

This section introduces the concept of design life for closure, and the implications on criteria for aspects such as water quality, geotechnical stability and the design of hydraulic structures.

16.2.2.1 Design life

At some point in the development of the closure and reclamation plan, it may be necessary to state a 'design life' for closure. There is relatively little agreement about what constitutes an appropriate design life for closure, and considerable variation in the understanding of the concept.

In talking about design life for waste rock dump closure and reclamation, it can be useful to separate the discussion into the following three concepts:

- the design life of the overall closure system
- the design life of a specific closure component (e.g. concrete used in a spillway)
- the recurrence interval of a design event (e.g. a 1:200 or 1:1000 year storm).

It is possible that all three of these can be different for a single closure and rehabilitation plan. For example, a closure period of 1000 years may have been selected as the design life for the waste rock facility. The design storm used to size water diversion studies may be the 1:200 year,

24 h storm, based on an analysis that shows that if the design storm flow is exceeded, the overflow of the diversion system will have acceptable consequences. At the same time, a concrete structure needed for some aspect of closure may be constructed with concrete that has a design life of 50 years, with the understanding that this structure will need to be periodically renovated throughout the design life of the overall system.

There is no consensus on what is an appropriate design life for the overall closure. This arises from the tension between the limits on what we can reasonably design based on engineering experience and the self-evident reality that a waste rock facility can be expected to remain more or less where it has been placed for the foreseeable future, barring a massive relocation program or the action of geological processes over thousands (or tens of thousands) of years. When a closure period that seems somewhat credible based on engineering experience (such as 100 or 200 years) is postulated, it will fall short when compared to the reality of geological time scales. When a period that attempts to take into account the geological time scale (1000 or 10 000 years, for example) is postulated, we are left without proven engineering tools.

While set periods such as 1000 years or more have entered into practice (or at least regulations), there are very few engineering experiences available with which to meaningfully evaluate such design periods. This is especially true for designs that make use of material such as geosynthetics that have only existed for a few decades. And while closure periods of over 10 000 years have been discussed, this is beyond the reach of what we can meaningfully analyse with models calibrated against real world experience. In periods of time of the order of 10 000 years, one can quite reasonably expect the impact of geological processes and significant climate change on any closure design, especially taking as context that the last major period of glaciation (or 'ice age') ended ~11 000 years ago.

Jurisdictional mandates for design life need to be taken into account, although the reasonableness of definitions such as 'perpetuity' are questionable. In the absence of another specific formulation, one approach that has gained some acceptance was developed for the US *Uranium Mill Tailings Radiation Control Act* of 1978, which required closure measures to be effective 'for up to 1000 years to the extent reasonably achievable and, in any case, for at least 200 years' (EPA 1983). According to Logsden (2013), the EPA guideline was formulated to cover the 'periods over which climatological and geomorphic processes could reasonably be predicted, given current knowledge of earth sciences and engineering'. The 200 year period can be seen as being reasonably within reach of analytical or predictive engineering approaches, while the 1000 year span enters into the realm of qualitative evaluations (Logsden 2013).

Even if the *Uranium Mill Tailings Radiation Control Act* formulation is not explicitly being followed, that document outlines the limits of our current ability to design, independent of what is said in the governing regulations. Analytical design can be taken to its limit (which is probably of the order of 100–200 years), and longer periods can be considered in a qualitative way.

16.2.2.2 Water quality criteria

Water quality criteria for closure are generally the most straightforward of all closure criteria, at least in terms of theory, if not in application. In general terms, chemical constituents that are released from the waste rock dump to downstream surface water or groundwater should not endanger public, wildlife or environmental health and safety, and should not result in the inability to achieve the water quality objectives in the receiving environment over the long-term.

Water quality objectives are commonly defined by local regulations, and the closure design is obligated to work within that. There are two categories of effluent regulations: (1) end-of-pipe standards for the quality of the effluent at the point of discharge and (2) receiving water standards, which govern the quality of a natural water body downstream of a discharge point after a mixing zone. Many jurisdictions have regulations governing both. Receiving water quality standards recognise that the assimilative capacity of the environment is a factor in determining the design removal efficiency of a treatment system. A larger receiving water body may require less strict treatment requirements.

In the absence of local regulations, ecological and human health risk assessments provide a useful and defensible tool for the establishment of site-specific water quality goals. Alternatively, it is also common practice to adopt as a guideline the water quality regulations of a recognised jurisdiction.

Where possible, proper definition of baseline conditions in the areas upgradient and downgradient of the waste rock facility is highly recommended. The lack of such information can lead to demands to achieve water quality that may meet a regulatory guideline but is beyond what was observed under baseline conditions.

16.2.2.3 Geotechnical criteria

Typical closure criteria for waste rock facilities call for dumps to be physically stable. The interpretation of 'stable' can vary, but includes concepts such as the following:

- Slopes have an adequate factor of safety against significant failure. The nature of a 'significant' failure is tied to the use of the dump and the surrounding land post-closure (this may be evaluated semiquantitatively through risk assessment).

- The waste rock dump will not undergo deformations under the design seismic load that are significant (as per the previous definition of 'significant').
- The final surface does not erode significantly under natural events (such as storms or earthquakes) or other disruptive forces (e.g. all-terrain-vehicle use) after closure.

The required factors of safety for long-term geotechnical stability are frequently defined in regulations. In the absence of local regulations, global practices (such as those summarised in Chapter 8) should be referenced in the development of site-specific criteria.

For seismic stability, some jurisdictions mandate consideration of the maximum credible earthquake, and that the structures and slopes should sustain the maximum credible earthquake largely intact, and that designed landforms should be maintained without significant changes. Other jurisdictions require less stringent criteria, such as the 1:475 year return period event, or even the 1:100 year return period event. Acceptable deformation levels are less commonly defined, and should be evaluated based on site conditions and performance goals. Chapter 8 provides some guidance regarding acceptable deformation criteria under seismic loading.

Erosion criteria are less often stipulated in regulations, but are relevant for reclamation design in waste rock dumps where soil covers and/or revegetation is planned. There are now a considerable number of examples where high-intensity rainfall events have led to significant erosion or complete failure of soil covers. While some level of soil loss due to hydraulic processes is to be expected, this erosion rate should be balanced against long-term rates of soil generation.

16.2.2.4 Surface water management criteria

Water management features that are to remain following closure should be designed and built considering return periods prescribed by applicable regulations for closure and reclamation.

Design storm events are generally specified in terms of storm duration and a return period, such as (for example) a 24 h storm with a 1:200 year return period. Typically regulations specify only the return period, and the most critical duration will depend on the specific climate and catchment characteristics. Regulations typically either do not mention the event duration or specify that most critical durations should be considered in the design.

The return periods for the design criteria to be applied for the design life can be estimated on the basis of assuming a probability of exceedance. The probability of exceeding the design value during the design lifetime may be calculated assuming that the occurrence of events (e.g. seismic, floods) follows a Poisson process (ICOLD 2013):

$$P = 1 - e^{-t/T} \qquad \text{(Eqn 16.1)}$$

where:
P = probability of exceedance
t = design life
T = return period.

For example, using a closure design life of 1000 years, the return period calculated for a 5% probability of exceeding the design value would be ~20 000 years. Similarly, a 10% probability of exceeding the design criteria would result in a return period of ~10 000 years. Numerical techniques are also often used to develop a probable maximum flood (PMF).

Closure works are often designed to accommodate the PMF, the rationale presumably being that given the very long design life, the probability of any design storm occurring approaches 100% ($P = 1$). This approach may be enshrined in regulations. However, it presents at least two conceptual difficulties. The first is that, in some instances, the consequences of sporadically exceeding the design flow in a control structure may be manageable. It may be that the net present value of periodic repair and maintenance works is far more acceptable than the capital expenditures implied by the construction of works sized to handle the PMF.

A second issue is the reliability of PMF estimates. Especially when working at remote sites or in developing regions, the period of historical rainfall data may be quite limited. It is not uncommon to work on design at remote mine sites where 10 or 20 years of rainfall data would be considered outstanding. In contrast, obtaining a prediction of just the 25 year storm with 90% confidence would require 59 years of data. The statistical validity of even a 1:200 year estimate developed based on 10 or 20 years of data is questionable. Using these limited data to estimate a PMF, and then investing millions of dollars in structures sized on these extrapolated numbers is a practice that should be carefully considered.

This issue has not been adequately resolved in current practice. In the absence of clear design practices, it is recommended to consider the site-specific closure conditions/requirements and limitations on precision of estimates when developing hydrological criteria.

16.2.2.5 Other criteria

A wide range of other criteria may be defined, depending on the specific needs of the closure design. For instance, covers may be designed to maintain a minimum level of saturation to prevent oxygen entering the waste rock. The compliance with these criteria can be monitored with properly maintained probes that measure the degree of saturation in the soil cover, and with oxygen probes measuring oxygen percentages in the waste rock. If vegetation is part of the final rehabilitation plan, a variety of criteria can be used to evaluate the development of vegetation, such as measurements of the density of plants per square metre, or biodiversity measurements.

The definition of these criteria forms part of the development of the closure plan for the waste rock facility and needs to be established based on the closure measures selected.

16.3 Geochemical stability

As discussed in Chapter 14, avoiding unacceptable impacts on downgradient water quality that result from ARD/ML can be a critical issue in the closure design of waste rock facilities. Where ARD and/or ML are present or have the potential to develop, closure measures are needed to prevent their development (where possible), to minimise them, or to treat them. These three approaches are described briefly below.

16.3.1 Acid rock drainage/metal leaching prevention

The techniques commonly applied that have the potential to completely prevent the generation of ARD/ML were previously described in Chapter 14, and include subaqueous disposal, blending and encapsulation. Each of these techniques has a greater chance of success if incorporated into the early design and construction of the waste rock facility. Several of the innovative approaches identified in Chapter 15 also have the potential to prevent ARD/ML generation. These innovative approaches also generally require early incorporation for the most cost-effective implementation.

16.3.2 Acid rock drainage/metal leaching reduction

Even where control techniques fail to result in complete elimination of ARD/ML, they may result in the reduction of the volumes generated. Mixing potentially acid forming waste with limestone or non-acid forming material that does not result in complete neutralisation or elimination of ARD may still result in the treatment or elimination of a large portion of the net reactivity.

Barriers and seals (discussed in Chapter 14) are considered viable techniques for the reduction of ARD/ML as the various types of barrier covers have generally been shown to reduce net infiltration into waste disposal piles, but not eliminate it. Some percentage of percolation through the cover is to be expected, with the net percolation being a function of the overall cover system design, including soil profiles, climate and site topography/exposure. As discussed in Chapter 15, reduction in infiltration rates may not be sufficient to achieve the desired benefits from a cover. Lower volumes of

infiltration may have higher concentrations, resulting in equivalent mass loadings of the contaminants of concern to downstream receivers, when compared to the loadings that would have resulted from a higher flow at lesser concentrations. For this reason, geochemical characterisation and prediction of mass loading is key in the evaluation of designs for ARD/ML reduction.

16.3.3 Acid rock drainage/metal leaching treatment

Acid rock drainage/metal leaching treatment will, in general, be required at sites where (1) there is a net positive water balance and (2) the geochemical stability following application of prevention and reduction measures is not sufficient to ensure that the excess water meets the required criteria for discharge to the environment. The need for an integrated management approach to waste rock handling and excess water treatment is increasing. This need is highlighted by increasingly higher costs for compliance with the required criteria, including an increasing number of constituents of concern in discharges and stricter regulations, as well as an increase in competing uses for receiving water bodies, which reduces the assimilative capacity of the environment and increases regulatory pressure. For example, the removal of heavy metals, acid or excess cyanide in mining effluents has long been practised, but the regulations governing these discharges are becoming stricter. Furthermore, as recently as a decade ago, the treatment of waste rock leachate or runoff for the removal of selected constituents such as sulphate or selenate was rare. Today, we see specific cases of treatment plants for sulphate or selenate removal having capital costs of over one hundred million US dollars, and the implementation of stricter regulations governing these constituents is becoming increasingly common. These costs may be debilitating to mine operators and reinforce the need for a holistic approach and a careful analysis of the trade-offs between treatment and prevention. An area of recent research involves the incorporation of treatment chemistry into waste dump design, such as incorporation of reactive elements into the design of waste dump toes and the use of saturated granular media as a biochemical reactor.

While a wide range of proven treatment options exist, all of them imply a level of ongoing care and maintenance, resulting in ongoing operating costs after closure. A review of case studies of treatment practices for 110 mine sites (Zinck and Griffith 2013) is provided by Mine Environment Neutral Drainage, a program funded by the Canadian Ministry of Natural Resources and the Mining Association of Canada. An alternative review of treatment theory is provided in *The Global Acid Rock Drainage (GARD) Guide*, funded by International Network for Acid

Prevention (INAP 2009). The available treatment systems are often broadly categorised as active, passive and hybrid (combined active and passive).

Active treatment refers to a range of facilities that require the input of energy and/or treatment chemicals on an ongoing basis and require ongoing supervision to ensure correct functioning. Active treatment plants can be sized to accommodate most flow rates and acidity loadings, with costs being a function of factors such as flow, flow variability, mass loadings, types of contaminants to be removed and final desired discharge concentration. Typical treatment plants for ARD/ML include low-density sludge and high-density sludge plants, where the collected water is dosed with an alkaline reagent to neutralise pH and promote metal precipitation in the form of hydroxides. Typical alkaline reagents include lime, caustic soda, soda ash and ammonia. The precipitates form a sludge, which will require collection and disposal on an ongoing basis. There exists considerable documentation in the literature on experiences with the treatment of ARD and ML with active treatment systems (e.g. Skousen *et al.* 2000).

With increasing treatment costs, and higher metal values, ARD/ML treatment operators have implemented systems designed to recover metals. Such systems can reduce net operating costs through the sale of the recovered metals, and in some cases operate treatment systems at a profit (Bratty *et al.* 2006). Metal recovery systems are normally implemented for waste streams with relatively high levels of metals, in cases where relatively little prevention effort has been implemented. However, case studies exist where metal recovery is economically justified at trace levels.

While generally reliable and effective, the operating expenses for an active treatment plant need to be properly dimensioned, taking into account the requirements for energy, reagents, labour and sludge disposal. In addition, the needed treatment period will very likely exceed the design life of many treatment plant components. As a result, long-term operating expense estimates should include periodic refurbishment of the plant. Because of the cost implications of these considerations, there can be a strong motivation to design waste rock dumps in such a way as to avoid active treatment.

Passive treatment involves the use of natural geochemical, biological and physical reactions that result in the improvement of water quality. Since, by definition, passive systems run unattended for extended periods, such systems typically exhibit variability in process parameters and in effluent results. Therefore, passive systems are ideally suited to applications where load reduction takes preference over meeting high removal efficiency specifications.

Water flow through passive systems is typically by gravity drainage, and can offer greatly reduced level of complexity and greatly reduced staffing levels and skills; as a result, it potentially offers cost savings. Passive treatment systems are designed for low maintenance, with the possible exception of periodic reconstruction or replenishment of active components over the long-term (normally periods of the order of years to decades between replenishments). Some of the typical passive water treatment systems include wetlands (both aerobic and anaerobic), biochemical sulphate-reducing reactors, anoxic limestone drains, vertical flow wetlands, limestone leach beds and open limestone channels.

Passive treatment of ARD is no longer an emerging technology, and there exists a considerable base of practical experience to guide designs. Passive treatment research and development have been carried out over a period of decades. Skousen and Ziemkiewicz (2005) provide a review of the performance of 116 passive treatment systems for acid drainage. In general, passive treatment requires a much larger land area than active treatment and, as a result, the process has not been frequently applied to large flow rates. An important drawback of certain passive treatment systems is the undesired release of nutrients from the reactors.

The costs associated with both passive and active treatment systems can be significant. The long-term costs, especially for active treatment systems, have the potential to exceed all other closure costs for a given mine site. As a result, the requirement for treatment systems should be carefully evaluated and mechanisms to avoid the need for water treatment given full consideration.

16.4 Physical stability

The approach to geotechnical stability analysis of waste rock facilities has been described in depth elsewhere in this book. The applicable analytical techniques are described in Chapter 8, and the applicable criteria have been discussed earlier in this chapter. It is increasingly common that waste rock dumps are designed for closure, at least in terms of using stability criteria in the design that are suitable for closure. However, additional, long-term stabilisation measures for waste rock facilities are required in the following cases:

- The original design or 'as-built condition' of the waste rock facility does not meet suitable closure criteria for long-term stability (e.g. the selected design earthquake is smaller than that required for closure purposes).
- Changes of material properties that could affect long-term behaviour of the facility are expected (e.g. the presence of residual waste that may be subject to

weathering or breakdown in the long term, or thawing of previously frozen soil), and were not taken into account in the original design.

- Settlements that would adversely affect drainage patterns and/or long-term landforms are expected.

Closure activities to ensure long-term stability of waste rock facilities generally consist of reshaping or buttressing of the facility slopes to achieve an adequate factor of safety under the selected closure criteria for long-term stability. The design of adequate buttressing or reshaped slope geometry to ensure slope stability in the long term should be assessed based on the results of stability analysis. This includes addressing factors such as weathering of the materials within the facility and resulting reductions in strength expected in the long term.

16.5 Land forms and erosion control

The traditional construction of waste rock facilities by the loose end dumping of waste rock from the tip heads of multiple lifts results in uniform slopes at the angle of repose of the material (typically 37° or ~1.3H:1V), separated by flat benches. As part of waste rock facility closure and reclamation, the angle of repose slopes may be flattened, contour drains constructed on the benches, downslope drains constructed to deliver rainfall runoff to the toe, and the slope surface topsoiled and revegetated. Typical motives for this are stated as being improved geotechnical stability, safe access, facilitation of revegetation efforts and post-mining land use, and meeting stakeholder and regulatory demands. However, the flattening of the angle of repose slopes that are made up of durable waste rock is not typically required to ensure

adequate geotechnical stability, and will very likely exacerbate erosion due to the longer slope length, smooth flattened surface and placement of topsoil (Williams 2016). Depending on the differences between the constructed configuration and the desired final landform, slope flattening may also involve significant expense for earth moving and increase the facility footprint. As such, it should not be undertaken lightly, or simply to meet vaguely defined perceptions.

From a construction point of view, slope angles of the order of 2.5H:1V or 3H:1V (of the order of 20°) represent a practical limit on the steepness of slope on which heavy equipment can be used to place soil for revegetation. As a result, some facilities have been designed with benches set back such that, at closure, successive angle of repose lifts can be flattened out to an overall 20° slope without affecting the final facility footprint.

For the long-term stability of vegetated waste rock dump covers, the geomorphic approach has gained increasing attention since the mid 2000s, arising as an alternative to the more traditional 'structural' approach. In both approaches, the importance of surface runoff as an erosional agent is recognised. In the structural approach, water movement is controlled and channelled over the waste rock facility using benches, berms and other traditional water control structures. In the geomorphic approach, natural analogues are sought out through observation of the landscape in the area of the facility, with hill slope forms and gradients, watershed characteristics and vegetation observed and measured. These data are then used to develop surface contours and systems that, using the facilities existing configuration as a base, will to some degree mimic the natural landforms and provide a structure that will permit discharge without excessive erosion.

Figure 16.1: Waste dump with structural closure on south side and geomorphic approach on north (Washington, USA). Source: Google Earth Pro (2015)

The contrast between these two approaches is illustrated in a single waste dump at TransAlta's Centralia Mine in Washington State, USA. As shown in Fig. 16.1, the south side of the dump was closed using the structural approach, with linear structural features clearly visible in the satellite image. The northern and north-east side of the facility were designed using the geomorphic approach, with curved surfaces and dendritic drainage patterns.

The current approach for final landform design and planning has been refined as a result of observation of the performance of historical reclamation practices and the development of numerical models with the ability to simulate the evolution of landforms under the action of runoff and erosion over long periods of time. Some general design considerations to optimise waste rock facility configurations at closure, based on observations at various sites, are as follows (Environment Australia 1998; Sawatsky 2004; Ayres *et al.* 2006):

- Where possible, it is preferable that final geometry resemble a 'mature' landform, which involves measures such as the following:
  - → designing the final landform using natural analogues (geometries and natural landforms of the surroundings)
  - → avoiding benches, terraces, contour banks and abrupt changes in topography
  - → avoiding man-made materials (pipes, concrete, gabions)
  - → using a 'spur-end' shape in plan with a concave–convex profile if feasible
  - → providing appropriate distribution and quantity of drainage features (function of climate, 'soils' and slope)
  - → situating watercourses in 'valleys' as opposed to banks
  - → establishing vegetation progressively.
- The top surface should be sloped and minimised if practicable. A large flat surface at the top of a waste facility may act as a temporary pond for incidental rainfall. Especially in wet climates, ponding is a concern as accumulated water may find a low point on the rim and discharge, leading to the formation of deep erosion gullies on the slopes. If a sloped or domed shape is impractical, construction of small containment berms (bunding) on the top of the facilities may be required to retain runoff.
- When practicable, it may be desirable to include the construction of small lakes and wetlands upstream of final surface water discharge points, provided they are geomorphically compatible and stable. Such features will attenuate surface runoff to reduce peak flows and increase sedimentation before reaching receiving streams.

As presented in earlier sections of this chapter, addressing closure issues during design phases can be beneficial in ensuring successful closure. Where possible, waste rock dumping should be planned to minimise material rehandling, controlling closure costs. If the desired landform is identified early through the application of geomorphic principals, construction can be planned accordingly. This may involve early definition of bench location, so that benches form a stepped outer profile that can be easily reshaped into a stable landform at closure. A waste rock facility constructed using this approach may only require cut/fill movement of the waste rock between the setbacks, whereas a facility constructed without such advance planning may require significantly more earthworks to achieve the desired closure configuration.

16.6   Revegetation

Where technically and economically feasible, it is normally either desired or legally mandated to revegetate the closed facility surface. There are exceptions to this, such as in extremely arid climates or cold climates, where it may be standard practice to leave the closed waste rock facility surface with no vegetation.

Options for revegetation include direct revegetation of the waste rock (Fig. 16.2) or revegetation of some final cover surface (typically a soil or soil blend) that has been placed over the waste rock, either with or without a cover system.

Considerable experience in the revegetation of mine waste facilities has been amassed in the last 50 years. It is now common practice to salvage any existing topsoil at waste rock facility sites before construction and to either use it to support the reclamation of existing waste rock dumps (progressive closure) or stockpile it for use in future closure works. If stockpiled, the soil storage should be designed and managed to maximise the long-term viability of the soil. Management to promote viability includes protecting the stockpiled topsoil from erosion and maintaining microbial communities, which are necessary to provide nutrients to plants. This includes maintaining good aeration by avoiding compaction during stockpiling, and discing or ripping to loosen the surface. It is also important to establish native vegetation on the stockpiles, since many of the soil organisms need to be in contact with plant roots to survive (e.g. mycorrhiza).

Where topsoil (either previously stockpiled or from borrow sources) is not available, or the impacts of taking topsoil from other sites are unacceptable, there may be workable alternatives, such as the application of organic matter (manure, compost, biosolids), chemical amendments, soil conditioners, mulches and nitrogen fixing plant species to promote revegetation efforts.

Figure 16.2: Vegetation growing directly on a waste rock facility surface in an arid climate (135 mm precipitation average per year). Source: B Weeks

While a variety of actions (including irrigation) may be used to promote the initial establishment of vegetation on the facility surface, the normal goal is to attain a self-sustaining ('walk away') ecosystem. As a general rule, the ecosystems in the surrounding area provide an indication of the attainable final state for the cover. Vegetation densities in the surrounding area indicate the maximum vegetation density that could reasonably be attained over the longer term, although there may be some variations in attainable densities and species distributions as a function of factors such as substrate thickness and the direction of exposure.

The establishment of self-sustaining vegetation on a waste rock dump may take years. Factoring in species succession at the site, it may be decades before the plant communities reach their long-term 'steady state'. The species that dominate immediately after revegetation efforts are generally replaced by other species over time. Fortunately, plant species succession is a natural process and in many instances does not require any human intervention to proceed (although weed control may be necessary). On the other hand, if the natural progression results in conditions that would be contrary to the overall closure design, controlling natural succession may generate an ongoing care and maintenance need. The common example of this is vegetation on covers that have an underlying barrier layer. While the vegetation provides an important component of the cover (promoting evaporation), tree growth is generally not desirable due

to concerns that tree roots will penetrate the barrier layer and in time degrade the layer's effectiveness. This has necessitated ongoing tree removal programs at sites where volunteer species would otherwise generate a succession of trees over the planted grasslands.

Erosion control, as alluded to in the previous section, is a key consideration in effective revegetation. There are many instances of covers that have failed due to erosion of the surface soil, occurring either because of design inadequacy or because of a low-frequency storm occurring before the establishment of vegetation that would have otherwise prevented the soil loss.

Species selection for revegetation is a subject for site-specific, specialist analysis. While it is not uncommon to see conceptual closure plans (or even regulations) that specify revegetation with native species, careful consideration should be given to both native and non-native species. Especially in the early stages of revegetation, the lands to be rehabilitated are likely to exhibit many conditions that are distinct from the surrounding lands, and there may be strong arguments for the use of non-native species to meet specific design goals.

The development of revegetation programs for rehabilitation is an area that benefits greatly from research work conducted well in advance of closure. Test plots and progressive closure give an opportunity to test alternative approaches; evaluate seed mixes, seeding methods and seedbed preparation; and undertake preparation of the seed banks needed for closure.

APPENDIX 1
SUMMARY OF BRITISH COLUMBIA MINE WASTE DUMP INCIDENTS, 1968–2005

Reference number								
BCMEM (2012) number	BCMWRPRC (1995) event number	BCMWRPRC (1992a) record number	Date of event	Dump name	Failed volume (m³)	Runout distance from toe (m)	Dump height (m)	Comments
2	40	–	24 Nov 1968	No. 2	150 000	620	180	The weak glaciolacustrine sediments present at the lower portion of the dump failed when a combination of high load and high piezometric pressures were incurred. The second stage of failure occurred due to loss of support as a result of failure of toe. Failure along foundation contact in toe zone causing loss of support and failure through dump along plane of weakness. Failure debris crossed Hwy 3 and buried a car, killing two people inside.
3	42	–	5 May 1971	A29-N	2 200 000	2600	260	No warning. Clayey till and pore water pressure.
–	43	–	13 May 1971	HKNOB		870	120	Mudflow went afterwards.
4	–	–	1 May 1972	South Pit Dump		1000		Very small volume debris flow from toe of spoil in upper drainage which slumped into creek. High fines, water content and mobility. Three people on a CP Rail crew were killed; they had stopped by the creek for their lunch in lower drainage.
9	19	–	5 May 1972	Clode/2-SPOIL	50 000	275	120	Very heavy snowfall. But groundwater not felt important.
10	20	–	27 May 1972	Clode/2-SPOIL	160 000	270	135	Very heavy snowfall. But groundwater not felt important.
11	–	–	12 Nov 1972	2 Spoil			220	
5	44	–	25 May 1973	A29-S	110 000	500	120	Steep toe topography.
12	24	–	8 Nov 1974	2-SPOIL	450 000	270	220	Rapid.
27	27	–	12 Nov 1974	2-SPOIL	750 000	250	110	Organic foundation soils. Organic soil displaced ahead.
6	38	–	8 Feb 1976	5975 (alternative A29S?)		335	260	
1	–	3	1 Nov 1979	A – active at time of failure	10 500	45	45	Rapid loading of soft weak saturated foundation soils causing excess pore pressure generation; foundation seepage pressures. Lateral yielding of foundation soils at toe region and basal sliding along foundation contact in a slow creeping movement.

Reference number								
BCMEM (2012) number	BCMWRPRC (1995) event number	BCMWRPRC (1992a) record number	Date of event	Dump name	Failed volume (m³)	Runout distance from toe (m)	Dump height (m)	Comments
–	35	–	18 Nov 1980	H2B3		260	230	Abnormal on 13 Sept, 2 months prior, spatial.
48	–	30	24 Apr 1981	G – active at time of failure	80 000 dump, 250 000 foundation 320 000 total		20	Exhibited heaving of natural soils below dump toe instead of large displacement. Increased foundation pore pressure due to groundwater recharge; increased load due to saturation of fines in dump. Slumping in dump material was rotational, in foundation was translational.
7	36	1	29 Jun 1982	A40-C1	575 000	415	150	Steep foundation slope with unconfined dump toe; snow in dump forming weak plane; precipitation elevating internal pore pressure and increasing foundation seepage pressure; poor-quality material with high fines content. Failure along foundation contact in toe zone (unconfined), causing loss of support and planar failure through dump along plane of weakness. Spatial variations measured, channeled runout.
40	62	28	1 Jul 1982	1966 W	250 000	490	136	Failure was a two stage event including a relatively small rotational failure through dump material, followed by a much larger disturbance of natural soils which liquefied due to the sudden load. Rotational slump within dump material. Steep toe slope.
31	75	18	20 Mar 1983	East Dump	400 000	700	160	High fines content causing over-steepening; snow and saturated soils at foundation contact; foundation seepage; limited toe support. Rotational failure of dump toe and foundation material causing loss of support for mass above and basal sliding along foundation. Steep toe; poor waste; over-steep face.
13	–	5	8 May 1983	A – active at time of failure	140 000	36	220	Fine-grained poor-quality material; crest over-steepening due to apparent cohesion of fines. Rotational failure through dump material.
32	76	19	11 May 1983	East Dump	720 000	320	170	High foundation pore pressure; high dump pore pressure; redistribution stresses. Yielding toe/basal along foundation contact. Failure characterised by slow and gradual deformation over period of several days. Quasi-circular. Excess foundation pore water pressure. Not sturzstrom. Poor waste.
49	–	41	27 May 1983	1660 WN Dump	100 000	350	200	High loading rate and associated strain rate; captured snow forming continuous plane of weakness; spring meltwater causing infiltration and groundwater recharge; redistribution of stresses; toe failure causing loss of support for mass above and planar sliding along plane of weakness.
–	162	–	1 Jun 1983	BROWNIE		400	170	Steep foundation.
14	1	–	11 Jun 1983	BROWNIE	110 000	420	235	Nearly 10 days of 0.9 m/d displacement.

Reference number			Date of event	Dump name	Failed volume (m³)	Runout distance from toe (m)	Dump height (m)	Comments
BCMEM (2012) number	BCMWRPRC (1995) event number	BCMWRPRC (1992a) record number						
15	9	4	30 Jan 1984	2-SPOIL	775 000	140	220	Foundation frozen during construction forming an impermeable boundary below dump, with resultant increase in foundation pore pressures during groundwater recharge; slow creeping movement occurring over a period of several hours. Basal sliding. Pore water pressure under frozen crust.
41	–	27	1 May 1984	D – active at time of failure	300 000	200	200	Rapid loading of relatively fine material including rehandle and fine coal material; over-steepening due to apparent cohesion; failed along zone of poor-quality material. Sliver failure within dump material. Meltwater is considered to be significant. Visual observation of face bulging and over-steepening at the crest were made prior to failure.
16	5	–	24 Jul 1984	BROWNIE	300 000	470	260	Groundwater seen flowing from toe a few days earlier. Did not reach crest.
50	112	37	17 Aug 1984	1690 Marmot North Dump	1 500 000	300	250	Two dozers retreated just prior to failure due to crest movement. Monitor had been pulled to dress dump up. High percentage of wet fine-grained material and readily degradable carbonaceous mudstone; large strains due to rapid loading; dump height. Series of three or four rotational failures within dump material. Regressive with several distinct slumps.
17	3	11	21 Sep 1984	Brownie	400 000	900	215	Primarily related to steep foundation topography and high strains due to loading rate. Basal sliding along foundation contact. Crest deformation rates in excess of 1400 mm/d 4 days prior to failure. Dumping stopped at 600 mm/d.
33	–	17	31 May 1985	B – active at time of failure	2 500 000	600	140	Fine material leading to over-steepening; promontory developed with little support; saturated, weak foundation soils; foundation seepage pressure; additional load applied after initial deformation. Toe failure due to yielding foundation causing loss of support and non-circular rotation within dump material.
18	16	–	17 Jun 1985	Blackfill	55 000	475	140	Delayed failure; rapid; suspect bulging face.
51	–	–	18 Jun 1985	1705 WN Dump	100 000			Due to waste dumped rapidly on a new lift; small cracks and bulges noticed; at 10:30 am monitor was replaced on lift; at 11:45 pm dozer operator reported crest failure, dump closed at 01:00 am when dump failed again.
19	7	–	29 Jun 1985	Brownie	50 000	360	200	High rates and sliver failures all month.
34	89	21	1 Jul 1985	NORTH – Active at time of failure	200 000	500	275	Over-steepening due to percentage of fine grained material in dump; soft foundation soils at toe resulting in liquefaction upon rapid loading. Sliver failure causing sudden loading of weak soils below dump toe resulting in flow of debris. Steep foundation. Poor rock; high load rate.

Reference number								
BCMEM (2012) number	BCMWRPRC (1995) event number	BCMWRPRC (1992a) record number	Date of event	Dump name	Failed volume (m³)	Runout distance from toe (m)	Dump height (m)	Comments
20	6	–	16 Aug 1985	Brownie	140 000	360	215	Very steep upper foundation.
52	54	36	9 Sep 1985	1660 Marmot North Dump	3 000 000	2000	245	Elevated dump pore pressure due to infiltration; snow contained in dump forming a continuous weak plane; wet fine material; foundation seepage pressure; dump height. Damage to powerline and sedimentation pond. Rotational toe failure causing loss of support for mass above and failure along a plane of weakness within dump. Failed within dump due to wet fines.
21	11	12	16 Sep 1985	Brownie	225 000	900	215	Primarily related to steep foundation topography and loading rate. Basal sliding along foundation contact. Crest deformation rates in excess of 1400 mm/d 4 days prior to failure. Dumping stopped at 400 mm/d.
53	113	38	4 Oct 1985	1690 Marmot North Dump	2 200 000	1200	250	Three drainage courses existed beneath dump. Large percentage of fines; internal pore pressure caused by melting snow/ice within dump; continuous layer of snow along winter construction slope. Rotational slump of toe material causing loss of support for mass above and planar sliding along weak plane. Heavy precipitation in last 2 months noted. Poor rock. Previous slip.
54	–	–	16 Oct 1985	1705 WN Dump	300 000	250		A 200–300 m runout, 100 m crest length.
55	–	–	10 Nov 1985	1705 WN Dump	40 000	200		Due to high loading rate of dump. Lift put on in one area for four shifts. Large bulge visible prior to failure.
35	–	16	11 Nov 1985	A – active at time of failure	200 000	500	150	Poor-quality material including coal stringer material; high loading rate; additional load applied after initial deformation. Failure characterised by a rotational slump, indicative of relatively homogeneous fine grained soils. Foundation not a factor in failure.
–	15	–	16 Nov 1985	B Stone			300	Delayed failure.
–	149	–	16 Jun 1986	Blaine		250	230	
56	48	35	21 Jun 1986	1640 Marmot Dump	450 000	300	200	High loading rate and a history of large strain deformation; high precipitation; redistribution of dump stresses causing rapid loading of foundation and the generation of pore pressure. Toe failure causing loss of support for mass above and planar failure along a plane of weakness within the dump. High rates over past 3 months. Toe bulge. Foundation pore water pressure.
57	49	44	30 Jun 1986	1705 WN Dump	1 200 000	1700	180	Pore pressure within dump material caused by infiltration; steep foundation slopes, redistribution of stresses, history of high deformation and slides in previous year. Rotational, through dump material. Toe 'exploded', debris flow type velocity of 70–90 km/h.

Reference number								
BCMEM (2012) number	BCMWRPRC (1995) event number	BCMWRPRC (1992a) record number	Date of event	Dump name	Failed volume (m³)	Runout distance from toe (m)	Dump height (m)	Comments
36	79	22	12 Jul 1986	East Dump	240 000	400	155	Over-steepening due to percentage of fine grained material in dump; soft foundation soils at toe resulting in liquefaction upon rapid loading. Sliver failure causing sudden loading of weak soils below the dump toe resulting in flow of debris.
22	–	8	16 Jul 1986	Blain	4 000 000	350	230	Foundation frozen during construction forming an impermeable boundary below dump toe; increase in foundation pore pressures due to frozen boundary and spring melt causing groundwater recharge. No dumping for 3 weeks prior to failure. Toe failure causing loss of support for dump mass above.
42	–	26	11 Aug 1986	C – active at time of failure	600 000	200	220	High fines content and precipitation leading to over-steepening of crest due to apparent cohesion; rapid loading. Series of sliver failures within dump material. Total of 10 mm precipitation during days prior to failure.
–	12	–	14 Aug 1986	13 Seam	250 000	250	180	Groundwater seepage on face, inactive 3 months, toe not involved.
–	70	–	29 Aug 1986	Brownie			215	Dump height may be higher.
23	14	–	5 Oct 1986	Blackstone	115 000	515	300	Failure did not involve toe. Inactive 4 months. Blasting.
–	13	–	23 Nov 1986	Brownie	45 000		290	Crest unaffected; inactive 2 weeks; snowmelt; high load rate.
–	–	2	24 Nov 1986	B – active at time of failure	150 000	550	210	
37	80	–	12 Feb 1987	East	580 000	250	119	Poor-quality debris at toe; steep terrain; possibly affected by previous slips.
–	81	–	7 Mar 1987	East	285 000	170	85	Poor-quality debris at toe; frozen foundation pushed ahead.
24	–	6	15 Aug 1987	A – active at time of failure	500 000	240	220	Foundation frozen during construction forming a boundary below dump; high precipitation with increase in foundation pore pressure; slow creeping movement. Dumping was suspended 5 days prior to failure since rates were above 600 mm/d; 7000 mm/d prior to failure. Non-circular rotational slump.
58	–	32	16 Aug 1987	1615 Marmot	100 000	250	175	Loading rate; pore pressure in foundation due to redistribution of stress; rain water infiltration into dump; seepage pressure at toe area due to groundwater recharge. Slow and gradual plane failure along plane of weakness within dump material.
25	–	7	10 Sep 1987	A – active at time of failure	1 000 000	190	205	Pore pressure generation associated with rapid loading of previous failure debris from the 15 August 1987 failure and in the saturated soils of the foundation. Non-circular rotational slump. Movement rates exceeded 600 mm/d 2 days prior and exceeded 1200 mm/d for 20 hours prior to failure.

BCMEM (2012) number	BCMWRPRC (1995) event number	BCMWRPRC (1992a) record number	Date of event	Dump name	Failed volume (m³)	Runout distance from toe (m)	Dump height (m)	Comments
59	–	–	22 Sep 1987	1660 WN Dump	100 000			Due to steep topography and poor dump material. A 350 m runout, 100 m of crest length. Closed 3 days prior to failure. Last reading was 69 cm/h.
30	–	9	21 Oct 1987	B – active at time of failure	1 000 000		270	Runout contained by toe dike, lateral spreading. Pore pressure response in foundation and 1986 failure debris, induced by strain rate and increased rate of loading. Yielding toe/planar type failure. Recent approval to operation at 1200 mm/d from 600 mm/d.
60	157	42	7 Nov 1987	1660 WN Dump	5 600 000	2000	255	Loading rate causing increased pore pressure generation in foundation material; reduced shear strength of dump material due to elevated phreatic surface; water disposal from mine facilities. Toe failure along foundation contact causing loss of support and rotational slump through dump material. Runout channeled in stream bed, over and past previous failure.
61	–	33	5 Dec 1987	1615 Marmot Dump	200 000	100	175	High loading rate causing the generation of pore pressure, thick zone of saturated fines at dump face; loading remnant bulge of previous failure plane. Creeping plane failure.
62	–	–	12 May 1988	1675 WS Dump	80 000	800	200	High percentage of fine grained material causing over-steepening; lack of toe support; spring melt adding to material moisture content and foundation seepage pressures. Toe failure causing loss of support for mass above and failure along plane of weakness.
–	158	–	13 May 1988	1680 WW	80 000	405	240	Pore pressure in colluvium.
–	–	43	21 May 1988	H – active at time of failure	80 000	800	200	
43	–	31	23 May 1988	G – active at time of failure	230 000		20	Exhibited heaving of natural soils below dump toe instead of large displacement. Increased foundation pore pressure due to groundwater recharge; saturated fines in dump. Slumping in dump material was rotational, in foundation was transitional.
44	–	29	20 Aug 1988	F – active at time of failure	90 000	100	150	Increased pore pressure within the dump itself; high fines content; captured snow adding to moisture; foundation seepage zone in natural gully; over-steepening of crest; steep foundation especially in toe region. Rotational slump within dump material.
63	–	34	7 Sep 1988	B – active at time of failure	450 000	100	190	Wet fine material placed on dump face 2 months previously; high loading rate; continued loading to maintain platform grade. Sliver failure.
26	–	16	9 Oct 1988	Brownie	8 000 000		350	The failure can be attributed to toe failure brought about by an increase in foundation pore pressures at the toe, causing a loss of support for the mass above. Due to rapid redistribution of stresses within the dump. Double wedge, or yielding foundation soils at the toe foundation and loss of support for mass above permitting basal sliding along foundation.

Reference number								
BCMEM (2012) number	BCMWRPRC (1995) event number	BCMWRPRC (1992a) record number	Date of event	Dump name	Failed volume (m³)	Runout distance from toe (m)	Dump height (m)	Comments
64	–	–	4 Jun 1989	835 Shikano South Dump	10 000	75		Due to poor foundation material, high dumping rates and unfavorable dump advance direction.
38	–	–	1 Jul 1989	North Dump?				This event inferred from report on 22 November 1989 event.
65	–	–	28 Sep 1989	1650 WS Dump	120 000	250		Due to overloading short crest and fines over-steepening face.
27	–	15	26 Oct 1989	South Spoil	2 500 000	1200	435	High deformation leading to strain softening along foundation contact; overestimation of foundation shear strengths and 3D confining effects; high loading rate during winter months and toe advance over frozen ground. Yielding toe/basal along foundation contact.
–	155	–	29 Oct 1989	South	2 500 000	500	390	Review report.
66	–	–	8 Nov 1989	1570 P1 Dump	85 000	200		Due to poor foundation conditions.
39	156	20	22 Nov 1989	North Dump (2000 in South Cataract Creek)	2 000 000	1750	250	Liquefaction of soft saturated organics and debris at toe; high loading and strain rate; poor material quality. The cause of the failure appears to be due to the presence of saturated soils and debris at the toe. Lateral yielding of foundation toe soils at the contact causing loss of support for the mass above and rotational slump through dump material. Excess pore water pressure in foundation organics.
67	–	39	13 Mar 1990	Shikano South Dump	350 000	100	25	Strain-induced pore pressure in clay silt; poor pore pressure dissipation in clay-silt; groundwater recharge. Failed into Sediment Pond 1. Dozer operator sensed movements on dump and retreated. Rotational – series of rotation slumps through dump and foundation. Dump was located on floodplain of river.
45	165	23	24 Mar 1990	A – active at time of failure	200 000	500	170	High pore pressure in toe of adjacent dump, overloaded suddenly by new dump construction; steep slopes; fine material; high loading rates. Toe failure of adjacent dump, progressive non-circular rotational slump within new dump construction, saturated flow failure. Steep foundation, poor waste, 1 m/d crest advance.
68	–	–	19 Apr 1990	1580 WN Dump	40 000			Due to steep topography in combination with weak soil mantle. Developed quickly, monitors on either side showed low movement, no monitor in centre. DZ118 caught in slide, operator walked down valley to escape.
46	166	24	2 May 1990	A – active at time of failure	200 000	750	170	High pore pressure in toe of adjacent dump, overloaded suddenly by new dump construction; steep slopes; fine material; high loading rates. Toe failure of adjacent dump, progressive non-circular rotational slump within new dump construction, saturated flow failure.

BCMEM (2012) number	BCMWRPRC (1995) event number	BCMWRPRC (1992a) record number	Date of event	Dump name	Failed volume (m³)	Runout distance from toe (m)	Dump height (m)	Comments
	Reference number							
69	–	40	5 May 1990	Shikano South Dump	1 500 000	600	25	Liquefaction within sand layers due to strain generated pore pressure; groundwater recharge. Spring meltwater contributing to groundwater recharge and increased foundation seepage pressure. Horizontal translation along sand parting layer. Failure was 0.5 km long and 0.5 km wide. About 5–10% moved into and across Murray River. No flow for 8 hours.
70	–	–	14 May 1990	1570 P1 Dump	250 000	75		Due to high loading rates, build-up of fines due to dump cutting, and high rainfall.
28	153	14	29 May 1990	South Spoil – roadway fill access	300 000	300	130	Caused by high pore pressure due to excess precipitation and runoff (initially under frozen foundation, and later in dump material); loading failure with excess material to bring up roadway grade after initial slump. Basal sliding. Steep foundation; possible pore water pressure in colluvium; fines in waste: possible sliver failure.
–	154	–	7 Jul 1990	SS STG 2	300 000	350	140	Steep toe; wet weather causes excess pore water pressure in colluvium.
71	–	–	12 Sep 1990	1570 P1 Dump				Due to poor foundation and fines in dump.
47	–	25	29 Jan 1991	B – active at time of failure	400 000		195	High loading rate; winter construction. Initially rotational slump, progressing to a larger plane failure within dump material.
72	–	–	2 Feb 1991	Shikano 770 Haul Road	3500			Due to dumping on flood plain of low shear strength material. Dozer was pushing hole in berm for rainwater runoff.
29		10	6 Mar 1991	Brownie Spoil	30 000 000		300	Failure caused uplift and thickening of previous debris. Caused by generation of pore pressures within the toe foundation contact zone, due to rapid redistribution of stresses within the dump, or strain-induced pore pressure of the foundation. Double wedge, or yielding foundation soils at the toe foundation and loss of support for mass above, permitting basal sliding along foundation.
74	–	–	27 May 1991	1530 P1 Dump	250 000			Due to spring runoff, poor foundation conditions, and dump cutting. Occurred without warning. Concave dump profile preceding failure. Failed at toe, dump platform remained stable.
75	–	–	1 Jan 1992	Brownie Spoil Stage V	30 000	200		Small relatively new spoil, about 200 m high in steep foundations greater than 30 degrees, material ran down two gullies.
76	–	–	1 Mar 1992	2116	100 000			Low–normal range mobility.
77	164	–	11 May 1992	North Cougar Access Rd	150 000	270	88	Not permitted; normal range mobility; one fatality – a mechanic who was on way to a shovel at shift change; access road crossed toe of dump. Mechanic's vehicle swept off road and buried; dump was inactive ~1 year. Steep foundation, poor material, dormant, fatality.

BCMEM (2012) number	BCMWRPRC (1995) event number	BCMWRPRC (1992a) record number	Date of event	Dump name	Failed volume (m³)	Runout distance from toe (m)	Dump height (m)	Comments
78	–	–	7 Jun 1992	Taylor 1857 Access	2000			Crest slough after heavy rain.
79	–	–	1 Jul 1992	S or T 19	5000			Low dump over peat bog.
80	–	–	1 Aug 1992	H1	50 000			Recontouring till/dump onto weak foundation.
81	–	–	1 Sep 1992	Ingerbell	100 000			Dump was inactive ~10 years; suspect pore pressure in glacier-lacustrine, terrace in foundation.
82	–	–	24 Dec 1992	2116	750 000			Low–normal range mobility, runout over old debris.
83	–	–	10 Feb 1993	Brownie Stage 2	300 000	100		Low–normal range mobility, runout over old debris.
84	–	–	30 Apr 1993	Cataract North Access	50 000			Spoil was closed at beginning of day shift when cracking and berm settlement was observed during the first check; at 2:35 pm the north 15 m of the access road slumped and ran out 40 m.
85	163	–	31 May 1993	South Stage 1	8 000 000	375	385	Normal range mobility; flow stopped by toe dykes. Pore water pressure is surficial found soils.
86	–	–	9 Jun 1993	1434 Ph I	150 000	400		The dump was closed on June 8 due to excessive movement rates and had failed some ten hours later in morning of June 9; approx. 150 000 bcm of waste rock moved down the dump face, running out to 400 m beyond the original dump toe.
87	–	–	20 Aug 1993	Beach Dump	250 000			Marine dump; sections 24.5–26.5.
88	–	–	9 Sep 1993	Hennetta East	7 000	70		Waste moved 40 m but foundation moved 70 from toe.
89	–	–	7 Oct 1993	Brownie Stage 1	800 000	200		Runout over old debris.
90	–	–	1 Nov 1993	Cataract South 2060	75 000			Rapid slump/creep.
91	–	–	10 Nov 1993	Mesa 1630/40	700 000			No records on file.
92	–	–	12 Nov 1993	Cataract South 2060	20 000			Crest slump/creep.
93	–	–	30 Dec 1993	Wolverine South 1533	750 000	1000		Last used January 1993 (140 bcm) then September 1992; lots of fines, steep foundation 270, pore pressure? Failure between Dec 30, 1993, and Jan 15, 1994.
94	–	–	24 Apr 1994	Cataract North 2060	1 000 000	1500		Not permitted to elev. 2,060 steep foundation pore pressure in foundation colluvium; shearing in bedrock foundation.
95	–	–	27 Apr 1994	Cataract North 2075	100 000			Rapid loading; weak soil foundation?
96	–	–	9 May 1994	Cataract North 2060	1 000 000	1500		Adjacent to April 24 failure; runout in 2 drainages; south lobe over previous runout in North Fork Cataract Creek.
97	–	–	1 Jun 1994	Mesa North 1570	500 000			Low mobility.

Reference number								
BCMEM (2012) number	BCMWRPRC (1995) event number	BCMWRPRC (1992a) record number	Date of event	Dump name	Failed volume (m³)	Runout distance from toe (m)	Dump height (m)	Comments
98	–	–	7 Jun 1994	Cataract North 2075	250 000			High mobility; confined, stopped by Swift Creek. Crossover/rock drain; flowed 35 m up 370 face; scoured weak foundation along runout path.
99	–	–	12 Jun 1994	Cataract North 2075	100 000			Normal mobility.
100	–	–	26 Jun 1994	Shikano 803 Road	30 000	145		The 803 roadside pad failed due to the presence of clay-silt material in foundation; a 58 m length of crest broke away and traveled 145 m towards the west, overrunning material from previous slide.
101	–	–	29 Jun 1994	Wilson Creek	10 000			Entrance to rock drain plugged by debris; washout in heavy rain – 650, 665 platforms.
102	–	–	30 Jun 1994	Cataract North 2060, 2075	250 000			After a day and a half of steadily increasing rates on #26 monitor, the northeast corner of the spoil ran into the valley below and about 400 m down the South Fork of Swift Creek drainage.
103	–	–	1 Jul 1994	Cataract South 2050	750 000			After a day and a half of steadily increasing rates on #11 and #15 monitors, the southeast portion of the spoil ran into the South Cataract Creek area.
104	–	–	10 Jul 1994	Cataract North 2000 Access	75 000			Sliver failure.
105	–	–	16 Aug 1994	South Stage 2	250 000			Sliver failure blocked drainage cross in previous debris ditch.
106	–	–	22 Nov 1994	Mesa 1518 P2	1 600 000	300		The 1518P2 dump failed and slid for 275 m to the north, over material from a previous slide; dump was closed for 3.5 days prior to the failure.
107	–	–	10 Dec 1994	Road to 1685 North	90 000	20		Rapid loading, previously disturbed wet and weak foundation; foundation heave beyond toe.
108	–	–	22 Jan 1995	Cataract South 1995 Access	500 000			An 80 m long by 15 m deep section of crest failed; runout stopped 300 m from the impact barrier.
109	–	–	6 Apr 1995	Wolverine South 1473	10 000	150		Runout down steep slope.
110	–	–	1 May 1995	Henretta				Henretta dragline pad failure – wiped out diversion and killed all fish in creek.
111	–	–	18 May 1995	Mesa 1570 P2	800 000	500		Included 67% waste, 33% foundation.
112	–	–	11 Jun 1995	Raven North	700 000			Slump, no runout.
113	–	–	21 Aug 1995	Cataract North 2075 (into Swift Ck)	700 000	500		Normal range mobility; ran up opposite hill, stopped 250 m short of rock drain.
114	–	–	1 Oct 1995	Lindsay Creek CRD	1500			Failure in clayey silt foundation excavation.

BCMEM (2012) number	BCMWRPRC (1995) event number	BCMWRPRC (1992a) record number	Date of event	Dump name	Failed volume (m³)	Runout distance from toe (m)	Dump height (m)	Comments
115	–	–	8 Nov 1995	UNNCK 2038				Initial platform failed; normal range mobility; covered road at 1855 switchback; road closure procedures not followed.
116	–	–	9 Nov 1995	Lindsay Creek CRD	1500			Second failure liquefied and covered backhoe.
117	–	–	29 Nov 1995	Tinkerbell	1000			Dump above main mine access surface gully from heavy precipitation/ponding on dump.
118	–	–	17 May 1996	UNNCK 2038	568 000	2800		Very mobile; confined, dynamic friction angle ~9°; travelled to Station 0; procedures not adequate for this mobility.
119	–	–	22 May 1996	Brownie Stage 2A	600 000	600		Runout over old debris; dump was inactive since Nov 1995.
120	–	–	3 Jun 1996	MSA North	1000			Gully erosion, small volume initiated by blocked culvert; very mobile finest water.
121	–	–	4 Jun 1996	West side	20 000			Very mobile, erosion in same gully as last year's flood (second gully south), rain, freshet, pore pressure in gully bottom.
122	–	–	5 Jun 1996	Brownie Stage 4	800 000			Rapid creep – 140 m/d.
123	–	–	10 Sep 1996	East 2028 (7 Pit)	370 000	700		Two events – first triggered second in rapid succession; dump predicted to fail with average mobility; two fatalities in a logging crew who were working in the valley bottom.
124	–	–	5 Oct 1996	South end, west side				Debris flows south end of west side; high sediment levels in ditch.
125	–	–	18 Oct 1996	Mesa 1353 P1	25 000	50		Spoil failure small; detected by visual inspection only.
126	–	–	27 Oct 1996	Brownie Stage 3	300 000	100		Creep failure over several days.
127	–	–	4 Nov 1996	Brownie 4	1 900 000	400		Creep failure over several days.
128	–	–	6 Nov 1996	Wolverine 1475	200 000	450		Run up and change of direction in runout, toe heave, failure in foundation.
129	–	–	30 Nov 1996	East 2028 (7 Pit)	80 000			No injuries/equipment damage; spoil monitors indicated high and increasing rates; spoil closed at time of failure.
130	–	–	10 Dec 1996	East 2028 (7 Pit)	90 000			Same location as 10 Sep 1996 event.
131	–	–	13 Dec 1996	East 2028 (7 Pit)	20 000			Extreme south end.
132	–	–	30 Jan 1997	East 2028 (7 Pit)	1 000 000	800		No injuries/equipment damage; spoil monitors indicated high and increasing rates; spoil closed at time of failure.
133	–	–	31 Jan 1997	East 2028 (7 Pit)	700 000	1000		No records on file.
134	–	–	19 Feb 1997	Bodie 9	8000	10		Failed when toe crossed over steep bank of Bodie Creek; to be expected; no damage to rock drain or future dump stability.
135	–	–	1 Mar 1997	East 2028 (7 Pit)	700 000	1000		No injuries or equipment damage; spoil monitors indicated high and increased rates, monitor rates 1–2 weeks prior without failure; spoil closed.

BCMEM (2012) number	BCMWRPRC (1995) event number	BCMWRPRC (1992a) record number	Date of event	Dump name	Failed volume (m³)	Runout distance from toe (m)	Dump height (m)	Comments
	Reference number							
136	–	–	1 Apr 1997	Main waste dump	100 000	500		Rapid creep of wet overburden and peat over several days.
137	–	–	1 Apr 1997	TSF Ov Dump in pond	10 000	50		Several failures (three to five) of wet overburden from TSF borrow areas, on weak foundations including peat and organic.
138	–	–	6 Apr 1997	East 2028 (7 Pit)	100 000	600		From north portion.
139	–	–	9 Apr 1997	Bodie 3	25 000	70		No impacts, failed within limits and away from western toe; failed through foundation silt deposit of Pod I.
140	–	–	30 Apr 1997	East 1992 Spoil (7 Pit)	400 000	1200		No injuries/equipment damage; large sliver failure ~150 m wide × 10 m deep; new access to lower platform; ran over top of 1720 at Southend slide across Corbin Creek.
141	–	–	28 May 1997	North Horseshoe R. 2122	150 000	500		Contractor dozer operator on dump; inadequate monitoring and inadequate training and awareness of dump hazards.
142	–	–	31 May 1997	Mesa P1 1380	100 000	300		Sliver failure.
143	–	–	9 Jun 1997	Bodie 9	600			No impacts or runout; normally not considered a reportable incident.
144	–	–	10 Jun 1997	East 1992 (7 Pit)	1 400 000	1450		Lots of advance warning; two failures, one ran across valley, the other ran down valley to 1700 platform.
145	–	–	11 Jun 1997	Cataract North 1975	800 000	1500		Into Swift Creek and blocked entrance to rock drain. Impacts?
146	–	–	14 Jun 1997	Babcock Little Windy 1546	30 000	150		Inadequate contractor training and awareness of dump hazards; inadequate visual monitoring as wirelines inappropriate for early stage dump development.
147	–	–	17 Apr 1998	Baldy 6	5000	40		Dump failure occurred over crest length of 40 m on south side of Baldy 6 dump; Bodie Creek not impeded; failure caused by previous dozer cutting, which loaded the face with fine material, dumping of fine material and poor foundation.
148	–	–	17 Apr 1998	Baldy	30 000	10		Progressive silver failure; fine saturated material, possibly high loading rates for this material.
149	–	–	1 May 1998	Upper Waste Dump	50 000	100		Wet, mixed organic soil, overburden and waste rock circular failures or slump and flow, probably more than once.
150	–	–	24 Jun 1998	Brownie Spoil Stage V				No injuries/equipment damage; creep type spoil failure at approx. 10 pm; height of the spoil approx. 145 m and failure runout angle was 29°; toe of runout remains within permitted spoil limits; high monitor rates triggered evacuation prior.
151	–	–	4 Jul 1998	Baldy 6	80 000			No injuries/equipment damage; classic foundation failure through wet surface organic soil up to 3 m thick; small dump; less than 50 m high with short crest length and high loading rate from P&H 2800 shovel; dump failed quickly after one high monitor reading.

Reference number								
BCMEM (2012) number	BCMWRPRC (1995) event number	BCMWRPRC (1992a) record number	Date of event	Dump name	Failed volume (m³)	Runout distance from toe (m)	Dump height (m)	Comments
152	–	–	14 Aug 1998	Brownie Spoil Stage V	100 000			No injuries/equipment damage; creep type spoil involving 100 000 bcm of material; last load dumped on spoil 8:30 am Aug 13; rates gradually increased to +12 000 mm/d by 7 am on 14 Aug; most of toe displacement occurred 5–10 am on 14 Aug.
153	–	–	24 Aug 1998	Henretta Toe Berm Spoil				No injuries/equipment damage; small spoil failure at 2:15 am; small volume involved and did not travel beyond exfiltration ditch; Henretta Creek not impacted.
154	–	–	29 Sep 1998	West Line 1762	50 000			There are no plans to conduct any rehabilitation work in this area at this time as the dump is scheduled to be extended in the future as part of the long range plan.
155	–	–	22 Nov 1998	Bodie 5 Dump				No injuries/equipment damage; a 75 m crest length of dump sloughed 2–3 m back into the dump face; three monitors on the dump crest didn't pick up any movement prior to incident, which indicated that the failure was localised.
156	–	–	17 May 1999	Cougar North 1885 Spoil	400 000			No injuries/equipment damage; spring freshet high groundwater, concentrated point dumping and poor-quality material; failure contained by natural topography; high settlement rates recorded on 12th, increasing to max. rates 21 and 31 m/d; no impacts on mine plan.
157	–	–	19 May 1999	Horseshoe Ridge		150		No injuries/equipment damage; small dump failure toe of dump, runout material moving approx. 150 m; main portion of dump OK; resumed spoiling on section of dump which toes out in different direction.
158	–	–	28 May 1999	1540 Big Windy dump	75 000	500		No injuries/equipment damage; warming trend 5 days previous caused increased snowmelt and wet toe conditions; progressive shearing in water softened foundation (till); movements accelerated over a 7 hour period on date of failure.
159	–	–	9 Jun 1999	Big Windy Dump (Babcock)	7500	100		No injuries/equipment damage; dump developed toe bulge above bottom of drainage; toe had began to rise up opposite slope; bulge then failed taking small portion of face down drainage path; attributed to wet overburden fines in failed face.
160	–	–	10 Jun 1999	1980 spoil waste dump	200 000	200		No injuries/equipment damage; 200 000 bcm loose material in 1980 spoil failed and flowed into Corbin Valley approx. 200 m from original toe of spoil; access to area below spoil blocked since construction began.
161	–	–	12 Jun 1999	South Spoil 1970 West Spoil	800 000	243		No injuries/equipment damage; waste failed from south-west corner of South Spoil from 1970 spoil elevation; rail access to property closed temporarily to June 13 and vehicular access diverted.
162	–	–	15 Jun 1999	Surface waste dump	20 000			No injuries/equipment damage; waste dump failed, 20 000 tonnes, wet weak foundation soils.

Reference number								
BCMEM (2012) number	BCMWRPRC (1995) event number	BCMWRPRC (1992a) record number	Date of event	Dump name	Failed volume (m³)	Runout distance from toe (m)	Dump height (m)	Comments
163	–	–	15 Jul 1999	1980 Spoil	100 000	200		No injuries/equipment damage; approx. 100 000 lcm of material in 1956 spoil failed and material flowed into the Corbin Valley, approx. 200 m for original toe of spoil.
164	–	–	12 Aug 1999	North Face of Henretta Ridge	30 000	550		No injuries; equipment damage; RS#1 blast triggered slide ~30 000 bcm of mine waste and original ground on north face of Henretta Ridge; debris flowed ~550 m and blocked north access road; no evidence of instability prior to blast; 25 mm rain had accumulated.
165	–	–	7 Nov 1999	South End of Brownie Stage 5 Spoil	50 000	27		No injuries/equipment damage; 50 000 lcm waste failed in a creep even from south crest of Brownie Stage 5 spoil; no impact to water quality.
166	–	–	3 Dec 1999	1980 Spoil	130 000	100		No injuries/equipment damage; 130 000 bcm material in 1980 spoil failed; movement in form of silver failure; affected dump length approx. 160 m and 15 m into dumping platform itself at widest point.
167	–	–	11 Jan 2000	Waste spoil	144			No injuries/equipment damage; dozer pushing off wet till and snow; section of dump crest worked on by dozer cracked and slumped down approx. 3 m and excavator required to dig/pull out dozer.
168	–	–	11 Feb 2000	Babcock LW1600 Dump	300 000	300		No injuries/equipment damage; approx. 150 m crest length failed, ran out 300 m at toe; portion of failed dump closed and bermed off; portions from original crest continue to scarp off.
169	–	–	28 Mar 2000	Bodie 5 Dump, 1440 Platform				Deformation of toe of Bodie Dump – cracks were visible in areas previously covered by snow; area adjacent to toe not been closely inspected during winter and pre-existence of cracks not ruled out; monitoring frequency increased and regrade plans made.
170	–	–	1 May 2000	Brownie Stage 5 Spoil South end	200 000			No injuries/equipment damage; no impact to water quality; 200 000 bcm of waste failed in a creep event; adequate warning of event provided by monitoring instrumentation; future events expected until more toe support is provided. Contributing factors: steep topography and excessive, strain-induced pore water pressures.
171	–	–	19 Jun 2000	2000 spoil waste dump	700 000			No injuries/equipment damage; approx. 700 000 bcm material in 2000 backfill spoil failed and flowed into mined out south portion of 14 Pit approx. 100 m from original spoil; dumping face is being re-established by dump short and push.
172	–	–	27 Jul 2000	Cougar North 1945 Spoil: Upper Swift Creek	270 000	150		No injuries/equipment damage; initially the debris moved north then turned east down the Swift Creek drainage, terminating approx. 150 m short of Swift Creek rock drain; spoil inactive 58 hours prior to failure.

Reference number								
BCMEM (2012) number	BCMWRPRC (1995) event number	BCMWRPRC (1992a) record number	Date of event	Dump name	Failed volume (m³)	Runout distance from toe (m)	Dump height (m)	Comments
173	–	–	12 Aug 2000	Brownie Stage 5 Spoil	450 000	250		No injuries/equipment damage; 450 000 lcm of waste failed from the spoil crest in a creep event; majority of material resulted in mid-slope bulge rather than displacements of toe; failed section of crest was 250 m long by 40 m deep; no impact on water quality.
174	–	–	10 Nov 2000	1956 Spoil	3 000 000	900		No injuries/equipment damage; approx. 3 000 000 lcm of material failed and flowed into Corbin Valley approx. 900 m from original toe of spoil; access to crest has been bermed off; spoil to be re-established.
175	–	–	26 Apr 2001	Horseshoe Ridge	500 000			The 2014 dump platform began accelerated settling rates on 26 April. By 29 April, access to the platform was cut off with platform having settled 45 m by the morning of 30 April. No damage; dumping was transferred to another area off the north limit of the Horseshoe Pit.
176	–	–	2 Jul 2001	Brownie 2A North Spoil: 2210 Platform	1 200 000			An estimated 1.2 million bcm of waste failed from the Brownie Stage 2A North in a creep event. No personnel were injured; no damage to equipment of infrastructure resulted. There was no impact on water quality.
178	–	–	18 Jul 2001	Wolverine North	9000	1200		Debris flow (loose 'overbank material') from pit castings was triggered by a storm event affecting south corner of the 1680 floor. Trees and a local drainage were buried. Actions to be taken by mine: the daylight contact at 1680 is to be crest rounded and sloped to direct drainage away from the affected crest; large boulders along the crest are to be pulled back 15 m.
179	–	–	5 Dec 2001	1956 Spoil	500 000			Spoil failed on the 1956 platform, and was a circular type failure. The section of platform that was affected was approx. 200 m in length and approx. 50 m of platform was affected. The face height was estimated at 200 m. No injuries or damage to equipment. Access to the crest has been bermed off. Dumping face will be re-established by dump short and push from the north end of the failure.
180	–	–	12 Aug 2002	1980 and 1956 East Spoils	1 000 000			Approx. 175 m of the most recently active 1980 spoil crest and about 200 m of the most recently active 1956 crest failed. The mine was on a regularly scheduled maintenance/vacation shutdown at the time of the failure. The failures are completely contained in the Corbin Valley south of the lower east spoils. No injuries or damage to equipment resulted from the movement.
181	–	–	19 Nov 2002	1795 Toe Berm	10 000			Material in the 1795 toe berm failed to the northwest. The toe of the failed material moved approx. 30–40 m further downhill ahead of the existing dump toe. The failure scarp was approx. 70 m in width and the face height was estimated at 80 m. There were no injuries or damage to equipment as a result of the failure.

Reference number								
BCMEM (2012) number	BCMWRPRC (1995) event number	BCMWRPRC (1992a) record number	Date of event	Dump name	Failed volume (m³)	Runout distance from toe (m)	Dump height (m)	Comments
182	–	–	1 Dec 2002	W1985CS North Access Road	14 000			Material in the W1985CS North Access Road failed to the northwest. The toe of the failed material moved approx. 100 m downhill, but came to rest 30 m short of a previous failure (8 Nov 2002). The failure scarp was approx. 70 m in width; approx. 70 m of platform was affected by the failure. No injuries or damage to equipment. Access to the crest has been bermed off. The slope will be cut with a dozer, then the dumping face will be re-established by dump short and push from the north end of the failure.
183	–	–	4 Dec 2002	1975 Toe Berm	10 000			Less than 10 000 lcm moved approx. 85 m downhill, but came to rest well within the design foot print of the original toe berm. The failure scarp was approx. 50 m wide. No injuries or damage to equipment resulted from the failure. Activity was stopped and the dumping face will be re-established by dump short and push from the north end of the failure.
184	–	–	21 Apr 2003	W1985CS North Access Road	3000	400		Material in the W1985CS North Access Road failed to the northwest during a period of elevated moisture and snow. The material moved approx. 400 m downhill, but came to rest within the west spoil toe limit. No injuries or damage to equipment. Action by mine: access restricted temporarily, installed wire line monitor, data collected regularly. No further movement observed.
185	–	–	27 Apr 2003	W1985CS North Access Road	15 000	1000		Material in the W1985CS North Access Road failed down the North Thompson drainage during night shift on 27 April 2003. The toe of the failed material moved approximately 1 km downhill and came to rest at the Upper Thompson settlement ponds. No injuries or damage to equipment resulted from the failure; however, the mud flow plugged the outlet of the North Thompson settlement pond.
186	–	–	17 May 2003	1956 Waste Rock Spoil	700 000	700		Spoil failure was circular, breaking back 30 m from the crest. Monitor showed movement up to 11 000 mm/d on 16 May. The failure appears to be caused by the build-up of water due to rain and melting snow. No injuries or damage to equipment. An extensometer was set up after the failure to monitor any continued movement. Access to the area above the failure was bermed off. Geotechnical investigation to be carried out.
187	–	–	28 May 2003	1980 Waste Rock Spoil	100 000	100		Spoil failure breaking back 6 m from the crest. Movement rates exceeded 11 000 mm/d on 27 May. Failure was likely due to a build-up of pore water pressure near the toe, causing the toe to move and drag the higher sliver failure. No injuries or damage to equipment. An extensometer was set up after the failure to monitor any continued movement. Access to the area above the failure was bermed off. Dump short and push guidelines in effect until a proper dump face is re-established.

Reference number								
BCMEM (2012) number	BCMWRPRC (1995) event number	BCMWRPRC (1992a) record number	Date of event	Dump name	Failed volume (m³)	Runout distance from toe (m)	Dump height (m)	Comments
188	–	–	22 Jun 2003	W1985CS North Access Road	15 000	600		Approx. 15 000 lcm of material in the W1985CS North Access Road failed to the west at approx. 11:20 am, after pre-existing foundation cracks were observed to be opening. The toe of the failed material moved approx. 600 m downhill but came to rest within the toe berm footprint. No injuries or damage to equipment resulted from the failure. Access to crest has been bermed off. Dumping face will be re-established by dump short and push from north end of the failure.
189	–	–	26 Jun 2003	1944 Waste Rock Spoil	30 000	100		Spoil failed to the east. The failure plane appears to be along a 3 m deep sliver. Movement rates prior to failure reached 6600 mm/d. Failure was probably due to a build-up of new material on a poor base. No injuries or damage to equipment. An extensometer was set up after the failure to monitor any continued movement. Access to the area above the failure was bermed off. Dump short and push guidelines will be in effect until a proper dump face is re-established.
190	–	–	3 Jul 2003	W1985CS North Access Road	25 000	700		Material in the W1985CS North Access Road failed down the North Thompson drainage after spoil was observed to be settling at rates up to 5025 mm/d. No injuries or damage to equipment. Area was roped off and access restricted. Area was re-sloped and a wire line monitor installed. The scarp and re-slope have been filled in with coarse rock. To alleviate high pore water pressures in the foundation material, ditching or trenching the original ground ahead of the spoil advance will be undertaken.
191	–	–	16 Oct 2003	RA2170SP/ Raven Spoil	300 000	100		Material in the RA2170SP spoil failed down the Greenhills drainage. Settlement rates increased to 1000 mm/d starting on 13 Oct. A large bulge that remained from a prior slump was the primary cause of the failure. No injuries or damage to equipment. Ongoing visual monitoring and wire line monitors already in place. Area has been bermed off and signs posted; rehabilitation of the failure scarp will begin from the north and south ends, advancing towards centre. The vertical scarp will be cut with a dozer.
192	–	–	25 Jun 2004	South Toe Berm Access Road	4000	30		Material on the slope above the south toe berm access road sloughed down onto the access road. No injuries or damage to equipment. Inspection of slope material showed a fine coaly waste and some evidence of water. Access to the south access road was blocked off. An extensometer was placed above the slough area to monitor potential movement. The sloughed waste on the road will be cleared off, and an abutment will be built with pit run waste.

Reference number								
BCMEM (2012) number	BCMWRPRC (1995) event number	BCMWRPRC (1992a) record number	Date of event	Dump name	Failed volume (m³)	Runout distance from toe (m)	Dump height (m)	Comments
193	–	–	26 Jun 2004	Pollyanna Pit	2500			Coarse rock failed along a plane within a zone of mixed rock and overburden at the Pollyanna Pit – Short Haul Dump. The waste rock was placed on a steep zone or rock and wet overburden (mainly glacial till). A weak layer in the foundation contributed to the failure. Heavy rain was present leading up to the failure, beginning one day prior to the failure. No injuries or damage to equipment. The failure zone was bermed off and dumping was moved to a new area in the same dump.
194	–	–	30 Jun 2004	Pollyanna Pit	2500			Waste rock failed at the Pollyanna Pit – Short Haul Dump due to a foundation failure. Dump loading caused the weak foundation material (a layer of organic and inorganic lacustrine silt) to 'kick out' and deposit the rock mass to the west of the June 26 failure heap. Dumping was not taking place at the time of failure. No injuries or damage to equipment. The dump was bermed off, dumping stopped, and movement monitoring initiated.
195	–	–	13 Apr 2005	Raven Spoil – South End	340 000	700		Material slid down from the crest of the RA2145 spoil. Material dumped on April 10 contained many coal fines. The fines hung up near the top of the face and eventually settled down towards the toe. In-place extensometers measured settlement levels above 1000 mm/d on April 11. No injuries or damage to equipment. Area was closed off to dumping prior to the failure. The area will continue to be monitored until all material is down at the toe. In the future, an effort will be made to mix in coarse material with the fines.
196	–	–	3 Jun 2005	Topsoil Stockpile	600			A topsoil stockpile constructed in March 2005 experienced superficial failure at the downslope toe with saturated material flowing slowly down towards Tributary 1. Failure appears due to soil thawing and the melting of snow material mixed within. No injuries or damage to equipment. A water pump was installed to reduce the water level on top of the soil pile and isolate the site from further influx of runoff. A remediation plan was developed.
197	–	–	13 Jun 2005	South Pit Waste Haul Road	100 000	125		Road embankment slumped into Milligan Creek and blocked the channel to the mid-level sediment pond after several days of heavy rain. The road was not in operation at the time of failure. Till material on the north side of the road failed due to the heavy rain. Permitted TSS loads were temporarily exceeded. Road access to the scarp and mid-pond were closed until the site was inspected. Repairs are in progress and include a new culvert to direct water north of the debris.

– = no data, d = day

APPENDIX 2
SUMMARY OF THE 2013 MINE WASTE DUMP SURVEY

The Large Open Pit (LOP) project, an international research and technology transfer project on the stability of large slopes associated with open pit mines, provided funding for the development of a reference book entitled *Guidelines for Mine Waste Dump and Stockpile Design* (hereafter 'the Guidelines'). The Guidelines covering the investigation, analyses, design, operation and monitoring of mine waste dumps, stockpile and dragline spoils were jointly developed by the Project Team, which includes:

- Piteau Associates Engineering Ltd
- Golder Associates Ltd
- Schlumberger Water Services
- Dr Oldrich Hungr
- Dr Ward Wilson
- Sherwood Geotechnical and Research Services.

The Guidelines are intended to reflect the current state of practice with respect to the investigation, design and management of mine waste dumps and stockpiles, and to be a practical reference text for both geotechnical practitioners and mine operators. As an important first step in capturing the current state of practice in waste dumps, stockpiles and dragline spoils, the Project Team conducted an online survey. The primary purpose of the survey was to obtain statistical data on waste dump and stockpile geometries, conditions and performance to support development of the Guidelines. The survey was run between April and August 2013 and was populated by a staff from a range of the LOP sponsors organisations and various consultants. Data from 69 waste dumps was provided and the following presents a summary of the results received. Data from only three dragline spoils were entered into the survey and these results are not included in this summary.

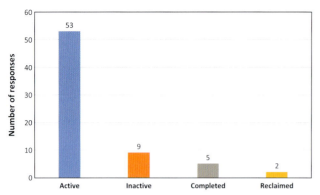

Figure A2.1: Summary of survey responses showing the status of the dump at the time the survey was completed in 2013

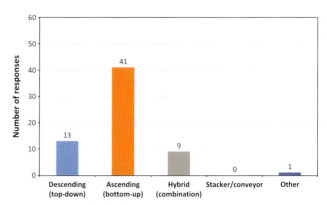

Figure A2.4: Method of dump construction

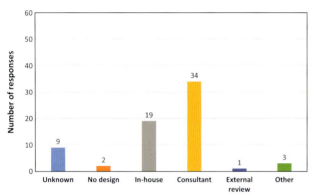

Figure A2.2: Summary of survey responses showing who the dump was designed by

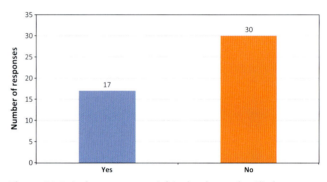

Figure A2.5: Is the waste material in the dump classified according to quality?

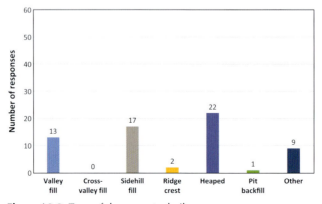

Figure A2.3: Type of dump or stockpile

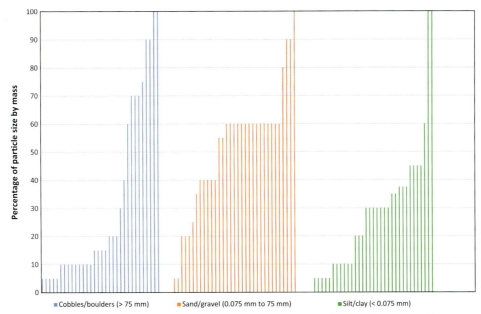

Figure A2.6: Summary of survey responses showing the distribution of the gradation of waste dump material

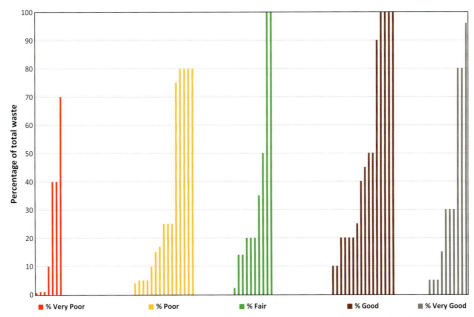

Figure A2.7: Summary of survey responses showing the distribution of the waste dump material quality by classification from Very Poor to Very Good

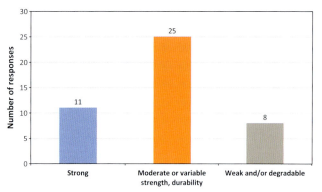

Figure A2.8: Distribution of the waste material intact strength and durability

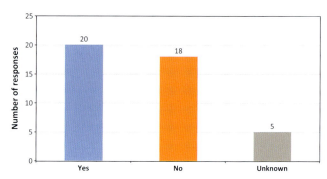

Figure A2.11: Does the waste have an acid rock drainage potential?

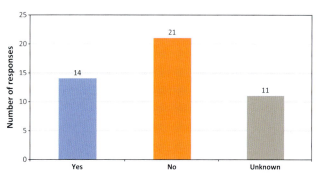

Figure A2.9: Has the potential for liquefaction of the waste material been assessed?

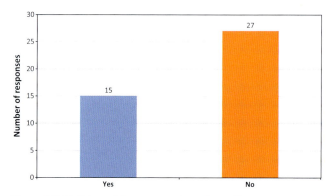

Figure A2.12: Was a risk assessment conducted?

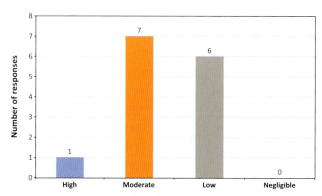

Figure A2.10: What is the liquefaction potential?

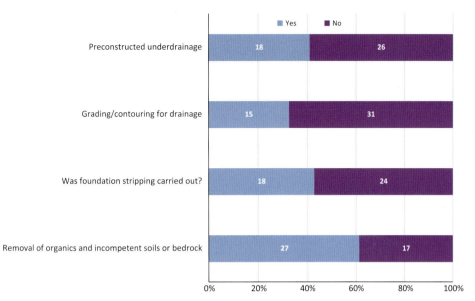

Figure A2.13: What is the method of foundation preparation for the dump?

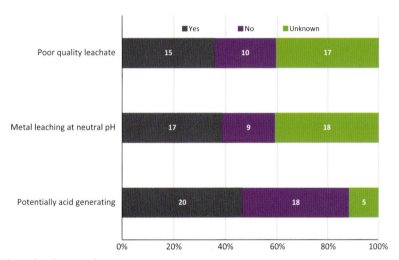

Figure A2.14: What is the dump leachate quality?

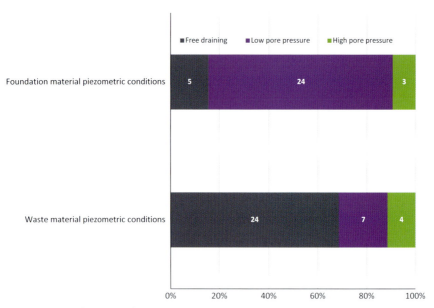

Figure A2.15: What is the dump groundwater conditions?

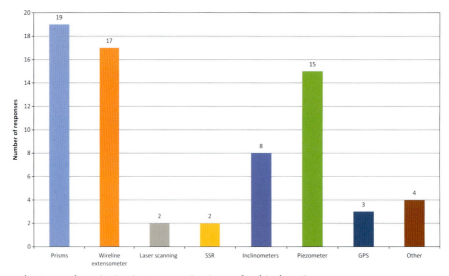

Figure A2.16: What are the types of monitoring instrumentation in use for this dump?

Table A2.1: Summary of the 2013 LOP Waste Dump Survey Part 1 – mass, geometry and average foundation slope angles

Survey number	2013 Status of dump	Dump was designed by	Type of dump	Method of construction	Current volume (bank cubic metres)	Planned volume (bank cubic metres)	Current bulking factor	Planned bulking factor	Average lift height (m)–current	Average lift height (m)–planned	Maximum lift height (m)–current	Maximum lift height (m)–planned	Maximum overall slope angle (°)–current	Maximum overall slope angle (°)–planned	Maximum vertical thickness (m)–current	Maximum vertical thickness (m)–planned	Initial (in dump initiation zone)	Upper 1/3 of foundation	Middle 1/3 of foundation	Lower 1/3 of foundation (toe)
1	Active	Other	Valley fill	Ascending (bottom-up)	–	–	1.3	1.5	100	100	200	300	29	24	100	300	–	–	–	–
2	Active	In-house	Ridge Crest	Hybrid (combination)	–	–	–	–	30	–	40	–	21	–	75	–	–	–	–	–
3	Active	Consultant	Sidehill fill	Hybrid (combination)	261 983 085	367 566 667	1.36	1.3	150	150	345	345	24.5	30	150	210	37	30	27	37
4	Active	Consultant	Ridge crest	Ascending (bottom-up)	59 500 000	128 520 000	1.7	1.68	10	10	20	10	23	21.5	60	140	4	6	6	4
5	Active	In-house	Heaped	Ascending (bottom-up)	2 000 000	2 000 000	1.2	1.2	10	10	15	15	15	15	30	30	60	–	–	–
6	Completed	Consultant	Other	Descending (top-down)	2 412 200	2 412 200	1.4	1.4	130	130	140	140	27	27	65	65	46	42	57	47
7	Active	Consultant	Heaped	Ascending (bottom-up)	74 000 000	600 000 000	–	–	15	15	45	30	31	12	60	86	8	8	7	5
8	Active	–	Heaped	Descending (top-down)	–	–	–	–	–	–	–	–	–	–	–	–	45	45	45	0
9	Active	In-house	Other	Ascending (bottom-up)	–	–	–	–	–	–	–	–	–	–	–	–	–	–	–	–
10	Active	In-house	Heaped	Ascending (bottom-up)	183 671 497	200 000 000	1.2	1.2	20	20	30	30	37	37	90	140	–	–	–	–
11	Active	No design	Heaped	Ascending (bottom-up)	–	–	–	1.3	–	20	–	–	–	–	100	–	–	–	–	–
12	Active	No design	Heaped	Ascending (bottom-up)	–	–	–	–	–	–	–	–	–	–	100	–	–	–	–	–
13	Active	In-house	Sidehill fill	Descending (top-down)	–	–	–	–	–	–	–	–	–	–	–	–	–	–	–	–
14	Completed	Consultant	Other	Ascending (bottom-up)	–	–	1.4	1.4	130	130	140	140	27	27	65	65	–	–	–	–
15	Active	Unknown	Heaped	Ascending (bottom-up)	–	–	–	–	–	–	–	–	–	–	–	–	–	–	–	–
16	Active	Consultant	Heaped	Ascending (bottom-up)	20 000 000	–	1.4	–	12	12	12	36	22	22	–	36	22	–	–	–
17	Active	Consultant	Valley fill	Ascending (bottom-up)	–	–	1.6	1.6	10	10	–	–	20	20	150	170	–	–	–	–
18	Active	Consultant	Heaped	Ascending (bottom-up)	4 619 860	–	1.9	–	15	–	15	–	32	–	100	–	0	35	25	15
19	Active	In-house	Heaped	Ascending (bottom-up)	6 627 831	–	1.8	–	12	–	12	–	30	–	50	–	0	30	20	10
20	Completed	In-house	Other	Ascending (bottom-up)	13 781 333	–	1.9	–	12	–	12	–	30	–	75	–	0	0	0	0

Survey number	2013 Status of dump	Dump was designed by	Type of dump	Method of construction	Mass and geometry												Average foundation slope angle (°)			
					Current volume (bank cubic metres)	Planned volume (bank cubic metres)	Current bulking factor	Planned bulking factor	Average lift height (m) – current	Average lift height (m) – planned	Maximum lift height (m) – current	Maximum lift height (m) – planned	Maximum overall slope angle (°) – current	Maximum overall slope angle (°) – planned	Maximum vertical thickness (m) – current	Maximum vertical thickness (m) – planned	Initial (in dump initiation zone)	Upper 1/3 of foundation	Middle 1/3 of foundation	Lower 1/3 of foundation (toe)
21	Completed	Consultant	Heaped	Ascending (bottom-up)	27 312 500	–	1.9	–	15	–	15	–	27	–	75	–	0	30	5	0
22	Active	Consultant	Heaped	Descending (top-down)	–	–	–	–	–	9	–	9	–	18	–	72	–	–	–	–
23	Inactive	Consultant	Sidehill fill	Ascending (bottom-up)	–	–	–	2	–	90	–	90	–	26	–	300	10	30	17	10
24	Inactive	Unknown	Sidehill fill	Descending (top-down)	–	–	2	–	225	–	225	–	35	–	200	–	–	–	–	–
25	Active	Unknown	Sidehill fill	Descending (top-down)	–	–	–	–	100	–	100	–	26	–	100	–	30	30	22	17
26	Inactive	Unknown	Valley fill	Descending (top-down)	–	–	–	–	–	–	–	–	–	–	–	–	20	20	15	10
27	Inactive	Consultant	Valley fill	Other	–	–	–	2	–	60	–	60	–	26	–	375	12–5	25–30	15–20	12–5
28	Inactive	Consultant	Valley fill	Ascending (bottom-up)	–	–	–	2	–	60	–	60	–	26	–	375	12–5	25–45	25–12	12–5
29	Inactive	Unknown	Sidehill fill	Descending (top-down)	–	–	–	–	–	–	–	–	–	–	–	–	25	17–25	17–10	10–5
30	Active	Unknown	Sidehill fill	Descending (top-down)	–	–	–	–	300	–	300	–	35	–	–	–	30	20–30	20–10	10–3
31	Active	Consultant	Sidehill fill	Ascending (bottom-up)	–	23 000 000	–	–	22–25	–	25	–	26–28	–	–	–	5	20	20	20–5
32	Active	Unknown	Sidehill fill	Descending (top-down)	–	–	–	–	–	–	–	–	–	–	–	–	–	–	–	–
33	Active	In-house	Valley fill	Ascending (bottom-up)	–	–	–	–	45	45	45	45	19	20	150	300	37	26	26	26
34	Active	In-house	Sidehill fill	Ascending (bottom-up)	176 785 200	351 897 183	1.3	1.3	20	20	20	20	35	20	127	207	35	35	35	35
35	Active	In-house	Sidehill fill	Ascending (bottom-up)	–	–	–	–	–	–	–	–	–	–	–	–	–	–	–	–
36	Active	External Review	Sidehill fill	Ascending (bottom-up)	–	–	–	–	10	–	–	–	21	–	235	–	–	–	–	–
37	Active	Consultant	Heaped	Descending (top-down)	20 000 000	50 000 000	1.4	–	12	12	12	12	–	22	–	36	22	–	–	–
38	Active	Consultant	Other	Hybrid (combination)	–	–	–	–	–	–	–	–	–	–	–	–	–	–	–	–
39	Active	Consultant	Sidehill fill	Ascending (bottom-up)	–	–	1.6	–	5	–	10	–	45	–	20	–	–	–	–	–
40	Completed	Consultant	Heaped	Ascending (bottom-up)	27 312 500	–	1.9	–	15	–	15	–	27	–	75	–	0	30	5	0
41	Active	Consultant	Valley fill	Descending (top-down)	–	–	–	–	–	–	–	–	–	–	–	–	–	–	–	–
42	Active	Unknown	–	–	–	–	–	–	–	–	–	–	–	–	–	–	–	–	–	–
43	Active	Other	Valley fill	Ascending (bottom-up)	–	–	–	–	45	45	45	45	30	20	370	430	–	–	–	–
44	Inactive	Consultant	Valley fill	Ascending (bottom-up)	–	–	–	–	45	–	60	–	24	–	90	–	10	3	10	10

Mass and geometry

Average foundation slope angle (°)

Survey number	2013 Status of dump	Dump was designed by	Type of dump	Method of construction	Current volume (bank cubic metres)	Planned volume (bank cubic metres)	Current bulking factor	Planned bulking factor	Average lift height (m)–current	Average lift height (m)–planned	Maximum lift height (m)–current	Maximum lift height (m)–planned	Maximum overall slope angle (°)–current	Maximum overall slope angle (°)–planned	Maximum vertical thickness (m)–current	Maximum vertical thickness (m)–planned	Initial (in dump initiation zone)	Upper 1/3 of foundation	Middle 1/3 of foundation	Lower 1/3 of foundation (toe)
45	Reclaimed	Consultant	Heaped	Ascending (bottom-up)	–	–	–	–	–	15	–	15	–	28.5	–	60	0	0	0	0
46	Active	Consultant	Valley fill	Hybrid (combination)	–	–	–	–	140	–	280	–	25	–	280	–	35	35	6	2
47	Active	Consultant	Sidehill fill	Hybrid (combination)	–	–	–	–	60	60	90	–	25	–	300	–	15	25	3.5	12
48	Active	In-house	Sidehill fill	Ascending (bottom-up)	–	–	–	–	–	–	–	–	–	–	–	–	–	–	–	–
49	Active	Consultant	Valley fill	Ascending (bottom-up)	–	–	–	–	25	–	50	–	17	–	60	–	10	10	10	10
50	Inactive	Consultant	Valley fill	Ascending (bottom-up)	–	–	–	–	25	–	60	–	21	–	220	–	14	19	5	14
51	Active	Consultant	Other	Ascending (bottom-up)	–	–	–	–	25	–	25	–	18	–	150	–	0	39	0	0
52	Inactive	Consultant	Other	Ascending (bottom-up)	–	–	–	–	30	–	40	–	23	–	80	–	0	39	0	0
53	Reclaimed	Consultant	Sidehill fill	Descending (top-down)	–	–	–	–	100	–	100	–	30	–	45	–	17	17	17	17
54	Active	In-house	Other	Hybrid (combination)	–	–	–	–	–	–	–	–	–	–	–	–	–	–	–	–
55	Active	In-house	Pit backfill	Hybrid (combination)	–	–	–	–	–	–	–	–	–	–	–	–	–	–	–	–
56	Active	Unknown	–	–	–	–	–	–	–	–	–	–	–	–	–	–	–	–	–	–
57	Active	In-house	Heaped	Ascending (bottom-up)	–	–	–	–	–	–	–	–	–	–	–	–	–	–	–	–
58	Active	Consultant	Heaped	Ascending (bottom-up)	–	–	–	–	15	15	15	30	11	22	140	170	–	–	–	–
59	Active	In-house	–	–	–	–	–	–	–	–	–	–	–	–	–	–	–	–	–	–
60	Active	Consultant	Heaped	Ascending (bottom-up)	–	–	–	–	–	–	–	–	–	–	–	–	–	–	–	–
61	Active	In-house	Other	Hybrid (combination)	98 882 000	247 205 000	18.5 TFC	18.5 TFC	85	100	105	120	25	25	340	340	0	0	0	0
62	Active	Other	Sidehill fill	Hybrid (combination)	–	–	–	–	–	–	–	–	–	–	–	–	–	–	–	–
63	Active	Consultant	Heaped	Ascending (bottom-up)	3 715 000	12 700 000	1.3	–	4	8	4	8	37	–	16	–	37	–	–	–
64	Active	Consultant	Heaped	Ascending (bottom-up)	3 858 000	4 480 000	1.25	–	4	4	4	4	25	–	12	–	25	–	–	–
65	Active	Consultant	Heaped	Ascending (bottom-up)	–	–	–	–	18	18	18	18	22	22	36–90	36–90	–	–	–	–
66	Active	In-house	–	–	–	–	–	–	–	–	–	–	–	–	–	–	–	–	–	–
67	Active	In-house	Valley fill	Ascending (bottom-up)	–	–	–	–	45	45	45	45	19	20	150	300	37	26	26	26
68	Active	In-house	–	–	–	–	–	–	–	–	–	–	–	–	–	–	–	–	–	–
69	Active	Consultant	Heaped	Ascending (bottom-up)	189 000 000	319 000 000	–	–	9	9	9	9	18	18	8	8	37	18	18	18

Table A2.2: Summary of the 2013 LOP Waste Dump Survey Part 2 – Overburden type, thickness, depth to bedrock and bedrock conditions

| Survey number | Typical overburden type and thickness (m) | | | | Depth to bedrock | Adverse bedrock conditions/structure? (Describe) |
	Organics (topsoil/peat)	Coarse grained soils (talus/ cobbles, gravel, sand)	Mixed-grained soils (colluvium/ moraine)	Fine-grained soils (clays/ silts)		
1	0	10	50	50	10	Strong schistosity and more than two cleavage directions
2	–	–	5	–	–	Some located within major geological structures
3	0	0	20	0	20	No
4	1	3	3	1	6	None
5	–	–	10	40	50	–
6	2	0	0	51	53	Nil
7	Top soil = 0.15	Gravel = 1	Hard cap = 3	B-zone = 4	35	No adverse bedrock conditions or structure identified
8	–	–	–	–	–	–
9	–	–	–	–	–	–
10	–	–	–	–	–	–
11	–	15	–	–	30	Sand and calcrete
12	–	–	–	–	–	–
13	–	–	–	–	–	–
14	–	–	–	–	–	–
15	–	–	–	–	–	–
16	–	–	–	–	–	Highly weathered Mudstone
17	–	–	–	–	–	–
18	0.0–0.5	–	0.5–3	–	3	Elevated pore pressure in altered clays
19	0.0–0.5	–	–	0.5–40	40	Elevated pore pressure in altered clay
20	0	0	0	0	–	None
21	0.0–0.5	–	0.5–10.0	–	10	Elevated pore pressure in altered clays
22	–	–	–	–	–	Highly competent
23	1	–	5	–	5–10	Adverse orientation of bedding possible
24	–	–	–	–	–	–
25	1	–	5–10	–	10	Bedding dip could be adversely oriented
26	–	–	–	–	–	Faulting
27	–	–	4.5	–	> 5	No – GSI used
28	–	–	5	–	> 5	–
29	–	–	–	–	–	–
30	–	–	–	–	–	–
31	–	–	–	–	–	–
32	–	–	–	–	–	–
33	–	–	–	–	10	Cohesion 200 kPa; friction angle 45°
34	–	–	–	50	50	No
35	–	–	–	–	–	–
36	–	–	–	–	–	–

Survey number	Typical overburden type and thickness (m)					Adverse bedrock conditions/structure? (Describe)
	Organics (topsoil/peat)	Coarse grained soils (talus/ cobbles, gravel, sand)	Mixed-grained soils (colluvium/ moraine)	Fine-grained soils (clays/ silts)	Depth to bedrock	
37	0.3	–	–	–	1.5	Weathered mudstone
38	–	–	–	–	–	–
39	–	–	–	–	–	–
40	0.0–0.5	–	0.5–10.0	–	10	Elevated pore pressure in altered clays
41	–	–	–	–	–	–
42	–	–	–	–	–	–
43	–	–	–	–	–	–
44	0.5	–	5	5	10	No
45	0	0	0	0	0	None
46	0.5	–	5	5	10	Unknown
47	0.5	–	5	5	10	None
48	–	–	–	–	–	–
49	0.5	–	5	–	5	Argillic/clay alteration
50	2	–	7	–	0–7	Argillic alteration
51	–	–	–	–	0	Argillic alteration
52	–	–	–	–	0	Argillic alteration, fault zones
53	0.5	–	5	–	10	–
54	–	–	–	–	–	–
55	–	–	–	–	–	–
56	–	–	–	–	–	–
57	–	–	–	–	–	–
58	–	–	–	–	–	–
59	–	–	–	–	–	–
60	–	–	–	–	–	–
61	NA	NA	NA	NA	0	None
62	–	–	–	–	–	–
63	Muskeg (2–4)	–	–	Clay	15–25	None
64	Muskeg (2–4)	–	–	Clay	15–25	None
65	–	–	–	10–15	15	–
66	–	–	–	–	–	–
67	–	–	–	–	10	None
68	–	–	–	–	–	–
69	2	–	–	–	–2	Flat, shallow to bedrock

Table A2.3: Summary of the 2013 LOP Waste Dump Survey Part 3 – Foundation strength, depth to groundwater and foundation preparation

Survey number	Design foundation shear strength parameters	Depth to water table (typical, below original ground surface) (m)	Foundation preparation		
			Removal of organics and incompetent soils or bedrock	Was foundation stripping carried out?	If yes, provide average depth of stripping (m)
1	–	0	No	No	–
2	Some not known	30–50	No	Yes	–
3	Bedrock: ϕ = 50°, c = 240 kPa, unit weight 22.56 kN/m^3	No	No	No	–
4	ϕ = 32°, c = 0	1	Yes	Yes	1.5
5	–	20–30 below surface	Yes	No	–
6	ϕ = 67°, c = 1300 kPa	55	Yes	No	–
7	Laterite (gravel and clay) density 21 kN/m^3, friction 35°, c = 10 kPa Cemented clay/low strength basement rock density 22kN/m^3, friction 50°, c = 45 kPa	25	Yes	Yes	1.5
8	–	–	Yes	Yes	–
9	–	–	–	–	–
10	–	–	–	–	–
11	Not known	Varies	Yes	Yes	–
12	–	–	Yes	Yes	–
13	–	–	–	–	–
14	–	–	–	–	–
15	–	–	–	–	–
16	None	54	Yes	–	0.5
17	–	–	Yes	No	–
18	ϕ = 15–20°	0	Yes	Yes	–
19	ϕ = 15–20°	0	Yes	Yes	–
20	–	0	No	No	–
21	ϕ = 15–20°	0	No	No	–
22	–	–	Yes	–	–
23	ϕ = 20–25° with no cohesion	0–25	Yes	No	–
24	–	–	–	No	–
25	ϕ = 20–25° with no cohesion	5–50	No	No	–
26	–	–	No	No	–
27	ϕ = 20–35° with no cohesion	0	–	–	–
28	GSI/RMR = 48	0	–	–	–
29	–	–	No	No	–
30	–	–	No	No	–
31	–	–	Yes	Yes	1–3 m
32	Founded on old spoil run-out	–	No	No	–
33	ϕ = 10°, C = 30 kPa	0	Yes	Yes	3
34	Unknown	30	Yes	Yes	3
35	–	–	–	–	–

Survey number	Design foundation shear strength parameters	Depth to water table (typical, below original ground surface) (m)	Foundation preparation		
			Removal of organics and incompetent soils or bedrock	Was foundation stripping carried out?	If yes, provide average depth of stripping (m)
36	–	0.6	Yes	No	–
37	–	54	Yes	Yes	0.5
38	–	–	–	–	–
39	–	–	Yes	No	–
40	$\phi = 15–20°$	0	No	No	–
41	–	–	–	–	–
42	–	–	–	–	–
43	–	–	–	–	–
44	$\phi = 32°$	0	Yes	Yes	5
45	$\phi = 45°$, $C = 1000$ kPa	20	No	No	In-pit dump
46	$\phi = 32°$, $c = 60$ kPa; $\phi = 45°$, $c = 1000$ kPa	0	Yes	No	–
47	$\phi = 32°$, $c = 40$ kPa; $\phi = 45°$, $c = 1000$ kPa	0	Yes	Yes	0.5
48	–	–	–	–	–
49	Variable based on rock mass and depth below surface due to weathering front. Lower bound is 20.5°, 115 kPa for near surface argillically altered bedrock	2	Yes	Yes	1
50	$\phi = 20.5°$, $c = 115$ kPa	2	Yes	Yes	2
51	$\phi = 29°$, $c = 165$ kPa	0	No	No	–
52	Bedrock: $\phi = 29°$, $c = 165$ kPa Faults: $\phi = 18°$, $c = 50$ kPa	0	No	No	–
53	$\phi = 32°$, $c = 150$ kPa	0	Yes	–	1
54	–	–	–	–	–
55	–	–	–	–	–
56	–	–	–	–	–
57	–	–	–	–	–
58	–	–	–	–	–
59	–	–	–	–	–
60	–	–	–	–	–
61	Exposed unaltered igneous rocks located in the bottom of Cut-21 pit have an *in situ* rock strength of 5400 lb/ft^2 cohesion and an average friction angle of 42°	5	No	No	–
62	–	–	–	–	–
63	No foundation	2	No	No	–
64	No foundation	2	No	No	–
65	Duracrust foundation: $\phi = 25°$, $c = 50$ Kpa	–	–	–	–
66	–	–	–	–	–
67	Bedrock $\phi = 45°$, $c = 200$ kPa; soil $\phi = 10°$, $c = 30$ kPa	0	Yes	Yes	3
68	–	–	–	–	–
69	–	60	Yes	Yes	1–2

Table A2.4: Summary of the 2013 LOP Waste Dump Survey Part 4 – Foundation preparation and waste material gradation

Survey number	Foundation preparation		Lithology	Alteration	Waste Material Gradation			Waste material intact strength and durability
	Grading/ contouring for drainage	Pre-constructed underdrainage			Cobbles/ boulders (> 75 mm)	Sand/gravel (0.075– 75 mm)	Silt/Clay (<0.075 mm)	
1	Yes	Yes	Iron formations waste material (itabirites and mafic intrusive dike, quartzites and phylites)	Highly weathered –class V	5	35	60	Moderate or variable strength, durability
2	Yes	Yes	Laterite soil	Minimum	–	–	–	Weak and /or degradable
3	No	No	Tobas y brechas	Silice, silice-alunita, alunitasilice, argilica, propilitica, coluvio, steamheat	70	20	10	Strong
4	Yes	Yes	Andesite, coal, mudstone and sandstone	Argilization	10	80	10	Moderate or variable strength, durability
5	Yes	No	Quartz-feldspar gneiss – garnet gneiss	–	–	–	–	Strong
6	No	Yes	Dolerite, altered andesites, porphyry	–	75	20	5	Strong
7	No	No	Andesite, diorite, dolerite	Sericite, clinozoicite and actinolite veins	70	20	10	Strong
8	No	No	–	–	–	–	–	Strong
9	–	–	–	–	–	–	–	–
10	–	–	–	–	–	–	–	–
11	No	No	Shales, dolerite	Carbonaceous shales affected	–	–	–	Moderate or variable strength, durability
12	No	No	–	–	–	–	–	Moderate or variable strength, durability
13	–	–	–	–	–	–	–	–
14	–	–	–	–	–	–	–	–
15	–	–	–	–	–	–	–	–
16	No	No	Mudstone and basalt	–	–	–	–	–
17	No	Yes	Phylites, poor iron formations and clay soils	–	20	60	20	Weak and /or degradable
18	Yes	Yes	Argillic	Yes	5	30–90	15–60	Moderate or variable strength, durability
19	No	Yes	Argillic	Yes	0	0	100	Moderate or variable strength, durability
20	No	No	Argillic	No	100	0	0	Moderate or variable strength, durability
21	No	No	Argillic	Yes	5	30–90	15–60	Moderate or variable strength, durability
22	No	No	–	–	–	–	–	–
23	No	No	Sedimentary strata	N/A	5	60	35	Moderate or variable strength, durability
24	No	No	Sedimentary strata	–	10	60	30	Moderate or variable strength, durability
25	No	No	Sedimentary strata	–	10	60	30	Moderate or variable strength, durability

Survey number	Foundation preparation		Lithology	Alteration	Waste Material Gradation			Waste material intact strength and durability
	Grading/ contouring for drainage	Pre-constructed underdrainage			Cobbles/ boulders (> 75 mm)	Sand/gravel (0.075– 75 mm)	Silt/Clay (<0.075 mm)	
26	No	No	Sedimentary strata	–	10	60	30	Moderate or variable strength, durability
27	–	–	Sedimentary	–	10	60	30	Moderate or variable strength, durability
28	–	–	Sedimentary strata	–	10	60	30	Moderate or variable strength, durability
29	No	No	Sedimentary strata	–	10	60	30	Moderate or variable strength, durability
30	No	No	Sedimentary	–	10	60	30	Moderate or variable strength, durability
31	Yes	Yes	Coal coal refuse	–	–	100	–	Moderate or variable strength, durability
32	No	No	Sedimentary strata	–	10	55	35	Moderate or variable strength, durability
33	Yes	Yes	Volcanic, diorite, tonalite	Secondary biotite, pale green mica, chlorite-epidote, feldspar destructive	90	5	5	Strong
34	Yes	No	Hard rock	–	100	–	–	Strong
35	–	–	–	–	–	–	–	–
36	No	Yes	Poor itabirites, friable siliceous itabirites, phyllites	–	20	60	20	Weak and /or degradable
37	No	No	Mudstone and basalt	–	–	–	–	–
38	–	–	–	–	–	–	–	–
39	No	–	–	–	–	–	–	–
40	No	No	Argillic	Yes	5	30–90	15–60	Moderate or variable strength, durability
41	–	–	–	–	–	–	–	–
42	–	–	–	–	–	–	–	–
43	–	–	–	–	–	–	–	–
44	Yes	Yes	Sulphides	None	0	90	10	Moderate or variable strength, durability
45	No	No	Sulphide	None	–	90	10	Moderate or variable strength, durability
46	No	No	Limestone	None	40	55	5	Strong
47	No	Yes	Limestone, intrusive, skarn	–	30	60	10	Moderate or variable strength, durability
48	–	–	–	–	–	–	–	–
49	No	Yes	Argillically altered andesite, quartz alunite, fresh andesite (toe buttress), vuggy silica	Argillic/clay	15	40	45	Weak and /or degradable
50	Yes	Yes	Argillicaly altered andesite, quartz alunite and vuggy silica	Argillic	15	40	45	Weak and /or degradable

Survey number	Foundation preparation		Lithology	Alteration	Waste Material Gradation			Waste material intact strength and durability
	Grading/ contouring for drainage	Pre-constructed underdrainage			Cobbles/ boulders (> 75 mm)	Sand/gravel (0.075– 75 mm)	Silt/Clay (<0.075 mm)	
51	Yes	Yes	Argillically altered andesite, vuggy silica and quartz alunite	Argillic	15	40	45	Weak and /or degradable
52	No	Yes	Argillically altered andesite	Argillic/clay	15	40	45	Weak and /or degradable
53	Yes	Yes	Andesite, schist	Clay	20	60	20	Moderate or variable strength, durability
54	–	–	–	–	–	–	–	–
55	–	–	–	–	–	–	–	–
56	–	–	–	–	–	–	–	–
57	–	–	–	–	–	–	–	–
58	–	–	–	–	–	–	–	–
59	–	–	–	–	–	–	–	–
60	–	–	–	–	–	–	–	–
61	No	No	Igneous diabase mixed with thin bedded calcareous siltstone	Minor clay alteration in igneous rocks and decalcification in the Calc. siltstone rocks	70	25	5	Moderate or variable strength, durability
62	–	–	–	–	–	–	–	–
63	No	No	Rock – limestone	Moderate	60	40	–	Moderate or variable strength, durability
64	No	No	Clay	None	–	–	100	Weak and /or degradable
65	Yes	–	Primary rock and oxide rock	–	–	–	–	Strong
66	–	–	–	–	–	–	–	–
67	Yes	Yes	Volcanic, diorite, tonalite	Secondary biotite, pale green mica, chlorite-epidote, feldspar destructive	90	5	5	Strong
68	–	–	–	–	–	–	–	–
69	Yes	No	Metamorphic rocks	–	–	–	–	Strong

Table A2.5: Summary of the 2013 LOP Waste Dump Survey Part 5 – Waste material quality distribution, shear strength, liquefaction potential and geochemistry

Survey number	Is the waste/ stockpile material classified according to quality?	Waste/stockpile material quality distribution					Design waste material shear strength parameters	Has the potential for liquefaction of the waste material been assessed?	Liquefaction potential	Geochemistry			
		% Very Poor	% Poor	% Fair	% Good	% Very Good				Potentially acid generating	Metal leaching at neutral pH	Poor-quality leachate	Poor-quality leachate (item of concern)
1	No	70	25	–	–	5	–	No	–	No	No	Unknown	–
2	Yes	–	75	–	–	–	Variable $\phi = 15°$, c = 0	Unknown	–	Yes	Yes	Yes	–
3	Yes	0.6	17	2.2	–	80	40.2	No	–	No	No	No	–
4	Yes	10	15	50	25	–	Good = 37°; Fair = 32°; Poor = 25°, 10 kPa; Very Poor = 17.5°, 6 kPa; Peat = 10°, 2 kPa	No	–	Yes	Unknown	Unknown	–
5	Yes	–	–	–	90	–	–	No	–	–	–	–	–
6	Yes	–	–	–	20	80	$\phi = 38°$, c = 36 kPa	Unknown	–	No	Unknown	No	–
7	Yes	–	10	20	40	30	Unit weight: 20 kN/m³, $\phi = 40°$, c = 0 kPa	No	–	No	Unknown	Unknown	–
8	Yes	–	–	–	–	–	–	–	–	–	–	–	–
9	–	–	–	–	–	–	–	–	–	–	–	–	–
10	–	–	–	–	–	–	–	–	–	–	–	–	–
11	No	–	–	–	–	–	–	No	–	Yes	Unknown	Unknown	–
12	No	–	–	–	–	–	–	No	–	Unknown	No	Unknown	–
13	–	–	–	–	–	–	–	–	–	–	–	–	–
14	–	–	–	–	–	–	–	–	–	–	–	–	–
15	–	–	–	–	–	–	–	–	–	–	–	–	–
16	Yes	–	–	–	–	–	–	No	–	Unknown	Unknown	Unknown	–
17	No	40	25	20	10	5	–	No	–	No	No	No	–
18	No	N/A	N/A	N/A	N/A	N/A	$\phi = 35$	Yes	Moderate	Yes	Yes	Yes	–
19	No	N/A	N/A	N/A	N/A	N/A	$\phi = 30$	Yes	Moderate	Yes	Yes	No	–
20	No	N/A	N/A	N/A	N/A	N/A	$\phi = 35$	Yes	Low	Yes	Yes	No	–
21	No	N/A	N/A	N/A	N/A	N/A	$\phi = 35$	Yes	Low	Yes	Yes	Yes	–
22	Yes	–	–	–	–	–	–	Yes	Moderate	Unknown	Unknown	Unknown	–
23	No	–	–	–	–	–	Shear normal strength function	No	–	No	Yes	No	Selenium
24	No	–	–	–	–	–	Frictional, unknown value	Unknown	–	No	Yes	Yes	Selenium/nitrates
25	No	–	–	–	–	–	Frictional?	Unknown	–	No	Yes	No	Selenium/nitrates
26	No	–	–	–	–	–	–	Unknown	–	No	Yes	No	Selenium/nitrates
27	Yes	–	–	100	–	–	Non-linear strength curve	No	–	No	Yes	Yes	Selenium
28	Yes	–	–	100	–	–	Non-linear strength curve	No	–	–	Yes	Yes	Selenium

Survey number	Is the waste/stockpile material classified according to quality?	% Very Poor	% Poor	% Fair	% Good	% Very Good	Design waste material shear strength parameters	Has the potential for liquefaction of the waste material been assessed?	Liquefaction potential	Potentially acid generating	Metal leaching at neutral pH	Poor-quality leachate	Poor-quality leachate (item of concern)
29	No	–	–	–	–	–	–	Unknown	–	No	Yes	Yes	Selenium
30	No	–	–	–	–	–	–	No	–	No	Yes	Yes	Selenium
31	No	–	–	–	–	–	–	Unknown	–	No	No	No	Selenium
32	No	–	–	–	–	–	–	No	–	No	Yes	Yes	Selenium/Nitrates
33	No	1	5	14	50	30	$\phi = 33°, c = 0$	Yes	Low	Yes	Yes	Unknown	–
34	No	–	–	–	100	–	$\phi = 35°, c = 0$	Unknown	–	Yes	Unknown	Unknown	–
35	–	–	–	–	–	–	–	–	–	–	–	–	–
36	No	40	25	20	10	5	–	Yes	Low	Unknown	Unknown	Unknown	–
37	No	–	–	–	–	–	–	Unknown	–	Unknown	Unknown	Unknown	–
38	–	–	–	–	–	–	–	–	–	–	–	–	–
39	–	–	–	–	–	–	–	–	–	–	–	–	–
40	No	N/A	N/A	N/A	N/A	N/A	$\phi = 35$	Yes	Low	Yes	Yes	Yes	–
41	–	–	–	–	–	–	–	–	–	–	–	–	–
42	–	–	–	–	–	–	–	–	–	–	–	–	–
43	–	–	–	–	–	–	–	–	–	–	–	–	–
44	No	–	–	–	100	–	34–36°	No	–	Yes	Unknown	Yes	–
45	No	–	–	–	100	–	34–36°	No	–	Yes	Unknown	Unknown	–
46	No	–	–	–	100	–	37°	No	–	No	No	No	–
47	No	–	–	–	–	–	37°	No	–	Yes	Unknown	Unknown	–
48	–	–	–	–	–	–	–	–	–	–	–	–	–
49	Yes	–	80	–	0	–	27°, 20 kPa	Yes	Moderate	Yes	Unknown	Unknown	–
50	Yes	–	80	–	20	–	27°, 20 kPa	Yes	Moderate	Yes	Unknown	Unknown	–
51	Yes	–	80	–	20	–	27°, 20 kPa	Yes	Moderate	Yes	Unknown	Unknown	–
52	Yes	–	80	–	20	–	27°, 20 kPa	Yes	Moderate	Yes	Unknown	Unknown	–
53	No	–	–	–	–	–	35°	No	–	Yes	Unknown	Yes	–
54	–	–	–	–	–	–	–	–	–	–	–	–	–
55	–	–	–	–	–	–	–	–	–	–	–	–	–
56	–	–	–	–	–	–	–	–	–	–	–	–	–
57	–	–	–	–	–	–	–	–	–	–	–	–	–

Survey number	Is the waste/ stockpile material classified according to quality?	Waste/stockpile material quality distribution					Design waste material shear strength parameters	Has the potential for liquefaction of the waste material been assessed?	Liquefaction potential	Geochemistry			Poor-quality leachate (item of concern)
		% Very Poor	% Poor	% Fair	% Good	% Very Good				Potentially acid generating	Metal leaching at neutral pH	Poor-quality leachate	
58	–	–	–	–	–	–	–	–	–	–	–	–	–
59	–	–	–	–	–	–	–	–	–	–	–	–	–
60	–	–	–	–	–	–	–	–	–	–	–	–	–
61	Yes	0	5	35	45	15	Unit weight = 129 lb/ft³, ϕ = 33°, c = 0	No	–	Yes	Unknown	No	Carbonate and sulphide waste tonnes are tracked to balance the pH of the future pit lake
62	–	–	–	–	–	–	–	–	–	–	–	–	–
63	No	–	–	–	–	–	–	No	–	No	No	No	–
64	No	–	–	–	–	–	–	Yes	High	–	–	–	–
65	Yes	–	4	–	–	96	–	Unknown	–	No	No	–	–
66	–	–	–	–	–	–	–	–	–	–	–	–	–
67	No	1	5	14	50	30	ϕ = 33°, C = 0	Yes	Low	Yes	Yes	–	–
68	–	–	–	–	–	–	–	–	–	–	–	–	–
69	No	–	–	–	–	–	–	Unknown	–	No	No	No	–

Table A2.6: Summary of the 2013 LOP Waste Dump Survey Part 6 – Downslope buffer zone width, impact berm size, construction equipment and crest advance

Survey number	Width of designated buffer zone downslope from toe (m)	Impact/deflection berm		Construction – typical values			Average crest advance rate (m/day)	
		Height (m)	Distance downslope from toe (m)	Type and size of equipment	Mass loading rate (tonnes/ day)	Average active crest length (m)	Planned	Actual
1	–	7	–	Dozers D-10	–	–	–	–
2	30	1.5	5	Dozers	100 000	10	5	3
3	70 a 100 metros en sur, 300 metros en botadero norte	15	60	Cat 793, 240 tonnes	175 000	350	388	370
4	50	–	–	Komatsu 730E & Bulldozer D375A	10 000	80	–	6
5	45/25 guidline	–	–	793 to dump, D10 to rehab. slopes	100 000	100	50	–
6	–	5	4	Cat 793 trucks and EX3600 digger	–	180	5	–
7	50	–	–	Cat 793	4400	200	–	–
8	–	–	–	–	–	–	–	–
9	–	–	–	–	–	–	–	–
10	–	–	–	–	–	–	–	–
11	–	–	20	–	–	–	–	–
12	–	0	0	–	–	–	–	–
13	–	–	–	–	–	–	–	–
14	–	–	–	–	–	–	–	–
15	–	–	–	–	–	–	–	–
16	No buffer zone	–	–	D9 track dozer	–	100	–	10
17	–	–	–	Dozers	–	20	–	–
18	40	6	20	Cat 785	50 000	200	> 2	> 3
19	40	0	0	Cat 785	10 000	80	0.75	1.75
20	0	N/A	N/A	Cat 785	20 000	80	> 2	> 2
21	40	3	6	Cat 785	50 000	200	> 2	> 2
22	–	–	–	2	130 000	–	–	–
23	22–1000	–	–	–	–	–	–	–
24	N/a	–	–	–	0	0	0	0
25	Unknown	–	–	–	–	–	–	–
26	–	–	–	–	–	–	–	–
27	60–225	–	–	–	–	–	–	–
28	60–225	–	–	–	–	–	–	–
29	–	–	–	–	100 000	–	–	–
30	–	–	–	–	67 000	–	–	–
31	–	–	–	–	14 000	–	–	–
32	–	–	–	–	40 000	–	–	–
33	330	10	80	HT-793, DZ-11	240 000	–	–	–

Survey number	Width of designated buffer zone downslope from toe (m)	Impact/deflection berm		Construction – typical values			Average crest advance rate (m/day)	
		Height (m)	Distance downslope from toe (m)	Type and size of equipment	Mass loading rate (tonnes/ day)	Average active crest length (m)	Planned	Actual
34	Unknown	–	–	Cat 785	60 000	Paddock tip	Paddock tip	Paddock tip
35	–	–	–	–	–	–	–	–
36	–	–	–	Dozers	–	20	–	–
37	None	–	–	D9 track dozer	–	100	–	–
38	–	–	–	–	–	–	–	–
39	–	–	–	–	–	–	–	–
40	40	3	6	Cat 785	50 000	200	> 2	> 2
41	–	–	–	–	–	–	–	–
42	–	–	–	–	–	–	–	–
43	–	–	–	–	–	–	–	–
44	200	15	70	993	16 800	225	–	–
45	0	0	0	993/997	–	100	–	–
46	0	45	0	993/997	75 000	750	0.5	0.5
47	0	15	0	993/997	–	–	0.5	–
48	–	–	–	–	–	–	–	–
49	0	0	0	240 tonnes	–	–	–	–
50	0	–	–	240 tonnes	–	–	–	–
51	0	–	–	240 tonnes	–	–	–	–
52	–	–	–	240 tonnes	–	–	–	–
53	0	3	0	–	–	–	–	–
54	–	–	–	–	–	–	–	–
55	–	–	–	–	–	–	–	–
56	–	–	–	–	–	–	–	–
57	–	–	–	–	–	–	–	–
58	–	–	–	–	–	–	–	–
59	–	–	–	–	–	–	–	–
60	–	–	–	–	–	–	–	–
61	35 m	25 m	35 m	Cat 793D 215 tonne trucks	28 000	150	1.0	0.47
62	–	–	–	–	–	–	–	–
63	10	–	–	777(100t), D-9, D-6	15 000	300	–	–
64	10	–	–	773(50t), 735, 777, D-6, D-9	6000	300	–	–
65	60	1.5	–	Bulldozer	–	–	–	–
66	–	–	–	–	–	–	–	–
67	330	10	80	HT-793	240 000	–	NA	NA
68	–	–	–	–	–	–	–	–
69	–	2	–	793 trucks + D-10 dozers	110 000	> 1000	–	–

Table A2.7: Summary of the 2013 LOP Waste Dump Survey Part 7 – Site climate, site seismicity and single lift stability

Survey number	Total annual precipitation (mm)			Annual temperature (°C)			Peak horizontal ground acceleration		Single lift – static				Single lift – seismic				Seismic loading was based on:	Seismic loading was based on:
	Annual rainfall	Maximum 24 hour	Annual snowfall	Minimum	Maximum	Average	1:475 return period	Maximum credible earthquake	Operations – FoS	Operations – PoF	Closure – FoS	Closure – PoF	Operations – FoS	Operations – PoF	Closure – FoS	Closure – PoF		
1	1500	90	0	7	38	26	–	–	1.3	–	1.5	–	–	–	–	–	–	–
2	3500	150	Nil	26	40	32	Not known	–	2	0.1	1.5	<5	NA	NA	NA	–	Other	N/a
3	193	–	3	-30	22	10	–	0.3 g	1.13	–	1.13	–	0.97	–	0.87	–	1.475 year return period	–
4	1440	76	–	-3.6	16	7.6	0.11 g	0.21 g	–	–	1.1	–	–	–	0.9	–	1.475 year return period	–
5	200	20	0	0	50	37	–	–	–	–	–	–	–	–	–	–	–	–
6	230	140	0	0	50	33	–	–	–	–	–	–	–	–	–	–	–	–
7	800	40	–	-2	43	20	0.035 g	0.14 g	1.4	–	3.1	–	1.1	–	2.4	–	Other	ANCOLD
8	–	–	–	–	–	–	–	–	–	–	–	–	–	–	–	–	–	–
9	–	–	–	–	–	–	–	–	–	–	–	–	–	–	–	–	–	–
10	–	–	–	–	–	–	–	–	–	–	–	–	–	–	–	–	–	–
11	300	50	0	0	40	30	–	–	–	–	–	–	–	–	–	–	–	–
12	–	–	–	–	–	–	–	–	–	–	–	–	–	–	–	–	–	–
13	–	–	–	–	–	–	–	–	–	–	–	–	–	–	–	–	–	–
14	–	–	–	–	–	–	–	–	–	–	–	–	–	–	–	–	–	–
15	–	–	–	–	–	–	–	–	–	–	–	–	–	–	–	–	–	–
16	535.6	117.6	–	–	–	–	0.03 g	–	1.3	–	1.5	–	–	–	–	–	Other	–
17	1500	400	0	5	38	30	–	–	–	–	–	–	–	–	–	–	–	–
18	4277	277	0	21.9	33.5	26.7	4.6 m/s^2	3.92 m/s^2	1.38	–	N/A	–	0.92	–	N/A	–	MCE	–
19	4277	277	0	21.9	33.5	26.7	4.6 m/s^2	3.92 m/s^2	0.9	–	N/A	–	0.71	–	N/A	–	Other	1:95 year return period
20	4277	277	0	21.9	33.5	26.7	4.6 m/s^2	3.92 m/s^2	–	–	–	–	–	–	–	–	–	–
21	4277	277	0	21.9	33.5	26.7	4.6 m/s^2	3.92 m/s^2	–	–	–	–	–	–	–	–	–	–
22	–	–	–	–	–	–	–	–	–	–	–	–	–	–	–	–	–	–
23	–	–	–	–	–	–	$0.0415 = k_h$	–	1.1	–	–	–	–	–	–	–	Other	1 in 1000
24	–	–	–	–	–	–	–	–	–	–	–	–	–	–	–	–	–	–
25	–	–	–	–	–	–	–	–	–	–	–	–	–	–	–	–	–	–
26	–	–	–	–	–	–	–	–	–	–	–	–	–	–	–	–	–	–
27	–	–	–	–	–	–	–	–	1	–	1.3	–	1	–	1.1	–	1.475 year return period	–
28	–	–	–	–	–	–	–	–	1	–	1.2	–	–	–	–	–	1.475 year return period	–
29	–	–	–	–	–	–	–	–	–	–	–	–	–	–	–	–	–	–
30	–	–	–	–	–	–	–	–	–	–	–	–	–	–	–	–	–	–
31	–	–	–	–	–	–	–	–	–	–	–	–	–	–	–	–	–	–
32	–	–	–	–	–	–	–	–	–	–	–	–	–	–	–	–	–	–
33	3146	180	0	15.6	31.4	19.3	–	2.94 m/s^2	1.25	–	–	–	–	–	–	–	MCE	–
34	980	50	0	16	32	Unknown	1.17 m/s^2	M5	–	–	–	–	–	–	–	–	MCE	–
35	–	–	–	–	–	–	–	–	–	–	–	–	–	–	–	–	–	–
36	1743.3	196.2	–	16.5	22.9	20.5	–	–	1.3	–	1.5	–	–	–	–	–	–	–

Survey number	Total annual precipitation (mm)			Annual temperature (°C)			Peak horizontal ground acceleration		Single lift – static				Single lift – seismic				Seismic loading was based on:	Seismic loading was based on:
	Annual rainfall	Maximum 24 hour	Annual snowfall	Minimum	Maximum	Average	1:475 return period	Maximum credible earthquake	Operations –FoS	Operations –PoF	Closure –FoS	Closure –PoF	Operations –FoS	Operations –PoF	Closure –FoS	Closure –PoF		
37	535.6	117.6	–	–	–	–	–	–	–	–	–	–	–	–	–	–	–	–
38	–	–	–	–	–	–	–	–	–	–	–	–	–	–	–	–	–	–
39	–	–	–	–	–	–	–	–	–	–	–	–	–	–	–	–	–	–
40	4277	277	0	21.9	33.5	26.7	4.6 m/s²	3.92 m/s²	–	–	–	–	–	–	–	–	–	–
41	–	–	–	–	–	–	–	–	–	–	–	–	–	–	–	–	–	–
42	–	–	–	–	–	–	–	–	–	–	–	–	–	–	–	–	–	–
43	–	–	–	–	–	–	–	–	–	–	–	–	–	–	–	–	–	–
44	1350	32.3	–	2	11.3	5.6	0.28 g	0.48 g	1.36	–	–	–	1.09	–	–	–	1.475 year return period	–
45	1350	32.3	–	2	5.6	11.3	2.75 m/s²	4.71 m/s²	1.25	–	–	–	0.95	–	–	–	–	–
46	1350	32.3	–	2	11.3	5.6	2.75 m/s²	4.71 m/s²	–	–	1.25	–	–	–	0.95	–	1.475 year return period	–
47	1350	32.3	–	2	5.6	11.3	2.75 m/s²	4.71 m/s²	1.25	–	–	–	0.95	–	–	–	–	–
48	–	–	–	–	–	–	–	–	–	–	–	–	–	–	–	–	–	–
49	900	–	–	–	–	–	3.43 m/s²	–	1.3	–	–	–	0.98	–	–	–	1.475 year return period	–
50	900	36	–	–	–	–	3.43 m/s²	–	1.3	–	–	–	0.98	–	–	–	1.475 year return period	–
51	900	36	–	–	–	–	3.43 m/s²	–	1.08	–	1.08	–	0.82	–	0.82	–	1.475 year return period	–
52	900	36	–	–	–	–	3.43 m/s²	–	1.08	–	1.08	–	0.82	–	0.82	–	1.475 year return period	–
53	–	–	–	–	–	–	0.059 g	–	–	–	1.42	–	–	–	1.32	–	1.475 year return period	–
54	–	–	–	–	–	–	–	–	–	–	–	–	–	–	–	–	–	–
55	–	–	–	–	–	–	–	–	–	–	–	–	–	–	–	–	–	–
56	–	–	–	–	–	–	–	–	–	–	–	–	–	–	–	–	–	–
57	–	–	–	–	–	–	–	–	–	–	–	–	–	–	–	–	–	–
58	–	–	–	–	–	–	–	–	–	–	–	–	–	–	–	–	–	–
59	–	–	–	–	–	–	–	–	–	–	–	–	–	–	–	–	–	–
60	–	–	–	–	–	–	–	–	–	–	–	–	–	–	–	–	–	–
61	215	32	584	-31.6	40	9.5	0.98 m/s²	1.18 m/s²	1.23	–	2.15	–	1.01	–	1.59	–	1.475 year return period	–
62	–	–	–	–	–	–	–	–	–	–	–	–	–	–	–	–	–	–
63	488.7	–	241.7	-33.8	23.2	-2	–	–	–	–	–	–	–	–	–	–	–	–
64	488.7	–	241.7	-33.8	23.2	-2	–	–	–	–	–	–	–	–	–	–	–	–
65	1208	100 (return period 5 years)	0	24	30	26	–	0.08 g	1.3	10	–	–	1	50	–	–	MCE	–
66	–	–	–	–	–	–	–	–	–	–	–	–	–	–	–	–	–	–
67	3146	180	0	15.6	31.4	19.3	–	2.94 m/s²	1.25	–	–	–	–	–	–	–	MCE	–
68	–	–	–	–	–	–	–	–	–	–	–	–	–	–	–	–	–	–
69	–	–	–	–	–	–	–	–	–	–	–	–	–	–	–	–	–	–

Table A2.8: Summary of the 2013 LOP Waste Dump Survey Part 8 – Overall stability and risk assessment

Survey number	Overall – static				Overall – seismic				Other criteria (e.g. maximum allowable deformation)	Was a risk assessment conducted?	Type of risk assessment	Results of risk assessment	Mitigation measures
	Operations –FoS	Operations –PoF	Closure– FoS	Closure– PoF	Operations –FoS	Operations –PoF	Closure– FoS	Closure– PoF					
1	1.3	–	1.5	–	–	–	–	–	–	Yes	Monitoring; visual inspections, plurennial auditing	–	–
2	–	–	–	–	–	–	–	–	–	Yes	Waste Dump Failure risks assessment	Could fail if not well managed	Good dewatering systems and proper dumping sequence
3	–	–	–	–	–	–	–	–	–	Yes	Evaluacion de riezgo formal, evaluacion de riezgo a nivel del terreno, inspecciones diarias	Se realiza una evaluacion interna considerando si el riezgo es bajo, medio o alto	Dependiendo del nivel de riezgo se toman medidas correctivas a tal punto de cerrar el botadero
4	–	–	1.35	–	–	–	1.05	–	–	No	–	–	–
5	–	–	–	–	–	–	–	–	–	No	–	–	–
6	1.21	–	–	–	–	–	–	–	–	No	–	–	–
7	1.6–2.9	–	2.4–3.1	1.5E-07	1.5–2.5	–	1.5–2.4	1.1E-07	–	Yes	Semi-quantitative	Likelihood: rare Consequence: moderate Risk level = '6' or moderate	None
8	–	–	–	–	–	–	–	–	–	–	–	–	–
9	–	–	–	–	–	–	–	–	–	–	–	–	–
10	–	–	–	–	–	–	–	–	–	–	–	–	–
11	–	–	–	–	–	–	–	–	–	No	–	–	–
12	–	–	–	–	–	–	–	–	–	–	–	–	–
13	–	–	–	–	–	–	–	–	–	–	–	–	–
14	–	–	–	–	–	–	–	–	–	–	–	–	–
15	–	–	–	–	–	–	–	–	–	–	–	–	–
16	–	–	–	–	–	–	–	–	–	Yes	–	–	–
17	–	–	–	–	–	–	–	–	Saturation index or water level	Yes	Monitoring control (piezometers and prisms); visual control, periodical monitoring	Monthly reports; in-house database monitoring system	Backfill or discharging
18	1.38	–	N/A	–	0.92	–	N/A	–	–	Yes	SQRA	Foundation Failure	Control pore pressure-underdrains
19	0.9	–	N/A	–	0.71	–	N/A	–	–	Yes	FOS	Foundation Failure	Control pore pressure-underdrains
20	>1.2	–	N/A	–	1	–	N/A	–	–	Yes	FOS	/	/
21	>1.2	–	N/A	–	1	–	N/A	–	–	No	–	–	–
22	–	–	–	–	–	–	–	–	–	–	–	–	–
23	1.3	–	–	–	1	–	–	–	–	Yes	Qualitative risk completed on the design	High risk design	revise design
24	–	–	–	–	–	–	–	–	–	No	–	–	–
25	–	–	–	–	–	–	–	–	–	No	–	–	–
26	–	–	–	–	–	–	–	–	–	No	–	–	–
27	1.3	–	1.5	–	–	–	1.1	–	–	No	–	–	–
28	1.3	–	1.5	–	–	–	1.1	–	–	No	–	–	–
29	–	–	–	–	–	–	–	–	–	No	–	–	–
30	–	–	–	–	–	–	–	–	–	No	–	–	–

Survey number	Overall – static				Overall – seismic				Other criteria (e.g. maximum allowable deformation)	Was a risk assessment conducted?	Type of risk assessment	Results of risk assessment	Mitigation measures
	Operations –FoS	Operations –PoF	Closure– FoS	Closure –PoF	Operations –FoS	Operations –PoF	Closure– FoS	Closure– PoF					
31	–	–	–	–	–	–	–	–	–	No	–	–	–
32	–	–	–	–	–	–	–	–	–	No	–	–	–
33	–	–	1.69	–	–	–	1.15	–		Yes	Semi Quantitative	Medium Risk	Monitoring performance of waste dump, geotechnical hazard call up procedure
34	1.2	0.1	1.3	<1%	–	–	–	–	–	No	–	–	–
35	–	–	–	–	–	–	–	–	–		–	–	–
36	1.3	–	1.5	–	–	–	–	–	–	Yes	Monitoring control (piezometers and prisms); visual control; periodical monitoring	Monthly reports; in-house database monitoring system	Backfill or discharging
37	–	–	–	–	–	–	–	–	–	–	–	–	–
38	–	–	–	–	–	–	–	–	–	–	–	–	–
39	>1.2	–	–	–	–	–	–	–	–	–	–	–	–
40	–	–	N/A	–	1	–	N/A	–	–	No	–	–	–
41	–	–	–	–	–	–	–	–	–	–	–	–	–
42	–	–	–	–	–	–	–	–	–	–	–	–	–
43	–	–	–	–	–	–	–	–	–	–	–	–	–
44	1.64	–	–	–	1.22	–	–	–	1% of dump height	No	–	–	–
45	1.38	–	–	–	1	–	–	–	1% of maximum height	No	–	–	–
46	1.96	–	1.96	–	1.38	–	1.38	–	1% of maximum failure height	No	–	–	–
47	1.96	–	–	–	1.38	–	–	–	1% of failure height	No	–	–	–
48	–	–	–	–	0.88	–	–	–	–	No	–	–	–
49	1.35	–	–	–	0.88	–	–	–	1% of failure height	No	–	–	–
50	1.35	–	–	–	0.9	–	–	–	1% of failure height	No	–	–	–
51	1.42	–	1.3	–	0.9	–	0.82	–	–	No	–	–	–
52	1.42	–	1.3	–	0.9	–	0.82	–	1% of maximum failure height	No	–	–	–
53	–	–	–	–	–	–	–	–	–	No	–	–	–
54	–	–	–	–	–	–	–	–	–	–	–	–	–
55	–	–	–	–	–	–	–	–	–	–	–	–	–
56	–	–	–	–	–	–	–	–	–	–	–	–	–
57	–	–	–	–	–	–	–	–	–	–	–	–	–
58	–	–	–	–	–	–	–	–	–	–	–	–	–
59	–	–	–	–	–	–	–	–	–	–	–	–	–

| Survey number | Overall – static | | | | Overall – seismic | | | | | Other criteria (e.g. maximum allowable deformation) | Was a risk assessment conducted? | Type of risk assessment | Results of risk assessment | Mitigation measures |
	Operations –FoS	Operations –PoF	Closure– FoS	Closure –PoF	Operations –FoS	Operations –PoF	Closure– FoS	Closure– PoF					
60	–	–	–	–	–	–	–	–	–	–	–	–	–	–
61	1.23	–	2.34	–	1.01	–	1.72	–	–	Yes	Fault tree analysis and incident control analysis	Catastrophic, unpredicted and uncontrolled failure– likelihood rating of a 2. Unlikely	Controls: (1) Manage the material characteristics of the dump fill material; (2) limit the inter-dump lift height; (3) limit the dump fill rate; (4) monitor water levels in the dump with piezometers
62	–	–	–	–	–	–	–	–	–	–	–	–	–
63	–	–	–	–	–	–	–	–	–	No	–	–	–
64	–	–	–	–	–	–	–	–	–	No	–	–	–
65	1.3	10	–	–	1	50	–	–	–	Yes	–	–	–
66	–	–	–	–	–	–	–	–	–	–	–	–	–
67	–	–	1.69	–	–	–	1.15	–	–	Yes	–	–	–
68	–	–	–	–	–	–	–	–	–	–	–	–	–
69	–	–	–	–	–	–	–	–	–	–	–	–	–

Table A2.9: Summary of the 2013 LOP Waste Dump Survey Part 9 – Monitoring, triggers and groundwater conditions

Survey Number	Monitoring Type: select all that apply								Monitoring Frequency	Criterion that triggers temporary closure	Criterion to re-open	Special procedures to re-open	General stability conditions	Groundwater Conditions	
	Prisms	Wireline Extensometers	Laser Scanning	SSR	Inclinometers	Piezometers	GPS	Other (describe)						Waste material piezometric conditions	Foundation material piezometric conditions
1	Yes	–	–	–	–	Yes	–	–	Monthly	–	–	–	Stable	Free draining	Low pore pressure
2	–	Yes	–	–	–	–	–	–	Once a week	N/A	N/A	N/A	Stable	free draining	low pore pressure
3	Yes	Yes	–	–	Yes	Yes	Yes	–	Daria, seminal y mensural	Si el movimiento detectado por el instrumental supera un determinado valor se cierra el botadero. Esto esta prosedimentado	Se realiza saneo del botadero y se re abre	–	Stable	–	–
4	Yes	Yes	–	–	–	Yes	–	–	Monthly	–	–	–	Stable	Low pore pressure	Free draining
5	Yes	–	Yes	–	–	Yes	–	–	Monthly	–	–	–	–	–	–
6	Yes	–	–	Yes	–	–	–	–	Daily, hourly	Rainfall	Inspection and rainfall ceases	Inspection and rainfall ceases	Stable	Free draining	Low pore pressure
7	Yes	Yes	–	–	–	Yes	–	–	Real time for radar; hourly for automated prisms; daily for extensometers	Failure or open cracks observed, extensometers showed movement > = 2 mm/day	All remedial works completed; assessment completed and cleared by geotechnical engineer	Slope hazard response procedure; waste dump and stockpile tip heads	Stable	Free draining	Low to high pore pressure
8	–	–	–	–	–	–	–	–	–	–	–	–	–	–	–
9	–	–	–	–	–	–	–	–	–	–	–	–	–	–	–
10	–	–	–	–	–	–	–	–	–	–	–	–	–	–	–
11	–	–	–	–	–	–	–	Satellite and visual inspections	Every quarter on satellites and every shift by Foreman	None	None	None	Stable	Free draining	Low pore pressure
12	–	–	–	–	–	–	–	–	–	–	–	–	–	–	–
13	–	–	–	–	–	–	–	–	–	–	–	–	–	–	–
14	–	–	–	–	–	–	–	–	–	–	–	–	–	–	–
15	–	–	–	–	–	–	–	–	–	–	–	–	Stable	Low pore pressure	–
16	–	–	–	–	–	–	–	–	No monitoring is carried out in Voorspoed Waste Rock Dump	–	–	–	–	–	–

Survey Number	Monitoring Type: select all that apply								Monitoring Frequency	Criterion that triggers temporary closure	Criterion to re-open	Special procedures to re-open	General stability conditions	Groundwater Conditions	
	Prisms	Wireline Extensometers	Laser Scanning	SSR	Inclinometers	Piezometers	GPS	Other (describe)						Waste material piezometric conditions	Foundation material piezometric conditions
17	Yes	–	–	–	–	Yes	–	–	Monthly	Global failure risk; access interruption	–	–	Stable	–	Free draining
18	Yes	–	–	–	Yes	Yes	–	–	Daily	Prism movement/pore pressure exceeds target	Pressure drops/movement stabilises	Group RA	Stable	–	Low pore pressure
19	Yes	–	–	–	Yes	Yes	–	–	–	Prism movement/pore pressure exceeds target	Pressure drops/movement stabilises	Group RA	Metastable Satisfactory	Low pore pressure	High pore pressure
20	–	–	–	–	–	–	–	–	–	Prism movement exceeds target	Movement stabilises	Group RA	Stable	Free draining	Low pore pressure
21	Yes	–	–	–	Yes	Yes	–	–	Daily	Prism movement exceeds target	Movement stabilises	Group RA	Stable	Low pore pressure	–
22	–	–	–	–	–	–	–	–	–	–	–	–	Stable	–	–
23	–	Yes	–	–	–	–	–	–	–	–	–	–	Metastable Satisfactory	–	–
24	–	Yes	–	–	–	–	–	–	Daily when active	600 mm/h movement	–	–	Stable	Free draining	Low pore pressure
25	–	Yes	–	–	–	–	–	–	Daily	600 mm/h	Rates fall below 600 mm/h criteria	–	Metastable satisfactory	–	Low pore pressure
26	–	Yes	–	–	Yes	–	–	–	Hourly/daily	–	–	–	Metastable Satisfactory	–	Low pore pressure
27	–	–	–	–	–	–	–	–	–	–	–	–	Stable	Free draining	Low pore pressure
28	–	Yes	–	–	–	–	–	–	–	–	–	–	Stable	Free draining	Low pore pressure
29	–	Yes	–	–	–	–	–	–	Daily/hourly	600–1200 mm/h (dump dependent)	< 600–1200 mm/hr	–	Stable	Free draining	Low pore pressure
30	–	Yes	–	–	–	–	–	–	Daily/hourly	1200 mm/day	< 1200 mm/day	–	Stable	Free draining	–
31	Yes	–	–	–	–	Yes	–	–	Dataloggers on piezos, bi-monthly on prisms	–	–	–	Stable	Free draining	Low pore pressure
32	–	Yes	–	–	–	Yes	–	–	Hourly/daily	–	–	–	Metastable Satisfactory	Free draining	Low pore pressure

| Survey Number | Monitoring Type: select all that apply | | | | | | | | Monitoring Frequency | Criterion that triggers temporary closure | Criterion to re-open | Special procedures to re-open | General stability conditions | Groundwater Conditions | |
	Prisms	Wireline Extensometers	Laser Scanning	SSR	Inclinometers	Piezometers	GPS	Other (describe)						Waste material piezometric conditions	Foundation material piezometric conditions
33	Yes	–	–	–	Yes	Yes	Yes	–	Monthly	Crack on bench	Area has been remediated	Conduct coordination meeting and approval from geotech, mine engineering, mine operation and safety	Stable	Free draining	Low pore pressure
34	–	–	–	–	–	–	–	Visual checks for tension cracks	Monthly	Tension cracks > 10 mm	Risk assessment/controls put in place to buttress the lift if required	–	Stable	Free draining	–
35	–	–	–	–	–	–	–	–	–	–	–	–	–	–	–
36	Yes	–	–	–	–	Yes	–	–	Monthly	Global failure risk; access interruption	–	–	Stable	–	–
37	–	–	–	–	–	–	–	–	–	–	–	–	–	–	–
38	–	–	–	–	–	–	–	–	–	–	–	–	–	–	–
39	–	–	–	–	–	–	–	–	–	–	–	–	–	–	–
40	Yes	–	–	–	Yes	–	–	–	Daily	Prism movement exceeds target	Movement stabilises	Group RA	–	–	–
41	–	–	–	–	–	–	–	–	–	–	–	–	–	–	–
42	–	–	–	–	–	–	–	–	–	–	–	–	–	–	–
43	–	–	–	–	–	–	–	–	–	–	–	–	–	–	–
44	–	Yes	–	–	–	–	–	–	4 h	2.5 cm/h	1 cm/h	–	Stable	Free draining	Low pore pressure
45	–	Yes	–	–	–	–	–	–	4 h	2.5 cm/h	1 cm/h	–	Stable	–	Free draining
46	–	Yes	–	–	–	–	–	–	4 h	2.5 cm/h	1 cm/h	–	Metastable Satisfactory	Free draining	Low to high pore pressure
47	–	Yes	–	–	–	–	–	–	4 h	4 cm/h	2 cm/h	–	Stable	Free draining	Low pore pressure
48	–	–	–	–	–	–	–	–	–	–	–	–	–	–	–
49	Yes	–	–	Yes	–	–	–	–	Weekly, real time	Varies	–	–	Stable	High pore pressure	Free draining
50	Yes	–	–	–	–	–	–	–	–	–	–	–	Stable	Low pore pressure	Free draining

Survey Number	Monitoring Type: select all that apply								Monitoring Frequency	Criterion that triggers temporary closure	Criterion to re-open	Special procedures to re-open	General stability conditions	Groundwater Conditions	
	Prisms	Wireline Extensometers	Laser Scanning	SSR	Inclinometers	Piezometers	GPS	Other (describe)						Waste material piezometric conditions	Foundation material piezometric conditions
51	–	–	–	–	–	–	–	–	–	–	–	–	Stable	High pore pressure	Low pore pressure
52	–	–	–	–	–	–	–	Visual	Annually	–	–	–	Stable	High pore pressure	Low pore pressure
53	Yes	–	–	–	–	–	–	–	Weekly	–	–	–	Stable	Free draining	Low pore pressure
54	–	–	–	–	–	–	–	–	–	–	–	–	–	–	–
55	–	–	–	–	–	–	–	–	–	–	–	–	–	–	–
56	–	–	–	–	–	–	–	–	–	–	–	–	–	–	–
57	–	–	–	–	–	–	–	–	–	–	–	–	–	–	–
58	–	–	–	–	–	–	–	–	–	–	–	–	–	–	–
59	–	–	–	–	–	–	–	–	–	–	–	–	–	–	–
60	–	–	–	–	–	–	–	–	–	–	–	–	–	–	–
61	Yes	Yes	Yes	–	–	Yes	–	–	Quarterly	Visible extension cracks, deformation in the dump face, and loss of berm height.	Inspection by trained pit foreman and/ore geotechnical personnel	Large deformation in the dump face will require laser scan of the dump	Stable	Free draining	–
62	–	–	–	–	–	–	–	–	None	–	–	–	Stable	–	–
63	–	–	–	–	–	–	–	–	None	–	–	–	Stable	Free draining	–
64	–	–	–	–	–	–	–	Visual inspections	Weekly	–	–	–	Stable	–	–
65	–	–	–	–	–	–	–	–	Weekly	Cracks near to crest, settlement	Compacted by bulldozer	Sometimes short dump load pushing with bulldozer to the crest	Stable	Free draining	–
66	–	–	–	–	–	–	–	–	–	–	–	–	–	–	–
67	Yes	–	–	–	Yes	Yes	Yes	–	Monthly	Crack on bench, failure on slope	Area has been remediate	Conduct coordination and approval from geotech, mine engineering and mine operation	Stable	Free draining	–
68	–	–	–	–	–	–	–	–	–	–	–	–	–	–	–
69	–	–	–	–	–	–	–	–	–	–	–	–	–	–	–

List of symbols

The following tables present symbols (in alphabetical order of first Roman followed by Greek characters), including the dimensions and the name of the property represented, the preferred unit, the preferred unit reduced to SI base units and some other commonly used metric units.

Symbol	Dimensions	Property	Preferred unit	SI base unit	Other common units
A	L^2	Area	m^2	m	mm^2, cm^2, km^2, ha
$B, \bar{B}$	–	Pore pressure parameter	–	–	–
c	$M/L.T^2$	Cohesion	Pa	$kg/m.s^2$	N/m^2; kPa, MPa
C	–	Hazen's constant for estimating hydraulic conductivity	–	–	–
C_α	–	Secondary compression index	–	–	–
C_C	–	Coefficient of curvature	–	–	–
C_c	–	Compression index	–	–	–
C_r	–	Recompression index	–	–	–
C_U	–	Coefficient of uniformity	–	–	–
c_u	$M/L.T^2$	Undrained shear strength	Pa	$kg/m.s^2$	N/m^2; kPa, MPa
c_v	L^2/T	Coefficient of consolidation	m^2/s	–	–
D	L	Distance or displacement	m	m	mm, cm
D_{10}, d_{10}	L	Particle diameter corresponding to 10% of the sample passing by weight	mm	m	m, cm
D_x, d_x	L	Particle diameter corresponding to x% of the sample passing by weight	mm	m	m, cm
e	–	Void ratio	–	–	–
E_s	L	Evaporation from surface	mm	m	m, cm
ET	L	Evaporation or evapotranspiration from the near-surface materials	mm	m	m, cm
f	–	Friction coefficient	–	–	–
F_d	$M.L/T^2$	Driving force	N (newton)	$kg.m/s^2$	
F_r	$M.L/T^2$	Resisting force	N (newton)	$kg.m/s^2$	
Gs	–	Specific gravity	–	–	–
H	L	Flow depth used in runout analysis	m	m	mm, cm
ha	L^2	Hectare	m^2	m	–
i	L/L	Gradient	m/m	m	–
k	L/T	Hydraulic conductivity or coefficient of permeability	m/s	cm/s	m/s, m/day, m/d
k_h	L/T	Horizontal coefficient of permeability	m/s	m/s	m/day, m/d
k_v	L/T	Vertical coefficient of permeability	m/s	m/s	m/day, m/d
L	L	Distance or displacement	m	m	mm, cm
m	L	Hydraulic radius	m	m	mm, cm
n	–	Porosity (in soil and rock)	–	–	–
n	–	Manning's roughness coefficient for hydraulic calculations	–	–	–
P	L	Precipitation	mm	m	m, cm
P	–	Probability of exceedance	–	–	–

(Continued)

Symbol	Dimensions	Property	Preferred unit	SI base unit	Other common units
p'	$M/L.T^2$	Mean effective stress	Pa	$kg/m.s^2$	N/m^2; kPa, MPa
Q	L^3/T	Volumetric rate of flow	m^3/s	m^3/s	m^3/day
q	$M/L.T^2$	Deviatoric stress	Pa	$kg/m.s^2$	N/m^2; kPa, MPa
q_u	$M/L.T^2$	Unconfined compressive strength	Pa	$kg/m.s^2$	N/m^2; kPa, MPa
R	L	Downward percolation	mm	–	–
r_u	–	Pore pressure ratio, between pore pressure and total normal stress	–	–	–
Ro	L	Surface water runoff	mm	m	m, cm
S	L	Potential maximum water retention	mm	m	m, cm
Si	$M/L.T^2$	Shear strength intercept of Mohr-Coulomb envelope	kPa	$kg/m.s^2$	N/m^2; kPa, MPa
Sr		Degree of saturation	%		
SW	L	Water storage of the near-surface materials	mm	m	m, cm
T	T	Return period for probability of exceedance	year	–	–
t	T	Design life for probability of exceedance	year	–	–
u	$M/L.T^2$	Pore pressure	kPa	$kg/m.s^2$	N/m^2; kPa, MPa
V	L/T	Velocity	m/s	m/s	mm/s, cm/s
V_{void}	L/T	Void velocity	m/s	m/s	mm/s, cm/s
W	%	Moisture content	%	–	–
W	$L^{0.5}/T$	Wilkins constant	$m^{0.5}/s$	$m^{0.5}/s$	–

Symbol	Dimensions	Property	Preferred unit	SI base unit	Other common units
α	Degrees	Travel distance angle	degrees	–	–
γ	$M/L^2.T^2$	Specific weight (or weight density) ($= \rho.g$)	N/m^3	$kg/m^2.s^2$	KN/m^3
γ_d	$M/L^2.T^2$	Unit weight in dry state	N/m^3	$kg/m^2.s^2$	KN/m^3
γ_{sat}	$M/L^2.T^2$	Unit weight in saturated state	N/m^3	$kg/m^2.s^2$	KN/m^3
ξ	L/T^2	Turbulence parameter	m/s^2	m/s^2	–
ρ_d	M/L^3	Density in dry state	kg/m^3	kg/m^3	–
ρ_{sat}	M/L^3	Density in saturated state	kg/m^3	kg/m^3	–
σ_c, σ_3	$M/L.T^2$	Confining stress	Pa	$kg/m.s^2$	N/m^2; kPa, MPa
σ, σ_n, σ_1	$M/L.T^2$	Normal stress	Pa	$kg/m.s^2$	N/m^2; kPa, MPa, GPa
σ'_p	$M/L.T^2$	Pre-consolidation stress	Pa	$kg/m.s^2$	N/m^2; kPa, MPa
τ	$M/L.T^2$	Shear stress	Pa	$kg/m.s^2$	N/m^2; kPa, MPa, GPa
ϕ	Degrees	Angle of internal friction	degrees	–	–
ϕ'	Degrees	Effective friction angle	degrees	–	–
ϕ_b	Degrees	For use in rock strength: basic friction angle of a smooth diamond saw-cut surface	degrees	–	–
ϕ_b	Degrees	For use in runout analysis: bulk friction coefficient	degrees	–	–
ϕ_i	Degrees	Internal friction coefficient	degrees	–	–
ϕ_m	Degrees	Mobilised friction	degrees	–	–

Note: M = mass, L = length, T = time

List of abbreviations

ABA	Acid–base accounting
ALS	Airborne laser scanning
AP	Acid potential
ARD	Acid rock drainage
AS/NZS	Standards Australia and Standards New Zealand
ASTM	American Society for Testing and Materials
BCMEM	British Columbia Ministry of Energy, Mines and Natural Gas
BCMWRPRC	British Columbia Mine Waste Rock Pile Research Committee
BPT	Becker penetration test
CANMET	Canadian Centre for Mining and Metallurgy
CN	Curve number
CPT	Cone penetration test
CU	Consolidated undrained
d	Day
DEM	Distinct element method
DPI	Design and performance index
DSGSD	Deep-seated gravitational slope deformations
DSHA	Deterministic seismic hazard analysis
DS	Direct shear
DSR	Dump stability rating
DSC	Dump stability class
EGI	Engineering geology index
EIA	Environmental impact assessment
FEM	Finite element method
FLAC	Fast Lagrangian Analysis of Continua
FMECA	Failure modes, effects and criticality analysis
FoS	Factor of Safety
FVT	Field vane test
GARD	Global acid rock drainage
GLE	Generalised limit equilibrium
GMPE	Ground motion prediction equation
GSHAP	Global seismic hazard assessment program
GSI	Geological strength index
GPS	Global positioning system
HDPE	High-density polyethylene
HEC-HMS	Hydraulic engineering centre hydraulic modelling system
HRDEM	High-resolution digital elevation model
ICOLD	International Commission on Large Dams
INAP	International Network for Acid Prevention
ISRM	International Society for Rock Mechanics
JCS	Joint wall compressive strength
JRC	Joint roughness coefficient
LDM	Laser distance measuring
LE	Limit equilibrium
LI	Liquidity index
LiDAR	Light detection and ranging
LL	Liquid limit
LOP project	Large open pit project
LPT	Large penetration test

(Continued)

List of abbreviations (Continued)

MEMS	Microelectromechanical systems
MESA	Mining Enforcement and Safety Administration
ML	Metal leaching
N/A	Not applicable
NAF	Non-acid forming
NAG	Net acid generation
NASA	National Aeronautics and Space Administration
NAVFAC	Naval facilities engineering command
NP	Neutralising potential
NPR	Net potential ratio
NRC	Natural Resources Canada
OSM	Office of Surface Mining Reclamation and Enforcement
PAF	Potentially acid forming
PI	Plasticity index
PL	Plastic limit
PMF	Probable maximum flood
PoF	Probability of failure
PSD	Particle size distribution
PSHA	Probabilistic seismic hazard analysis
Q	Tunnelling quality index
QA/QC	Quality assurance/quality control
RMR	Rock mass rating
RQD	Rock quality index
RTS	Robotic total station
SCS	Soil conservation service
SL	Shrinkage limit
SMD	Soil moisture deficit
SME	Society for Mining, Metallurgy, and Exploration
SPT	Standard penetration test
SRF	Strength reduction factor
SSR	Slope stability radar
TARP	Trigger action response plan
TDR	Time domain reflectometry
UC	Unconfined compression
UCS	Unconfined compressive strength
UDEC	Universal distinct element code
USACE	US Army Corps of Engineers
USBM	US Bureau of Mines
USCS	Unified Soil Classification System
USDA NRCS	US Department of Agriculture Natural Resources Conservation Service
USGS	US Geological Survey
VW	Vibrating wire
WHC	Waste dump and stockpile hazard class
WSR	Waste dump and stockpile stability rating
WSRHC	Waste dump and stockpile stability rating and hazard classification
2D	Two-dimensional
3D	Three-dimensional

Glossary

Active wedge = wedge of rockfill material in the active Rankine state of stress. When the rockfill is allowed to deform laterally or relax it will mobilise the active shear resistance.

Atterberg limits = named after Swedish agriculturist Albert Atterberg, who established consistency limits or Atterberg limits, which define methods to determine the water content when a soil passes to the next stage of state. The Atterberg limits generally in use include:

> **liquid limit** = the minimum water content at which a soil is still in a liquid or flowable state
>
> **plastic limit** = the minimum water content at which a soil begins to crumble or change from plastic to semi-solid state
>
> **shrinkage limit** = the maximum water content when a reduction in water content does not cause a reduction in volume or lowest water content a soil can be completely saturated.

Consequence = outcome or impact of an event.

Darcy's law = an empirical law that describes the flow of fluid through a porous medium. It states that the rate of flow of fluid is proportional to (1) the hydraulic gradient within the fluid, (2) the cross-sectional area through which flow is occurring, and (3) a property called the coefficient of permeability or hydraulic conductivity.

Drainable porosity = *see* Porosity, drainable.

Effective rainfall = the difference between total rainfall and actual evapotranspiration.

Effective stress = *see* Stress, effective.

Extinction depth = Maximum depth from which evapotranspiration can remove subsurface pore water.

Factor of Safety (FoS) = the capacity beyond the expected loads; in slope stability, the ratio of available shear resistance along a sliding surface to the driving shear forces required for equilibrium.

Finite difference = a numerical method for approximating the solutions to differential equations using finite-difference equations to approximate derivatives.

Finite element = a numerical method for approximating the solutions to differential equations using numerical integration.

Groundwater = the water contained in the pore spaces of a rock or soil mass in a saturated state (i.e. below the water table).

Hazard = source of potential harm; a potential occurrence or condition that could lead to injury, damage to the environment, delay or economic loss. *See also* Risk.

Homogeneous = having the same properties at every point; of a porous medium, one in which any two volumes large enough to be a representative elementary volume will have the same properties (e.g. of porosity or permeability).

Hydraulic conductivity = the volume of groundwater (or other fluid), at the prevailing viscosity, that will move through rock or soil fully saturated with that fluid in unit time under a unit hydraulic gradient through a unit area measured at right angles to the direction of flow. It depends on the properties of the fluid saturating the pores as well as the properties of the medium and has dimensions of L/T. (*See also* Permeability).

Hydraulic gradient = the rate of change of static head with distance.

Likelihood = the probability or frequency of occurrence of an event, described in qualitative or quantitative terms.

Limit equilibrium analysis = a technique used to calculate the Factor of Safety of a soil or rock slope based on solving one or more of the three conditions of static equilibrium (moment equilibrium, vertical force equilibrium and horizontal force equilibrium).

Overall dynamic stability = the expected global stability behaviour of a structure under seismic or earthquake loading conditions to withstand movement.

Overall static stability = the expected global stability behaviour of a structure under conditions of normal static (gravity) loading to withstand movement.

Particle size distribution = the graphical representation of a soil's range of particle sizes. Also referred to as a grain size distribution curve. Typically plotted as % passing, by mass, versus grain size in mm. The grain size distribution curve is typically plotted with a logarithmic scale for grain size, from a maximum of 100% down to the lowest available % passing, which is determined by the smallest sieve size used.

Passive wedge = wedge of rockfill material in a passive Rankine state of stress. When rockfill is compressed or deformed inward it will mobilise a passive shear resistance.

Perched groundwater zones = saturated parts of the rock or soil mass separated by unsaturated material from the main water table. Perched groundwater zones are normally caused by stratigraphic or structural horizons of low permeability that impede downward percolation of water. The vertical hydraulic gradient between the perched phreatic levels or water tables will be greater than unity.

Permeability = a term loosely used to mean hydraulic conductivity (with dimensions of L/T). In engineering practice, where there is usually little variation in the viscosity of the pore fluids, permeability is commonly used to mean hydraulic conductivity.

Phreatic surface = free water elevation, water table or groundwater table.

Porosity = the property of a rock or soil of possessing pores or voids; expressed quantitatively as the pore volume divided by the bulk volume.

Porosity, drainable = the fraction of the total volume of a rock or soil mass that can drain by gravity in response to lowering of the water table.

Porosity, total = the fraction of the total volume of a rock or soil mass occupied by void space, usually expressed as a percentage. These voids may be intergranular spaces or fractures. Total porosity (n) may include pores such as fluid inclusions or isolated fractures that are not connected with other pores.

Probability of failure = where there is variability or uncertainty in the shear strengths determined in the laboratory or in the field and the variability can be described by a probability distribution function, stability analyses can be carried out, assuming that the shear strength can vary according to the distribution function. The probability of failure is defined as the number of analyses with a FoS less than 1.0 divided by the total number of FoS results.

Risk = the effect of uncertainty on objectives; often expressed as a combination of the consequence of an event and its associated likelihood of occurrence resulting in the chance or probability of an adverse effect if exposed to a hazard.

Risk analysis = the systematic process used to understand the nature of and to deduce the level of risk.

Risk assessment = the overall process of risk identification, risk analysis and risk evaluation.

Rock drain = a zone of coarse, durable rockfill that is capable of transmitting hydraulic flows with low impedance.

Saturation = reached when the water content becomes equal to the porosity.

Saturated zone = the zone below the water table, where the pore water is everywhere at a pressure greater than atmospheric.

Seismic hazard analysis, deterministic = uses available historical seismic and geological data to generate discrete, single-value earthquake events to model earthquake ground motions at a particular site.

Seismic hazard analysis, probabilistic = uses the seismic source identification and characterisation elements of the deterministic seismic hazard analysis and adds an assessment of the likelihood that ground motions of specified amplitude will occur at a particular site.

Directly accounts for the probability or frequency of occurrence of different magnitude earthquakes on each significant seismic source and inherent uncertainties.

Shear strength = a common constitutive model for soils used to quantify shear strength based on Mohr-Coulomb failure criteria. This model defines the shear strength of cohesionless soils based on the friction and apparent cohesion between the soil particles. For cohesive soils, Mohr-Coulomb defines the shear strength based on a combination of inter-particle bonding (cohesion) and friction between the soil particles.

Soil moisture deficit = the amount of water that is needed at any given time to bring the water content of the near-surface zone back to field capacity.

Specific gravity = the density of the solid soil particles. It is a dimensionless coefficient and is expressed as a ratio of the density of soil solids to the density of water.

Steady state condition = when seepage velocities or flows are constant and do not vary with time.

Stress = force per unit area. Stress therefore has the same units as pressure, but is normally used for a body's internal response to the externally applied pressure.

Stress, effective = the difference between the total stress (the total pressure experienced as a result of the weight of the overlying material) and the pore water pressure. The effective stress is the pressure with which the grains or blocks of the formation are held in contact. If the pore pressure is reduced by lowering the phreatic surface (water table or potentiometric surface), the effective stress will increase. If the total stress is reduced (e.g. by removal of overlying material) the effective stress will reduce.

Total porosity = *see* Porosity, total.

Travel distance angle = (fahrböschung in German) after a waste dump failure with runout, the angle determined by the ratio of the height (H) from original crest to runout toe over the length (L) defined from the original crest to the runout toe.

Underdrainage = the collection and conveyance of excess water (e.g. groundwater, meteoric water) or leachate through a network of interconnected drains or a drainage layer constructed in, or on, the foundation of a waste dump or stockpile.

Unsaturated zone = the zone between the ground surface and the water table, where the pore pressure is less than atmospheric. Some of the pores may, however, be completely filled with water. Also referred to as the vadose zone.

Void ratio = the ratio of pore volume to grain volume. *See also* Porosity, which is the ratio of pore volume to bulk volume.

Water table = the surface in a rock or soil mass at which the pore pressure is exactly equal to atmospheric pressure. Frequently used interchangeably with the term 'phreatic surface'.

Wireline extensometer an instrument that measures the change in distance between two points using a thin cable or wire; the most commonly used instrument for monitoring waste dump and stockpile crest deformation.

References

Abramson LW, Lee TS, Sharma S, Boyce GM (2002) *Slope Stability and Stabilization Methods.* John Wiley & Sons, New York, NY.

Abt SR, Ruff JF, Wittler RJ (1991) Estimating flow through riprap. *Journal of Hydraulic Engineering* **117**(5), 670–675. doi:10.1061/(ASCE)0733-9429(1991)117:5(670)

ACG (2008) *Proceedings of the First International Seminar on the Management of Rock Dumps, Stockpiles and Heap Leach Pads.* 5–6 March, Perth. (Ed. A Fourie). Australian Centre for Geomechanics, Nedlands.

Agliardi F, Crost GB, Frattini P (2012) Slow rock-slope deformation. In *Landslides: Type, Mechanism and Modeling.* (Eds JJ Clague, D Stead). pp. 201–221. Cambridge University Press, Cambridge.

Agliardi F, Crosta GB, Zanchi A (2001) Structural constraints on deep-seated slope deformation kinematics. *Engineering Geology* **59**, 83–102. doi:10.1016/S0013-7952(00)00066-1

Al-Hussaini M (1983) Effect of particle size and strain conditions on the strength of crushed basalt. *Canadian Geotechnical Journal* **20**(4), 706–717. doi:10.1139/t83-077

Albrecht B, Benson C (2001) Effect of desiccation on compacted natural clays. *Journal of Geotechnical and Geoenvironmental Engineering* **127**(1), 67–75. doi:10.1061/(ASCE)1090-0241(2001)127:1(67)

American Institute of Chemical Engineers (1989) *Guidelines for Chemical Process Quantitative Risk Analysis.* Center for Chemical Process Safety of the American Institute of Chemical Engineers, New York, NY.

Andrina J (2009) Physical and geochemical behavior of mine rock stockpiles in high rainfall environments. PhD thesis. Department of Mining Engineering, University of British Columbia, Vancouver.

Andrina J, Wilson GW, Miller S, Neale A (2006) Performance of the acid rock drainage mitigation waste rock trial dump at Grasberg Mine. In *Proceedings of the 7th International Conference on Acid Rock Drainage.* St Louis, MO. (Ed. RI Barnhisel). pp. 30–44. American Society of Mining and Reclamation, Lexington, KY.

Andrina J, Wilson GW, Miller SD (2012) Waste rock kinetic testing program: assessment of the scale up factor for sulphate and metal release rates. In *Proceedings of 9th International Conference on Acid Rock Drainage.* 20–26 May 2012, Ottawa, Ontario. (Eds WA Price, C Hogan, G Tremblay). pp. 882–893.

ANSI/IES (American National Standards Committee; Illuminating Engineering Society of North America) (1984) *American National Standard Practice for Industrial Lighting.* Illuminating Engineering Society of North America, Industrial Lighting Committee, New York, NY.

ANZMEC and MCA (Australian and New Zealand Minerals and Energy Council and Minerals Council of Australia) (2000) *Strategic Framework for Mine Closure.* Australian and New Zealand Minerals and Energy Council and Minerals Council of Australia, Canberra and Dickson.

Araújo NF, Correia AG (2011) The cyclic triaxial test as tool to quantify coarse non-conventional materials response. In *Proceedings of WASTES: Solutions, Treatments and Opportunities.* 12–14 September, Guimarães. pp. 179–189. Centro para a Valorização de Resíduos, Campus de Azurém, Guimarães.

ASTM (2005) *Standard Test Method for Mechanical Cone Penetration Tests of Soil (Withdrawn 2014)* (ASTM D3441–05). ASTM International, West Conshohocken, PA. <http://compass.astm.org/Standards/WITHDRAWN/D3441.htm>.

ASTM (2006a) *Standard Test Method for Resistance to Degradation of Small-Size Coarse Aggregate by Abrasion and Impact in the Los Angeles Machine* (ASTM C131/C131M-14). ASTM International, West Conshohocken, PA. <http://www.astm.org/cgi-bin/resolver.cgi?C131C131M>

ASTM (2006b) *Standard Test Method for Permeability of Granular Soils (Constant Head)* (ASTM D2434–68(2006); withdrawn 2015). ASTM International, West Conshohocken, PA. <http://compass.astm.org/Standards/WITHDRAWN/D2434.htm>.

ASTM (2007a) *Standard Test Method for Particle-Size Analysis of Soils* (ASTM D422–63(2007)e2; withdrawn 2016). ASTM International, West Conshohocken, PA. <http://compass.astm.org/Standards/WITHDRAWN/D422.htm>.

ASTM (2007b) *Standard Practice for Description of Frozen Soils (Visual-Manual Procedure)* (ASTM D4083–89(2007)). ASTM International, West Conshohocken, PA. <http://www.astm.org/cgi-bin/resolver.cgi?D4083>.

ASTM (2008a) *Standard Test Method for Determination of Water (Moisture) Content of Soil by Microwave Oven Heating* (ASTM D4643–08). ASTM International, West Conshohocken, PA. <http://www.astm.org/cgi-bin/resolver.cgi?D4643>.

ASTM (2008b) *Standard Test Method for Determining Rock Quality Designation (RQD) of Rock Core* (ASTM D6032–08). ASTM International, West Conshohocken, PA. <http://www.astm.org/cgi-bin/resolver.cgi?D6032>.

ASTM (2008c) *Standard Test Method for Determination of the Point Load Strength Index of Rock and Application to Rock Strength Classifications* (ASTM D5731–08). ASTM International, West Conshohocken, PA. <http://www.astm.org/cgi-bin/resolver.cgi?D5731>.

ASTM (2008d) *Standard Test Method for Splitting Tensile Strength of Intact Rock Core Specimens* (ASTM D3967–08). ASTM International, West Conshohocken, PA. <http://www.astm.org/cgi-bin/resolver.cgi?D3967>.

ASTM (2008e) *Standard Test Method for Performing Laboratory Direct Shear Strength Tests of Rock Specimens Under Constant Normal Force* (ASTM D5607–08). ASTM International, West Conshohocken, PA. <http://www.astm.org/cgi-bin/resolver.cgi?D5607>.

ASTM (2008f) *Standard Test Method for Slake Durability of Shales and Similar Weak Rocks* (ASTM D4644–08). ASTM International, West Conshohocken, PA. <http://www.astm.org/cgi-bin/resolver.cgi?D4644>.

ASTM (2009a) *Standard Practice for Soil Exploration and Sampling by Auger Borings* (ASTM D1452–09). ASTM International, West Conshohocken, PA. <http://www.astm.org/cgi-bin/resolver.cgi?D1452>.

ASTM (2009b) *Standard Practice for Description and Identification of Soils (Visual-Manual Procedure)* (ASTM D2488–09a). ASTM International, West Conshohocken, PA. <http://www.astm.org/cgi-bin/resolver.cgi?D2488>.

ASTM (2009c) *Standard Test Methods for Particle-Size Distribution (Gradation) of Soils Using Sieve Analysis* (ASTM D6913–04(2009)e1). ASTM International, West Conshohocken, PA. <http://www.astm.org/cgi-bin/resolver.cgi?D6913>.

ASTM (2010a) *Standard Guide for Planning and Conducting Borehole Geophysical Logging* (ASTM D5753–05[2010]). ASTM International, West Conshohocken, PA. < http://www.astm.org/cgi-bin/resolver.cgi?D5753>.

ASTM (2010b) *Standard Guide for Comparison of Field Methods for Determining Hydraulic Conductivity in Vadose Zone* (ASTM D5126/D5126M–90[2010]e1). ASTM International, West Conshohocken, PA. <http://www.astm.org/cgi-bin/resolver.cgi?D5126D5126M>.

ASTM (2010c) *Standard Guide for Selection of Aquifer Test Method in Determining Hydraulic Properties by Well Techniques* (ASTM D4043–96[2010]e1) ASTM International, West Conshohocken, PA. < http://www.astm.org/cgi-bin/resolver.cgi?D4043>.

ASTM (2010d) *Standard Test Methods for Laboratory Determination of Water (Moisture) Content of Soil and Rock by Mass* (ASTM D2216–10). ASTM International, West Conshohocken, PA. <http://www.astm.org/cgi-bin/resolver.cgi?D2216>.

ASTM (2010e) *Standard Test Methods for Liquid Limit, Plastic Limit, and Plasticity Index of Soils* (ASTM D4318–10e1). ASTM International, West Conshohocken, PA. <http://www.astm.org/cgi-bin/resolver.cgi?D4318>.

ASTM (2011a) *Standard Practice for Classification of Soils for Engineering Purposes (Unified Soil Classification System).* (ASTM D2487–11). ASTM International, West Conshohocken, PA. <http://www.astm.org/cgi-bin/resolver.cgi?D2487>.

ASTM (2011b) *Standard Guide for Selecting Surface Geophysical Methods* (ASTM D6429–99[2011]e1). ASTM International, West Conshohocken, PA. < http://www.astm.org/cgi-bin/resolver.cgi?D6429>.

ASTM (2011c) *Standard Guide for Using the Seismic Refraction Method for Subsurface Investigation* (ASTM D5777–00[2011]e1). ASTM International, West Conshohocken, PA. < http://www.astm.org/cgi-bin/resolver.cgi?D5777>.

ASTM (2011d) *Standard Test Method for Standard Penetration Test (SPT) and Split-Barrel Sampling of Soils* (ASTM D1586–11). ASTM International, West Conshohocken, PA. <http://www.astm.org/cgi-bin/resolver.cgi?D1586>.

ASTM (2011e) *Standard Practice for Determining the Normalized Penetration Resistance of Sands for Evaluation of Liquefaction Potential* (ASTM D6066–11). ASTM International, West Conshohocken, PA. <http://www.astm.org/cgi-bin/resolver.cgi?D6066>.

ASTM (2011f) *Standard Test Method for Direct Shear Test of Soils Under Consolidated Drained Conditions* (ASTM D3080/D3080M–11). ASTM International, West Conshohocken, PA. <http://www.astm.org/cgi-bin/resolver.cgi?D3080D3080M>.

ASTM (2011g) *Method for Consolidated Drained Triaxial Compression Test for Soils* (ASTM D7181–11). ASTM International, West Conshohocken, PA. <http://www.astm.org/cgi-bin/resolver.cgi?D7181>.

ASTM (2011h) *Standard Test Method for Consolidated Undrained Triaxial Compression Test for Cohesive Soils* (ASTM D4767–11). ASTM International, West Conshohocken, PA. <http://www.astm.org/cgi-bin/resolver.cgi?D4767>.

ASTM (2011i) *Standard Test Methods for One-Dimensional Consolidation Properties of Soils Using Incremental Loading* (ASTM D2435/D2435M–11). ASTM International, West Conshohocken, PA. <http://www.astm.org/cgi-bin/resolver.cgi?D2435D2435M>.

ASTM (2011j) *Standard Test Method for Laboratory Determination of Creep Properties of Frozen Soil Samples by Uniaxial Compression* (ASTM D5520–11). ASTM International, West Conshohocken, PA. <http://www.astm.org/cgi-bin/resolver.cgi?D5520>.

ASTM (2012a) *Standard Practice for Estimating Peat Deposit Thickness* (ASTM D4544–12). ASTM International, West Conshohocken, PA. <http://www.astm.org/cgi-bin/resolver.cgi?D4544>.

ASTM (2012b) *Standard Test Method for Electronic Friction Cone and Piezocone Penetration Testing of Soils* (ASTM D5778–12). ASTM International, West Conshohocken, PA. <http://www.astm.org/cgi-bin/resolver.cgi?D5778>.

ASTM (2012c) *Standard Test Method for Resistance to Degradation of Large-Size Coarse Aggregate by Abrasion and Impact in the Los Angeles Machine* (ASTM C535–12). ASTM International, West Conshohocken, PA. <http://www.astm.org/cgi-bin/resolver.cgi?C535>.

ASTM (2013a) *Standard Test Method for Load Controlled Cyclic Triaxial Strength of Soil* (ASTM D5311/D5311M–13). ASTM International, West Conshohocken, PA. <http://www.astm.org/cgi-bin/resolver.cgi?D5311D5311M>.

ASTM (2013b) *Standard Test Method for Evaluation of Durability of Rock for Erosion Control Under Freezing and Thawing Conditions* (ASTM D5312/D5312M–12). ASTM International, West Conshohocken, PA <http://www.astm.org/cgi-bin/resolver.cgi?D5312D5312M>.

ASTM (2013c) *Standard Test Method for Soundness of Aggregates by Use of Sodium Sulfate or Magnesium Sulfate* (ASTM C88–13). ASTM International, West Conshohocken, PA. <http://www.astm.org/cgi-bin/resolver.cgi?C88>.

ASTM (2013d) *Standard Test Methods for Density of Soil and Rock in Place by the Water Replacement Method in a Test Pit* (ASTM D5030/D5030M–13a). ASTM International, West Conshohocken, PA. <http://www.astm.org/cgi-bin/resolver.cgi?D5030D5030M>.

ASTM (2013e) *Standard Test Method for Laboratory Weathering of Solid Materials Using a Humidity Cell* (ASTM D5744e1). ASTM International, West Conshohocken, PA. <http://www.astm.org/cgi-bin/resolver.cgi?D5744>.

ASTM (2014a) *Standard Test Methods for Crosshole Seismic Testing* (ASTM D4428/D4428M–14). ASTM International, West Conshohocken, PA. < http://www.astm.org/cgi-bin/resolver.cgi?D4428D4428M>.

ASTM (2014b) *Standard Test Method for (Field Procedure) for Withdrawal and Injection Well Testing for Determining Hydraulic Properties of Aquifer Systems* (ASTM D4050–14). ASTM International, West Conshohocken, PA. < http://www.astm.org/cgi-bin/resolver.cgi?D4050>.

ASTM (2014c) *Standard Practices for Preserving and Transporting Soil Samples* (ASTM D4220/D4220M–14). ASTM International, West Conshohocken, PA. <http://www.astm.org/cgi-bin/resolver.cgi?D4220D4220M>.

ASTM (2014d) *Standard Test Method for Sieve Analysis of Fine and Coarse Aggregates* (ASTM C136/C136M–14).

ASTM International, West Conshohocken, PA. <http://www.astm.org/cgi-bin/resolver.cgi?C136C136M>.

ASTM (2014e) *Standard Test Methods for Specific Gravity of Soil Solids by Water Pycnometer* (ASTM D854–14). ASTM International, West Conshohocken, PA. <http://www.astm.org/cgi-bin/resolver.cgi?D854>.

ASTM (2014f) *Standard Practice for Rock Core Drilling and Sampling of Rock for Site Exploration* (ASTM D2113–14). ASTM International, West Conshohocken, PA. <http://www.astm.org/cgi-bin/resolver.cgi?D2113>.

ASTM (2014g) *Standard Test Method for Compressive Strength and Elastic Moduli of Intact Rock Core Specimens under Varying States of Stress and Temperatures* (ASTM D7012–14). ASTM International, West Conshohocken, PA. <http://www.astm.org/cgi-bin/resolver.cgi?D7012>.

ASTM (2015a) *Standard Test Method for Field Vane Shear Test in Saturated Fine-Grained Soils* (ASTM D2573/D2573M–15). ASTM International, West Conshohocken, PA. <http://www.astm.org/cgi-bin/resolver.cgi?D2573D2573M>.

ASTM (2015b) *Standard Test Method for Performing the Flat Plate Dilatometer* (ASTM D6635–15). ASTM International, West Conshohocken, PA. <http://www.astm.org/cgi-bin/resolver.cgi?D6635>.

ASTM (2015c) *Standard Test Method for (Field Procedure) for Instantaneous Change in Head (Slug) Tests for Determining Hydraulic Properties of Aquifers* (ASTM D4044/D4044M–15). ASTM International, West Conshohocken, PA. <http://www.astm.org/cgi-bin/resolver.cgi?D4044D4044M>.

ASTM (2015d) *Standard Guide for Soil Sampling from the Vadose Zone* (ASTM D4700–15). ASTM International, West Conshohocken, PA. <http://www.astm.org/cgi-bin/resolver.cgi?D4700>.

ASTM (2015e) *Standard Practice for Thin-Walled Tube Sampling of Fine-Grained Soils for Geotechnical Purposes* (ASTM D1587/D1587M–15). ASTM International, West Conshohocken, PA. <http://www.astm.org/cgi-bin/resolver.cgi?D1587D1587M>.

ASTM (2015f) *Standard Test Method for Density, Relative Density (Specific Gravity), and Absorption of Coarse Aggregate* (ASTM C127–15). ASTM International, West Conshohocken, PA. <http://www.astm.org/cgi-bin/resolver.cgi?C127>.

ASTM (2015g) *Standard Test Method for Unconsolidated-Undrained Triaxial Compression Test on Cohesive Soils* (ASTM D2850–15). ASTM International, West Conshohocken, PA. <http://www.astm.org/cgi-bin/resolver.cgi?D2850>.

ASTM (2015h) *Standard Test Methods for Particle Size Analysis of Natural and Man-Made Riprap Materials* (ASTM D5519–15). ASTM International, West

Conshohocken, PA. <http://www.astm.org/cgi-bin/resolver.cgi?D5519>.

Aube BC, St-Arnaud LC, Payant SC, Yanful EK (1995) Laboratory evaluation of the effectiveness of water covers for preventing acid generation from pyritic rock. In *Proceedings of Sudbury '95: Mining and the Environment*. 18 May to 1 June, Sudbury, Canada. pp. 494–505. Laurentian University, Sudbury.

Aubertin M (2005) Coefficient of diffusion versus degree of saturation for saturated porous media. Figure from unpublished presentation, reproduced in *Global Acid Rock Drainage Guide*. Chapter 6. (INAP [International Network for Acid Prevention] 2014). <http://www.gardguide.com/index.php?title=Chapter_6>.

Ayres B, Dobchuk B, Christensen D, O'Kane M, Fawcett M (2006) Incorporation of natural slope features into the design of final landforms for waste rock stockpiles. In *Proceedings of 7th International Conference on Acid Rock Drainage*. 26–30 March, St Louis, MO. (Ed. RI Barnhisel). pp. 59–75. American Society of Mining and Reclamation, Lexington, KY.

Aziz ML, Ferguson KD (1997) Equity Silver Mine – integrated case study. In *Proceedings of the Fourth International Conference on Acid Rock Drainage*. 31 May to 6 June, Vancouver. pp. 181–196. Bitech Publishers, Richmond.

Barton N, Kjaernsli B (1981) Shear strength of rockfill. *Journal of the Geotechnical Engineering Division. Proceedings of the American Society of Civil Engineers* **107**(GT7), 873–891.

Barton N, Lien R, Lunde J (1974) Engineering classification of rock masses for design of tunnel support. *Rock Mechanics* **6**(4), 189–236. doi:10.1007/BF01239496

Barton NR, Choubey V (1977) The shear strength of rock joints in theory and practice. *Rock Mechanics* **10**(1–2), 1–54. doi:10.1007/BF01261801

BCMEM (British Columbia Ministry of Energy and Mines) (2012) Waste dump incidents pre92.doc, Waste dump incidents post92.xls. Summary files received by email on 1 November 2012 from George Warnock, BCMEM.

BCMWRPRC (1991a) *Investigation and Design Manual – Interim Guidelines*. Mine Rock and Overburden Piles 1. Prepared by Piteau Associates Engineering Ltd, May 1991. <http://mssi.nrs.gov.bc.ca/Geotechnical/minedrockoverburdenpile_investigationdesignmanual.pdf>.

BCMWRPRC (1991b) *Operating and Monitoring Manual – Interim Guidelines*. Mine Rock and Overburden Piles 2. Prepared by Klohn Leonoff Ltd, May 1991. <http://mssi.nrs.gov.bc.ca/Geotechnical/operationmonitoring_minedumps.pdf>.

BCMWRPRC (1992a) *Review and Evaluation of Failures, Interim Report*. Mine Rock and Overburden Piles 3. Prepared by Scott Broughton, March 1992.

BCMWRPRC (1992b) *Runout Characteristics of Debris from Dump Failures in Mountainous Terrain Stage 1 Data Collection*. Vol. I Text and Tables, Vol. II Drawings and Photographs. Mine Rock and Overburden Piles 4. Prepared by Golder Associates Ltd, March 1992. <http://www.gov.bc.ca/ener/popt/down/RunoutCharacteristicsofDebris_DumpFailuresinMountainousTerrain_Stage1_Vol1.pdf>.

BCMWRPRC (1992c) *Methods of Monitoring Waste Dumps Located in Mountainous Terrain*. Mine Rock and Overburden Piles 5. Prepared by HBT AGRA Limited, March 1992. <http://mssi.nrs.gov.bc.ca/Geotechnical/monitoring_waste_dumps_in_mountainous_terrain.pdf>.

BCMWRPRC (1994) *Consequence Assessment for Mine Waste Dump Failures, Interim Report*. Mine Rock and Overburden Piles 8. Prepared by Golder Associates Ltd, December 1994. <http://mssi.nrs.gov.bc.ca/Geotechnical/ConsequenceAssessment_MineWasteDumpFailures.pdf>.

BCMWRPRC (1995) *Runout Characteristics of Debris from Dump Failures in Mountainous Terrain Stage 2 analysis, modelling and prediction*. Mine Rock and Overburden Piles 9. Prepared by Golder Associates Ltd and O. Hungr Geotechnical Research Ltd, February 1995. <http://www.gov.bc.ca/ener/popt/down/RunoutCharacteristicsofDebris_DumpFailuresinMountainousTerrain_Stage2.pdf>.

BCMWRPRC (1997) *Rock Drain Research Program, Final Report*. Mine Rock and Overburden Piles 11. Prepared by Piteau Engineering, March 1997. <http://mssi.nrs.gov.bc.ca/Geotechnical/RockDrainResearchProgram_FinalReport.pdf>.

Beale G, Read J (2013) *Guidelines for Evaluating Water in Pit Slope Stability*. CSIRO Publishing, Melbourne, and CRC Press/Balkema, Rotterdam.

Becker E, Chan CK, Seed HB (1972) *Strength and Deformation Characteristics of Rockfill Materials in Plane Strain and Triaxial Compression Tests*. Report No. TE-72-3. Department of Civil Engineering, University of California, Berkeley, CA.

Beckie RD, Aranda C, Blackmore S, Peterson HE, Hirsche DT, Javadi M, Blaskovich R, Haupt C, Dockrey J, Conlan M, Bay D, Harrison B, Brienne S, Smith L, Klein B, Mayer KU (2011) A study of the mineralogical, hydrological and biogeochemical controls on drainage from waste rock at the Antamina Mine, Peru: an overview. In *Proceedings of the 15th International Conference on Tailings and Mine Waste*. 6–9 November, Vancouver. Norman B. Keevil Institute of Mining Engineering, University of British Columbia, Vancouver.

Benda LE, Cundy TW (1990) Predicting deposition of debris flows in mountain channels. *Canadian Geotechnical Journal* **27**, 409–417. doi:10.1139/t90-057

Biarez J, Hicher P-Y (1994) *Elementary Mechanics of Soil Behaviour: Saturated Remoulded Soils*. A.A. Balkema, Rotterdam.

Bieniawski ZT (1976) Rock mass classification in rock engineering. In *Exploration for Rock Engineering, Proceedings of the Symposium on Exploration for Rock Engineering*. 1–5 November. (Ed. ZT Bieniawski). pp. 97–106. Cape Town, Balkema.

Bieniawski ZT (1989) *Engineering Rock Mass Classifications*. John Wiley & Sons, New York, NY.

Bishop AW (1955) The use of the slip circle in the stability analysis of slopes. *Geotechnique* **5**(1), 7–17. doi:10.1680/geot.1955.5.1.7

Bishop AW (1973) The stability of tips and spoil heaps. *Quarterly Journal of Engineering Geology* **6**, 335–376. doi:10.1144/GSL.QJEG.1973.006.03.15

Bishop AW, Hutchinson JN, Penman ADM, Evans HE (1969) Geotechnical investigation into the causes and circumstances of the disaster of 21 October 1966. In *A Selection of Technical Reports Submitted to the Aberfan Tribunal*. pp. 1–80 (Item 1). HMSO, London, Welsh Office and The Stationary Office Ltd, Norwich.

Blake TF, Hollingsworth RA, Stewart JP (2002) *Recommended Procedures for Implementation of DMG Special Publication 117 – Guidelines for Analyzing and Mitigating Landslide Hazards in California*. Committee organized through the American Society of Civil Engineers, Los Angeles Section Geotechnical Group, Southern California Earthquake Center, Los Angeles.

Blight G (1981) Failure mode. In *Workshop on Design of Non-Impounding Mine Waste Dumps*. SME, AIME, Denver, CO.

Bradfield LR, Fityus SG, Simmons JV (2015) A trilinear shear strength envelope for coal mine spoil. In *Proceedings of the International Symposium in Mining and Civil Engineering 2016*. pp. 215–229. South African Institute of Mining and Metallurgy, Cape Town.

Bratty MP, Kratochvil D, Lawrence R (2006) Applications of biological H_2S production from elemental sulphur in the treatment of heavy metal pollution including acid rock drainage. In *Proceedings of the 7th International Conference on Acid Rock Drainage (ICARD)*. 26–30 March, St Louis, MO. (Ed. RI Barnhisel). pp. 271–281. American Society of Mining and Reclamation, Lexington, KY.

Breitenbach AJ (1993) Rockfill placement and compaction guidelines. *Geotechnical Testing Journal* **16**(1), 76–84. doi:10.1520/GTJ10270J

British Standards Institution (2002) *Geotechnical Investigation and Testing – Identification and Classification of Soil – Part 1: Identification and Description* (BS EN ISO 14688-1). BSI Standards, London.

British Standards Institution (2004) *Geotechnical Investigation and Testing – Identification and Classification of Soil – Part 2: Principles for a Classification* (BS EN ISO 14688-2). BSI Standards, London.

British Standards Institution (2015) *Code of Practice for Ground Investigations* (BS5930:2015). BSI Standards, London.

Brown J, Ferrians OJ, Jr, Heginbottom JA, Melnikov ES (Eds) (1997) *Circum-Arctic Map of Permafrost and Ground-Ice Conditions*. US Geological Survey in Cooperation with the Circum-Pacific Council for Energy and Mineral Resources, Washington, DC. Circum-Pacific Map Series CP-45, scale 1:10,000,000, 1 sheet.

Bryant S, Duncan JM, Seed HB (1983) *Application of Tailings Flow Analyses to Field Conditions*. Report No. UCB/GT/83-01. Department of Civil Engineering, University of California, Berkeley, CA.

Caldwell JA, Moss ASE (1981) Simplified stability analysis. In *Workshop on Design of Non-impounding Mine Waste Dumps*. SME, AIME, Denver, CO.

Campbell DB (1981) Construction and performance in mountainous terrain. In *Workshop on Design of Non-impounding Mine Waste Dumps*. SME, AIME, Denver, CO.

Campbell DB (1986) Stability and performance of waste dumps on steeply sloping terrain. In *Proceedings of International Symposium on Geotechnical Stability in Surface Mining*. Calgary.

Campbell DB (2000) The mechanism controlling angle-of-repose stability in waste rock embankments. In *Proceedings, 4th International Conference on Stability in Open Pit Mining, Society for Mining, Metallurgy, and Exploration Inc.* Denver. pp. 285–291. SME, Littleton, CO.

Campbell DB, Kent A (1993) High mine rock piles in mountain terrain: Current design trends. In *Proceedings, International Congress on Mine Design*. Kingston, Ontario. pp. 67–82. A.A. Balkema, Rotterdam.

Canadian Geotechnical Society (2006) *Canadian Foundation Engineering Manual*. 4th edn. Canadian Geotechnical Society, Bitech Publishers Ltd, Vancouver.

CANMET (1977) Waste embankments. In *Pit Slope Manual*. Chapter 9. CANMET Report 77-01. Minerals Research Program, Mining Research Laboratories, Ottawa.

CANMET (1992) *Instability Mechanisms Initiating Flow Failures in Mountainous Mine Waste Dumps Phase I*. Mine Rock and Overburden Piles 6. CANMET and the University of Alberta, Edmonton.

CANMET (1994) *Liquefaction Flowslides in Western Canadian Coal Mine Waste Dumps Phase II Case Histories.* Mine Rock and Overburden Piles 7. CANMET and the University of Alberta, Edmonton.

CANMET (1995) *Liquefaction Flowslides in Western Canadian Coal Mine Waste Dumps Summary Report Phase III Volumes I and II.* Mine Rock and Overburden Piles 10. CANMET and University of Alberta, Edmonton.

Cannon SH (1993) An empirical model for the volume-change behaviour of debris flows. In *Proceedings, Hydraulic Engineering '93.* Vol. 2, pp. 1768–1777. American Society of Civil Engineers, San Francisco.

Carey SK, Barbour SL, Hendry MJ (2005) Evaporation from a waste-rock surface, Key Lake, Saskatchewan. *Canadian Geotechnical Journal* **42**, 1189–1199. doi:10.1139/t05-033

Carter M, Bentley SP (1991) *Correlations of Soil Properties.* Pentech Press Publishers, London.

Casagrande A (1940) *Characteristics of Cohesionless Soils Affecting the Stability of Slopes and Earth Fills. Contributions to Soil Mechanics, 1925 to 1940.* Boston Society of Civil Engineers, Boston, pp. 257–276.

Casagrande A (1948) Classification and identification of soils. *American Society of Civil Engineers (ASCE) Journal for Soil Mechanics and Foundation Engineering Transactions* **113**, 901–930.

Casagrande A, Fadum RE (1940) *Notes on Soil Testing for Engineering Purposes.* Publication 268. Harvard University Graduate School of Engineering, Cambridge, MA.

CCME (Canadian Council of Ministers of the Environment) (2003) *Climate, Nature, People: Indicators of Canada's Changing Climate.* CCME, Winnipeg.

CEN (European Committee for Standardization) (2004) *Eurocode 7: Geotechnical Design – Part 1: General rules. EN 1997-1: 2004.* BSI Standards, London.

CEN (European Committee for Standardization) (2007) *Eurocode 7: Geotechnical Design – Part 2: Ground investigation and testing. EN 1997-2: 2007.* BSI Standards, London.

Cetin K, Seed R, Der Kiureghian A, Tokimatsu K, Harder L, Jr, Kayen R, Moss R (2004) Standard penetration test-based probabilistic and deterministic assessment of seismic soil liquefaction potential. *Journal of Geotechnical and Geoenvironmental Engineering* **130**, 1314–1340. doi:10.1061/(ASCE)1090-0241(2004)130:12(1314)

CGS (Canadian Geotechnical Society) (2006) *Canadian Foundation Engineering Manual.* 4th edn. Canadian Geotechnical Society, Vancouver.

Chakraborty D, Choudhury D (2013) Pseudo-static and pseudo-dynamic stability analysis of tailings dam under seismic conditions. *Proceedings of the National Academy of Sciences, India Section A: Physical Sciences.* **83**(1), 63–71.

Charles JA, Watts KS (1980) The influence of confining pressure on shear strength of compacted rockfill. *Geotechnique* **30**(4), 353–367. doi:10.1680/geot.1980.30.4.353

Cheney R, Chassie R (2000) *Soils and Foundations Workshop Reference Manual.* National Highway Institute Publication NHI-00-045. pp. 3–8. Federal Highway Administration, Washington, DC.

Chowdhury RN, Flentje PN, Bhattacharya G (2009) *Geotechnical Slope Analysis.* CRC Press – Taylor & Francis Group, Boca Raton, FL.

Claridge FB, Nichols RS, Stewart AF (1986) Mine waste dumps constructed in mountain valleys. *Canadian Institute of Mining and Metallurgy Bulletin* **79**(892), 79–87.

Clayton CRI, Matthews MC, Simons NE (1995) *Site Investigation.* 2nd edn. <www.geotechnique.info>.

Clover Technology (2015) *Galena Slope Stability Analysis System.* Version 6.1. <www.galenasoftware.com>.

Contreras L (2011) Comportamiento friccionante de materiales granulares gruesos. MSc thesis. Department of Civil Engineering, University of Chile, Santiago [in Spanish].

Corominas J (1996) The angle of reach as a mobility index for small and large landslides. *Canadian Geotechnical Journal* **33**, 260–271. doi:10.1139/t96-005

Cornforth DH, Worth EG, Wright WL (1974) Observations and analysis of a flow slide in sand fill. In *Proceedings of the Symposium on Field Instrumentation in Geotechnical Engineering, UK.* 30 May to 1 June 1973, London. (Ed. British Geotechnical Society). pp. 136–151. Butterworths, London.

Crosta GB, Cucchiaro S, Frattini P (2003) Validation of semi-empirical relationship for the definition of debris flow behavior in granular materials. In *Debris-Flow Hazards Mitigation.* (Eds D Rickenmann and CL Chen). pp. 821–831. Millpress, Rotterdam.

Crosta GB, Imposimato S, Roddeman D (2009) Numerical modeling of 2-D granular step collapse on erodible and nonerodible surface. *Journal of Geophysical Research* **114**, F03020. doi:10.1029 /2008JF001186

Crova R, Jamiolkowski M, Lancellota R, Lo Presti DCF (1993) Geotechnical characterization of gravelly soils at Messina site: selected topics. In *Predictive Soil Mechanics: Proceedings of the Wroth Memorial Symposium.* 27–29 July 1992, St Catherine's College, Oxford. (Eds GT Houlsby and AN Schofield). pp. 199–218. Thomas Telford, London, UK.

Cubrinovski M, Ishihara K (2000) Flow potential of sandy soils with different grain compositions. *Soil and Foundation* **40**, 103–119. doi:10.3208/sandf.40.4_103

Daniel CR, Howie JA, Sy A (2003) A method for correlating large penetration test (LPT) to standard penetration test (SPT) blow counts. *Canadian Geotechnical Journal* **40**, 66–77. doi:10.1139/t02-094

Darling P (2011) *SME Mining Engineering Handbook*. 3rd edn. Society for Mining, Metallurgy, and Exploration, Inc., Denver, CO.

Das BM (2002) *Principles of Geotechnical Engineering*. 5th edn. Thomson Brooks/Cole, Pacific Grove.

Davis LA, Neuman SP (1983) *Documentation and User's Guide: UNSAT2 – Variably Saturated Flow Model (Including 4 Example Problems)*. Final Report. Water, Waste and Land, Inc., Fort Collins.

Dawson EM, Roth WH, Drescher A (1999) Slope stability analysis by strength reduction. *Geotechnique* **49**(6), 835–840. doi:10.1680/geot.1999.49.6.835

Dawson RF (1994) Mine waste geotechnics. PhD thesis. University of Alberta, Edmonton.

Dawson RF, Morgenstern NR, Stokes AW (1998) Liquefaction flowslides in Rocky Mountain coal mine waste dumps. *Canadian Geotechnical Journal* **35**, 328–343. doi:10.1139/t98-009

De la Hoz K (2007) Estimación de los parámetros de resistencia al corte en suelos granulares gruesos. MSc thesis. Department of Civil Engineering, University of Chile, Santiago [in Spanish].

De Mello V (1977) Reflection on design decisions of practical significance to embankment dams. *Geotechnique* **27**(3), 281–355. doi:10.1680/geot.1977.27.3.281

Deere DU (1963) Technical description of rock cores for engineering purposes. *Rock Mechanics and Engineering Geology* **1**(1), 16–22.

Department of the Navy (1971) *Design Manual – Soil Mechanics, Foundations and Earth Structures (NAVFAC DM-7)*. Naval Facilities Engineering Command, Alexandria.

DHI (2016) *FEFLOW*. Version 7.0. <http://www.mikepoweredbydhi.com/products/feflow>.

Donaghe RT, Torrey VH (1979) Scalping and replacement effects on strength parameters of earth-rock mixtures. In *Proceedings of Design Parameters in Geotechnical Engineering* **2**, 29–34. London.

Dorador L (2010) Análisis experimental de las metodologías de curvas homotéticas y corte en la evaluación de propiedades geotécnicas de suelos gruesos. MSc thesis. Department of Civil Engineering, University of Chile, Santiago [in Spanish].

Douglas KJ (2002) The shear strength of rock masses. PhD thesis. University of New South Wales, School of Civil and Environmental Engineering, Sydney.

Dowding CH, Dussud ML, Kane WF, O'Connor KM (2003) Monitoring deformation in rock and soil with TDR sensor cables. *Geotechnical Instrumentation News* **June 2003**, 51–59.

Dumbleton MJ (1981) *The British Soil Classification System for Engineering Purposes: Its Development and Relation to Other Comparable Systems*. TRRL Report LR 1030. Transport and Road Research Laboratory, Crowthorne.

Duran A (2012) Spoil piles – comparison of stability analysis methods. *Australian Geomechanics* **47**(4), 33–42.

Duran A (2013) Undrained behaviour in spoil piles. In *Proceedings, International Symposium on Slope Stability in Mining and Civil Engineering 2013*. pp. 853–865. Australian Centre for Geomechanics, Nedlands.

Eberhardt E (2006) *Sea to Sky Geotechnique 2006 Short Course – Course Notes*. October 1.

Eckersley JD (1990) Instrumented laboratory flowslides. *Geotechnique* **40**, 489–502. doi:10.1680/geot.1990.40.3.489

Environment Australia (1998) Landform design for rehabilitation. In *Best Practice Environmental Management in Mining*. Environment Australia, Kingston.

Environment Canada (2011) *Guidelines for the Assessment of Alternatives for Mine Waste Disposal*. <http://ec.gc.ca/pollution/default.asp?lang=En&n=125349F7-1>

EPA (Environmental Protection Agency) (1983) Standards for remedial action as at inactive uranium processing sites. *Federal Register* **48**(3), 590–606.

Eurocode 7 (2007) *Geotechnical design – Part 2: Ground investigation and testing*. CEN (European Committee for Standardization), Brussels.

Fannin RJ, Wise MP (2001) An empirical-statistical model for debris flow travel distance. *Canadian Geotechnical Journal* **38**, 982–994. doi:10.1139/t01-030

Fell R, Fry JJ (2007) *Internal Erosion of Dams and Their Foundations*. Taylor & Francis/Balkema, Rotterdam.

Fell R, MacGregor P, Stapledon D, Bell G (2005) *Geotechnical Engineering of Dams*. A.A. Balkema, Rotterdam.

Fellenius W (1936) Calculation of the stability of earth dams. In *Transactions of the 2nd Congress on Large Dams*. Vol. 4. pp. 445–459. International Commission on Large Dams, Washington, DC.

Fines P, Wilson GW, Williams DJ, Tran AB, Miller S (2003) Field characterization of two full-scale rock piles. In *Proceedings of the Sixth International Conference on Acid Rock Drainage*. 14–17 July, Cairns. pp. 903–910. Australasian Institute of Mining and Metallurgy, Melbourne.

Finn LWD, Kramer SL, O'Rourke TD (Chair), Youd TL (2010) *Re: Final Report: Technical Issues in Dispute with EERI MNO-12, Soil Liquefaction During Earthquakes*. Report prepared for the Earthquake Engineering Research Institute Ad Hoc Committee on Soil Liquefaction During Earthquakes, 16 August.

Fredlund DG, Krahn J (1977) Comparison of slope stability methods of analysis. *Canadian Geotechnical Journal* **14**, 429–439.

Fredlund DG, Krahn J, Pufahl DE (1981) The relationship between limit equilibrium slope stability methods. In *Proceedings of the International Conference on Soil Mechanics and Foundation Engineering*. Stockholm. Vol. 3. pp. 409–416. Balkema, Rotterdam.

GeoAnalysis (2000) *Soil Cover*. University of Saskatchewan, Saskatoon.

Geoffrey Walton Practice and Great Britain, Department of the Environment (1991) *Handbook on the Design of Tips and Related Structures*. Her Majesty's Stationery Office, London.

Geo-Slope (2012a) *VADOSE/W*. Geo-Slope International, Calgary. <http://www.geo-slope.com/>.

Geo-Slope (2012b) *SEEP/W*. Geo-Slope International, Calgary. <http://www.geo-slope.com/>.

Geo-Slope (2012c) *SLOPE/W*. Geo-Slope International, Calgary. <http://www.geo-slope.com/>.

Gitrana G, Santos M, Fredlund MD (2008) Three dimensional analysis of the Lodalen Landslide. In *Proceedings of Geocongress 2008: Geosustainability and Geohazard Mitigation*. 9–12 March, New Orleans. pp. 186–190. American Society of Civil Engineers, Reston, VA.

Golder Associates Ltd (1987) *Regional Study of Coal Mine Waste Dumps in British Columbia, Stage II*. Report prepared for Canadian British Columbia Mineral Development Agreement. November 1987 (862–1231).

Gómez P, Díaz M, Lorig L (2002) Stability analysis of waste dumps at Chuquicamata Mine, Chile. *Gluckauf-Forschungshefte* **64**(3), 93–98.

Gonano LP (1980) *An Integrated Report on Slope Failure Mechanisms at Goonyella – November 1976*. CSIRO Division of Applied Geomechanics, Technical Report Number 114. Commonwealth Scientific and Industrial Research Organisation, Institute of Earth Resources, Division of Applied Geomechanics, Melbourne.

Goodman RE (1989) *Introduction to Rock Mechanics*. 2nd edn. John Wiley & Sons, New York, NY.

Google Earth Pro (2015) *Google Earth Pro V 7.1.5.1557*. Washington, DC. 45 45-05N, 122 49′04″W, Eye alt 2.00 km. July 16, 2014. <http://www.earth.google.com>.

Gowan M, Lee M, Williams DJ (2010) Co-disposal techniques that may mitigate risks associated with storage and management of potentially acid generating wastes. In *Proceedings of Mine Waste 2010*. 29 September to 1 October, Perth. (Eds A Fourie, R Jewell) pp. 389–404. Australian Centre for Geomechanics, Nedlands.

Griffiths DV, Lane PA (1999) Slope stability analysis by finite elements. *Geotechnique* **49**(3), 387–403. doi:10.1680/geot.1999.49.3.387

GSHAP (Global Seismic Hazard Assessment Program) (1999) *Global Seismic Hazard Assessment Program*. <http://www.seismo.ethz.ch/GSHAP>.

Haan CT, Barfield BJ, Hayes JC (1994) *Design Hydrology and Sedimentology for Small Catchments*. Academic Press Inc., San Diego.

Harder LF, Jr, Seed HB (1986) *Determination of Penetration Resistance for Coarse-Grained Soils using the Becker Hammer Drill*. Rep. UCB/EERC-86/06, Earthquake Engineering Research Center, University of California at Berkeley, Berkeley.

Hawley PM (2000) Site selection, characterization, and assessment. In *Proceedings, Slope Stability in Surface Mining*. pp. 267–274. Society for Mining Metallurgy, and Exploration, Denver, CO.

Hawley PM, Ochoa X, Sharon R (2002) Design of the Pierina Waste Dump. *CIM Bulletin* **96**(1073), 76–83.

Herasymiuk G (1996) Hydrogeology of a sulfide waste rock dump. MSc thesis. Department of Civil Engineering, University of Saskatchewan, Saskatoon.

HGL (HydroGeologic, Inc.) (2013) *MODFLOW-SURFACT*. HydroGeologic, Inc., Reston, VA. <https://www.hgl.com/expertise/modeling-and-optimization/software-tools/modflow-surfact/>.

HKIE (Hong Kong Institution of Engineers) (1998) *Soil Nails in Loose Fill: A Preliminary Study (Draft Report)*. Hong Kong Institution of Engineers, Geotechnical Division, Hong Kong.

Hoek E (2007) *Practical Rock Engineering*. <www.rocscience.com/education/hoeks_corner>

Hoek E, Brown ET (1980) Empirical strength criterion for rock masses. *Journal of the Geotechnical Engineering Division* **106**(GT9), 1013–1035.

Hoek E, Brown ET (1997) Practical estimates of rock mass strength. *International Journal of Rock Mechanics and Mining Sciences & Geomechanics Abstracts* **34**(8), 1165–1186. doi:10.1016/S1365-1609(97)80069-X

Hoek E, Carranza-Torres C, Corkum B (2002) Hoek-Brown failure criterion. 2002 edn. In *Mining and Tunnelling Innovation and Opportunity. Proceedings of 5th North American Rock Mechanics Symposium and 17th Tunnelling Association of Canada Conference*. 7–10 July, Toronto, ON (Eds R Hammah, W Bawden, J Curran, M Telesnicki). Vol. 1, pp. 267–273.

Hoek E, Carter TG, Diederichs MS (2013) Quantification of the geological strength index chart. *The 47th US Rock Mechnaics/Geomechnics Symposium*. 23–26 June, San Francisco, CA. (Ed. LA Pyrak-Nolte). American Rock Mechanics Association, Alexandria.

Hoek E, Diederichs MS (2005) Empirical estimation of rock mass modulus. *International Journal of Rock Mechanics and Mining Sciences* **43**(2006) 203–215.

Hoek E, Kaiser PK, Bawden WF (1995) *Support of Underground Excavations in Hard Rock*. A.A. Balkema, Rotterdam.

Holdridge LR (1971) *Forest Environments in Tropical Life Zones: A Pilot Study*. Pergamon Press, Oxford/New York, NY.

Hough BK (1957) *Basic Soil Engineering*. Ronald Press Company, New York, NY.

Howes DE, Kenk E (1997) *Terrain Classification System for B.C. (Version 2)*. BC Ministry of Environment, Recreational Fisheries Branch and BC Ministry of Crown Lands, Surveys and Resource Mapping Branch, Victoria.

Hsü KJ (1975) Catastrophic debris stream (sturzstroms) generated by rockfalls. *Geological Society of America Bulletin* **86**, 129–140. doi:10.1130/0016-7606(1975)86<129:CDSSGB>2.0.CO;2

Hungr O (1995) A model for the runout analysis of rapid flow slides, debris flows and avalanches. *Canadian Geotechnical Journal* **32**(4), 610–623. doi:10.1139/t95-063

Hungr O, Dawson R, Kent A, Campbell D, Morgenstern NR (2002) Rapid flow slides of coal mine waste in British Columbia, Canada. In *Catastrophic Landslides*. (Ed. SG Evans). pp. 191–208. Geological Society of America, Boulder, CO.

Hungr O, McDougall S (2009) Two numerical models for landslide dynamic analysis. Computers & Geoscience 35(5), 978–992.

Hungr O, McDougall S, Bovis M (2005) Entrainment of material by debris flows. In *Debris Flow Hazards and Related Phenomena*. (Eds M Jakob, O Hungr). pp. 135–158. Springer Verlag, Heidelberg.

Hungr O, Morgan GC, Kellerhals R (1984) Quantities analysis of debris torrent hazards for design of remedial measures. *Canadian Geotechnical Journal* **21**(4), 663–677. doi:10.1139/t84-073

Hungr O, Morgenstern NR (1984) Experiments in high velocity open channel flow of granular materials. *Geotechnique* **34**, 405–413. doi:10.1680/geot.1984.34.3.405

Hungr O, Morgenstern NR, Wong HN (2007) Review of benchmarking exercise on landslide debris runout and mobility modelling. In *Proceedings of the International Forum on Landslide Disaster Management*. 10–12 December, Hong Kong. Vol. II, pp. 755–812. Geotechnical Division, Hong Kong Institution of Engineers, Hong Kong.

Hungr O, Picarelli L, Leroueil S (2014) The Varnes classification of landslides, an update. *Landslides* **11**(2), 167–194. doi:10.1007/s10346-013-0436-y

Hunter G, Fell R (2003) Travel distance angle for 'rapid' landslides in constructed and natural slopes. *Canadian Geotechnical Journal* **40**, 1123–1141. doi:10.1139/t03-061

Hustrulid WA, McCarter MK, Van Zyl DJA (Eds) (2000) *Slope Stability in Surface Mining*. Society for Mining, Metallurgy, and Exploration, Inc., Littleton, CO.

Hutchinson JN (1986) A sliding-consolidation model for flow slides. Canadian Geotechnical Journal 23, 115–126.

Hutchinson JN, Bhandari RK (1971) Undrained loading: a fundamental mechanism of mudflows and other mass movements. *Geotechnique* **21**(4), 353–358. doi:10.1680/geot.1971.21.4.353

Hynes-Griffin ME, Franklin AG (1984) Rationalizing the seismic coefficient method. Department of the Army, US Army Corps of Engineers, Waterways Experimental Station, Miscellaneous Paper GL-84-13. Vicksburg, MS.

ICMM (International Council on Mining & Metals) (2008) *Planning for Integrated Mine Closure: Toolkit*. International Council on Mining and Metals, London.

ICOLD (International Commission on Large Dams) (2013) Sustainable design and post-closure performance of tailings dams. Bulletin 153. International Commission on Large Dams, Paris.

Imran J, Parker G, Locat J, Lee H (2001) A 1-D numerical model of muddy subaqueous and subaerial debris flows. *Journal of Hydraulic Engineering* **127**(11), 959–968. doi:10.1061/(ASCE)0733-9429(2001)127:11(959)

INAP (International Network for Acid Prevention) (2009) *The Global Acid Rock Drainage (GARD) Guide*. <http://www.gardguide.com>.

INAP (International Network for Acid Prevention) (2014) *Global Acid Rock Drainage Guide* Rev. 1. International Network for Acid Prevention and Global Alliance. <http://www.gardguide.com/index.php?title=Main_Page>.

ISRM (1974) *Suggested Methods for Determining Shear Strength*. Document No. 1. International Society for Rock Mechanics, Commission on Standardization of Laboratory and Field Tests, Lisbon.

ISRM (1978a) Suggested methods for the quantitative description of discontinuities in rock masses. *International Journal of Rock Mechanics and Mining Sciences & Geomechanics Abstracts* **15**, 319–368.

ISRM (1978b) Suggested methods for determining tensile strength of rock materials. Part 2: suggested method for determining indirect tensile strength by the Brazil Test. *International Journal of Rock Mechanics and Mining Sciences* **15**, 99–103. doi:10.1016/0148-9062(78)90003-7

ISRM (1981a) Rock characterization testing and monitoring. In *ISRM Suggested Methods*. (Ed. ET Brown). pp. 3–52. Pergamon, Oxford.

ISRM (1981b) Basic geotechnical description of rock masses. *International Journal of Rock Mechanics and Mining Sciences* **18**(1), 85–110.

ISRM (1985) Suggested methods for determining point load strength. *International Journal of Rock Mechanics and Mining Sciences & Geomechanics Abstracts* **22**(2), 51–60. doi:10.1016/0148-9062(85)92327-7

ISRM (2007) *The Complete ISRM Suggested Methods for Rock Characterisation, Testing and Monitoring: 1974–2006.* (Eds R Ulusay, JA Hudson). International Society for Rock Mechanics, Commission on Testing Methods, Ankara.

ISSMFE Technical Committee on Penetration Testing of Soils–TC 16 (1989) *Report of the ISSMFE Technical Committee on Penetration Testing of Soils – TC16 with Reference Test Procedures CPT-SPT-DP-WST.* Swedish Geotechnical Society, Linkpin.

Itasca (2011) *FLAC/Slope.* Version 7. <http://www.itascacg.com/>.

Itasca (2012) *FLAC 3D.* Version 5. <http://www.itascacg.com/>.

Itasca (2014) *UDEC.* Version 6. <http://www.itascacg.com/>.

Itasca (2016) *FLAC.* Version 8. <http://www.itascacg.com/>.

Iverson RM (1997) The physics of debris flows. *Reviews of Geophysics* **35**(3), 245–296. doi:10.1029/97RG00426

Iverson RM, Denlinger RP (2001) Flow of variably fluidized granular masses across three-dimensional terrain. 1. Coulomb mixture theory. *Journal of Geophysical Research* **106**, 537–552. doi:10.1029/2000JB900329

Jakob M, Hungr O (2005) *Debris-flow Hazards and Related Phenomena.* Springer-Praxis, Chichester.

Janbu N, Bjerrum L, Kjaernsli B (1956) *Soil Mechanics Applied to Some Engineering Problems.* Norwegian Geotechnical Institute, Oslo.

Jefferies M, Lorig L, Alvarez C (2008) Influence of rock-strength spatial variability on slope stability in continuum and distinct element numerical modeling in geo-engineering. In *Proceedings of the 1st International FLAC/DEM Symposium.* August, Minneapolis. Paper No. 05-01. (Eds R. Hart *et al.*) Itasca Consulting Group, Inc., Minneapolis.

Jibson RW (2011) Methods for assessing the stability of slopes during earthquakes – a retrospective. *Engineering Geology* **122**, 43–50. doi:10.1016/j.enggeo.2010.09.017

Kaito T, Sakaguchi S, Nishigaka Y, Miki K, Yukami H (1971) Large penetration test. *Tsuchi-to-Kiso* **19**(7), 15–21 [in Japanese].

Kayen R, Moss R, Thompson E, Seed R, Cetin K, Kiureghian A, Tanaka Y, Tokimatsu K (2013) Shear-wave velocity–based probabilistic and deterministic assessment of seismic soil liquefaction potential. *Journal of Geotechnical and Geoenvironmental Engineering* **139**, 407–419. doi:10.1061/(ASCE) GT.1943-5606.0000743.

Kehew AE (2006) *Geology for Engineers and Environmental Scientists.* 3rd edn. Pearson Prentice Hall, Upper Saddle River, NJ.

Kemeny J, Devgan A, Hagaman R, Wu X (1993) Analysis of rock fragmentation using digital image processing. *Journal of Geotechnical Engineering* **119**, 1144–1160. doi:10.1061/(ASCE)0733-9410(1993)119:7(1144)

Kho AK, Williams DJ, Kaneka N, Smith NJW (2013) Shear strength parameters for assessing geotechnical slope stability of open pit coal mine spoil based on laboratory tests. In *Proceedings International Symposium on Slope Stability in Mining and Civil Engineering 2013.* pp. 867–880. Australian Centre for Geomechanics, Nedlands.

Koloski JW, Schwarz SD, Tubbs DW (1989) Geotechnical properties of geologic materials. In *Engineering Geology in Washington.* Vol. 1. *Washington Division of Geology and Earth Resources Bulletin* **78**, 19–26. <http://www.tubbs.com/geotech/geotech.htm>.

Koopejan AW, Wamelen BM, Weinberg LJH (1948) Coastal flow slides in the Dutch provinces of Zeeland. In *Proceedings of the 2nd International Conference on Soil Mechanics and Foundation Engineering.* 21–30 June, Rotterdam. Vol. 5, pp. 89–96. American Association of Civil Engineers, New York, NY.

Körner HJ (1980) The energy line method in the mechanics of avalanches. *Journal of Glaciology* **26**, 501–505.

Krahn J (2003) The 2001 R.M. Hardy Lecture: the limits of limit equilibrium analyses. *Canadian Geotechnical Journal* **40**, 643–660.

Kramer SL (1988) Triggering of liquefaction flow slides in coastal soil deposits. *Engineering Geology* 26, 17–31.

Kramer SL (1996) *Geotechnical Earthquake Engineering.* Prentice-Hall Inc., Upper Saddle River.

Krumbein WC, Sloss LL (1963) *Stratigraphy and Sedimentation.* 2nd edn. WH Freeman and Company, San Francisco.

Kulhawy FH, Mayne PW (1990) *Manual on Estimating Soil Properties for Foundation Design.* Report EL-6800, Electric Power Research Institute, Palo Alto.

Laigle D, Coussot P (1997) Numerical modelling of mudflows. *Journal of Hydraulic Engineering* **123**(7), 617–623. doi:10.1061/(ASCE)0733-9429(1997)123:7(617)

Lambe W, Whitman RV (1969) *Soil Mechanics.* John Wiley & Sons, New York, NY.

Lamontagne A, Fortin S, Poulin R, Tasse N, Lefebvre R (2000) Layered co-mingling for the construction of waste rock piles as a method to mitigate acid mine drainage – laboratory investigations. In *Proceedings of the Fifth International Conference on Acid Rock Drainage.* 21–24 May, St Louis, MO. pp. 779–788. Society for Mining, Metallurgy, and Exploration, Denver, CO.

Lancaster ST, Hayes SK, Grant GE (2003) Effects of wood on debris flow runout in small mountain watersheds.

Water Resources Research **39**(6), 1168. doi:10.1029/2001WR001227

La Rosa D, Girdner K, Valery W, Abramson S (2001) *Recent Applications of the Split-Online Image Analysis System.* VI SHMMT/XVIII ENTMME, Rio de Janeiro.

Leps TM (1970) Review of shearing strength of rockfill. *Journal of the Soil Mechanics and Foundations Division. Proceedings of the American Society of Civil Engineers* **96**(SM4), 1159–1170.

Leps TM (1973) Flow through rockfill. In *Embankment Dam Engineering: Casagrande Volume.* pp. 87–107. John Wiley & Sons, New York, NY.

Lied K, Bakkehöi S (1980) Empirical calculations of snow avalanche runout distance based on topographic parameters. *Journal of Glaciology* **26**, 165–177.

Linero S, Palma C, Apablaza R (2006) Geotechnical characterisation of waste material in very high dumps with large scale triaxial testing. In *Proceedings International Symposium on Rock Slope Stability in Open Pit Mining and Civil Engineering 2007.* (Ed. Y Potvin). pp. 59–75. Australian Centre for Geomechanics, Nedlands.

Lo DOK (2000) *Review of Natural Terrain Landslide Debris-resisting Barrier Design.* GEO Report No. 104. Geotechnical Engineering Office, Hong Kong.

Logsden MJ (2013) What does 'perpetual' management and treatment mean? Toward a framework for determining an appropriate period-of-performance for management of reactive, sulphide-bearing mine wastes. In *Proceedings of the International Mine Water Association (IMWA) Conference 2013.* Golden, Colorado. pp. 53–58.

Lowe J (1964) Shear strength of coarse embankment dam materials. In *Proceedings of the International Congress on Large Dams.* pp. 745–761.

Lunne T, Berre T, Andersen KH, Strandvik S, Sjursen M (2006) Effects of sample disturbance and consolidation procedures on measured shear strength of soft marine Norwegian clays. *Canadian Geotechnical Journal* **43**, 726–750. doi:10.1139/t06-040

Lunne T, Berre T, Strandvik S (1997) Sample disturbance effects in soft low plastic Norwegian clay. In *Proceedings of the Conference on Recent developments in Soil and Pavement Mechanics.* 25–27 June, Rio de Janeiro. pp. 81–102. A.A. Balkema, Rotterdam.

Maerz NH, Palangio TC, Franklin JA (1996) WipFrag image based granulometry system. In *Proceedings of the FRAGBLAST 5 Workshop on Measurement of Blast Fragmentation*, Montreal, Quebec, Canada. (Eds JA Franklin, T Katsabanis). pp. 91–99. A.A. Balkema, Rotterdam.

Maksimović M (1996) A family of nonlinear failure envelopes for non-cemented soils and rock discontinuities. *The Electronic Journal of Geotechnical Engineering* **1**, 1–15.

Mancarella D, Hungr O (2010) Analysis of run-up of granular avalanches against steep, adverse slopes and protective barriers. *Canadian Geotechnical Journal* **47**, 827–841. doi:10.1139/T09-143

Marachi ND, Chan CK, Seed HB, Duncan JM (1969) *Strength and Deformation Characteristics of Rockfills Materials.* Report No. TE-69-5. Department of Civil Engineering, University of California, Berkeley, CA.

Marcuson WF, Franklin AG (1983) Seismic design, analysis, and remedial measures to improve the stability of existing earth dams – Corps of Engineers approach. In *Seismic Design of Embankments and Caverns* (Ed. TR Howard). pp. 65–78. ASCE, New York, NY.

Marsal R (1973) Mechanical properties of rockfill. In *Embankment-dam Engineering: Casagrande Volume.* (Eds R Hirschfeld, S Poulos). John Wiley & Sons, New York, NY.

Marsal R, Moreno E, Núñes A, Cuéllar R, Moreno YR (1965) *Research on the Behaviour of Granular Materials and Rockfill Samples.* Comision Federal de Electricidad Mexico, Mexico City.

Matsuda H, Andre PH, Ishikura R, Kawahara S (2011) Effective stress change and post-earthquake settlement properties of granular materials subjected to multi-directional cyclic simple shear. *Soil and Foundation* **51**(5), 873–884. doi:10.3208/sandf.51.873

Matsuda H, Nhan TT, Ishikura R, Inazawa T, Andre PH (2012) New criterion for the liquefaction resistance under strain-controlled multi-directional cyclic shear. In *Proceedings of the 15th World Conference on Earthquake Engineering.* 24–28 September, Lisbon. Paper No. 4149. pp. 23386–23395. Sociedade Portuguesa de Engenharia Sismica, Lisbon.

Mayne PW (2007) *Synthesis 368: Cone Penetration Testing.* National Cooperative Highway Research Program (NCHRP), Transportation Research Board, National Academies, Washington, DC.

McCarter MK (1976) *Monitoring Stability of High Waste Dumps.* SME–AIME Fall Meeting & Exhibit, Denver, Colorado, September 1–3.

McClung D (2001) Extreme avalanche runout: a comparison of empirical models. *Canadian Geotechnical Journal* **38**, 1254–1265. doi:10.1139/t01-041

McClung D, Mears AL, Schaerer P (1989) Extreme avalanche runout: data from four mountain ranges. *Annals of Glaciology* **13**, 180–184.

McClung D, Schaerer P (2006) *The Avalanche Handbook.* Mountaineers Books, Seattle.

McDougall S, Hungr O (2003) Objectives for the development of an integrated three-dimensional continuum model for the analysis of landslide runout. In *Proceedings of the 3rd International Conference on Debris Flows.* 10–12 September, Davos. (Ed. D Rickenmann, C-L Chen). pp. 647–657. Millpress, Rotterdam.

McDougall S, Hungr O (2004) A model for the analysis of rapid landslide runout motion across three-dimensional terrain. *Canadian Geotechnical Journal* **41**, 1084–1097. doi:10.1139/t04-052

McDougall S, Hungr O (2005) Modelling of landslides which entrain material from the path. *Canadian Geotechnical Journal* **42**, 1437–1448. doi:10.1139/t05-064

McKenna GT (1995) Grouted-in installation of piezometers in boreholes. *Canadian Geotechnical Journal* **32**, 355–363.

McLemore V, Donahue K, Dunbar N, Heizler L (2009) Characterization of weathering of mine rock piles: example from the Questa Mine, New Mexico, USA. In *Proceedings of the 8th International Conference on Acid Rock Drainage*. 22–26 June, Skelleftea. pp. 501–510. Swedish Association of Mines, Mineral and Metal Producers, Stockholm.

Melo C, Sharma S (2004) Seismic coefficients for pseudostatic slop analysis. In *Proceedings of the 13th World Conference on Earthquake Engineering*. 1–6 August 2004, Vancouver. Paper No. 369.

MEND (Mine Environment Neutral Drainage) (2001a) *MEND MANUAL Volume 1 – Summary MEND 5.4.2a*. Natural Resources Canada (CANMET), Ottawa.

MEND (Mine Environment Neutral Drainage) (2001b) *MEND MANUAL Volume 2 – Sampling and Analysis MEND 5.4.2b*. Natural Resources Canada (CANMET), Ottawa.

MESA (Mining Enforcement and Safety Administration) (1975) *Engineering and Design Manual – Coal Refuse Disposal Facilities*. Report prepared for US Dept. Int., Mining Enforcement and Safety Administration by E. D'Appolonia Consulting Engineers, Inc. USGPO, Washington.

Meyerhof GG (1957) The mechanism of flow slides in cohesive soils. *Geotechnique* **7**, 41–49. doi:10.1680/geot.1957.7.1.41

Mihai S, Deak St, Deak Gy, Oancea I, Petrescu A (2008) Tailings dams and waste-rock dumps safety assessment using 3D numerical modelling of geotechnical and geophysical data. In *Proceedings of the 12th International Conference of the International Association for Computer Methods and Advances in Geomechanics (IACMAG)*. 1–6 October, Goa. pp. 42212–42221. International Association for Computer Methods and Advances in Geomechanics, Mumbai.

Mikkelsen PE and Slope Indicator (2000) *Grouting-in Piezometers*. <http://www.slopeindicator.com/>.

Miller S, Rowles T, Millgate J, Pellicer J, Morris L, Gaunt J (2012) Integrated acid rock drainage management at the Phu Kham Copper Gold Operation in Lao PDR. In *Proceedings of the Ninth International Conference on Acid Rock Drainage*. 20–26 May, Ottawa. (Eds WA Price, C Hogan, G Tremblay) pp. 615–627. Golder Associates Ltd, Kanata.

Mirabediny H, Baafi E (1998) Dragline digging methods in Australian strip mines. In *Coal 1998: Coal Operators' Conference*. (Ed. N Aziz). pp. 313–324. University of Wollongong & the Australasian Institute of Mining and Metallurgy, Wollongong.

Mitchell JK, Soga K (2005) *Fundamentals of Soil Behavior*. 3rd edn. John Wiley & Sons, New York, NY.

Morgenstern NR (1992) The evaluation of slope stability: a 25 year perspective. In *Stability and Performance of Slopes and Embankments*. (Eds RB Seed, RW Boulanger). ASCE Geotechnical Special Publication **31**(1), 1–26.

Morgenstern NR, Price VE (1965) The analysis of the stability of general slip surfaces. *Geotechnique* **15**(1), 79–93. doi:10.1680/geot.1965.15.1.79

Morgenstern NR, Vick SG, Van Zyl D (2015) *Report on Mount Polley Tailings Storage Facility Breach*. Government of British Columbia, Ministry of Energy and Mines, Victoria.

Moss RES, Seed RB, Kayen RE, Stewart JP, Der Kiureghian A, Cetin KO (2006) CPT-based probabilistic and deterministic assessment of in situ seismic soil liquefaction potential. *Journal of Geotechnical and Geoenvironmental Engineering* **132**(8), 1032–1051. doi:10.1061/(ASCE)1090-0241(2006)132:8(1032)

MSHA (Mine Safety and Health Administration) (2009) *Engineering and Design Manual – Coal Refuse Disposal Facilities*. 2nd edn. Report prepared for US Dept. Int., Mine Safety and Health Administration by E. D'Appolonia Engineering. US Dept. Int., Mine Safety and Health Administration, Pittsburgh.

MVLWB/AANDC (Mackenzie Valley Land and Water Board/Aboriginal Affairs and Northern Development Canada) (2013) *Guidelines for the Closure and Reclamation of Advanced Mineral Exploration and Mine Sites in the Northwest Territories*. Aboriginal Affairs and Northern Development Canada, Ottawa.

National Academy of Sciences (1968–1973) *The Great Alaska Earthquake of 1964*. Vol. 1, Geology, and Vol. 3, Hydrology. National Academy of Sciences Printing and Publishing Office, Washington, DC.

NAVFAC (Naval Facilities Engineering Command) (1986) *Soil Mechanics Design Manual DM 7.01*. Department of the Navy Naval Facilities Engineering Command, Alexandria, VA; US Government Printing Office, Washington, DC.

Newmark NM (1965) Effects of earthquakes on dams and embankments. *Geotechnique* **15**(2), 139–160. doi:10.1680/geot.1965.15.2.139

Nichols RR (1987) Rock segregation in waste dumps. In *Proceedings of the International Symposium on Flow-*

Through Rock Drains. 8–11 September 1986, Inn of the South, Cranbrook. British BioTech, Vancouver.

NRC (Natural Resources Canada) (2013) *Earthquake Hazard*. <http://earthquakescanada.nrcan.gc.ca/hazard-alea>.

NWSRFS (National Weather Service River Forecast System) (2016) *NWSRFS User Manual Documentation*. <http://www.nws.noaa.gov/ohd/hrl/nwsrfs/users_manual/htm/xrfsdocpdf.php>.

O'Brien JS, Julien PY, Fullerton WT (1993) Two-dimensional water flood and mudflow simulation. *Journal of Hydraulic Engineering* **119**(HY2), 244–261. doi:10.1061/(ASCE)0733-9429(1993)119:2(244)

Oasys (2014) *Slope*. <http://www.oasys-software.com/products/engineering/slope.html>.

O. Hungr Geotechnical Research (2009) *Clara-W*. <http://www.clara-w.com/>.

OSM (Office of Surface Mining) (1982) *Surface Mining Water Diversion Design Manual*. OSM/TR-32/2. Contract report (No. J5101050) prepared for US Department of the Interior, Office of Surface Mining by Simons, Li & Associates, Inc. Office of Surface Mining, Washington, DC.

OSM (Office of Surface Mining Reclamation and Enforcement) (1989) *Engineering Design Manual for Disposal Of Excess Spoil*. Contract report (No. J5110084) prepared for US Department of the Interior, Office of Surface Mining Reclamation and Enforcement by CTL/Thompson Inc. Office of Surface Mining, Washington, DC.

Palma C, Linero S, Apablaza R (2009) Caracterización geotécnica de materiales de lastre en botaderos de gran altura mediante ensayos triaxiales y odométricos de gran tamaño. In *Proceedings of III Young South-American Geotechnical Conference*. 30 March to 1 April, Córdoba, Argentina [in Spanish].

Parkhurst DL, Appelo CAJ (1999) *User's Guide To PHREEQC (Version 2) – A Computer Program For Speciation, Batch-Reduction, One-Dimensional Transport, and Inverse Geochemical Calculations*. Water-Resources Investigations Report 99-4259. US Department of the Interior, US Geological Society, Denver, CO. <ftp://brrftp.cr.usgs.gov/pub/charlton/phreeqc/Phreeqc_2_1999_manual.pdf>.

Pastor M, Blanc T, Haddad B, Petrone S, Sanchez Morles M, Drempetic V, Issler D, Crosta GM, Cascini L, Sorbino G, Cuomo S (2014) Application of a SPH depth-integrated model to landslide run-out analysis. *Landslides* **11**(5), 793–812. doi:10.1007/s10346-014-0484-y

Patel SK, Sanghvi CS (2012) Seismic slope stability analysis of Kaswati earth dam. *International Journal of Advanced Engineering Research and Studies* **1**(3), 305–308.

Patton FD (1966) Multiple modes of shear failure in rock. In *1st Congress of the International Society of Rock Mechanics*. 25 September to 1 October, Lisbon. Vol. 1, pp. 509–513. Laboratório Nacional de Engenharia Civil, Lisbon.

PC-Progress (2011) *HYDRUS*. Version 2. PC-Progress, Prague.

Peng S, Zhang J (2007) *Engineering Geology for Underground Rocks*. Springer, Berlin and New York.

Perla R, Cheng TT, McClung DM (1980) A two-parameter model of snow avalanche motion. *Journal of Glaciology* **26**, 197–207.

Pernichele AD, Kahle MB (1971) Stability of waste dumps at Kennecott's Bingham Canyon Mine. *SME Transactions* **250**, 363–367.

Pihlainen JA, Johnston GH (1963) *Guide to a Field Description of Permafrost for Engineering Purposes*. National Research Council of Canada (NRCC) Document NRCC-7576. National Research Council, Canada, Associate Committee on Soil and Snow Mechanics, Ottawa.

Plaxis (2014) *Plaxis*. <http://www.plaxis.nl/>.

Pollet N, Schneider JLM (2004) Dynamic disintegration processes accompanying transport of the Holocene Flims sturzstrom (Swiss Alps). *Earth and Planetary Science Letters* **221**, 433–448. doi:10.1016/S0012-821X(04)00071-8

Poulos HG, Davis EH (1974) *Elastic Solutions for Soil and Rock Mechanics*. John Wiley & Sons, New York, NY.

Powers MC (1953) A new roundness scale for sedimentary particles. *Journal of Sedimentary Petrology* **23**, 117–119. doi:10.1306/D4269567-2B26-11D7-8648000102C1865D

Pusch R (1995) *Rock Mechanics on a Geological Base*. Elsevier, Amsterdam.

Read J, Stacey P (2009) *Guidelines for Open Pit Slope Design*. CSIRO Publishing, Melbourne, and CRC Press/Balkema, Rotterdam.

RIC (Resources Inventory Committee) (1996) *Terrain Stability Mapping in BC: A Review and Suggested Methods for Landslide Hazard and Risk Mapping*. Crown Publications, Victoria.

Richards BG, Coulthard MA, Toh CT (1981) Analysis of slope stability at Goonyella Mine. *Canadian Geotechnical Journal* **18**(2), 179–194. doi:10.1139/t81-023

Rickenmann D (1999) Empirical relationships for debris flows. *Natural Hazards* **19**, 47–77. doi:10.1023/A:1008064220727

Rizkalla MFA (1983) Stability of open pit mine spoils. MEng thesis. University of Alberta, Edmonton.

Robertson PK, Cabal KL (2015) *Guide to Cone Penetration Testing for Geotechnical Engineering*. 6th edn. Gregg Drilling & Testing Inc., Signal Hill, CA.

RocScience (2016a) *RS²*. <http://www.rocscience.com/>.

RocScience Inc (2016b) *Settle3D Code for Settlement and Consolidation Analysis*. <www.rocscience.com>.

RocScience Inc (2016c) *Slide Code for Two Dimensional Slope Stability Analysis*. <www.rocscience.com>.

Sabatini PJ, Bachus RC, Mayne PW, Schneider JA, Zettler TE (2002) *Geotechnical Engineering Circular No. 5, Evaluation of Soil and Rock Properties.* Rep. FHWA-IF-02-034, Federal Highway Administration, Washington, DC.

Sarma SK (1973) Stability analysis of embankments and slopes. *Geotechnique* **23**(3), 423–433. doi:10.1680/geot.1973.23.3.423

Sassa K (1985) The mechanism of debris flows. In *Proceedings of XI International Conference on Soil Mechanics and Foundation Engineering.* 12–16 August, San Francisco, CA. Vol. 3, pp. 1173–1176. A.A. Balkema, Rotterdam.

Sassa K (2000) Mechanism of flows in granular soils. In *Proceedings of the International Conference of Geotechnical and Geological Engineering, GEOENG2000.* 19–24 November, Melbourne. Vol. 1, pp. 1671–1702. Technomic, Lancaster, PA.

Sassa K, Fukuoka H, Wang GH, Ishikawa N (2004) Undrained dynamic-loading ring-shear apparatus and its application to landslide dynamics. *Landslides* **1**, 7–19. doi:10.1007/s10346-003-0004-y

Savage SB, Hutter K (1989) The motion of a finite mass of granular material down a rough incline. *Journal of Fluid Mechanics* **199**, 177–215. doi:10.1017/S0022112089000340

Sawatsky L (2004) Reclamation strategies that address mine closure drainage. In *Proceedings of the Canadian Institute Mine Reclamation Conference.* 8–9 November 2004, Vancouver.

Scheidegger A (1973) On the prediction of the reach and velocity of catastrophic landslides. *Rock Mechanics* **5**, 231–236. doi:10.1007/BF01301796

Schleifer J, Tessier B (1996) FRAGSCAN: A tool to measure fragmentation of blasted rock. In *Proceedings of the FRAGBLAST 5 Workshop on Measurement of Blast Fragmentation*, 23–24 August, Montreal. pp. 73–78. A.A. Balkema, Rotterdam.

Seed HB (1967) Slope stability during earthquakes. *Journal of the Soil Mechanics and Foundations Division* **93**(SM4), 299–323.

Seed HB (1979) Considerations in the earthquake-resistant design of earth and rockfill dams. *Geotechnique* **29**(3), 215–263. doi:10.1680/geot.1979.29.3.215

Seed RB, Cetin KO, Moss RES, Kammerer AM, Wu J, Pestana JM, Riemer MF, Sancio RB, Bray JD, Kayen RE, Faris A (2003) Recent advances in soil liquefaction engineering: a unified and consistent framework. In *Proceedings of the 26th Annual ASCE Los Angeles Geotechnical Spring Seminar.* Keynote presentation. 30 April, H.M.S. Queen Mary, Long Beach, CA.

Seif El Dine B, Dupla JC, Frank R, Canou J (2010) Mechanical characterization of matrix coarse-grained soils with a large-sized triaxial device. *Canadian Geotechnical Journal* **47**(4), 425–438. doi:10.1139/T09-113

Siddiqi FH (1984) Strength evaluation of cohesionless soils with oversize particles. PhD thesis, University of California, Davis.

Simmons JV, Fityus SG (2016) *The stability of very high spoil piles.* Report for Australian Coal Association Research Program Project C20019. <www.acarp.com.au>.

Simmons J, Fityus S, Donnelly B, Hammond A (2015) Moisture conditions in mine spoil dumps. In *Bowen Basin Symposium 2015 – Bowen Basin and Beyond.* (Ed. JW Beeston). pp. 429–439. Geological Society of Australia Inc. Coal Geology Group and the Bowen Basin Geologists Group, Brisbane.

Simmons JV, McManus DA (2004) Shear strength framework for design of dumped spoil slopes for open pit coal mines. In *Proceedings, Advances in Geotechnical Engineering: The Skempton Conference*, London, March 2004. (Eds R J Jardine, DM Potts, KG Higgins). Vol. 2, pp. 981–991. Thomas Telford Limited, London.

Skousen J, Sexstone A, Ziemkiewicz P (2000) Acid mine drainage control and treatment. In *Reclamation of Drastically Disturbed Lands.* (Eds RI Barnhisel, RG Darmody, WL Daniels). Agronomy Monograph 41. pp. 131–168. American Society of Agronomy, Madison, WI.

Skousen J, Ziemkiewicz P (2005) Performance of 116 passive treatment systems for acid mine drainage. National Meeting of the American Society of Mining and Reclamation. 19–23 June, Breckenridge. ASMR, Lexington, KY.

Sladen JA, Hewitt KJ (1989) Influence of placement method on the *in situ* density of hydraulic sand fills. Canadian Geotechnical Journal 26, 453–466.

SME (1985) Design of non-impounding mine waste dumps. In *Proceedings of Workshop held in Conjunction with Annual Fall Meeting of the Society of Mining Engineers.* November 1981. (Ed. MK McCarter).

Smith M (2002) Liquefaction in dump leaching. *Mining Magazine (London)*, July 2002.

SoilVision (2009) *SVSlope 3D Slope Stability Software Package.* <www.soilvision.com >.

SoilVision (2015) *SVFLUX.* SoilVision Systems Ltd, Saskatoon. <www.soilvision.com>.

Sousa J, Voight B (1991) Continuum simulation of flow failures. *Geotechnique* **41**, 515–538. doi:10.1680/geot.1991.41.4.515

Spencer E (1967) A method of analysis of embankments assuming parallel interslice forces. *Geotechnique* **17**(1), 11–26. doi:10.1680/geot.1967.17.1.11

Standards Australia (1993) *Geotechnical Site Investigations* (AS 1726:1993). Standards Australia, Homebush.

Standards Australia and Standards New Zealand (2004) *Risk Management Standard* (AS/ANZ 4360:2004).

Standards Australia, Sydney; Standards New Zealand, Wellington.

Standards Australia and Standards New Zealand (2009) *Risk Management – Principles and Guidelines* (AS/ANZ ISO 31000:2009). Standards Australia, Sydney; Standards New Zealand, Wellington.

Stark P (1997) Earthquake prediction: the null hypothesis. *Geophysical Journal International* **131**, 495–499. doi:10.1111/j.1365-246X.1997.tb06593.x

Strelkoff T (1970) Numerical solution of St.Venant equations. *Journal of Hydraulic Engineering* **96**, 223–251.

Suter GW, Luxmoore RJ, Smith ED (1993) Compacted soil barriers at abandoned landfill sites are likely to fail in the long term. *Journal of Environmental Quality* **22**(2), 217–226. doi:10.2134/jeq1993.00472425002200020001x

Swaisgood JT (2003) Embankment dam deformations caused by earthquakes. In *Proceedings of the 2003 Pacific Conference on Earthquake Engineering.* 13–15 February, Christchurch. Paper No. 014. New Zealand National Society for Earthquake Engineers, Wellington.

Sy A, Campanella RG (1994) Becker and standard penetration tests (BPT-SPT) correlations with consideration of casing friction. *Canadian Geotechnical Journal* **31**(3), 343–356. doi:10.1139/t94-042

Takahashi T (1991) *Debris Flow.* IAHR Monograph. A.A. Balkema, Rotterdam.

Takahashi T (2007) *Debris Flow: Mechanism, Prediction and Countermeasures.* Taylor and Francis, London.

Tapia A, Contreras LF, Jefferies M, Steffan O (2007) Risk evaluation of slope failure at the Chuquicamata mine. In *Slope Stability 2007, Proceedings of the 2007 International Symposium on Rock Slope Stability in Open Pit Mining and Civil Engineering.* (Ed. Y Potvin). pp. 477–495. Australian Centre for Geomechanics, Perth.

Taylor RK (1984) *Composition and Engineering Properties of British Colliery Discharge. Report to National Coal Board, Mining Department.* The National Coal Board, London.

Terzaghi K, Peck RB (1967) *Soil Mechanics in Engineering Practice.* 2nd edn. John Wiley & Sons, New York, NY.

Terzaghi K, Peck RB, Mesri G (1996) *Soil Mechanics in Engineering Practice.* 3rd edn. John Wiley & Sons, New York, NY.

Terzaghi K (1950) Mechanisms of landslides. In *Engineering Geology (Berkeley) Volume.* (Ed. S Paige). pp. 83–123. Harvard University, Cambridge.

Tran AB, Miller S, Williams DJ, Fines P, Wilson GW (2003) Geochemical and mineralogical characterisation of two contrasting waste rock dumps – the INAP waste rock dump characterisation project. In *Proceedings of the Sixth International Conference on Acid Rock Drainage.* 14–17 July, Cairns. pp. 939–947. Australasian Institute of Mining and Metallurgy, Melbourne.

Urrutia P, Wilson W, Aranda C, Peterson H, Blackmore S, Sifuentes F, Sanchez M (2011) Design and construction of field-scale lysimeters for the evaluation of cover systems at the Antamina Mine, Peru. In *Proceedings of the 15th International Conference on Tailings and Mine Waste.* 6–9 November, Vancouver. Norman B. Keevil Institute of Mining Engineering, University of British Columbia, Vancouver.

Urrutia VP (2012) Assessment of cover systems for waste rock in the Antamina Mine, Peru. MSc thesis. Department of Mining Engineering, University of British Columbia, Vancouver, BC.

USACE (US Army Corps of Engineers) (2013) *Geology.* Manual No. TM 3-34.61, Department of the Army, US Army Corps of Engineers, Washington, DC.

USACE (US Army Corps of Engineers) (1982) *Slope Stability Manual.* EM-1110-2-1902. Department of the Army, Office of the Chief of Engineers, Washington, DC.

USACE (US Army Corps of Engineers) (2001a) *Geotechnical Investigation.* Manual No. EM 1110-1-1804, Department of the Army, US Army Corps of Engineers, Washington, DC.

USACE (US Army Corps of Engineers) (2001b) *Hydrologic Modeling System HEC-HMS.* US Army Corps of Engineers Hydrologic Engineering Center, Davis, CA.

USBM (US Department of Interior, Bureau of Mines) (1982) *Development of Systematic Waste Disposal Plans for Metal and Nonmetal Mines.* Minerals research contract report (Contract No. J0208033) prepared for US Department of Interior, Bureau of Mines by Goodson and Associates, Inc. USBM Open File Report. pp. 183–82.

USDA NRCS (2004) Hydrologic soil-cover complexes. In *National Engineering Handbook.* Part 630 Hydrology. 210-VI-NEH. <http://www.wcc.nrcs.usda.gov/ftpref/wntsc/H&H/NEHhydrology/ch9.pdf>.

USDA NRCS (United States Department of Agriculture National Resources Conservation Service) (1986) TR-55 Urban Hydrology for Small Watersheds. Technical Release 55. 210-VI-TR-55. 2nd edn. <http://www.nrcs.usda.gov/Internet/FSE_DOCUMENTS/stelprdb1044171.pdf>.

US Department of Interior Bureau of Reclamation (1998) *Engineering Geology Field Manual.* Vol. 1. 2nd edn. Reprinted 2001. US Government Printing Office, Washington, DC.

USGS (United States Geological Survey) (2015) *Hazards.* <http://earthquake.usgs.gov/hazards>.

Valenzuela L, Bard E, Campana J, Anabalon ME (2008) High waste rock dumps – challenges and developments. In *Proceedings of Rock Dumps 2008.* Perth. pp. 65–78. Australia Centre for Geomechanics, Nedlands.

VanDine DF (1985) Debris flows and debris torrents in the Southern Canadian Cordillera. *Canadian Geotechnical Journal* **22**(1), 44–68. doi:10.1139/t85-006

VanDine DF (1996) Debris flow control structures for forest engineering. Ministry of Forest Research Program, Working Paper 22/ 1996. Province of British Columbia, Victoria.

Vardanega PJ, Haigh SK (2014) The undrained strength–liquidity index relationship. *Canadian Geotechnical Journal* **51**(9), 1073–1086. doi:10.1139/cgj-2013-0169.

Verdugo R, De la Hoz K (2006) Strength and stiffness of coarse granular soils. In *Soil Stress-Strain Behavior: Measurement, Modeling and Analysis Geotechnical Symposium in Rome*. 16–17 March. pp. 243–252. Springer, Dordrecht.

Voellmy A (1955) Uber die Zerstorung von Lawinen. *Sweizeriche Bauzeitung* **73**, 212–285.

Wahler WA (1979) A perspective – mine waste disposal structures – mine dumps, and mill and plant impoundments. In *Proceedings of the 6th Panamerican Conference on Soil Mechanics and Foundation Engineering, Vol. III*. Lima. International Society of Soil Mechanics and Geotechnical Engineering (ISSMGE), Lima.

Wickland BE (2006) Volume change and permeability of mixtures of waste rock and fine tailings. PhD thesis. Department of Mining Engineering, University of British Columbia, Vancouver.

Wickland BE, Wilson GW, Wijewickreme D, Klein B (2006) Design and evaluation of mixtures of mine waste rock and tailings. *Canadian Geotechnical Journal* **43**(9), 928–945. doi:10.1139/t06-058

Wilkins JK (1956) The flow of water through rockfill and its application to the design of dams. In *Proceedings of the 2nd Australia–New Zealand Conference on Soil Mechanics and Foundation Engineering*. pp. 141–149. Technical Publications for the New Zealand Institution of Engineers, Wellington.

Williams DJ (2016) Mine rehabilitation – are we reinventing the wrong wheel? In *Proceedings of the 11th International Seminar on Mine Closure*. 15–17 March, Perth. (Eds A Fourie and M Tibbett). pp. 595–608. Australian Centre for Geomechanics, Nedlands.

Williams DJ, Currey NA, Ritchie P, Wilson GW (2003) Kidston waste rock dump design and 'store and release' cover performance seven years on. In *Proceedings of the Sixth International Conference on Acid Rock Drainage*. 14–17 July, Cairns. pp. 419–426. Australasian Institute of Mining and Metallurgy, Melbourne.

Wilson GW (2000) Embankment hydrology and unsaturated flow in waste rock. In *Slope Stability in Surface Mining*. (Eds WA Hustrulid, MC McCarter, DJA Van Zyl). Chapter 33. p. 305. Society for Mining, Metallurgy, and Exploration, Denver, CO.

Wilson GW (2008) Why are we still battling ARD. In *Proceedings of the Sixth Australian Workshop on Acid and Metalliferous Drainage*. Bernie, Tasmania. (Ed. LC Bell, BMD Barrie, B Braddock, RW McLean). pp. 101–112. ACMER, St Lucia.

Wilson GW (2011) Mine waste cover systems for mine closure – meeting expectations. Invited Plenary Address. In *Proceedings 6th International Mine Closure Conference*. Lake Louise, Canada. (Ed. AB Fourie). Australian Centre for Geomechanics, Nedlands.

Wilson RC, Keefer DK (1985) Predicting aerial limits of earthquake-induced landsliding. In *Evaluating earthquake hazards in the Los Angeles region – An Earth-Science perspective*. pp. 316–345. USGS Professional Paper 1360. US GPO, Washington, DC.

Wolter A, Havaej M, Zorzi L, Stead D, Clague JJ, Ghirotti M, Genevois R (2013) Exploration of the kinematics of the 1963 Vajont Slide, Italy, using a numerical modelling toolbox. *Italian Journal of Engineering Geology and Environment* 2013, 599–612. doi:10.4408/IJEGE. 2013-06.B-58

Wolter A, Stead D, Ward BC, Clague JJ, Ghirotti M (2016) Engineering geomorphological characterisation of the Vajont Slide, Italy, and a new interpretation of the chronology and evolution of the slide. *Landslides* **13**, 1067. doi:10.1007/s10346-015-0668-0

Wyllie DC, Mah CW (2005) *Rock Slope Engineering*. 4th edn. Spon Press, New York, NY.

Yazdani J, Barbour L, Wilson W (2000) Soil water characteristic curve for mine waste rock containing coarse material. In: *Proceedings of the 6th Environmental Engineering Speciality Conference of the CSCE & the 2nd Spring Conference of the Geoenvironmental Division of the Canadian Geotechnical Society*. 7–10 June, London, Ontario. pp. 198–202.

Zinck J, Griffith W (2013) *Review of Mine Drainage Treatment and Sludge Management Operations*. MEND (Mine Environment Neutral Drainage) Report 3.43.1. Natural Resources Canada, Ottawa.

Zorzi L, Massironi M, Surian N, Genevois R, Floris M (2014) How multiple foliations may control large gravitational phenomena: a case study from the Cismon Valley, Eastern Alps, Italy. *Geomorphology* **207**, 149–160. doi:10.1016/j.geomorph.2013.11.001

Index